Handbook of Natural Gas Analysis

Handbook of Natural Gas Analysis

James G. Speight

CD&W Inc.
Laramie, WY, USA

WILEY

This edition first published 2018

Registered Office
John Wiley & Sons, Inc., 111 River Street, Hoboken, NJ 07030, USA

Editorial Office
111 River Street, Hoboken, NJ 07030, USA

For details of our global editorial offices, customer services, and more information about Wiley products visit us at www.wiley.com.

Wiley also publishes its books in a variety of electronic formats and by print-on-demand. Some content that appears in standard print versions of this book may not be available in other formats.

Library of Congress Cataloging-in-Publication Data

Names: Speight, James G., author.
Title: Handbook of natural gas analysis / by Dr. James G. Speight.
Description: Hoboken, NJ : John Wiley & Sons, 2018. |
Includes bibliographical references and index. |
Identifiers: LCCN 2018010730 (print) | LCCN 2018014739 (ebook) |
ISBN 9781119240303 (pdf) | ISBN 9781119240310 (epub) | ISBN 9781119240280
Subjects: LCSH: Natural gas–Analysis–Handbooks, manuals, etc.
Classification: LCC TN880.A3 (ebook) | LCC TN880.A3 S74 2018 (print) | DDC 665.7/3–dc23
LC record available at https://lccn.loc.gov/2018010730

Cover Design: Wiley
Cover Image: © suns07butterfly/Shutterstock

Set in 10/12pt Warnock by SPi Global, Pondicherry, India

V10002350_071418

Contents

About the Author

Dr James G. Speight has a BSc and PhD in Chemistry; he also holds a DSc in Geological Sciences and a PhD in Petroleum Engineering. He has more than 50 years of experience in areas associated with (i) the properties, recovery, and refining of reservoir fluids, conventional petroleum, heavy oil, and tar sand bitumen; (ii) the properties and refining of natural gas and gaseous fuels; and (iii) the properties and refining of biomass, biofuels, and biogas and the generation of bioenergy. His work has also focused on environmental effects, environmental remediation, and safety issues associated with the production and use of fuels and biofuels. He is the author (and coauthor) of more than 75 books in petroleum science and engineering, biomass, biofuels, and environmental sciences.

Although he has always worked in the private industry that focused on contract-based work, Dr Speight has served as adjunct professor in the Department of Chemical and Fuels Engineering at the University of Utah and in the Departments of Chemistry and Chemical and Petroleum Engineering at the University of Wyoming. In addition, he was a visiting professor in the College of Science, University of Mosul (Iraq), and has also been a visiting professor in chemical engineering at the following universities: University of Akron (Ohio), University of Missouri, Technical University of Denmark, and University of Trinidad and Tobago. He has served as a thesis examiner for more than 30 theses and has been an advisor/mentor to MSc and PhD students.

Dr Speight has been honored as the recipient of the following awards:

- Diploma of Honor, United States National Petroleum Engineering Society. For outstanding contributions to the petroleum industry, 1995.
- Gold Medal of the Russian Academy of Sciences. For outstanding work in petroleum science, 1996.
- Einstein Medal of the Russian Academy of Sciences. In recognition of outstanding contributions and service in the field of geologic sciences, 2001.

- Gold Medal – Scientists without Frontiers, Russian Academy of Sciences. In recognition of his continuous encouragement of scientists to work together across international borders, 2005.
- Methanex Distinguished Professor, University of Trinidad and Tobago. In recognition of excellence in research, 2006.
- Gold Medal – Giants of Science and Engineering, Russian Academy of Sciences. In recognition of continued excellence in science and engineering, 2006.

Preface

Natural gas is a flammable gaseous mixture that usually occurs with petroleum in reservoirs as well as in gas reservoirs. It is predominantly methane (CH_4) but does contain higher molecular weight hydrocarbon derivatives, such as paraffins (C_nH_{2n+2}), generally containing up to eight carbon atoms that may also be present in small quantities. In some gases, benzene and low molecular weight aromatic carbon derivative may also be present. The hydrocarbon constituents of natural gas are combustible, but nonflammable nonhydrocarbon components such as carbon dioxide (CO_2), hydrogen sulfide (H_2S), nitrogen (N_2), and helium (He) are often also present, which detract from the heating value of natural gas. In certain natural gases where the concentration of the nonhydrocarbon constituents is relatively high, they may be extracted as added-value products.

Effective gas recovery, transportation, storage, and processing operations require that analytical resources are optimized because the application of the relevant analytical methods meets the objectives required at each stage of gas handling. Method validation, as required by the local, state, or national regulatory agencies at the various stages of the approval process, necessitates the need for demonstrating that the analytical procedures applied to the sales gas are suitable for their intended use in terms of defining the sales specifications of the gas. Where appropriate, the results of the analytical test methods are incorporated into MSDS documents that provide valuable information to purchasers. In addition, regulations require chemical manufacturers to prepare and distribute an MSDS for various products, especially those products that might be harmful to the user and to the environment. This includes natural gas products (and natural gas itself) that are flammable, corrosive, explosive, or toxic.

However, to fully understand the results of the analytical test methods as applied to production to sales of natural gas and the various products, it is necessary to understand the composition of natural gas as it is related to its formation and character. For example, analytical data are the media for transmitting information related to the effectiveness of gas processing operations at the wellhead and at the refinery. Prior to transportation in pipelines, it is essential that any corrosive constituents in the gas (such as water, carbon dioxide, and hydrogen sulfide) are separated from natural gas.

Thus, analytical methods are employed to establish the identity, purity, physical characteristics, and potency of natural gas and the associated products. Methods have been developed to support testing against specifications during manufacturing and quality release operations, as well as during long-term stability studies. Furthermore, the validation of an analytical method demonstrates the reliability of the measurement. It is required to varying extents throughout the regulatory process, and the validation practice demonstrates that an analytical method measures (in the context of natural gas) the amount of the constituent and it is in the allowable range for that constituents.

As a result, it should not be a surprise that at each stage of natural gas production, wellhead treating, transportation, and processing, analysis of the gas to determine its composition and properties by standard test methods is an essential part of the chemistry of natural gas and related technology. Use of analytical methods offers vital information about the behavior of natural gas during recovery, wellhead processing, transportation, gas processing, and use. The data produced from the test methods are based on the criteria involving the suitability of the gas for use and the potential for interference with the environment.

Thus, it is the purpose of this book to identify and describe the criteria that are appropriate to the analysis and testing of natural gas and related gas streams. For this reason, there is reference to the relevant standard test method that falls under the umbrella of ASTM International, an organization that is recognized globally across borders, disciplines, and industries and works to improve performance in manufacturing and materials, products and processes, and systems and services.

This book presents the various aspects of the origin and analysis of natural gas and will provide a detailed explanation of the necessary standard test methods and procedures that are applicable to products to help predefine predictability of gas composition and behavior during gas cleaning operations and use. This information allows the analyst to understand the behavior of the method and to establish the performance limits of the method according to the origin of the sample received by the analytical laboratory. Where appropriate, the book also references process gas (also called refinery gas), coalbed methane, gas from tight formations, gas from gas hydrates, coal gas, biogas, landfill gas, and flue gas.

Each chapter is written as a stand-alone chapter so that all of the relevant information is at hand especially where there are tests that can be applied to several products. Where this was not possible, cross-references to the pertinent chapter are included. Several general references are listed for the reader to consult and obtain a more detailed description of natural gas properties, and the focus is to cite the relevant test methods that are applied to natural gas and its constituents.

The book is intended for use within analytical laboratories that specialize on the analysis of natural gas and for managers, professionals, and technicians working in the gas industry as well as gas processing scientists and engineers as a guide for the analysis of gas from other sources. The book will help the reader to understand where the various standard test methods are relevant and how these test methods fit into the technology of natural gas. In summary, it is not merely a matter of knowing which test methods to apply to the analysis of natural gas, but at which point analysis should be applied in the gas train that commences with the origin of the gas and ends with the sales of the gas to the consumer.

This book will serve as a valuable reference work for the application of analysis in the natural gas industry and will introduce the analytical chemist to the origin and production of natural gas and the natural gas engineer to the various methods of analysis that are required (by regulation) as well as the points of application of the relevant test methods.

January, 2018

Dr. James G. Speight
Laramie, WY, USA

Part I

Origin and Properties

1

History and Background

1.1 Introduction

Although the terminology and definitions involved with the natural gas technology are quite succinct, there may be those readers that find the terminology and definitions somewhat confusing. *Terminology* is the means by which various subjects are named so that reference can be made in conversations and in writings and the meaning is passed on. *Definitions* are the means by which scientists and engineers communicate the nature of a material to each other either through the spoken or through the written word. Thus, the terminology and definitions applied to natural gas (and, for that matter, to other gaseous products and fuels) are extremely important and have a profound influence on the manner by which the technical community and the public perceive that gaseous fuel. For the purposes of this book, natural gas and those products that are isolated from natural gas during recovery (such as natural gas liquids [NGLs], gas condensate, and natural gasoline) are a necessary part of this text.

Thus, the term *natural gas* is the generic term that is applied to the mixture of gases and low-boiling liquid hydrocarbon derivatives (typically up to and including hydrocarbon derivatives such as *n*-octane, $CH_3(CH_2)_6CH_3$, boiling point 125.1–126.1 °C, 257.1–258.9 °F) (Table 1.1) that is commonly associated with petroliferous (petroleum-producing, petroleum-containing) geologic formations (Mokhatab et al., 2006; Speight, 2007, 2014a) and that has been extended to gases and liquids from the recently developed shale formations (Speight, 2017b) as well as gas (biogas) produced from biological sources (John and Singh, 2011; Ramroop Singh, 2011; Singh and Sastry, 2011).

For clarification, natural gas (also called marsh gas and swamp gas in older texts and more recently landfill gas) is not the same as town gas, which is manufactured from coal, and the terms coal gas, manufactured gas,

Handbook of Natural Gas Analysis, First Edition. James G. Speight.

Table 1.1 Constituents of natural gas.

Name	Formula	Vol. %
Methane	CH_4	>85
Ethane	C_2H_6	3–8
Propane	C_3H_8	1–5
Butane	C_4H_{10}	1–2
Pentane$^+$	C_5H_{12}	1–5
Carbon dioxide	CO_2	1–2
Hydrogen sulfide	H_2S	1–2
Nitrogen	N_2	1–5
Helium	He	<0.5

Pentane$^+$: pentane and higher molecular weight hydrocarbon derivatives up to octane as well as benzene and toluene.

producer gas, and syngas (synthetic natural gas [SNG]) are also in regular use for gases produced from coal (Speight, 2013b). Also, by way of definition and clarification, town gas is a flammable gaseous fuel made by the destructive distillation of coal. It contains a variety of calorific gases including hydrogen, carbon monoxide, methane, and other volatile hydrocarbon derivatives, together with small quantities of noncalorific gases such as carbon dioxide and nitrogen. Town gas, although not used to any great extent in the United States, is still generated and used in some countries and is used in a similar way to natural gas. This is a historical technology and is not usually economically and environmentally competitive with modern sources of natural gas.

Most town gas-generating plants located in the Eastern United States in the late nineteenth century and early twentieth century were ovens that heated bituminous coal in airtight chambers to produce coke through the carbonization process. The gas driven off from the coal was collected and distributed through networks of pipes to residences and other buildings where it was supplied to industrial and domestic users – natural gas did not come into widespread use until the last half of the twentieth century. The coal tar collected in the bottoms of the gashouse ovens was often used for roofing and other waterproofing purposes, and when mixed with sand and gravel (aggregate), it was used for paving streets (road asphalt). The coal tar asphalt has been replaced by asphalt produced from crude oil (Speight, 2014a, 2015b). Thus, prior to the development of resources, virtually all fuel and lighting gas was manufactured from coal, and the history of and analysis of natural gas has its roots in town gas analysis and use (Speight, 2013b). Thus, with the onset of industrial expansion after World War II, natural gas has become one of the most important raw materials consumed by modern industries to provide raw materials for the ubiquitous plastics and other products as well as feedstocks for the energy and transportation industries.

From a chemical standpoint, natural gas is a mixture of hydrocarbon compounds and nonhydrocarbon compounds with crude oil being much more complex than natural gas (Mokhatab et al., 2006; Speight, 2007, 2012, 2014a). The fuels that are derived from this natural product supply more than one quarter of the total world energy supply. The more efficient use of natural gas is of paramount importance, and the technology involved in processing both feedstocks will supply the industrialized nations of the world for (at least) the next five decades until suitable alternative forms of energy (such as biogas and other nonhydrocarbon fuels) are readily available (Boyle, 1996; Ramage, 1997; Speight, 2008, 2011a, b, c; Rasi et al., 2011). Any gas sold, however, to an industrial or domestic consumer must meet the designated specification that is designed according to the use of the gas.

As a result, it should not be a surprise that at each stage of natural gas production, wellhead treating, transportation, and processing, analysis of the gas to determine the composition and properties of the gas by standard test methods is an essential part of the chemistry and technology of natural gas. Use of analytical methods offers vital information about the behavior of natural gas during recovery, wellhead processing, transportation, gas processing, and use. The data produced from the test methods are the criteria by means of which the suitability of the gas for use and the potential for interference with the environment.

1.2 History, Use, and the Need for Analysis

Natural gas is a versatile, clean-burning, and efficient fuel that is used in a wide variety of applications. In the late nineteenth century and in the early twentieth century, natural gas played a subsidiary role to coal gas insofar as coal gas was used for street lighting and for building lighting and provided what was known as gaslight (Mokhatab et al., 2006; Speight, 2013b). However, as the twenty-first century progresses, the discovery of large reserves of natural gas in various countries as well as improved distribution of gas has made possible a wide variety of uses in homes, businesses, factories, and power plants. The fastest-growing use of natural gas is for the generation of electric power and, to a large extent, has been a replacement fuel of many formerly coal-fired power plants and oil-fired power plants. Natural gas power plants usually generate electricity in gas turbines (which are derived from jet engines), directly using the hot exhaust gases from the combustion process.

Natural gas-fired plants are currently among the cheapest power plants to construct, which is a reversal of previous trends where operating costs were generally higher than those of coal-fired power plants because of the relatively high cost of natural gas. In addition, natural gas-fired plants have greater operational flexibility than coal-fired power plants because they can be fired up and turned down rapidly. Because of this, many natural gas plants in the United States were originally used to provide additional capacity (peak capacity) at times when electricity demand was especially high, such as the summer months when air conditioning is widely used. During much of the year, these natural gas peak plants were idle, while coal-fired power plants typically provided base load power. However, since 2008, natural gas prices in the United States have fallen significantly, and natural gas is now increasingly used as for base load power as well as for intermediate load power source in many cities. Natural gas can also be used to produce both heat and electricity simultaneously (cogeneration or combined heat and power [CHP]). Cogeneration systems are highly efficient, able to put 75–80% of the energy in gas to use. Trigeneration systems, which provide electricity, heating, and cooling, can reach even higher efficiencies using natural gas.

Natural gas also has a broad range of other uses in industry, not only as a source of both heat and power but also as a source of valuable hydrogen that is necessary for crude oil refining as well as for producing plastics and chemicals. Most hydrogen gas (H_2) production, for example, comes from reacting high-temperature water vapor (steam) with methane – steam-methane reforming reaction followed by the water–gas shift reaction:

$$CH_4 + H_2O \rightarrow CO + 3H_2 \left(\text{steam-methane reforming reaction}\right)$$

$$CO + H_2O \rightarrow CO_2 + H_2 \left(\text{water} - \text{gas shift reaction}\right)$$

Furthermore, the hydrogen produced from natural gas can itself be used as a fuel. The most efficient way to convert hydrogen into electricity is by using a fuel cell, which combines hydrogen with oxygen to produce electricity, water, and heat. Although the process of reforming natural gas to produce hydrogen still has associated carbon dioxide emissions, the amount released for each unit of electricity generated is much lower than for a combustion turbine.

As part of the industrial use of natural gas, there is the need for associated (or contract) laboratory operations to perform the necessary high numbers of analyses before the products (in this context, the gaseous products) are used by industrial and domestic consumers. Detection of even the slightest amounts of impurities can be an indication of process inefficiency and whether or not the gas is suitable for the designated use. In fact, one of the most important tasks in gas technology, especially in the context of petroleum-related natural gas, is the need for reliable values of the volumetric and thermodynamic properties for pure low-boiling hydrocarbon derivatives and their mixtures. These properties are important in the design and operation of much of the processing equipment (Poling et al., 2001).

For example, reservoir engineers and process engineers analyze pressure–volume–temperature (PVT) relationships and phase behavior of reservoir fluids (i) to estimate the amount of oil or gas in a reservoir, (ii) to develop a recovery process for a crude oil or gas field, (iii) to determine an optimum operating condition in a gas–liquid separator unit, (iv) to determine the need for a wellhead processing system to protect a pipeline from corrosion, and (v) to design suitable gas processing options. However, the most advanced design approaches or the most sophisticated simulation experiments cannot guarantee the optimum design or operation of a unit (or protection of a pipeline) if the physical properties as determined in an analytical laboratory dictate otherwise. For these reasons accurate knowledge of the properties of the gas is an extremely increasingly important aspect of gas technology. Analysis of the gas during production operations is also an essential practice.

Typically, in field operations, the composition of natural the gas (which affects the specific gravity), especially of associated gas, can vary significantly as the product flowing out of the well can change with variability of the production conditions as well as the change of pressure as gas is removed from the reservoir (Burruss and Ryder, 2003, 2014). Constituents of the gas that were in the liquid phase under the pressure of the reservoir can revert to the gas phase as the reservoir pressure is reduced by gas removal. As a result, it is necessary to collect representative samples of gas from high-pressure cylinders for analysis by gas chromatography in the laboratory (Chapter 6). In terms of gas analysis, the concept of obtaining a representative sample of the gas for analysis cannot be overstressed (Chapter 6).

1.2.1 History

Natural gas has been known for many centuries, but initial use for the gas was more for religious purposes rather than as a fuel. For example, gas wells were an important aspect of religious life in ancient Persia because of the importance of fire in their religion. In classical times these wells were often flared and must have been, to say the least, awe inspiring (Forbes, 1964). In the current contact, it is perhaps at least as awe inspiring – considering the history of the use of other fossil fuels such as coal and crude oil during the twentieth century – that the use of natural gas is superseding the use of crude oil and coal in many countries. During that time, natural gas was generally flared as a product of limited use until the depletion of crude oil reserves in the late twentieth century caused a back-and-forth concern about the future lack of energy-producing fuels (Speight, 2011a, 2014a; Speight and Islam, 2016).

After the discovery by the Chinese more than 2000 years ago that the energy in natural gas could be harnessed and used as a heat source, the use of natural gas has grown (Mokhatab et al., 2006; Speight, 2007, 2014a). However, the uses of natural gas did not necessarily parallel its discovery, and during recorded historical time, there was little or no understanding of what natural gas was; it posed somewhat of a mystery to man. Sometimes, events such as lightning strikes would ignite natural gas that was escaping from vents through the crust of the Earth. This would create a fire coming from the earth, burning the natural gas as it seeped out from underground. These fires puzzled most early civilizations and were the root of much myth and superstition. One of the most famous of these types of flames was found in ancient Greece, on Mount Parnassus approximately 1000 BC. The (realistic or legendary) story is that a goat herdsman came across what looked like a *burning spring*, a flame rising from a fissure in the rock. The Greeks, believing it to be of divine origin, built a temple on the flame. This temple housed a priestess who was known as the Oracle of Delphi, giving out prophecies she claimed were inspired by the flame.

These types of gas leaks became prominent in the religions of India, Greece, and Persia where the inhabitants of the region were unable to explain the origin of the fires and regarded the origin of the flames as divine, or supernatural, or both. As a result, the energy value of natural gas was not recognized until approximately 900 BC in China, and the Chinese drilled the first known natural gas well in 211 BC. Crude pipelines (probably state-of-the-art pipelines at the time) were constructed from bamboo stems to transport the gas, where it was used to boil seawater, removing the salt as a residue product, after which the water was condensed and, therefore, drinkable (Abbott, 2016).

Natural gas was discovered and identified in America as early as 1620, when French explorers discovered natives igniting gases that were seeping into and around Lake Erie (Table 1.2). However, Britain was the first country to commercialize the use of natural gas, and in 1785 natural gas produced from coal was used to light houses, as well as streetlights. Manufactured natural gas of this type (as opposed to naturally occurring gas) was first brought to the United States in 1816, when it was used to light the streets of Baltimore, Maryland. This manufactured gas was much less efficient, and less environmentally friendly, than modern natural gas that comes from underground.

In 1821 in Fredonia, United States, residents observed gas bubbles rising to the surface from a creek. William Hart, considered as America's *father of natural gas*, dug there the first natural gas well in North America (Speight, 2007). In 1859, Colonel Edwin Drake, a former railroad conductor (the origin of the title "Colonel" is unknown but seemed to impress the townspeople), dug the first well. Drake found crude oil and natural gas at 69 ft below the surface of the Earth. More recently, natural gas was discovered because of prospecting for crude oil. However, the gas was often an unwelcome

Table 1.2 Abbreviated timeline for the use of natural gas.

1620	French missionaries recorded that Indians in what is now New York state ignited gases in the shallows of Lake Erie and in the streams flowing into the lake
1785	HousNNatural gas is introduced for home lighting and streetlighting
1803	Gas lighting system patented in London by Frederick Winsor
1812	First gas company founded in London
1815	Metering for households, invented in 1815 by Samuel Clegg
1816	First US gas company (using manufactured gas) founded in Baltimore
1817	First natural gas from the wellhead used in Fredonia, NY, for house lighting
1840	Fifty or more US cities were burning public utility gas
1850	Thomas Edison postulated replacing gas lighting by electric lighting
1859	Carl Auer von Welsbach in Germany developed a practical gas mantle
1885	Depleted reservoirs are used for the first time to store gas

by-product because, as any gas-containing reservoirs were tapped during the drilling process, the drillers were forced to discontinue the drilling operations to allow the gas to vent freely into the air. Currently, and particularly after the crude oil shortages of the 1970s, natural gas has become an important source of energy in the world.

1.2.2 Use

Throughout the nineteenth century, natural gas was used almost exclusively as source of light, but because of the lack of any transport infrastructure that prevented the transportation of the gas to distant markets, the use was localized to areas close by where the gas was discovered. However, in 1890 with the invention of leakproof pipeline coupling, transportation of natural gas to long-distance customers became possible and finally achieved practicality in the early 1920s with additional advances in pipeline technology. Finally, after World War II, the use of natural gas underwent a marked increase as pipeline networks and natural gas storage systems underwent rapid development.

Once the transportation of natural gas was possible over considerable distances, the increased use of natural gas led to innovations from the discovery of new uses for natural gas that included the use of natural gas by industrial consumers. As the use of natural gas has increased and diversified, the need for analysis related to gas composition has also increased. Natural gas has many applications: for domestic use, industrial use, and transportation. In addition, natural gas is also a raw material for many common products such as paints, fertilizer, plastics, antifreeze, dyes, photographic film, medicines, and explosives. Along with these newer uses, there has been an increased need not only for the compositional analysis of natural gas but also for analytical data that provide other information about the behavior of natural gas.

1.2.3 The Need for Analysis

To satisfy the modern use of natural gas, analytical methods are applied to all aspects of gas production and use and include (i) gas in the reservoirs, (ii) associated gas, (iii) nonassociated gas, (iv) unconventional gas, including coalbed methane (CBM), tight shale gas, and gas from gas hydrates, (v) determining the quality of gas reserves; (vi) determining the suitability of the gas for use, (vii) the ability of the gas to conform to regulation, and (viii) the effect of natural gas on the environment.

However, the primary need for analysis of natural gas came with the recognition by the producers and the consumers that natural gas – a product of nature – is a gaseous mixture and that the composition of the gas will differ according to the reservoir from which the gas was produced. While natural gas consists predominantly of methane, with higher molecular weight (MW) hydrocarbon derivatives such as ethane, propane, and butane present as minor constituents, there are also the nonflammable (inert) constituents such as nitrogen, carbon dioxide, water (vapor), and helium. It is these other non-methane constituents that give rise to the need not only for compositional analysis of the gas but also for the determination of other properties that are relevant to the ultimate use of the gas.

Up to the introduction of gas chromatography in the 1950s, the analysis of natural gas and other fuel gases, such as coal gas (Speight, 2013b) and refinery process gas (Speight, 2014a), was performed using chemical absorption (chemisorption) and/or combustion methods (Speight, 2015a). These methods relied upon the selective absorption of different constituents (either in their original form or after reaction with other chemicals), e.g. carbon dioxide was determined by absorption in a solution of potassium hydroxide (chemisorption) followed by titration to determine the amount of the gas absorbed:

$$CO_2 + 2KOH = K_2CO_3 + H_2O$$

But in the early 1960s, chromatography had replaced the adsorption methods as the major analytical method applied to natural gas and other fuel gases.

Although the specification of the natural gas from a source is typically negotiated between the supplier and the customer, the gas must meet specification that is provided by various governmental and environmental bodies. The main aspects that should be covered include, but are not limited to, (i) properties such as gas quality, heating value, and Wobbe index; (ii) impurities, such as oxygen, inert gases, carbon dioxide, and sulfur compounds; (iii) hydrocarbon dew point, which prevents condensation of hydrocarbon liquids in the pipeline; and (iv) water content, which prevents water dropout and hydrate formation and corrosion in the pipeline.

Each of the categories enumerated in the preceding paragraph requires detailed analysis (and thence knowledge) of the properties of natural gas, and to understand the applicability of the analytical methods, it is helpful for the reader to understand (i) the origin of natural gas; (ii) the production of natural gas; (iii) the transportation of natural gas; (iv) the refining of natural gas, including gas cleaning; and (v) the influence of natural gas on the environment.

First and foremost, although covered in name by a generic term, natural gas varies in composition, and therefore quality, as well as other properties depending on the source of the gas. Therefore, it is essential that gas composition and properties should be determined by a series of standard test methods (Chapter 6) in order to choose and predict a suitable use for the gas. For example, it is often assumed that gases with the same heat content (calorific value [CV], heating value) are interchangeable, but this is not necessarily the case. Since all gas-fired equipment is designed to operate within a particular range of gas specification, the use of gases outside this range can lead to a range of issues from poor quality combustion to equipment damage and ultimately dangerous operation. As a result, gas interchangeability relates to more than just a parameter for the heat content. In fact, interchangeability is governed by a general umbrella property often referred to as *gas quality*, which is, in turn, a function of gas composition and specific gravity as well as any other property (or properties) that are relevant to the use of the gas. Thus, the essence of natural gas interchangeability relies on knowledge of the necessary properties of the gas, which are derived from application of a series of standard test methods that are accepted on an international basis.

The various purposes to be served by the analytical data and the necessary accuracy with which each constituent of each type of gas must be known in order to serve each specific purpose are then estimated. These estimates afford the first criterion by means of which the suitability of analytical methods and apparatus may be judged, but they are subject to revision when more is known about the limiting attainable accuracies of the analytical methods (Shepherd, 1947). Thus, because of the worldwide use of natural gas and the potential for gas from different wells to vary in composition, there is a need to apply a series of standard test to the gas in order that it can meet sales specifications (Table 1.3).

Gases analyzed include hydrocarbon derivatives (C_1–C_{6+}) such as methane, ethane, propane, isobutane, *n*-butane, isopentane, *n*-pentane, and hexane, plus higher MW hydrocarbon derivative. Nonhydrocarbon impurities include hydrogen, nitrogen, carbon monoxide, carbon dioxide, oxygen, mercury, sulfur-containing compounds such as hydrogen sulfide and thiols (also called

Table 1.3 Examples of standard test methods for natural gas (ASTM, 2017).

ASTM D1070 Standard Test Methods for Relative Density of Gaseous Fuels
ASTM D1071 Standard Test Methods for Volumetric Measurement of Gaseous Fuel Samples
ASTM D1072 Standard Test Method for Total Sulfur in Fuel Gases
ASTM D1142 Standard Test Method for Water Vapor Content of Gaseous Fuels by Measurement of Dew-Point Temperature
ASTM D1826 Standard Test Method for Calorific (Heating) Value of Gases in Natural Gas Range by Continuous Recording Calorimeter
ASTM D1945 Standard Test Method for Analysis of Natural Gas by Gas Chromatography
ASTM D1946 Standard Practice for Analysis of Reformed Gas by Gas Chromatography
ASTM D1988 Standard Test Method for Mercaptans in Natural Gas Using Length-of-Stain Detector Tubes
ASTM D3588 Standard Practice for Calculating Heat Value, Compressibility Factor, and Relative Density of Gaseous Fuels
ASTM D4084 Standard Test Method for Analysis of Hydrogen Sulfide in Gaseous Fuels (Lead Acetate Reaction Rate Method)
ASTM D4150 Standard Terminology Relating to Gaseous Fuels
ASTM D4810 Standard Test Method for Hydrogen Sulfide in Natural Gas
ASTM D4891 Standard Test Method for Heating Value of Gases in Natural Gas Range by Stoichiometric Combustion
ASTM D4984 Standard Test Method for Carbon Dioxide in Natural Gas Using Length-of-Stain Detector Tubes
ASTM D5287 Standard Practice for Automatic Sampling of Gaseous Fuels
ASTM D5454 Standard Test Method for Water Vapor Content of Gaseous Fuels Using Electronic Moisture Analyzers
ASTM D5503 Standard Practice for Natural Gas Sample-Handling and Conditioning Systems for Pipeline Instrumentation
ASTM D5504 Standard Test Method for Determination of Sulfur Compounds in Natural Gas and Gaseous Fuels by Gas Chromatography and Chemiluminescence
ASTM D5954 Standard Test Method for Mercury Sampling and Measurement in Natural Gas by Atomic Absorption Spectroscopy
ASTM D6228 Standard Test Method for Determination of Sulfur Compounds in Natural Gas and Gaseous Fuels by Gas Chromatography and Flame Photometric Detection
ASTM D6273 Standard Test Methods for Natural Gas Odor Intensity
ASTM D6350 Standard Test Method for Mercury Sampling and Analysis in Natural Gas by Atomic Fluorescence Spectroscopy

mercaptans, R—SH), and water. As an example of the need for a standard test method, natural gas composition analysis is applied to all phases of natural gas exploration and production, from the reservoir to recovery at the wellhead, initial processing prior to transportation, processing, storage, and distribution as well as liquefied natural gas-related activities. Natural gas composition includes testing for the following:

- Methane, CH_4
- Ethane, C_2H_6
- Propane, C_3H_8
- Butane, C_4H_{10}
- Carbon dioxide, CO_2
- Oxygen, O_2
- Nitrogen, N_2
- Hydrogen sulfide, H_2S
- Thiols (mercaptans), RSH
- Trace of rare gases
- Trace metals

Because of the presence of the nonmethane constituents, the energy density of the gas also differs. Since delivery of natural gas to the consumer is billed according to the energy quantities calculated from the gas analysis and the measured volume of gas, it is essential to analyze the gas composition as accurately as possible.

Moreover, the analysis of natural gas is carried out for many reasons: (i) identification of specific constituents, commonly minor constituents such as odorants, (ii) identification of the source of the gas either by fingerprinting or by molar composition, (iii) the application areas of increasing importance such as calculation (or estimation) of physical properties, and (iv) the measurement of gas quality when compared to the specification for use. However, the amount of detail that is required from the analysis depends upon the reason for the analysis, i.e. identification of source of the gas by molar composition might be done in some circumstances by an analysis up to pentane with a composite number for the presence of the higher MW hydrocarbon derivatives (such as the hexane and hydrocarbon derivatives up to, and including, C_8 or C_{10} hydrocarbon derivatives). Such data may also be suitable for estimation of the CV, assuming a single bulk contribution to the CV from this higher MW group (Chapter 6).

Direct methods of measurement of physical properties, particularly CV and relative density, are common and in some cases mandatory (Chapter 6). Chromatographic analysis followed by calculation of a property is a popular alternative for several reasons: (i) the apparatus is relatively inexpensive, (ii) the analysis will show why a change in property has occurred in addition to measuring the change, and (iii) a sufficiently detailed analysis allows several properties to be calculated at the same time. Nevertheless, the investigator should be aware of the risks of the use of *average values*, which may produce data that are not truly representative of the behavior of the gas mixture. Furthermore, in the case of property calculations related to phase changes, such as the hydrocarbon dew point, much more detail is needed about the distribution of the higher MW hydrocarbon derivatives. Briefly, the dew point is the temperature to which a given volume of gas must be cooled, at constant barometric pressure, for vapor to condense into liquid. Thus, the dew point is the saturation point that will vary as the carbon number of the hydrocarbon increases or decreases, thereby leading to results that are not consistent with the true behavior of the gas.

In summary, natural gas testing includes the analysis of conventional and shale gas, liquefied natural gas, and other hydrocarbon condensates and components (ASTM, 2017). The standard test methods that are cited within this book are, for convenience, the standard test methods developed by ASTM International (formerly the American Society for Testing and Materials) (ASTM, 2017). For example, standard test methods that might be used or consulted for comparison are (i) the British Standards (BS), which are the standards produced by BSI Group, located in Chiswick, London, which is incorporated under a Royal Charter and is formally designated as the national standards body (NSB) for the United Kingdom; (ii) the International Organization for Standardization (ISO), located in Geneva, Switzerland; and (iii) the German Institute for Standardization (DIN), located in Berlin, Germany, which publishes a variety of test methods that are applicable to determining the properties and behavior of natural gas and processed gases. Other organizations are available in several countries (Table 1.4) that might also be consulted as source of standard test methods.

However, the application of standard test methods need not (or does not) typically end with analysis of the gas. The mineralogy of the reservoir can (and often does) play a major role in the ability of the gas to be produced at the wellhead. Thus, there may be a need for the mineralogical analysis of samples of the reservoir rock.

1.3 Reservoirs

Natural gas is derived from aquatic plants and animals that lived and died hundreds of millions of years ago. Their remains mixed with mud and sand in layered deposits that, over the millennia, were geologically transformed into sedimentary rock. Gradually the organic matter decomposed and eventually formed petroleum (or a related precursor), which migrated from the original

Table 1.4 Examples of standards organizations in various countries (listed alphabetically).

Organization	Acronym	Country
American National Standards Institute	ANSI	United States
Badan Standardisasi Nasional	BSN	Indonesia
Brazilian National Standards Organization	ABNT	Brazil
British Standards Institution	BSI	United Kingdom
Bureau of Indian Standards	BIS	India
Bureau of Standards Jamaica	BSJ	Jamaica
Colombian Institute of Technical Standards and Certification	ICONTEC	Colombia
Deutsches Institut für Normung	DIN	Germany
Dirección General de Normas	DGN	Mexico
Ente nazionale italiano di unificazione	UNI	Italy
Estonian Centre for Standardisation	EVS	Estonia
Euro-Asian Council for Standardization, Metrology and Certification	GOST	Russia (Soviet Union)
Finnish Standards Association	SFS	Finland
French Association for Standardization	AFNOR	France
Instituto Argentino de Normalización y Certificación	IRAM	Argentina
Instituto Português da Qualidade	IPQ	Portugal
Japanese Industrial Standards Committee	JISC	Japan
Jabatan Standard Malaysia	DSM	Malaysia
Korean Agency for Technology and Standards	KATS	Korea (Republic)
Luxembourg Institute for Standardization	ILNAS	Luxembourg
Nederlandse Norm	NEN	Netherlands
Romanian Standards Association	ASRO	Romania
South African Bureau of Standards	SABS	South Africa
Spanish Association for Standardization and Certification	AENOR	Spain
Standardization Administration of China	SAC	China
Standards Australia	SAI	Australia
Standards Council of Canada	SCC	Canada
Standards New Zealand	SNZ	New Zealand
Standards Norway	SN	Norway
Standards Organisation of Nigeria	SON	Nigeria
Swedish Standards Institute	SIS	Sweden
Swiss Association for Standardization	SNV	Switzerland

source beds to more porous and permeable rocks, such as *sandstone* and *siltstone*, where it finally became entrapped. Such entrapped accumulations of petroleum are called *reservoirs*. A series of reservoirs within a common rock structure or a series of reservoirs in separate but neighboring formations is commonly referred to as an *oil field*. A group of fields is often found in a single geologic environment known as a *sedimentary basin* or *province*.

When a hydrocarbon reservoir is identified, it is important to also identify the types of fluids that are present, along with their main physicochemical characteristics. Generally, that information is obtained by performing a PVT analysis on a fluid sample of the reservoir. Conventional production measurements, such as a drill stem test (DST), are the typical parameters that can be measured almost immediately after a well is completed, as well as to obtain preliminary values of properties such as molar percentage of heptane and heavier components (% mole of C_{7+}), MW of the original fluid, maximum retrograde condensation (MRC), and dew-point pressure (P_d). Most of these properties are essential for exploitation of gas condensate reservoir, and their early availability will allow engineers to carry out reservoir studies that will ensure an efficient exploitation and maximize the final recovery of the liquids present in the reservoir. The only parameter needed to use these correlations is the value of the gas–condensate ratio (GCR) of the fluid during the early stage of production. These empirical equations should be valid for any gas condensate reservoir worldwide, although a range of usability is proposed for a better performance of the correlations. Moreover, just as crude oil can vary in composition, natural gas can also vary in composition. Differences in natural gas composition occur between different reservoirs, and two wells in the same field may also yield gaseous products that are different in composition (Speight, 1990; Mokhatab et al., 2006; Speight, 2007, 2014a).

Just as processes play a role in determining the composition and type of gaseous products, reservoirs (especially reservoir mineralogy) also play a major role in determining the composition and, hence, the behavior of natural gas. Different minerals adsorb gases at different rates, leading to a requirement for determining the mineralogical composition of the rock within the reservoir. The data related to mineralogical composition can lead to estimates of the gaseous constituents not produced through a well and the mean by which these constituents are held within the reservoir. In addition, the pressure within the reservoir may also be a major influence on the properties of the gas as it appears at the wellhead. For example, reservoir pressure varies depending on the depth, and because of the pressure, some of the typically gaseous constituents (at STP) of natural gas may be in liquid form and therefore appear at the surface as liquids (gas condensate) (Chapter 10).

In addition to the natural gas found in reservoirs that contain crude oil, there are also reservoirs in which natural gas may be the sole occupant. Again, the principal constituent of natural gas is methane, but other hydrocarbon derivatives, such as ethane, propane, and butane, may also be present but in lower proportions than those found in natural gas from crude oil reservoirs. Carbon dioxide is also a common constituent of natural gas. Trace amounts of rare gases, such as helium, may also occur, and certain natural gas reservoirs are a source of these rare gases.

On a worldwide scale, there are many reservoirs that produce gas condensate, and each reservoir will produce gas condensate with its own unique composition. However, in general, gas condensate has a specific gravity ranging from 0.5 to 0.8 and is composed of higher MW hydrocarbon derivatives such as pentane, hexane, heptane, and even octane, nonane, and decane in some cases. Propane and butane, normally gases at standard temperature and pressure, may also occur in gas condensate as gases soluble in the liquid hydrocarbon derivatives. In addition, gas condensate may contain additional impurities such as hydrogen sulfide, thiols (mercaptans, RSH), carbon dioxide, cyclohexane (C_6H_{12}), and low MW aromatics such as benzene (C_6H_6), toluene (C_6H5CH_3), ethylbenzene ($C_6H_5CH_2CH_3$), and xylene derivatives ($H_3CC_6H_4CH_3$) (Mokhatab et al., 2006; Speight, 2007, 2014a).

The composition of the gas condensate can have a major influence on the production of gas and condensate at the wellhead. For example, when condensation occurs in the reservoir, the phenomenon known as condensate blockage can halt flow of the liquids to the wellbore. Hydraulic fracturing is the most common mitigating technology in siliciclastic reservoirs (reservoirs composed of clastic rocks), and acidizing is used in reservoirs composed of carbonate mineral rocks (generally referred to as carbonate reservoirs) (Speight, 2016a). Briefly, clastic rocks are composed of fragments, or clasts, of preexisting minerals and rock. A clast is a fragment of geological detritus, chunks, and smaller grains of rock broken off other rocks by physical weathering. The geological term *clastic* is used with reference to sedimentary rocks as well as to particles in sediment transport whether in suspension or as bed load and in sedimentary deposits.

1.4 Conventional Gas

Natural gas resources, like crude oil resources, are typically divided into two categories: (i) conventional gas and (ii) unconventional gas (Mokhatab et al., 2006; Speight, 2007, 2014a). For the purposes of this text, the term unconventional gas resources also include CBM and natural gas from shale formations and from tight formations as well as biogas and landfill gas (John and Singh, 2011; Ramroop Singh, 2011; Singh and Sastry, 2011; Speight, 2011c, 2017b). Conventional gas is typically found in reservoirs with a permeability greater than 1 millidarcy (>1 mD) and can be extracted by means of traditional recovery methods. In contrast, unconventional gas is found in reservoirs with relatively low permeability (<1 mD) and hence cannot be extracted by conventional methods.

In addition, there are several general definitions that have been applied to natural gas from conventional formations that can also be applied to gas from tight formations. Thus, *lean* gas is gas in which methane is the predominant major constituent with other hydrocarbon constituents in the low minority, while *wet* gas contains considerable amounts of the higher MW hydrocarbon derivatives. *Sour* gas contains hydrogen sulfide and the equally odorous mercaptans, whereas *sweet* gas contains very little, if any, hydrogen sulfide or mercaptan. *Residue gas* is natural gas from which the higher MW hydrocarbon derivatives have been extracted, and *casinghead gas* is derived from crude oil but is separated at the separation facility at the wellhead. The term residue (as in residue gas) is used in relation to gas as a direct opposite as it is applied to crude oil in a refinery. In the refinery, the residue is the distillation residue of crude oil from which the lower MW constituents have been removed. In natural gas technology, residue gas is natural gas from which the higher MW constituents have been removed during gas processing operations (Chapter 4) to leave methane (the lower-boiling constituent) as residue gas.

Gas condensate (sometimes referred to as *condensate*) is a mixture of low-boiling hydrocarbon liquids obtained by condensation of the vapors of these hydrocarbon constituents either in the well or as the gas stream is emitted from the well. This is predominately pentane (C_5H_{12}) with varying amounts of higher-boiling hydrocarbon derivatives (up to C_8H_{18}) but relatively little methane or ethane; propane (C_3H_8) or butane (C_4H_{10}) may be present in condensate by dissolution in liquids. Depending upon the source of the condensate, benzene (C_6H_6), toluene ($C_6H_5CH_3$), xylene isomers ($CH_3C_6H_4CH_3$), and ethylbenzene ($C_6H_5C_2H_5$) may also be present (Mokhatab et al., 2006; Speight, 2007, 2014a).

The terms *condensate* and *distillate* are often used interchangeably to describe the liquid produced in tanks, but each term stands for a different material. Along with large volumes of gas, some wells produce a water-white or light straw-colored liquid that resembles low-boiling naphtha (Mokhatab et al., 2006; Speight, 2007, 2014a). The liquid has been called *distillate* because it resembles the products obtained from crude oil in refineries by distilling the volatile components from crude oil.

Lease condensate, so called because it is produced at the lease level from oil or gas wells, is the most common type of gas condensate and is typically a clear or translucent liquid. The API gravity of lease condensate ranges between 45 and 75°API, but, on the other hand, lease condensate with a lower API gravity can be black or near black color and, like crude oil, have higher concentrations of higher MW constituents. This condensate is generally recovered at atmospheric temperatures and pressures from wellhead gas production and can be produced along with large volumes of natural gas, and lease condensates with higher API gravity contains more NGLs, which include ethane, propane, and butane, but not many higher MW hydrocarbon derivatives.

Other terms applied to natural gas typically apply to the method by which the gas occurs in the reservoir. By way of explanation, natural gas is generated by any combination of (i) primary thermogenic degradation of organic matter, (ii) secondary thermogenic decomposition of petroleum, and (iii) biogenic degradation of organic matter. Gas generated by thermogenic and biogenic pathways may both exist in the same shale reservoir. After generation, the gas is stored in the reservoir formation in three different ways: (i) *adsorbed gas*, which is physically attached (adsorption) or chemically attached (chemisorption) to organic matter or to clay minerals; (ii) nonadsorbed gas, also known as *free gas* (also referred to as *nonassociated gas*), which occurs within the pore spaces in the reservoir rock or in spaces created by the rock cracking (fractures or microfractures); and (iii) solution gas, also referred to as *associated gas*, which is gas that exists in solution in liquids such as petroleum and heavy oil and (in the current context) in the gas condensate that occurs in some tight reservoirs with the gas (Speight, 2014a).

The amount of adsorbed gas component (typically methane) usually increases with an increase in organic matter or surface area of organic matter and/or clay. On the beneficial side, a higher free gas (nonassociated) content in unconventional tight reservoirs generally results in higher initial rates of production because the free gas resides in fractures and pores and, when production is commenced, moves easier through the fractures (induced channels) relative to any adsorbed gas. However, the high initial flow rate will decline rapidly to a low, steady rate as the nonassociated gas is produced, leaving the adsorbed gas to move to the well as it is slowly released from the shale.

1.4.1 Associated Gas

Crude oil cannot be produced without producing some associated gas, which consists of low-boiling hydrocarbon constituents that are emitted from solution in the crude oil as the pressure is reduced on the way to, and on, the surface. Well completion designs and reservoir management protocols are used to minimize the production of associated gas to retain the maximum energy in the reservoir and thus increase ultimate recovery of the crude oil (Parkash, 2003; Hsu and Robinson, 2006; Gary et al., 2007; Speight, 2014a, 2017a). Crude oil in the reservoir with minimal or no dissolved associated gas is rare and, as *dead crude oil*, is often difficult to produce as there is little reservoir energy to drive the oil into the production well and to the surface. Thus, *associated* or *dissolved* natural gas occurs either as free gas or as gas in solution in the petroleum. Gas that occurs as a solution in the petroleum is *dissolved* gas, whereas the gas that exists in contact with the petroleum is *associated* gas – the *gas cap* is an example of associated gas (Parkash, 2003; Hsu and Robinson, 2006; Mokhatab et al., 2006; Gary et al., 2007; Speight, 2014a, 2017a).

After the production fluids are brought to the surface, the gas is treated to separate out the higher MW NGLs, which are treated in a liquefied petroleum gas (LPG) processing (refining) plant to provide propane and butane, either separately or as a mixture of the two. By definition, NGLs include ethane, propane, butanes, pentanes, and higher MW hydrocarbon derivatives (C_{6+}). The higher MW hydrocarbon derivative product is commonly referred to as *natural gasoline* or *gas condensate.* The LPG is stored ready for transport, and the nonvolatile residue (i.e. nonvolatile under the conditions of the separation process), after propane and butane are removed, is gas condensate (or, simply, condensate), which is mixed with the crude oil or exported as a separate product (low-boiling naphtha) (Mokhatab et al., 2006; Speight, 2007, 2014a).

Rich gas has a high heating value and a high hydrocarbon dew point. However, the terms *rich gas* and *lean gas,* as used in the gas processing industry, are not precise indicators of gas quality but only indicate the relative amount of NGLs in the gas stream. When referring to NGLs in the natural gas stream, the term *gallons per thousand cubic feet* of gas is used as a measure of hydrocarbon richness.

Thus, in the case of associated gas, crude oil may be assisted up the wellbore by gas lift (Mokhatab et al., 2006; Speight, 2007, 2014a) in which the gas is compressed into the annulus of the well and then injected by means of a gas lift valve near the bottom of the well into the crude oil column in the tubing. At the top of the well, the crude oil and gas mixture passes into a separation plant (consisting of high-pressure and low-pressure separators) in which the gas pressure is reduced considerably in two stages. The crude oil and water exit the bottom of the lower-pressure separator, from where it is pumped to tanks for separation of the crude oil and water. The gas produced in the separators is recompressed, and the gas that comes out of solution with the produced crude oil (surplus gas) is then treated to separate out the NGLs that are treated in a gas plant to provide propane and butane or a mixture of the two (LPG) (Table 1.5). The higher-boiling residue, after propane and butane are removed, is the condensate, which is mixed with the crude oil or exported as a separate product. At each stage of this process (often referred to under the collective term *wellhead processing*), the composition of the gaseous and liquid products should be monitored to determine separator efficiency as well as for safety reasons.

The gas itself is then *dry* and, after compression, is suitable to be injected into the natural gas system where it substitutes for natural gas from the non-associated gas reservoir. Pretreated associated gas from other fields can also enter the system at this stage. Another use of the gas is as fuel for the gas

Table 1.5 General properties of unrefined natural gas (left-hand number) and refined natural gas (right-hand number).

Relative molar mass	20–16
Carbon content (% w/w)	73–75
Hydrogen content (% w/w)	27–25
Oxygen content (weight % w/w)	0.4–0
Hydrogen-to-hydrogen atomic ratio	3.5–4.0
Density relative to air at 15 °C	1.5–0.6
Boiling temperature (°C/1 atm)	−162
Autoignition temperature (°C)	540–560
Octane number	120–130
Methane number	69–99
Vapor flammability limits (% v/v)	5–15
Flammability limits	0.7–2.1
Lower heating/calorific value (Btu lb^{-1})	900
Methane concentration (% v/v)	100–80
Ethane concentration (% v/v)	5–0
Nitrogen concentration (% v/v)	15–0
Carbon dioxide concentration (% v/v)	5–0
Sulfur concentration (ppm, w/w)	5–0

turbines on-site. This gas is treated in a fuel gas plant to ensure it is clean and at the correct pressure. The start-up fuel gas supply will be from the main gas system, but facilities exist to collect and treat low-pressure gas from the various other plants as a more economical fuel source.

Other components such as carbon dioxide (CO_2), hydrogen sulfide (H_2S), and mercaptans (thiols; RSH), as well as trace amounts of other constituents, may also be present. Thus, there is no single composition of components that might be termed *typical natural gas* because of the variation in composition of the gas from different reservoirs, even from different wells from the same reservoir. Methane and ethane constitute the bulk of the combustible components; carbon dioxide (CO_2) and nitrogen (N_2) are the major noncombustible (inert) components.

1.4.2 Nonassociated Gas

In addition to the natural gas found in petroleum reservoirs, there are also those reservoirs in which natural gas is the sole occupant and is referred to as *nonassociated gas*. As with associated gas, the principal constituent of nonassociated gas is methane – higher MW hydrocarbon derivatives may also be present but in lower quantities than found in associated gas.

Thus, *nonassociated gas* (sometimes called *gas well gas*) is produced from geological formations that typically do not contain much, if any, crude oil or higher-boiling hydrocarbon derivatives (*gas liquids*) than methane. The nonassociated gas recovery system is somewhat simpler than the associated gas recovery system. The gas flows up the well under its own energy, through the wellhead control valves, and along the flow line to the treatment plant.

Processing of nonassociated gas is somewhat less complicated than processing of associated gas. Typically, nonassociated gas flows up the production well under the reservoir energy and then through the wellhead control valves and along the flow line to the wellhead processing plant. At this stage, the first processing option is to reduce the temperature of the gas to a point dependent upon the pressure in the pipeline so that the higher MW constituents that would exist as liquids at the temperature and pressure of the pipeline condense to a liquid phase and are removed in a separator. The temperature is reduced by expanding the gas through a Joule–Thomson valve, although other methods of removal do also exist (Mokhatab et al., 2006; Speight, 2007, 2014a). Briefly, the Joule–Thomson effect (also known as the Joule–Kelvin effect, the Kelvin–Joule effect, or the Joule–Thomson expansion) relates to the temperature change of a gas or liquid when it is forced through a valve while kept insulated so that no heat is exchanged with the environment.

Water in the gas stream must also be removed to mitigate the potential for the formation of gas hydrates that would block flow lines and have the potential for explosive dissociation. One method for water removal from the gas stream involves the injection of ethylene glycol ($HOCH_2CH_2OH$, also referred to as *glycol*), which combines with water and is later recovered in a glycol plant (Mokhatab et al., 2006; Speight, 2007, 2014a). The treated gas then passes from the top of the treatment vessel and into the pipeline. The water is treated in a glycol plant to recover the glycol, and the fraction of the natural gas stream that has been isolated as NGLs is sent, as additional feedstock, to the LGP plant. Alternatively, the lower-boiling constituents of the NGLs may be used as feedstock for the production of petrochemicals (Parkash, 2003; Hsu and Robinson, 2006; Gary et al., 2007; Speight, 2014a, 2017a).

Finally, another aspect of gas processing that requires attention (Chapter 4) and is worthy of mention here is the removal of sulfur from natural gas. The potential of sulfur-containing constituents, such as hydrogen sulfide (H_2S) and mercaptans (RSH) to corrode shipping equipment (such as pipelines), is high – especially in the presence of water (Speight, 2014b). Once the hydrogen sulfide has been removed by a suitable wellhead treatment process, it is environmentally undesirable to flare the hydrogen sulfide, so where there are significant quantities in the gas stream, it is converted into elemental sulfur and used for the manufacture of sulfuric acid and other products (Chapter 4). Sulfur can be transported over long distances by being pumped as a liquid at a temperature on the order of 120 °C (248 °F) through an insulated pipeline, which is maintained at this temperature by a counterflow of hot pressurized water.

1.4.3 Refinery Gas

The terms *refinery gas* and *petroleum gas* are often used to identify LPG or even gas that emanates from the top of a refinery distillation column. Refinery gas varies in composition and volume, depending on crude origin and on any additions to the crude made at the loading point – in this sense, it is probably the most difficult gas to analyze. Furthermore, it is not uncommon to reinject low MW hydrocarbon derivatives such as propane and butane into the crude before dispatch by tanker or pipeline. This results in a higher vapor pressure of the crude, but it allows one to increase the quantity of light products obtained at the refinery. Since light ends in most petroleum markets command a premium, while in the oil field itself propane and butane may have to be reinjected or flared, the practice of *spiking* crude oil with LPG is becoming common.

Thus, *refinery gas* (*fuel gas*) is the noncondensable gas that is obtained during distillation of crude oil or treatment (cracking, thermal decomposition) of petroleum. Refinery gas is produced in considerable quantities during the different refining processes and is used as fuel for the refinery itself and as an important feedstock to produce petrochemicals. It consists mainly of hydrogen (H_2), methane (CH_4), ethane (C_2H_6), propane (C_3H_8), butane (C_4H_{10}), and olefins ($RCH{=}CHR^1$, where R and R^1 can be hydrogen or a methyl group) and may also include off-gases from petrochemical processes. Olefins such as ethylene ($CH_2{=}CH_2$, boiling point: −104 °C, −155 °F), propene (propylene, $CH_3CH{=}CH_2$, boiling point: −47 °C, −53 °F), butene (butene-1, $CH_3CH_2CH{=}CH_2$, boiling point: −5 °C, 23 °F), *iso*-butylene ($(CH_3)_2C{=}CH_2$, −6 °C, 21 °F), *cis*- and *trans*-butene-2 ($CH_3CH{=}CHCH_3$, boiling point: ca. 1 °C, 30 °F), and butadiene ($CH_2{=}CHCH{=}CH_2$, boiling point: −4 °C, 24 °F) as well as higher-boiling olefins are produced by various refining processes.

Still gas is broad terminology for low-boiling hydrocarbon mixtures and is the lowest boiling fraction isolated from a distillation (*still*) unit in the refinery. If the distillation unit is separating light hydrocarbon fractions, the still gas will be almost entirely methane with only traces of ethane (CH_3CH_3) and ethylene ($CH_2{=}CH_2$). If the distillation unit is handling higher-boiling fractions, the still gas might also contain propane ($CH_3CH_2CH_3$), butane ($CH_3CH_2CH_2CH_3$), and their respective isomers. *Fuel gas* and still gas are terms that are often used interchangeably, but the term *fuel gas* is intended to denote the product's destination – to be used as a fuel for boilers, furnaces, or heaters.

In a series of processes commercialized as Platforming, Powerforming, Catforming, and Ultraforming, paraffinic and naphthenic (cyclic nonaromatic) hydrocarbon derivatives are converted into aromatics in the presence of hydrogen and a catalyst or isomerized to more highly branched hydrocarbon derivatives. Catalytic reforming processes thus not only result in the formation of a liquid product of higher octane number but also produce substantial quantities of gases. The latter are rich in hydrogen but also contain hydrocarbon derivatives from methane to butanes, with a preponderance of propane ($CH_3CH_2CH_3$), *n*-butane ($CH_3CH_2CH_2CH_3$), and isobutane [$(CH_3)_3CH$]. The composition of these gases varies in accordance with reforming severity and reformer feedstock. Since all catalytic reforming processes require substantial recycling of a hydrogen stream, it is normal to separate reformer gas into a propane ($CH_3CH_2CH_3$) and/or a butane [$CH_3CH_2CH_2CH_3/(CH_3)_3CH$] stream, which becomes part of the refinery LPG production, and a lighter gas fraction, part of which is recycled. In view of the excess of hydrogen in the gas, all products of catalytic reforming are saturated, and there are usually no olefin-type gases present in either gas stream.

A second group of refining operations that contributes to gas production is that of the catalytic cracking processes. These include Thermofor catalytic cracking and other variants in which heavy gas oils are converted into cracked gas, LPG, catalytic naphtha, fuel oil, and coke by contacting the heavy hydrocarbon with the hot catalyst. Both catalytic and thermal cracking processes, the latter being now largely used to produce chemical raw materials, result in the formation of unsaturated hydrocarbon derivatives, particularly ethylene ($CH_2{=}CH_2$), as well as propylene (propene, $CH_3CH{=}CH_2$), isobutylene [isobutene, $(CH_3)_2C{=}CH_2$], and the *n*-butenes ($CH_3CH_2CH{=}CH_2$, and $CH_3CH{=}CHCH_3$) in addition to hydrogen (H_2), methane (CH_4), and smaller quantities of ethane (CH_3CH_3), propane ($CH_3CH_2CH_3$), and butane isomers [$CH_3CH_2CH_2CH_3$, $(CH_3)_3CH$]. Diolefins such as butadiene ($CH_2{=}CHCH{=}CH_2$) are also present.

Additional gases are produced in refineries with coking or visbreaking facilities for the processing of their heaviest crude fractions. In the visbreaking process, fuel oil is passed through externally fired tubes and undergoes liquid-phase cracking reactions, which result in the formation of lighter fuel oil components. Oil viscosity is therefore reduced, and some gases, mainly hydrogen, methane, and ethane, are formed. Substantial quantities of both gas and carbon are also formed in coking (both fluid coking and delayed coking) in addition to the middle distillate and naphtha. When coking a residual fuel oil or heavy gas oil, the feedstock is preheated and contacted with hot carbon (coke), which causes extensive cracking of the feedstock constituents of higher MW to produce lower MW products ranging from methane, via LPG(s) and naphtha, to gas oil and heating oil. Products from coking processes tend to be unsaturated, and olefin-type components predominate in the tail gases from coking processes.

A further source of refinery gas is hydrocracking, a catalytic high-pressure pyrolysis process in the presence of fresh and recycled hydrogen. The feedstock is again heavy gas oil or residual fuel oil, and the process is mainly directed at the production of additional middle distillates and gasoline. Since hydrogen is to be recycled, the gases produced in this process again must be separated into lighter and heavier streams; any surplus recycle gas and the LPG from the hydrocracking process are both saturated.

Both hydrocracker and catalytic reformer tail gases are commonly used in catalytic desulfurization processes (Table 1.6). In the latter, feedstocks ranging from light to vacuum gas oils are passed at pressures of 500–1000 psi with hydrogen over a hydrofining catalyst. This results mainly in the conversion of organic sulfur compounds to hydrogen sulfide:

$$[S]_{feedstock} + H_2 \rightarrow H_2S + \text{hydrocarbon derivatives}$$

The process also produces some light hydrocarbon derivatives by hydrocracking.

The first and most important aspect of gaseous testing is the measurement of the volume of gas. However, in all cases, it is the composition of the gas in terms of *hydrocarbon type* that is more important in the context of application.

Table 1.6 Origin of petroleum-related gases.

Gas	Origin
Natural gas	Occurs naturally with or without the presence of crude oil in the reservoir
	A gaseous combination of low-boiling hydrocarbon derivatives
	Predominantly C_1 up to and including C_4 hydrocarbon derivatives
	May also contain gas condensate (C_{5+} hydrocarbon derivative) or natural gasoline (also C_{5+} hydrocarbon derivatives)
Refinery gas process gas	A combination of gases produced by distillation
	Products from the cracking of crude oil
	Consists of C_2 to C_4 hydrocarbon derivatives including olefin gases
	Boiling range of approximately −51−−1 °C (−60–30 °F)
Tail gas	A combination of hydrocarbon derivatives from the distillation of products from catalytically cracked feedstocks
	Predominantly of C_1 to C_4 hydrocarbon derivatives

For example, in petrochemical applications, the presence of propylene and butylene above 10% v/v can have an adverse effect on hydrodesulfurization prior to steam reforming. On the other hand, petrochemical processes, such as in the production of *iso*-octane from *iso*-butane and butylene, can require the exclusion of the saturated hydrocarbon derivatives. Furthermore, refinery gas specifications will vary depending on the gas quality available and the end use. For fuel uses, gas as specified above presents little difficulty used as supplied. Alternatively, a gas of constant Wobbe index, say, for gas turbine use, could readily be produced by the user. Part of the combustion air would be diverted into the gas stream by a Wobbe index controller, which would be preset to supply the gaseous fuel at the lowest Wobbe index of the undiluted gas.

Still gas is a term reserved for low-boiling hydrocarbon mixtures that represent the lowest boiling fraction isolated from an atmospheric distillation unit (*still*) in the refinery (Table 1.6). Typically, the still gas will contain methane, ethane (CH_3CH_3), propane ($CH_3CH_2CH_3$), butane ($CH_3CH_2CH_2CH_3$), and their respective isomers. If cracked products have been recycled to the atmospheric distillation unit, the still gas may also contain volatile olefins such as ethylene ($CH_2{=}CH_2$).

1.5 Unconventional Gas

There are several types of unconventional gas resources that are currently produced: (i) shale gas and tight natural gas, natural gas that occurs in low-permeability shale formations; (ii) geopressurized zones, natural underground formations that are under unusually high pressure for their depth; (iii) CBM, natural gas that occurs in conjunction with coal seams; and (iv) methane hydrates, natural gas that occurs at low-temperature and high-pressure regions such as the sea bed and is made up of a lattice of frozen water, which forms a *cage* around the methane (Mokhatab et al., 2006; Speight, 2007, 2013b, 2014a).

Typically, natural gas produced from shale reservoirs and other tight reservoirs has been classed under the general title *unconventional gas*. The production process requires stimulation by horizontal drilling coupled with hydraulic fracturing because of the extremely low permeability of the reservoir rocks that prohibits natural movement of the gas to a well. Moreover, maximization or optimization of reservoir producibility can only be achieved by a thorough understanding of the occurrence and properties of the shale gas resources as well as the producibility of the gas from the reservoir (Kundert and Mullen, 2009).

The boundary between conventional gas and unconventional gas resources is not well defined, because they result from a continuum of geologic conditions. Coal seam gas, more frequently called coalbed methane, is frequently referred to as unconventional gas. Tight shale gas and gas hydrates are also placed into the category of *unconventional gas*.

1.5.1 Gas in Geopressurized Zones

Geopressurized zones are natural underground formations that are under high pressure that is not always expected from the depth of the zone. These areas are formed by layers of clay minerals that are deposited and compacted on top of (or above) more porous, absorbent material such as sandstone or silt. Water and natural gas that are present in this clay formation are forced out by the rapid compression and enter the more porous sandstone or silt deposits. The natural gas or crude oil is deposited in this sandstone or silt under very high pressure (*geopressure*). Geopressurized zones are typically located at great depths – on the order of usually 10 000–25 000 ft below the surface of the Earth. The combination of the above factors makes the extraction of natural gas or crude oil located in geopressurized zones quite complicated (Speight, 2017b).

1.5.2 Coalbed Methane

Just as natural gas is often located in the same reservoir as with crude oil, it can also be found trapped within coal seams, where it is often referred to as coalbed methane (or *coal bed methane*, sometimes referred to as *coalmine methane*). Coalbed methane is a generic term for the methane found in most coal seams. To the purist, coalmine methane is the fraction of CBM that is released during the mining operation (referred to in the older literature as *firedamp* by miners because of its explosive nature). In practice, the terms coalbed methane and coalmine methane may usually refer to different sources of gas – both forms of gas, whatever the name, are equally dangerous to the miners. Methane in coalbeds is a gas formed as part of the geological process of coal generation and is contained in varying quantities within all coal. CBM is exceptionally pure compared with conventional natural gas, containing only very small proportions of higher MW hydrocarbon derivatives such as ethane and butane and other gases (such as hydrogen sulfide and carbon dioxide).

CBM is found within geological formations of coal deposits (Speight, 2013b). Because coal is a solid, very high carbon content mineral, there are usually no liquid hydrocarbon derivatives contained in the produced gas. The coalbed (coal seam) must first be dewatered to allow the trapped gas to flow through the formation to produce the gas. Consequently, CBM usually has a lower heating value and elevated levels of carbon dioxide, oxygen, and water that must be treated to an acceptable level, given the potential to be corrosive. However, the occurrence of methane in coal seams is not a new discovery, and methane (also called *firedamp*) was known to coal miners for at least 150 years (or more) before it was *rediscovered* and developed as CBM (Speight, 2013b). The gas occurs in the pores and cracks in the coal seam and is held there by underground water pressure. To extract the gas, a well is drilled into the coal seam, and the water is pumped out (*dewatering*), which allows the gas to be released from the coal and brought to the surface.

CBM is predominantly methane extracted from coalbeds (coal seams) and has become an important source of energy in the United States, Canada, and other countries. The gas is adsorbed into the solid matrix of the coal and is often referred to as *sweet gas* because of its lack of hydrogen sulfide. Unlike natural gas from conventional reservoirs, CBM contains very little higher MW hydrocarbon derivatives such as propane or butane or condensate. It often contains up to a few percent of carbon dioxide. Some coal seams contain little methane, with the predominant coal seam gas being carbon dioxide. In fact, CBM has been suggested with sufficient justification that the materials comprising a coalbed fall broadly into the following two categories: (i) volatile low MW materials (components) that can be liberated from the coal by pressure reduction, mild heating, or solvent extraction and (ii) materials that will remain in the solid state after the separation of volatile components.

Typically, with some exceptions, coalbed gas is typically in excess of 90% v/v methane and, subject to gas composition data, may be suitable for introduction into a commercial pipeline with little or no treatment (Rice, 1993; Levine, 1993; Mokhatab et al., 2006; Speight, 2007, 2013a). Methane within coalbeds is not structurally trapped by overlying geologic strata, as in the geologic environments typical of conventional gas deposits (Speight, 2007, 2013a, 2014a). Only a small amount (on the order of 5–10% v/v) of the CBM is present as free gas within the joints and cleats of coalbeds. Most of the CBM is contained within the coal itself (adsorbed to the sides of the small pores in the coal).

As the coal forms from the organic detritus (the precursor material), large quantities of methane-rich gas are produced and subsequently adsorbed onto the surface and within the pore system of the coal matrix. Because of its many natural cracks and fissures, as well as its porous nature, coal in the seam has a large internal surface area and can store much more gas than a conventional natural gas reservoir of similar rock volume. If a seam is disturbed, either during mining or by drilling into it before mining, methane is released from the surface and interior of the coal. This methane then leaks into any open spaces such as fractures in the coal seam (known as *cleats*). In these fractures, the coalmine methane mixes with nitrogen and carbon dioxide (CO_2). Boreholes or wells can be drilled into the seams to recover the methane. Large amounts of coal are found at shallow depths, where wells to recover the gas are relatively easy to drill at a relatively low cost. At greater depths, increased pressure may have closed the fractures, or minerals may have filled the fractures over time, lowering the permeability and thereby making it more difficult for the methane (and any other gases) to move through the coal seam.

The primary (or natural) permeability of coal is very low, typically ranging from 0.1 to 30 mD, and because coal is a very weak (low modulus) material and cannot take much stress without fracturing, it is almost always highly fractured and cleated. The resulting network of fractures commonly gives coalbeds a high secondary permeability (despite coal's typically low primary permeability). Groundwater, hydraulic fracturing fluids, and methane gas can more easily flow through the network of fractures. Because hydraulic fracturing generally enlarges preexisting fractures in addition to creating new fractures, this network of natural fractures is very important to the extraction of methane from the coal (Speight, 2013b, 2016b).

The gas from coal seams can be extracted by using technologies that are similar to those used to produce conventional gas, such as using wellbores. However, complexity arises from the fact that the coal seams are generally of low permeability and tend to have a lower flow rate (or permeability) than conventional gas systems, gas is only sourced from close to the well, and as such a higher density of wells is required to develop a CBM resource as an unconventional resource (such as tight gas) than a conventional gas resource.

Technologies such as horizontal and multilateral drilling with hydraulic fracturing are sometimes used to create longer, more open channels that enhance well productivity, but not all coal seam gas wells require application of this technique (Speight, 2013a, 2016b). Water present in the coal seam, either naturally occurring or introduced during the fracturing operation, is usually removed to reduce the pressure sufficiently to allow the gas to be released, which leads to additional operational requirements, increased investment, and environmental concerns. There are some horizontally drilled CBM wells and some that receive hydraulic fracturing treatments. However, some CBM reservoirs are also underground sources of drinking water, and as such, there are restrictions on hydraulic fracturing operations. CBM wells are mostly shallow, as the coal matrix does not have the strength to maintain porosity under the pressure of significant overburden thickness.

1.5.3 Gas in Tight Formations

The term *tight formation* refers to a formation consisting of extraordinarily impermeable, hard rock (Speight, 2013a). Tight formations are relatively low-permeability, nonshale, sedimentary formations that can contain oil and gas. A *tight reservoir* (*tight sands*) is a low-permeability sandstone reservoir that produces primarily dry natural gas. A tight gas reservoir is one that cannot be produced at economic flow rates or recover economic volumes of gas unless the well is stimulated by a large hydraulic fracture treatment and/or produced using horizontal wellbores. This definition also applies to CBM and tight carbonate reservoirs – shale gas reservoirs are also included by some observers (but not in this text). Typically, tight formations that formed under marine conditions contain less clay and are more brittle and thus more suitable for hydraulic fracturing than formations formed in freshwater that may contain more clay. The formations become more brittle with an increase in quartz content (SiO_2) and carbonate content (such as calcium carbonate, $CaCO_3$, or dolomite, $CaCO_3{\cdot}MgCO_3$).

By way of explanation and comparison, in a conventional sandstone reservoir, the pores are interconnected so that natural gas and crude oil can flow easily through the reservoir and to the production well. However, in tight sandstone formations, the pores are smaller and are poorly connected (if at all) by very narrow capillaries, which results in low permeability and immobility of the natural gas. Such sediments typically have an effective permeability of less than 1 millidarcy (<1 mD).

Shale is a sedimentary rock characterized by low permeability (Aguilera and Radetzki, 2014; Khosrokhavar et al., 2014), mainly compositing of mud, silts, and clay minerals, but, however, this composition varies with burial depth and tectonic stresses (Moslemizadeh et al., 2015). Shale reservoirs have a permeability that is three orders times less than tight gas reservoirs (Khosrokhavar et al., 2014). Natural gas and light tight oil (sometime erroneously referred to as shale oil – by way of definition, shale oil is the liquid product produced by the decomposition of the kerogen component of oil shale) are confined in the pore spaces of these impermeable rocks. The gas in a shale formation is present as a free gas in the pore spaces or is adsorbed by clay minerals and organic matters (Ross and Bustin, 2009). The free gas will be produced upon well completion, but the production of adsorbed gas depends on the pressure drop needed for desorption. It is, therefore, essential to know the relative amounts of free and adsorbed gas in shales. On the other hand, oil shale is a kerogen-rich petroleum source rock that was not buried under the correct maturation conditions to experience the temperatures required to generate oil and gas (Speight, 2014a, AAPG, 2015).

Natural gas from shale formations and other formations designated as and tight formations (i.e. formations having zero to extremely low permeability) offers additional energy-producing resources. This gas can be recovered from such formations that typically function as both the reservoir rock and the source rock. The tight formations that yield gas and oil are organic-rich shale formations that were previously regarded only as source rocks as well as seals for gas accumulating in the strata near sandstone and carbonate reservoirs of traditional onshore gas development.

In addition, the gas storage properties of such formations are quite different to conventional reservoirs. In the case of shale formations, the natural gas present in the matrix system of pores – similar to that found in conventional reservoir rocks – is accompanied by gas that is bound to, or adsorbed on, the surface of inorganic minerals the formation, especially clay minerals that are often present in shale. The relative contributions and combinations of free gas from matrix porosity and from desorption of adsorbed constituents is a key determinant of the production profile of the well. Thus, by definition, *shale gas* is the hydrocarbon gas present in organic-rich, fine-grained sedimentary rocks (shale and associated lithofacies). The gas is generated and stored *in situ* in gas shale as both adsorbed gas (on organic matter) and free gas (in fractures or pores). As such, shale containing gas is a self-sourced reservoir. Low-permeable shale requires extensive fractures (natural or induced) to produce commercial quantities of gas.

In terms of chemical makeup, shale gas is typically a dry gas composed primarily of methane (60–95% v/v), but some formations do produce wet gas (Table 1.7). The Antrim and New Albany plays have typically produced water and gas. Gas shale formations are organic-rich shale formations that were previously regarded only as source rocks and seals for gas accumulating in the strata near sandstone and carbonate reservoirs of traditional onshore gas development.

Tight gas is the fastest-growing natural gas resource in the United States and worldwide because of several recent developments. Advances in horizontal drilling technology allow a single well to pass through larger volumes of a shale gas reservoir and, thus, produce more gas. The development of hydraulic fracturing technology has also improved access to shale gas deposits. This process requires injecting large volumes of water mixed with sand and fluid chemicals into the well at high pressure to fracture the rock, increasing permeability and production rates.

Table 1.7 Composition of dry gas, wet gas, and gas condensate.

Component or property	Dry gas	Wet gas	Condensate
CO_2	0.10	1.41	2.37
N_2	2.07	0.25	0.31
C_1	86.12	92.46	73.19
C_2	5.91	3.18	7.80
C_3	3.58	1.01	3.55
$i\,C_4$	1.72	0.28	0.71
$n\,C_4$	—	0.24	1.45
$i\,C_5$	0.50	0.13	0.64
$n\,C_5$	—	0.08	0.68
C_6	—	0.14	1.09
C_{7+}	—	0.82	8.21
Molecular weight C_{7+}	—	132	184
Specific gravity C_{7+}	—	0.750	0.816

Briefly, to further define the terms *dry gas* and *wet gas* in quantitative measures, the term *dry* natural gas indicates that there is less than 0.1 gal (1 gal, US = 264.2 m^3) of gasoline vapor (higher MW paraffins) per 1000 ft^3 (1 ft^3 = 0.028 m^3). The term *wet natural gas* indicates that there are such paraffins present in the gas, in fact more than 0.1 gal/1000 ft^3.

The shale formations that yield gas and oil are organic-rich shale formations that were previously regarded only as source rocks and seals for gas accumulating in the strata near sandstone and carbonate reservoirs of traditional onshore gas development. On the other hand, the crude oil from shale formations is a light, highly volatile crude oil that contains more of the volatile hydrocarbon derivatives (C1–C4) than many conventional crude oils. In addition, the gas and oil storage properties of shale are quite different to conventional reservoirs. In the case of shale formations, the natural gas and crude oil present in the matrix system of pores – like the gas found in conventional reservoir rocks – is accompanied by gas or oil that is bound to, or adsorbed on, the surface of inorganic minerals in the shale. The relative contributions and combinations of free gas and oil from matrix porosity and from desorption of adsorbed constituents are a key determinant of the production profile of the well.

Natural gas from shale and tight formations offer additional energy-producing resources that can be recovered from shale formations that typically function as both the reservoir rock and the source rock. In terms of chemical makeup, shale gas is typically a dry gas of which methane is the primary constituent, but some formations do produce wet gas. The shale formations that yield gas are organic-rich shale formations that were previously regarded only as source rocks and seals for gas accumulating in the strata near sandstone and carbonate reservoirs of traditional onshore gas development. In the case of shale formations, the natural gas present in the matrix system of pores – like that found in conventional reservoir rocks – is accompanied by gas that is bound to, or adsorbed on, the surface of inorganic minerals in the shale. The relative contributions and combinations of free gas from matrix porosity and from desorption of adsorbed constituents is a key determinant of the production profile of the well.

1.5.4 Gas Hydrates

In addition to CBM (Section 1.5.2), another relatively new and possibly large source of methane that can be expected to extend the availability of natural gas is *methane hydrate* (also called *gas hydrate* and *methane ice*). Their production technologies have only recently been developed, and these sources are now becoming economically competitive.

Methane hydrates (also called *methane clathrates*) are a resource in which a large amount of methane is trapped within a crystal structure of water, forming a solid that resembles ice (Kvenvolden and McMenamin, 1980; Kvenvolden, 1995; Lorenson and Collett, 2000; Buffett and Archer, 2004; Gao et al., 2005; Gao, 2008; USGS, 2011). Natural gas hydrates are solids that form from a combination of water and one or more hydrocarbon or nonhydrocarbon gases. In physical appearance, gas hydrates resemble packed snow or ice. In a gas hydrate, the gas molecules (such as methane, hence the name *methane hydrates*) are trapped within a cagelike crystal structure composed of water molecules. Gas hydrates are stable only under specific conditions of pressure and temperature. Under an appropriate pressure, they can exist at temperatures significantly above the freezing point of water. The maximum temperature at which gas hydrate can exist depends on pressure and gas composition. For example, methane plus water at 600 psia forms hydrate at 5 °C (41 °F), while at the same pressure, methane with 1% v/v propane forms a gas hydrate at 9.4 °C (49 °F). Hydrate stability can also be influenced by other factors, such as salinity (Edmonds et al., 1996).

Estimates of the global inventory of methane clathrate have exceeded 10^{19} g of carbon (MacDonald, 1990a, b; Gornitz and Fung, 1994), which is comparable with estimates of potentially recoverable coal, oil, and natural gas. The proximity of this methane reservoir to the seafloor has motivated speculations about a release of methane in response to climate change (MacDonald, 1990b). Increases in temperature or decreases in pressure (through changes in sea level) tend to dissociate clathrate, releasing methane into the near-surface environment. Release of methane from clathrate has been invoked to explain abrupt increases in the atmospheric concentration of methane during the last glacial cycle.

On a unit volume basis, gas hydrates contain a high amount of gas. For example, 1 cubic yard of hydrate disassociates at atmospheric temperature and pressure to form approximately 160 cubic yards of natural gas plus 0.8 cubic yards of water (Kvenvolden, 1993). The natural gas component of gas hydrates is typically dominated by methane, but other natural gas components (e.g. ethane, propane, carbon dioxide) can also be incorporated into a hydrate. The origin of the methane in a hydrate can be either thermogenic gas or biogenic gas. Bacterial gas formed during early digenesis of organic matter can become part of a gas hydrate in continental shelf sediment. Similarly, thermogenic gas leaking to the surface from a deep thermogenic gas accumulation can form a gas hydrate in the same continental shelf sediment (Collett, 2002; Gupta, 2004; Collett et al., 2009; Demirbaş, 2010a, b, c; Chong et al., 2016).

Generally, methane clathrates are common constituents of the shallow marine geosphere, and they occur both in deep sedimentary structures and form outcrops on the ocean floor. Methane hydrates are believed to form by migration of gas from depth along geological faults, followed by precipitation, or crystallization, on contact of the rising gas stream with cold seawater. When drilling in petroleum-bearing and gas-bearing formations submerged in deep water, the reservoir gas may flow into the wellbore and form gas hydrates owing to the low temperatures and high pressures found during deepwater drilling. The gas hydrates may then flow upward with drilling mud or other discharged fluids. When the hydrates rise, the pressure in the annulus decreases and the hydrates dissociate into gas and water. The rapid gas expansion ejects fluid from the well, reducing the pressure further, which leads to more hydrate dissociation (sometimes explosive dissociation) and further fluid ejection.

As might be anticipated, the gas that is evolved from gas hydrates under laboratory conditions is composed of the following hydrocarbon gases: methane (C1), ethane (C2), and the sum C3+, composed of propane (C3), isobutane (*i*-C4), and *n*-butane (*n*-C4) (Lorenson and Collet, 2000). Other hydrocarbon gases present in some sediments include neopentane (neo—C5), isopentane (*i*-C5), *n*-pentane (*n*-C5), cyclopentane (cycloC5), neohexane (neo—C6), isohexane (*i*-C6), *n*-hexane (*n*-C6), isoheptane (*i*-C7), and methylcyclohexane (Lorenson and Collet, 2000). The composition of the gaseous product varies over a considerable range. Hydrogen sulfide (H_2S) has been detected in gas hydrates, but because hydrogen sulfide is water soluble and the gas hydrate dissociation measurements have the potential to be contaminated with some sediment and pore water, it is also possible that in some cases hydrogen sulfide may not have existed within the gas hydrate structure (Lorenson and Collet, 2000).

1.5.5 Coal Gas

Coal gas is any gaseous product that is produced by gasification of coal or by carbonization of coal. Most of these products have been replaced by natural gas (the city and town gas works is an industry of the past) but still find use in some parts of the world. The products of coal gasification are varied insofar as the gas composition varies with the system employed. It is emphasized that the gas product must be first freed from any pollutants such as particulate matter and sulfur compounds before further use, particularly when the intended use is a water–gas shift or methanation.

Low-heat-content gas (low-Btu gas) is produced when the oxygen is not separated from the air (ca. 150–300 Btu ft^{-3}). This gas is also the usual product of *in situ* gasification of coal, which is used essentially as a technique for obtaining energy from coal without the necessity of mining the coal. The process is a technique for utilization of coal that cannot be mined by other techniques. The nitrogen content of low-heat-content gas ranges from somewhat less than 33% v/v to slightly more than 50% v/v and cannot be removed by any reasonable means; the presence of nitrogen at these levels renders the product gas to be low heat content. The nitrogen also strongly limits the applicability of the gas to chemical synthesis. Two other noncombustible components (water, H_2O, and carbon dioxide, CO) further lower the heating value of the gas; water can be removed by condensation and carbon dioxide by relatively straightforward chemical means. The two major combustible components are hydrogen and carbon monoxide; the H_2/CO ratio varies from approximately 2 : 3 to 3 : 2, but methane may also make an appreciable contribution to the heat content of the gas. Of the minor components hydrogen sulfide is the most significant, and the amount produced is, in fact, proportional to the sulfur content of the feed coal. Any hydrogen sulfide present must be removed by one, or more, of several procedures. Low-heat-content gas is of interest to industry as a fuel gas or even, on occasion, as a raw material from which ammonia, methanol, and other compounds may be synthesized.

Medium-heat-content gas (medium-Btu gas) has a heating value in the range of 300–550 Btu ft^{-3}, and the composition is much like that of low-heat-content gas, except that there is virtually no nitrogen. The primary combustible gases in medium-heat-content gas are hydrogen and carbon monoxide. Medium-heat-content gas is considerably more versatile than low-heat-content gas; like low-heat-content gas, medium-heat-content gas may be used directly as a fuel to raise steam or used through a combined power cycle to drive a gas turbine, with the hot exhaust gases employed to raise steam, but medium-heat-content gas is especially amenable to synthesize methane (by methanation), higher hydrocarbon derivatives (by Fischer–Tropsch synthesis), methanol, and a variety of synthetic chemicals. The reactions used to produce medium-heat-content gas are the same as those employed for low-heat-content gas synthesis, the major difference being the application of a nitrogen barrier (such as the use of pure oxygen) to keep diluent nitrogen out of the system.

In medium-heat-content gas, the H_2/CO ratio varies from 2 : 3 to ca. 3 : 1, and the increased heating value correlates with higher methane and hydrogen contents as well as with lower carbon dioxide contents. Furthermore, the very nature of the gasification process used to produce the medium-heat-content gas has a marked effect upon the ease of subsequent processing. For example, the CO_2-acceptor product is quite amenable to use for methane production because it has (i) the desired H_2/CO ratio just exceeding 3 : 1, (ii) an initially high methane content, and (iii) relatively low water and carbon dioxide contents. Other gases may require appreciable shift reaction and removal of large quantities of water and carbon dioxide prior to methanation.

High-heat-content gas (high-Btu gas) is essentially pure methane and often referred to as *synthetic natural gas* or *substitute natural gas*. However, to qualify as substitute natural gas, a product must contain at least 95% methane; the energy content of SNG is 980–1080 Btu ft^{-3}. The commonly accepted approach to the synthesis of high-heat-content gas is the catalytic reaction of hydrogen and carbon monoxide:

$$3H_2 + CO \rightarrow CH_4 + H_2O$$

To avoid catalyst poisoning, the feed gases for this reaction must be quite pure, and, therefore, impurities in the product are rare. The large quantities of water produced are removed by condensation and recirculated as very pure water through the gasification system. The hydrogen is usually present in slight excess to ensure that the toxic carbon monoxide is reacted; this small quantity of hydrogen will lower the heat content to a small degree.

The carbon monoxide/hydrogen reaction is somewhat inefficient as a means of producing methane because the reaction liberates large quantities of heat. In addition, the methanation catalyst is troublesome and prone to poisoning by sulfur compounds, and the decomposition of metals can destroy the catalyst. Thus, hydrogasification may be employed to minimize the need for methanation:

$$C_{coal} + 2H_2 \rightarrow CH_4$$

The product of hydrogasification is far from pure methane, and additional methanation is required after hydrogen sulfide and other impurities are removed.

1.5.6 Biogas

On the other hand, biogas is gas produced from biological materials either by fermentation or by pyrolysis, while landfill gas is, as the name implied, gas produced during the *in situ* maturation of landfill materials, which can include municipal waste and industrial waste. Thus, *biomass* is the collective name for *renewable materials* that includes (i) energy crops grown specifically to be used as fuel, such as wood or various types of grass; (ii) agricultural residues and by-products, such as straw, sugarcane fiber, rice hulls, and animal waste; and (iii) residues from forestry, construction, and other wood-processing industries. Biomass is a renewable resource that has received great attention due to environmental considerations and the increasing demands of energy worldwide. It has a negligible content of sulfur, nitrogen, and ash that produces lower emissions of sulfur dioxide, nitrogen oxides, and soot than conventional fossil fuels. Following from this, a *biofuel* is any fuel that is derived from biomass. A biofuel has also been defined as any fuel with an 80% v/v minimum content of materials derived from living organisms harvested within the 10 years preceding its manufacture.

Biogas (often predominantly methane, also called biogenic gas and sometimes incorrectly known as swamp gas) typically refers to a biofuel gas produced by anaerobic digestion or fermentation of organic matter including manure, sewage sludge, municipal solid waste, biodegradable waste, or any other biodegradable feedstock under anaerobic conditions. Depending on where it is produced, biogas is also called swamp gas, marsh gas, landfill gas, and digester gas.

Biogas is produced by certain types of bacteria (methanogens) during the process of breaking down organic matter in an oxygen-free environment. Examples of biomass are (i) wood and wood-processing wastes; (ii) agricultural crops and waste materials; (iii) food, yard, and wood waste in garbage; and (iv) animal manure and human sewage, which are all potential sources of biogas (biogenic gas). Biogas production (typically an anaerobic process) is a multistep biological process where the originally complex and large organic solid wastes are progressively transformed in simpler and smaller-sized organic compounds by different bacteria strains up to have a final energetically worthwhile gaseous product and a semisolid material (digestate) that is rich in nutrients and thus suitable for its utilization in farming.

The composition of biogas varies depending upon the composition of the waste material in the landfill and the anaerobic digestion process. Landfill gas typically has methane concentration around 50%. Advanced waste treatment technologies can produce biogas with 55–75% v/v methane; often air is introduced (5% by volume) for microbiological desulfurization. For example, the constituents (by volume) of biogas generally are methane (50–75%), carbon dioxide (25–50%), nitrogen (0–10%), hydrogen (0–1%), hydrogen sulfide (0–3%), and oxygen (0–2%).

Biogas is characterized based on the chemical composition and the physical characteristics that result from it, including the methane content (Esposito et al., 2012). It is primarily a mixture of methane (CH_4) and inert carbonic gas (CO_2). However, the name biogas covers a wide variety of gases resulting from specific treatment processes, starting from various organic waste such as industrial waste, animal waste, or waste of domestic origin waste. Different sources of production lead to different specific compositions of the biogas (Table 1.8).

Table 1.8 Examples of biogas composition.

Constituents	Source-1	Source-2	Source-3
Methane, CH_4% v/v	50–60	60–75	60–75
Carbon dioxide, CO_2% v/v	38–34	33–19	33–19
Nitrogen, N_2% v/v	5–0	1–0	1–0
Oxygen, O_2% v/v	1–0	<0,5	<0,5
Water, H_2O% v/v	6 (40 °C)	6 (40 °C)	6 (40 °C)
Hydrogen sulfide, H_2S mg m^{-3}	100–900	1000–4000	3000–10 000
Ammonia, NH_3 mg m^{-3}	—	—	50–100

Source-1: Household waste.
Source-2: Wastewater treatment plant sludge.
Source-3: Agricultural waste.

Biogas is produced by methanogenic organisms in swamps and marshes by the decomposition of organic (both animal and plant) matter. The anaerobic (low or no oxygen present in the system) decomposition produces methane, carbon dioxide, a small amount of hydrogen, and a small amount of heat. Acidity, the carbon to nitrogen ratio of the feedstock, temperature, and percentage of solids are factors that influence the quality and quantity of biogas produced. Normally, cow manure, enriched by vegetable matter, is the ideal feedstock. The other principal type of biogas is wood gas, which is produced by gasification of wood or another biomass. This type of biogas is composed primarily of nitrogen, hydrogen, and carbon monoxide, with trace amounts of methane. Industrial anaerobic digesters and mechanical biological treatment systems are becoming increasingly popular. Methane is a potent greenhouse gas but is also a very clean, easy-burning fuel, and hence harnessing it from natural sources such as landfills is beneficial to the natural environment.

Biogas presents characteristics interesting to compare with natural gas and propane. It is a gas appreciably lighter than air; it produces twice as less calories by combustion with equal volume of natural gas. The presence of hydrogen sulfide, carbon dioxide, and water makes biogas very corrosive and requires the use of adapted materials. The composition of a gas issued from a digester depends on the substrate, its organic matter load, and the feeding rate of the digester.

When biogas is sent to processing for impurity removal, the characteristics of the resulting purified biogas are similar to that of a natural gas. In this instance the producer of the biogas can utilize the local gas distribution networks. The gas must be very clean to reach pipeline quality. Water (H_2O), hydrogen sulfide (H_2S), and particulates are removed if present at high levels or if the gas is to be completely cleaned. Carbon dioxide is less frequently removed, but it must also be separated to achieve pipeline quality gas. If the gas is to be used without extensive cleaning, it is sometimes cofired with natural gas to improve combustion. Biogas cleaned up to pipeline quality is called renewable natural gas. In this form the gas can be used in any application in which natural gas is used. The heating value of the raw biogas is on the order of 600 Btu ft^{-3} (5340 cal m^{-3}). The raw gas will burn in an engine, but corrosion can occur unless proper materials of construction are selected.

1.5.7 Landfill Gas

Landfill gas, which is often included under the umbrella definition of biogas, is also produced from the decay of organic wastes (such as municipal solid waste that contains organic materials), but these wastes may not be biomass-type materials (Lohila et al., 2007; Staley and Barlaz, 2009; Speight, 2011c). Landfills offer another underutilized source of biogas. When municipal waste is buried in a landfill, bacteria break down the organic material contained in garbage such as newspapers, cardboard, and food waste, producing gases such as carbon dioxide and methane. Rather than allowing these gases to go into the atmosphere, where they contribute to global warming, landfill gas facilities can capture them, separate the methane, and combust it to generate electricity, heat, or both.

Thus, landfill gas is another source of biogas generated from the decomposition of the biodegradable organic fraction of municipal solid waste by microorganisms. This generally occurs under semicontrolled conditions in a landfill; its constituents depend on the composition and age of the waste. Large variations may exist in the composition of landfill gas due to differences in sources of municipal solid waste and operating conditions at the landfill. The three main gas constituents, namely, methane (CH_4), carbon dioxide (CO_2), and hydrogen sulfide (H_2S), are used to characterize landfill gas.

The gas is a complex mix of different gases created by the action of microorganisms within a landfill. Typically, landfill gas is composed of 45–60% v/v methane, 40–60% v/v carbon dioxide, 0–1.0% v/v hydrogen sulfide, 0–0.2% v/v hydrogen (H_2), trace amounts of nitrogen (N_2), low MW hydrocarbon derivatives (dry volume basis), and water vapor (saturated). The specific gravity of landfill gas is approximately 1.02–1.06. Trace amounts of other volatile organic compounds comprise the remainder (typically, 1–2% v/v or less), and these trace gases include a large array of species, such as low MW hydrocarbon derivatives. Other minor components include hydrogen sulfide, nitrogen oxides, sulfur dioxide, nonmethane volatile organic compounds, polycyclic aromatic hydrocarbon derivatives, polychlorinated dibenzodioxin derivatives, and polychlorinated dibenzofuran derivatives (Brosseau, 1994; Rasi et al., 2007). All of the aforementioned agents are harmful to human health at high doses.

The gases produced within a landfill site can be collected for various uses, such as direct utilization on-site in a boiler or any type of combustion system, providing heat. Electricity can also be generated on-site through the use of microturbines, steam turbines, or fuel cells (Sullivan, 2010). The landfill gas can also be sold off-site and sent into natural gas pipelines. This approach requires the gas to be processed into pipeline quality, e.g. by removing various contaminants and components. As should be expected, the amount of methane that is produced varies significantly based on the composition of the waste (Staley and Barlaz, 2009). The efficiency of gas collection at landfills directly impacts the amount of energy that can be recovered – closed landfills (those no longer accepting waste) collect gas more efficiently than open landfills (those that are still accepting waste).

Landfill gas collection is typically accomplished through the installation of wells installed vertically and/or horizontally in the waste mass. Design heuristics for vertical wells call for about one well per acre of landfill surface, whereas horizontal wells are normally spaced about 50–200 ft apart on center. Efficient gas collection can be accomplished at both open and closed landfills, but closed landfills have systems that are more efficient, owing to greater deployment of collection infrastructure since active filling is not occurring. On average, closed landfills have gas collection systems that capture approximately 84% v/v of produced gas, compared with approximately 67% v/v for open landfills.

Landfill gas can also be extracted through horizontal trenches instead of vertical wells. Both systems are effective at collecting. Landfill gas is extracted and piped to a main collection header, where it is sent to be treated or flared. The main collection header can be connected to the leachate collection system to collect condensate forming in the pipes. A blower is needed to pull the gas from the collection wells to the collection header and further downstream. However, landfill gas cannot be distributed through utility natural gas pipelines unless it is cleaned up to less than 3% carbon dioxide and a few parts per million of hydrogen sulfide, because carbon dioxide and hydrogen sulfide corrode the pipelines (Speight, 2014b). The presence of carbon dioxide will lower the energy level of the gas below requirements for the pipeline. Siloxane derivatives in the gas will form deposits in gas burners and need to be removed prior to entry into any gas distribution or transmission system. Consequently, it may be more economical to burn the gas on-site or within a short distance of the landfill using a dedicated pipeline. Water vapor is often removed, even if the gas is burned on-site. If low temperatures condense water out of the gas, siloxane derivatives can be lowered as well because they tend to condense out with the water vapor. Other nonmethane components may also be removed to meet emission standards, to prevent fouling of the equipment, or for environmental considerations. Cofiring landfill gas with natural gas improves combustion, which lowers emissions.

Finally, although not strictly a gas, landfill gas condensate is a liquid that is produced in landfill gas collection systems. The condensate is removed as the gas is withdrawn from landfills. Production of condensate may be through natural or artificial cooling of the gas or through physical processes such as volume expansion. Condensate is composed principally of water and organic compounds. Often the organic compounds are not soluble in water, and the condensate separates into a watery (aqueous) phase and a floating organic (hydrocarbon) phase. This organic fraction may comprise up to 5% of the liquid.

The organic phase varies in volume and composition among sites and may range from less than 1 to 5% (by volume) of the total mixture. In general, the aqueous phase is mostly water and trace organic compounds. The organic phase consists principally of hydrocarbon derivatives, xylene isomers, chloroethane derivatives, chloroethylene derivatives, benzene, toluene, other priority pollutants, and trace moisture. However, a larger number of acid and base/neutral compounds are present in the aqueous than in the organic phase of the condensate. Concentration levels may be orders of magnitude higher in the organic phase than the aqueous phase and are dependent upon the types of compounds in the landfill (Briggs, 1988).

1.5.8 Flue Gas

Flue gas (sometimes called *exhaust gas* or *stack gas*) is gas that is generated through, in the current context, the combustion of a carbonaceous fuel. The steam generators in large power plants and the process furnaces in large refineries petrochemical plants, and incinerators, burn considerable amounts of carbonaceous fuels and therefore emit large amounts of flue gas to the ambient atmosphere. As with other gases, the composition of flue gas depends on the type of fuel and the combustion conditions, e.g. the air ratio value. Many flue gas components are air pollutants and must therefore, due to governmental regulations, be eliminated or minimized by special cleaning procedures before the gas is released to the atmosphere.

A typical flue gas from the combustion of a carbonaceous fuel contains varying amounts of nitrogen oxides (NOx), sulfur dioxide (SO_2), and particulate matter. The nitrogen oxides are derived from the nitrogen in the ambient air as well as from any nitrogen-containing compounds in the fuel. The sulfur dioxide is derived from any sulfur-containing compounds in the fuel. The particulate matter is composed of very small particles of solid materials and very small liquid droplets, which give flue gases their smoky appearance.

When burning coal, hydrogen chloride and hydrogen fluoride may be present in the flue gas as well as hydrocarbon derivatives and heavy metals in case of incineration of waste materials. In many countries, as part of a national environmental protection program, exhaust gases must comply with strict governmental regulations regarding the limit values of pollutants such as dust, sulfur and nitrogen oxides, and carbon monoxide. To meet these limit values, combustion plants are equipped with flue gas cleaning systems such as gas scrubbers and dust filters.

Thus, as environmental regulations regarding sulfur dioxide emissions have been enacted in many countries, sulfur dioxide is removed from flue gases by a variety of methods. Examples of common flue gas cleaning processes are (i) wet scrubbing process, which uses a slurry of alkaline sorbent, usually limestone or lime, or seawater to scrub the gases; (ii) spray-dry scrubbing process, which uses similar sorbent slurries as described in the first category; (iii) wet sulfuric acid process, which allows the recovery of sulfur in the form of commercial quality sulfuric acid; (iv) a process commonly referred to as a SNOX flue gas desulfurization process, which removes sulfur dioxide, nitrogen oxides, and particulate matter from flue gases; and (v) a dry sorbent injection process, which introduces powdered hydrated lime or other sorbent materials into the exhaust ducts to eliminate sulfur dioxide and sulfur trioxide from the process emissions.

1.6 Natural Gas and Energy Security

In terms of reserves, natural gas is produced on all continents except Antarctica. The proved reserves of natural gas are of the order of 6588.8 trillion cubic feet (1 Tcf = 1×10^{12} ft^3), of which approximately 374 Tcf exist in the United States and Canada (BP, 2017). It should also be remembered that the total gas resource base (like any fossil fuel or mineral resource base) is dictated by economics. Therefore, when resource data are quoted, some attention must be given to the cost of recovering those resources. Most importantly, the economics must also include a cost factor that reflects the willingness to secure total, or a specific degree of, energy independence. The total proved reserves of natural gas generally taken to be those quantities that geological and engineering information indicates with reasonable certainty can be recovered in the future from known reservoirs under existing economic and operating conditions.

A common misconception about natural gas is that resources are being depleted at an alarming rate and the supplies are quickly running out. In fact, there is a vast amount of natural gas estimated to still be retrieved from a variety of reservoirs. However, many proponents of the rapid-depletion theory believe that price spikes indicate that natural gas resources are depleted beyond the point of no return. However, price spikes of any commodity are not always caused by waning resources but can be the outcome of other forces at work in the marketplace.

Energy security is the continuous and uninterrupted availability of energy to a specific country or region. The security of energy supply conducts a crucial role in decisions that are related to the formulation of energy policy strategies. The economies of many countries are dependent on the energy imports in the notion that their balance of payments is affected by the magnitude of the vulnerability that the countries have in crude oil and natural gas (Speight, 2011b).

Energy security has been an on-again-off-again political issue in the United States since the first Arab oil embargo in 1973. Since that time, the speeches of various presidents and the Congress of the United States have continued to call for an end to the dependence on foreign oil and gas by the United States. The congressional rhetoric of energy security and energy independence continues, but meaningful suggestions of how to address this issue remain few and far between.

The energy literature and numerous statements by officials of oil-and-gas-producing and oil-and-gas-consuming countries indicate that the concept of energy security is elusive. Definitions of energy security range from uninterrupted oil supplies to the physical security of energy facilities to support for biofuels and renewable energy resources. Historically, experts and politicians referred to *security of oil supplies* as *energy security*. Only recently policy makers started to include natural gas supplies in the portfolio of energy definitions.

The security aspects of natural gas are similar, but not identical, to those of oil. Compared with oil imports, natural gas imports play a smaller role in most importing countries – mainly because it is less costly to transport liquid crude oil and petroleum products than natural gas. Natural gas is transported by pipeline over long distances because of the pressurization costs of transmission; the need to finance the cost of these pipelines encourages long-term contracts that dampen price volatility.

The past decade has yielded substantial change in the natural gas industry. Specifically, there has been rapid development of technology allowing the recovery of natural gas from shale formations. Since 2000, rapid growth in the production of natural gas from shale formations in North America has dramatically altered the global natural gas market landscape. Indeed, the emergence of shale gas is perhaps the most intriguing development in global energy markets in recent memory.

Beginning with the Barnett shale in northeast Texas, the application of innovative new techniques involving the use of horizontal drilling with hydraulic fracturing has resulted in the rapid growth in production of natural gas from shale. Knowledge of the shale gas resource is not new as geologists have long known about the existence of shale formations, and accessing those resources was long held in the geology community to be an issue of technology and cost. In the past decade, innovations have yielded substantial cost reductions, making shale gas production a commercial reality. In fact, shale gas production in the United States has increased from virtually nothing in 2000 to over 10 billion cubic feet per day (bcfd, 1×10^9 ft^3 per day) in 2010, and it is expected to more than quadruple by 2040, reaching 50% or more of total US natural gas production by the decade starting in 2030.

Natural gas – if not disadvantaged by government policies that protect competing fuels, such as coal – stands to play a very important role in the US energy mix for decades to come. Rising shale gas production has already delivered large beneficial impacts to the United States. Shale gas resources are generally located in close proximity to end-use markets where natural gas is utilized to fuel industry, generate electricity, and heat homes. This offers both security of supply and economic benefits.

The *Energy Independence and Security Act of 2007* (originally named the *Clean Energy Act of 2007*) is an Act of Congress concerning the energy policy of the United States. The stated purpose of the act is "to move the United States toward greater energy independence and energy security, to increase the production of clean renewable fuels, to protect consumers, to increase the efficiency of products, buildings, and vehicles, to promote research on and deploy greenhouse gas capture and storage options, and to improve the energy performance of the Federal Government, and for other purposes."

The bill originally sought to cut subsidies to the petroleum industry in order to promote petroleum independence and different forms of alternative energy. These tax changes were ultimately dropped after opposition in the Senate, and the final bill focused on automobile fuel economy, development of biofuels, and energy efficiency in public buildings and lighting. It was, and still is, felt by many observers that there should have been greater recognition of the role that natural gas can play in energy security. In fact, viewed from the perspective of the energy-importing countries as a whole, diversification in oil supplies has remained constant over the last decade, while diversification in natural gas supplies has steadily increased. Given the increasing importance of natural gas in world energy use, this is an indicator of an increase in overall energy security (Cohen et al., 2011).

However, natural gas is an attractive fuel, and its attraction is growing because of its clean-burning characteristics, compared with oil or coal, and because of its price advantage, on an energy equivalent basis, compared with oil. Accordingly, analysts predict significant future growth in natural gas consumption worldwide and growth in the trade of natural gas. Significant investments are being made to meet this future demand by bringing so-called stranded gas (including *shale gas*) to market.

Current trends suggest that natural gas will gradually become a global commodity with a single world market, just like oil, adjusted for transportation differences. The outcome of a global gas market is inevitable; once this occurs, the tendency will be toward a world price of natural gas, as with oil today, and the prices of oil and gas each will reach a global equivalence based on energy content (Deutch, 2010).

These trends also indicate the need for continuing development of methods for the analysis of natural gas.

1.7 Natural Gas Regulations

No matter what one can hear from various pundits, it is advantageous to accept that natural gas is dangerous – when mixed with air, it can form an explosive mixture at specific low concentrations, between a lower explosion limit (LEL) and an upper explosion limit (UEL). However, it can burn as a flame at the point of leakage. In an open land installation, the gas normally disperses quickly. However, in a closed system, simple explosions and boiling liquid expanding vapor explosions (BLEVE) can (will) cause serious damage. The latter is caused if a flame impinges on a vessel containing liquefied gas. Adequate ventilation is required by all safety codes for all enclosed spaces. It is possible to prevent a catastrophe by a chain of preventive measures. Insofar as the condition of the environment is important to floral and faunal life on Earth, any serious disturbance of the environment (pollution) can have serious consequences for continuation of this life.

Furthermore, all carbon-based fuels produce carbon dioxide and water when they are burned. Because of higher atomic ratio of hydrogen to carbon atom, natural gas produces less carbon dioxide per unit of energy than higher MW hydrocarbon derivatives (such as crude oil constituents) or coal. Also, natural gas impurities can be selectively and relatively easily removed and be completely converted by combustion, resulting in lower emissions of particulate matter. However, any releases of methane and ethane to the atmosphere should be minimized.

In the United States, the regulation of natural gas production has traditionally occurred primarily at the state level with most states that produce natural gas and crude oil issuing more rigorous standards that take primacy over federal regulations, as well as additional regulations that control areas not covered at the federal level, such as hydraulic fracturing. Within states, regulation is carried out by a range of agencies.

1.7.1 General Aspects

The specific regulations vary considerably among states, such as different depths for well casing, levels of disclosure on drilling and fracturing fluids, or requirements for water storage. Currently, many states that produce natural gas and crude oil have varying hydraulic fracturing regulations, specifically regulations related to (i) the disclosure of the components of hydraulic fracturing fluids, (ii) the proper casing of wells to prevent aquifer contamination, and (iii) management of wastewater from flow-back and produced water. The disposal of wastewater by underground injection has emerged as a concern for state regulators due to large interstate flows of wastewater to states with suitable geology for the water disposal as well as the potential for seismic activity near some well sites.

In terms of the development of unconventional natural gas and crude oil resources, a major environmental concern is the potential contamination of watercourses. Apart from water, which makes up 99.5% v/v of the hydraulic fracturing fluid, the fluid also contains chemical additives to improve the process performance. The additives are varied and can include acid, friction reducer, surfactant, gelling agent, and scale inhibitor (Speight, 2016b). The composition of the fracturing fluid is tailored to differing geological features and reservoir characteristics to address challenges, including scale buildup, bacterial growth, and proppant transport. Unfortunately, in the past many of the chemical compounds used in the hydraulic fracturing process lacked scientifically based maximum contaminant levels, making it more difficult to quantify their risk to the environment (Colborn et al., 2011). Moreover, uncertainty about the chemical makeup of fracturing fluids persists because of the limitations on required chemical disclosure (Centner, 2013; Maule et al., 2013; Centner and O'Connell, 2014).

The measures required by state and federal regulatory agencies in the exploration and production of natural gas from deep tight formations have been very effective, for example, in protecting drinking water aquifers from contamination. In fact, a series of federal laws govern most environmental aspects of natural gas development (Table 1.9). However, federal regulation may not always be the most effective way of assuring the desired level of environmental protection. Therefore, most of these federal laws have provisions for granting primacy to the state governments, which have usually developed their own sets of regulations. By statute, the different states may adopt these standards of their own, but they must be at least as protective as the federal principles they replace and, as a result, may be more protective to address local conditions.

State regulation of the environmental practices related to natural gas and crude oil development can more easily address the regional- and state-specific character of the activities, compared with a one-size-fits-all management by the federal level. Some of these factors include geology, hydrology, climate, topography, industry characteristics, development history, state legal structures, population density, and local economics, and, thus, the regulation of natural gas and crude oil production is a detailed monitoring of each stage of the development through the many controls at the state level. Each state has the necessary powers to regulate, permit, and enforce all activities – from drilling and fracturing of the well to production operations, to managing and disposing of wastes, to abandoning and plugging the production well(s). These state powers are a means of assuring that natural gas and crude oil operations do not have an adverse impact on the environment.

Moreover, because of the regulatory makeup of each state – which can vary from state to state – different states take different approaches to the regulation and enforcement of resource development, but the laws of each state generally give the state agency responsible for natural gas and crude oil development the discretion to require whatever is necessary to protect the environment, including human health. In addition, most have a general prohibition against pollution from natural gas and crude oil production. Most of the state requirements are written into rules or regulations, but some of the regulations may be added to permits on a case-by-case basis because of (i) environmental review, (ii) on-site inspections, (iii) commission hearings, and (iv) public comments.

Finally, the organization of regulatory agencies within the different states where natural gas and crude oil is produced varies considerably. Some states have several agencies that oversee different aspects (with some inevitable

Table 1.9 Examples of federal laws (*listed alphabetically*) in the United States to monitor hydraulic fracturing projects.

Act	Purpose
Clean Air Act	Limits air emissions from engines, gas processing equipment, and other sources associated with drilling and production
Clean Water Act	Regulation of surface discharges of water associated with natural gas and crude oil drilling and production, as well as storm water runoff from production sites
Energy Policy Act	Exempted hydraulic fracturing companies from some regulations; may disclose chemicals through a report submitted to the regulatory authority, but in some instances, chemical information may be exempt from disclosure to the public as trade secrets
NEPA[a]	Requires that exploration and production on federal lands be thoroughly analyzed for environmental impacts
NPDES[b]	Requires tracking of any toxic chemicals used in fracturing fluids
Oil Pollution Act	Regulation of ground pollution risks relating to spills of materials or hydrocarbon derivatives into the water table; also regulated under the Hazardous Materials Transportation Act
Safe Drinking Water Act	Directs the underground injection of fluids from natural gas and crude oil activities; disclosure of chemical content for underground injections; after 2005, see Energy Policy Act
TSCA[c]	Suggestion that this act be used to regulate the reporting of hydraulic fracturing fluid information

NB: The Fracturing Responsibility and Awareness of Chemicals Act (the FRAC Act) was an attempt to define hydraulic fracturing as a federally regulated activity under the Safe Drinking Water Act; no significant moves or passage of this act at the time of writing (https://www.congress.gov/bill/114th-congress/senate-bill/785/text).
[a] National Environmental Policy Act.
[b] National Pollutant Discharge Elimination System.
[c] TSCA, Toxic Substances Control Act.

overlap powers) of natural gas and crude oil operations, particularly the requirements that protect the environment. In different states, the various approaches have developed over time to create a structure that best serves the consumers, nonconsumers, and the various industries. The one constant is that in each state where natural gas and crude oil is produced, there is one agency that has the responsibility for issuing permits for gas and oil development projects. The permitting agencies work with other agencies in the regulatory process and often serve as a central organizing body and a useful source of information about activities related to natural gas and crude production.

1.7.2 Federal Regulation of the Industry

Since the Energy Policy Act passed in 2005, natural gas and crude oil production in the United States has grown significantly, and this rapid growth – along with continued reports of environmental effects – has led to renewed calls for the federal government to provide increased regulation or guidance. This pressure led to the introduction to Congress in 2009 of the Fracturing Responsibility and Awareness of Chemicals Act (the FRAC Act) to define hydraulic fracturing as a federally regulated activity under the Safe Drinking Water Act (Table 1.9). The proposed Act requires the energy industry to disclose the chemical additives used in the hydraulic fracturing fluid. The Act did not receive any action, was reintroduced in 2011, and appears to have been held in limbo since that time.

In the absence of new federal regulations, the various states that produce natural gas and crude oil by hydraulic fracturing have continued to use existing natural gas and crude oil and environmental regulations to manage natural gas and crude oil development, as well as introducing individual state regulations for hydraulic fracturing (Table 1.9). In fact, the current regulations comprise an overlapping collection of federal, state, and local regulations and permitting systems, and these regulations cover different aspects of the development and production of a natural gas and crude oil with the intention that the regulations combine to manage any potential impact on the surrounding environment, including any effects on water management. However, the hydraulic fracturing process has not previously been regulated under current laws, and, therefore, in terms of water management, emissions management, and site activity, the existing regulations are being (must be) reassessed for suitability for application to the hydraulic fracturing process. In the meantime, many states (including Wyoming, Arkansas, and Texas) have already implemented regulations requiring disclosure of the materials used in hydraulic fracturing fluids, and the US Department of the Interior has indicated an interest in requiring similar disclosure for sites on federal lands.

The Bureau of Land Management (BLM) of the Department of the Interior has proposed draft rules for natural gas and crude oil production on public lands, and these proposals would require disclosure of the chemical components used in hydraulic fracturing fluids. The proposed rule requires that an operations plan be submitted to the relevant authority – prior to initiation of the hydraulic fracturing project – that would allow BLM to evaluate groundwater protection designs based on (i) a review of the local geology, (ii) a review of any anticipated surface disturbance, and (iii) a review of the proposed management and disposal of project-related fluids. In addition, BLM would require submittal of the information necessary to confirm wellbore integrity before, during, and after the stimulation operation. Furthermore, before hydraulic fracturing commenced, the company would have to certify that the fluids comply with all applicable federal, state, and local laws, rules, and regulations. After the conclusion of the hydraulic fracturing stage of the project, a follow-up report would be required in which the actual events that occurred during fracturing activities would be summarized, and this report would have to include the specific chemical makeup of the hydraulic fracturing fluid.

On 17 April 2012, the US Environmental Protection Agency released new performance standards and national emission standards for hazardous air pollutants in the natural gas and crude oil industries. These rules include the first federal air standards for hydraulically fractured gas wells, along with requirements for other sources of pollution in the natural gas and crude oil industry that currently are not regulated at the federal level. These standards require either flaring or *green completion* on natural gas wells developed prior to 1 January 2015, with only green completions allowed for wells developed on and after that date.

Briefly, *green completion* requires natural gas companies to capture the gas at the wellhead immediately after well completion instead of releasing it into the atmosphere or flaring the gas. Thus, the green completion systems are systems to reduce methane losses during well completion. After a new well completion or workover, the wellbore and formation must be cleaned of debris and fracturing fluid. Conventional methods for doing this include producing the well into an open pit or tank to collect sand, cuttings, and reservoir fluids for disposal. Typically, the natural gas that is produced is vented or flared, and the large volume of natural gas that is lost may not only affect regional air quality. When using green completion systems, gas and hydrocarbon liquids are physically separated from other fluids (a form of wellhead gas processing) – there is no venting or flaring of the gas – and delivered directly into equipment that holds or transports the hydrocarbon derivatives for productive use. Furthermore, by using portable equipment to process natural gas and natural gas condensate, the recovered gas can be directed to a pipeline as sales gas. The use of truck-mounted or trailer-mounted portable systems can typically recover more than half of the total gas produced.

1.7.3 Natural Gas and the Environment

Natural gas, although often cited as a relatively clean fuel, is also capable of producing emissions that are detrimental to the environment. While the major constituent of natural gas is methane, there are components such as carbon dioxide (CO), hydrogen sulfide (H_2S), and mercaptans (thiols; R — SH), as well as trace amounts of sundry other emissions. The fact that methane has a foreseen and valuable end use makes it a desirable product, but in several other situations it is considered a pollutant, having been identified a greenhouse gas.

Also, a characteristic feature of natural gas that contains hydrogen sulfide is the presence of carbon dioxide (generally in the range of 1–4% v/v). In cases where the natural gas does not contain hydrogen sulfide, there may also be a relative lack of carbon dioxide. A sulfur removal process must be very precise, since natural gas contains only a small quantity of sulfur-containing compounds that must be reduced several orders of magnitude. Most consumers of natural gas require less than 4 ppm in the gas.

Acid rain occurs when the oxides of nitrogen and sulfur that are released to the atmosphere during the combustion of fossil fuels are deposited (as soluble acids) with rainfall, usually at some location remote from the source of the emissions. It is generally believed (the chemical thermodynamics are favorable) that acidic compounds are formed when sulfur dioxide and nitrogen oxide emissions are released from tall industrial stacks. Gases such as sulfur oxides (usually sulfur dioxide, SO_2) as well as nitrogen oxides (NO_x) react with the water in the atmosphere to form acids:

$$SO_2 + H_2O \rightarrow H_2SO_3$$

$$2SO_2 + O_2 \rightarrow 2SO_3$$

$$SO_3 + H_2O \rightarrow H_2SO_4$$

$$2NO + H_2O \rightarrow 2HNO_2$$

$$2NO + O_2 \rightarrow 2NO_2$$

$$NO_2 + H_2O \rightarrow HNO_3$$

Acid rain has a pH less than 5.0 and predominantly consists of sulfuric acid (H_2SO_4) and nitric acid (HNO_3). As a point of reference, in the absence of anthropogenic pollution sources, the average pH of rain is –6.0 (slightly acidic; neutral pH = 7.0). In summary, the sulfur dioxide that is produced during a variety of processes will react with oxygen and water in the atmosphere to yield environmentally detrimental sulfuric acid. Similarly, nitrogen oxides will also react to produce nitric acid.

Another acid gas, hydrogen chloride (HCl), although not usually considered to be a major emission, is produced from mineral matter and the brines that often accompany petroleum during production and is gaining increasing recognition as a contributor to acid rain. However, hydrogen chloride may exert severe local effects because it does not need to participate in any further chemical reaction to become an acid. Under atmospheric conditions that favor a buildup of stack emissions in the areas where hydrogen chloride is produced, the amount of hydrochloric acid in rainwater could be quite high.

In addition to hydrogen sulfide and carbon dioxide, natural gas may contain other contaminants, such as mercaptans (R—SH) and carbonyl sulfide (COS). The presence of these impurities may eliminate some of the sweetening processes since some processes remove large amounts of acid gas but not to a sufficiently low concentration. On the other hand, there are those processes that are not designed to remove (or are incapable of removing) large amounts of acid gases. However, these processes are also capable of removing the acid gas impurities to very low levels when the acid gases are there in low to medium concentrations in the gas.

At the global level there is concern that the increased use of any hydrocarbon-based fuels will ultimately raise the temperature of the planet (*global warming*), as carbon dioxide reflects the infrared or thermal emissions from the Earth, preventing them from escaping into space (*greenhouse effect*). Whether the potential for global warming becomes real will depend upon how emissions into the atmosphere are handled. There is considerable discussion about the merits and demerits of the global warming theory, and the discussion is likely to continue for some time. Be that as it may, the atmosphere can only tolerate pollutants up to a limiting value. And that value needs to be determined. In the meantime, efforts must be made to curtail the use of noxious and foreign (nonindigenous) materials into the air.

There are a variety of processes that are designed for sulfur dioxide removal from gas streams (Speight, 2014a), but scrubbing process utilizing limestone ($CaCO_3$) or lime [$Ca(OH)_2$] slurries has received more attention than other gas scrubbing processes. Most of the gas scrubbing processes are designed to remove sulfur dioxide from the gas streams; some processes show potential for removal of nitrogen oxide(s).

References

AAPG (2015). Unconventional energy resources: 2015 review. *Natural Resources Research* 24 (4): 443–508. American Association of Petroleum Geologists, Tulsa Oklahoma.

Abbott, M. (2016). *The Economics of the Gas Supply Industry*. New York: Routledge Publishers, Taylor & Francis Group.

Aguilera, R.F. and Radetzki, M. (2014). The shale revolution: global gas and oil markets under transformation. *Mineral Economics* 26 (3): 75–84.

ASTM (2017). *Annual Book of Standards*. West Conshohocken, Pennsylvania: ASTM International.

Boyle, G. ed. (1996). *Renewable Energy: Power for a Sustainable Future*. Oxford, UK: Oxford University Press.

BP (2017). *Statistical Review of World Energy 2016*. London, UK: British Petroleum. https://www.bp.com/content/dam/bp/en/corporate/pdf/energy-economics/statistical-review-2017/bp-statistical-review-of-world-energy-2017-full-report.pdf (accessed 18 August 2017).

Briggs, J. (1988). Municipal Landfill Gas Condensate. *Report No. EPA/600/S2-87/090*. Cincinnati, OH: Environmental Protection Agency Hazardous Waste Engineering Research Laboratory. February.

Brosseau, J. (1994). Trace gas compound emissions from municipal landfill sanitary sites. *Atmospheric-Environment* 28 (2): 285–293.

Buffett, B. and Archer, D. (2004). Global inventory of methane clathrate: sensitivity to changes in the deep ocean. *Earth and Planetary Science Letters* 227 (3–4): 185.

Burruss, R.C., and Ryder, R.T. (2003). Composition of Crude Oil and Natural Gas Produced from 14 Wells in the Lower Silurian "Clinton" Sandstone and Medina Group, Northeastern Ohio and Northwestern Pennsylvania. *Open-File Report 03–409*, Reston, VA: United States Geological Survey.

Burruss, R.C. and Ryder, R.T. (2014). Composition of natural gas and crude oil produced from 10 wells in the lower Silurian "Clinton" sandstone, Trumbull County, Ohio. In: *Coal and Petroleum Resources in the Appalachian Basin; Distribution, Geologic Framework, and Geochemical Character*, Professional Paper 1708 (ed. L.F. Ruppert and R.T. Ryder). Reston, Virginia: United States Geological Survey.

Centner, T.J. (2013). Oversight of shale gas production in the United States and the disclosure of toxic substances. *Resources Policy* 38: 233–240.

Centner, T.J. and O'Connell, L.K. (2014). Unfinished business in the regulation of shale gas production in the United States. *The Science of the Total Environment* 476–477: 359–367.

Chong, Z.R., Yang, S.H.B., Babu, P. et al. (2016). Review of natural gas hydrates as an energy resource: prospects and challenges. *Applied Energy* 162: 1633–1652.

Cohen, G., Joutz, F., and Loungani, P. (2011). Measuring energy security: trends in the diversification of oil and natural gas supplies. *IMF Working Paper WP/11/39*. Washington, DC: International monetary Fund.

Colborn, T., Kwiatkowski, C., Schultz, K., and Bachran, M. (2011). Natural gas operations from a public health perspective. *Human and Ecological Risk Assessment* 17: 1039–1056.

Collett, T.S. (2002). Energy resource potential of natural gas hydrates. *AAPG Bulletin* 86: 1971–1992.

Collett, T.S., Johnson, A.H., Knapp, C.C., and Boswell, R. (2009). Natural gas hydrates: a review. In: *Natural Gas Hydrates – Energy Resource Potential and Associated Geologic Hazards*, AAPG Memoir No. 89 (ed. T.S. Collett, A.H. Johnson, C.C. Knapp and R. Boswell), 146–219. Tulsa, Oklahoma: American Association of Petroleum Geologists.

Demirbaş, A. (2010a). Methane from gas hydrates in the Black Sea. *Energy Sources, Part A: Recovery, Utilization, and Environmental Effects* 32: 165–171.

Demirbaş, A. (2010b). Methane hydrates as potential energy resource: part 1-importance, resource and recovery facilities. *Energy Conversion and Management* 51: 1547–1561.

Demirbaş, A. (2010c). Methane hydrates as potential energy resource: part 2 – methane production processes from gas hydrates. *Energy Conversion and Management* 51: 1562–1571.

Deutch, J. (2010). *Oil and Gas Energy Security Issues*. Washington, DC: Resource for the Future, National Energy Policy Institute.

Edmonds, B., Moorwood, R., and Szczepanski, R. (1996). A practical model for the effect of salinity on gas hydrate formation. *Paper No. 35569*. Richardson, TX: Society of Petroleum Engineers.

Esposito, G., Frunzo, L., Liotta, F. et al. (2012). Bio-methane potential tests to measure the biogas production from the digestion and co-digestion of complex organic substrates. *The Open Environmental Engineering Journal* 5: 1–8.

Forbes, R.J. (1964). *Studies in Ancient Technology*. Leiden: E. J. Brill.

Gao, S. (2008). Investigation of interactions between gas hydrates and several other flow assurance elements. *Energy & Fuels* 22 (5): 3150–3153.

Gao, S., House, W., and Chapman, W.G. (2005). NMR MRI study of gas hydrate mechanisms. *Journal of Physical Chemistry B* 109 (41): 19090–19093.

Gary, J.G., Handwerk, G.E., and Kaiser, M.J. (2007). *Petroleum Refining: Technology and Economics*, 5e. Boca Raton, Florida: CRC Press, Taylor & Francis Group.

Gornitz, V. and Fung, I. (1994). Potential distribution of methane hydrate in the world's oceans. *Global Biogeochemical Cycles* 8: 335–347.

Gupta, A.K. (2004). Marine gas hydrates: their economic and environmental importance. *Current Science* 86: 1198–1199.

Hsu, C.S. and Robinson, P.R. (2006). *Practical Advances in Petroleum Processing*, vol. 1 and 2. New York: Springer.

John, E. and Singh, K. (2011). Production and properties of fuels from domestic and industrial waste. In: *The Biofuels Handbook* (ed. J.G. Speight), 333–376. London, UK: Royal Society of Chemistry.

Khosrokhavar, R., Griffiths, S., and Wolf, K.-H. (2014). Shale gas formations and their potential for carbon storage: opportunities and outlook. *Environmental Processes* 1 (4): 595–611.

Kundert, D., and Mullen, M. (2009). Proper evaluation of shale gas reservoirs leads to a more effective hydraulic-fracture stimulation. *Paper No. SPE 123586. Proceedings. SPE Rocky Mountain Petroleum Technology Conference*, Denver, Colorado (April 14–16).

Kvenvolden, K.A. (1993). Gas hydrates as a potential energy resource – a review of their methane content. In: *The Future of Energy Gases*, Professional Paper

No. 1570 (ed. D.G. Howell), 555–561. Washington, DC: United States Geological Survey.

Kvenvolden, K. (1995). A review of the geochemistry of methane in natural gas hydrate. *Organic Geochemistry* 23 (11–12): 997–1008.

Kvenvolden, K.A., and McMenamin, M.A. (1980). Hydrates of natural gas: a review of their geologic occurrence. *Circular No. 825*. United States Geological Survey, Reston, VA.

Levine, J.R. (1993). Coalification: the evolution of coal as a source rock and reservoir rock for oil and gas. *American Association of Petroleum Geologists, Studies in Geology* 38: 39–77.

Lohila, A., Laurila, T., Tuovinen, J.-P. et al. (2007). Micrometeorological measurements of methane and carbon dioxide fluxes at a municipal landfill. *Environmental Science & Technology* 41 (8): 2717–2722.

Lorenson, T.D. and Collett, T.S. (2000). Gas content and composition of gas hydrate from sediments of the southeastern north American continental margin. In: *Proceedings of the Ocean Drilling Program, Scientific Results* (ed. C.K. Paull, R. Matsumoto, P.J. Wallace, and W.P. Dillon), Vol 164, 37–46. http://www-odp.tamu.edu/Publications/164_SR/VOLUME/CHAPTERS/SR164_04.PDF (accessed 22 August 2017).

MacDonald, G.J. (1990a). The future of methane as an energy resource. *Annual Review of Energy* 15: 53–83.

MacDonald, G.J. (1990b). Role of methane clathrates in past and future climates. *Climate Change* 16: 247–281.

Maule, A.L., Makey, C.M., Benson, E.B. et al. (2013). Disclosure of hydraulic fracturing fluid chemical additives: analysis of regulations. *New Solutions* 23: 167–187.

Mokhatab, S., Poe, W.A., and Speight, J.G. (2006). *Handbook of Natural Gas Transmission and Processing*. Amsterdam: Elsevier.

Moslemizadeh, A., Shadizadeh, S.R., and Moomenie, M. (2015). Experimental investigation of the effect of henna extract on the swelling of sodium bentonite in aqueous solution. *Applied Clay Science* 105–106 (Supplement C): 78–88.

Parkash, S. (2003). *Refining Processes Handbook*. Amsterdam: Gulf Professional Publishing, Elsevier.

Poling, B.E., Prausnitz, J.M., and O'Connell, J.P. (2001). *The Properties of Gases and Liquids*, 5e. New York: McGraw-Hill.

Ramage, J. (1997). *Energy: A Guidebook*. Oxford, UK: Oxford University Press.

Ramroop Singh, N. (2011). Biofuels. In: *The Biofuels Handbook* (ed. J.G. Speight), 160–198. London, UK: Royal Society of Chemistry.

Rasi, S., Veijanen, A., and Rintala, J. (2007). Trace compounds of biogas from different biogas production plants. *Energy* 32: 1375–1380.

Rasi, S., Lantela, J., and Rintala, J. (2011). Trace compounds affecting biogas energy utilization – a review. *Energy Conversion and Management* 52 (12): 3369–3375.

Rice, D.D. (1993). Composition and origins of coalbed gas. *American Association of Petroleum Geologists Studies in Geology* 38: 159–184.
Ross, D.J.K. and Bustin, R.M. (2009). The importance of shale composition and pore structure upon gas storage potential of shale gas reservoirs. *Marine and Petroleum Geology* 26 (6): 916–927.
Shepherd, M. (1947). Analysis of a standard sample of natural gas by laboratories cooperating with the American Society for testing materials. *Journal of Research of the National Bureau of Standards* 38: 19–51.
Singh, K. and Sastry, M.K.S. (2011). Production of fuels from landfills. In: *The Biofuels Handbook* (ed. J.G. Speight), 408–453. London, UK: Royal Society of Chemistry.
Speight, J.G. ed. (1990). *Fuel Science and Technology Handbook*. New York: Marcel Dekker.
Speight, J.G. (2007). *Natural Gas: A Basic Handbook*. Houston, Texas: GPC Books, Gulf Publishing Company.
Speight, J.G. (2008). *Synthetic Fuels Handbook: Properties, Processes, and Performance*. New York: McGraw-Hill.
Speight, J.G. (2011a). *The Refinery of the Future*. Oxford, UK: Gulf Professional Publishing, Elsevier.
Speight, J.G. (2011b). *An Introduction to Petroleum Technology, Economics, and Politics*. Salem, Massachusetts: Scrivener Publishing.
Speight, J.G. ed. (2011c). *The Biofuels Handbook*. London, UK: Royal Society of Chemistry.
Speight, J.G. 2012. *Crude Oil Assay Database*. Knovel, New York. http://www.knovel.com/web/portal/browse/display?_EXT_KNOVEL_DISPLAY_bookid=5485&VerticalID=0 (accessed 15 December 2012).
Speight, J.G. (2013a). *Shale Gas Production Processes*. Oxford, UK: Gulf Professional Publishing, Elsevier.
Speight, J.G. (2013b). *The Chemistry and Technology of Coal*, 3e. Boca Raton, Florida: CRC Press, Taylor & Francis Group.
Speight, J.G. (2014a). *The Chemistry and Technology of Petroleum*, 5e. Boca Raton, Florida: CRC Press, Taylor & Francis Group.
Speight, J.G. (2014b). *Oil and Gas Corrosion Prevention*. Oxford, UK: Gulf Professional Publishing, Elsevier.
Speight, J.G. (2015a). *Handbook of Petroleum Product Analysis*, 2e. Hoboken, New Jersey: Wiley.
Speight, J.G. (2015b). *Asphalt Materials Science and Technology*. Oxford, UK: Butterworth-Heinemann, Elsevier.
Speight, J.G. (2016a). *Introduction to Enhanced Recovery Methods for Heavy Oil and Tar Sands*, 2e. Waltham, Massachusetts: Gulf Publishing Company, Taylor & Francis Group.
Speight, J.G. (2016b). *Handbook of Hydraulic Fracturing*. Hoboken, New Jersey: Wiley.

Speight, J.G. (2017a). *Handbook of Petroleum Refining*. Boca Raton, Florida: CRC Press, Taylor & Francis Group.

Speight, J.G. (2017b). *Deep Shale Oil and Gas*. Oxford, UK: Gulf Professional Publishing, Elsevier.

Speight, J.G. and Islam, M.R. (2016). *Peak Energy – Myth or Reality*. Beverly, Massachusetts: Scrivener Publishing.

Staley, B. and Barlaz, M.A. (2009). Composition of municipal solid waste in the United States and implications for carbon sequestration and methane yield. *Journal of Environmental Engineering* 135 (10): 901–909.

Sullivan, P. (2010). *The Importance of Landfill Gas Capture and Utilization in the U.S.* New York. 6 April. http://www.scsengineers.com/wp-content/uploads/2015/03/Sullivan_Importance_of_LFG_Capture_and_Utilization_in_the_US.pdf: Earth Engineering Center, Columbia University accessed 11 November 2017.

USGS. (2011). U.S. geological survey gas hydrates project: database of worldwide gas hydrates. http://woodshole.er.usgs.gov/project-pages/hydrates/database.html (accessed 7 October 2014).

2

Origin and Production

2.1 Introduction

Natural gas occurs deep beneath the Earth's surface and not only consists predominantly of methane (CH_4) but also contains small amounts of hydrocarbon gas liquids and nonhydrocarbon gases (Mokhatab et al., 2006; Speight, 2007, 2014a; Faramawy et al., 2016; Speight, 2017a). The formations that contain natural gas are varied in mineralogy and require considerable knowledge and expertise to produce the gas efficiently and economically. Thus, as the development of a natural gas commences, the mineralogy of the reservoir increases in importance, especially when the highly adsorbing clay minerals are present.

As a guideline, the four most important items to address during a mineralogical analysis of the reservoir rock are (i) the specific elements that are in the reservoir rock; (ii) the characteristics of these elements, such as the atomic weight and atomic number; (iii) the concentrations of the elements, % w/w and ppm w/w; and (iv) the minerals or mineral phases with which the elements are associated. These data are especially important for *in situ* analysis of the reservoir and the gas where knowledge of the phase relationships in the reservoir (involving determination of the dew point of the gas) is a first-order control on the setup of the analytical approach. The compilation of the data, obtained from the various ASTM test method procedures (ASTM, 2017), allows an accurate assessment of the properties of the natural gas present in the reservoir and also produces valuable data to assess the porosity, permeability, and saturation of the reservoir. Additionally, the geological data should describe any lateral and vertical variations of reservoir characteristics and how those characteristics will affect the production and composition of the natural gas.

The mineralogy of a rock determines its physical and chemical properties. In a reservoir rock, physical properties such as density, sonic velocity, and compressibility and wetting properties contribute to essential reservoir properties

Handbook of Natural Gas Analysis, First Edition. James G. Speight.

such as porosity, permeability, and fluid saturations. Also, the chemical properties of a reservoir rock are especially important in drilling and production operations in order to predict how the formation will react when exposed to foreign materials such as drilling mud and production chemicals. The knowledge of mineralogy also provides information about the deposition and diagenesis of reservoir rocks, which further helps in understanding flow characteristics in a reservoir. Thus, detailed reservoir characterization is essential to understanding how to optimally recover oil or gas reserves, and the relevant test methods should be applied on an as-needed basis (Table 2.1).

Furthermore, the identification of the lithology (i.e. identification of the rock types) is fundamental to all reservoir characterization because the physical and chemical properties of the rock that holds hydrocarbons and/or water affect the response of every tool used to measure formation properties. Understanding reservoir lithology is the foundation from which all other petrophysical calculations are made. In fact, the lithology of a reservoir impacts the petrophysical calculations in numerous ways. The depositional environment and sediments being deposited will define the grain size, its sorting, and its distribution within the reservoir interval. In most sandstone reservoirs, the depositional environment controls the porosity/permeability relationship – two essential factors that determine gas flow within and from the reservoir to the production well.

For carbonate formations, the rock formations typically consist of interbedded sequences of carbonates, dolomites, anhydrite, salt, and shale layers. The keys to reservoir development within the carbonate layers are the original grain size and how it has been altered by chemical diagenetic processes. As these chemical reactions take place, the pore-size distribution and porosity level will change (e.g. by dolomitization). Carbonate-reservoir porosity is also greatly enhanced by weathering, dissolution, and fracturing (Speight, 2016). The texture of clay minerals and their composition are requisite for understanding their influence on reservoir characteristics and the properties of the crude oil or natural gas (Archer, 1985; Hurst and Archer, 1986; Speight, 2014a; Saha et al., 2017). In fact, a detailed characterization of any crude oil or natural gas reservoir is an essential aspect to an understanding of the best method to optimally recover crude oil or gas reserves (Ballard, 2007). Furthermore, the knowledge of mineralogy also provides information about the deposition and diagenesis of reservoir rocks, which further helps in understanding flow characteristics in a reservoir. Another standard test method (ASTM D2113, 2017) presents the guidelines, requirements, and procedures for core drilling, coring, and sampling of the reservoir rock for the purposes of site exploration and also whether the borehole could be vertical, horizontal, or angled.

Successful reservoir description (Section 5.5) requires detailed quantitative sedimentological and mineralogical data fully integrated with a petrophysical interpretation (for example, ASTM D5890, 2017 and ASTM E1915, 2017).

Table 2.1 Test methods applied to determining the petrology and mineralogy of a reservoir.[a]

X-ray powder diffraction (XRD)

Used to determine the composition of rocks and other crystalline material. It is the best way to determine amounts and types of clay minerals:

- Bulk analysis: Material is ground to a fine powder. Results are reported in weight percent; useful for scales or rock samples where clay minerals are not present
- Formation rock composition: Includes a bulk analysis of all mineral components and analysis of the less than 4-μm clay-size material. The clay-size fraction is analyzed separately from bulk components and then recombined to provide the total composition of the rock. Reported results include all minerals present in the rock and includes types and amounts of clay minerals and are reported in weight percent
- Organic shale analysis: Analysis incorporates organic component (TOC) with inorganic constituents of rock. Reported results are in weight percent and volume percent and include grain density calculated from XRD data.
- Siliceous shale analysis (crystallinity index/silica diagenesis): Analysis included detailed analysis of quartz peaks to determine the index of crystallinity, which directly relates to the degree of silica diagenesis

Scanning electron microscopy (SEM)

Provides high magnification/high-resolution images of smallest features in rocks:

- Standard SEM analysis: Used to investigate authigenic clays and other cements associated with pore systems of conventional reservoir rocks. Analysis done in a low-pressure chamber using samples coated with gold or gold/palladium alloy
- Mudrocks: The method is used to investigate micro- and nanopore systems in unconventional reservoir rocks. Analysis done in a high-pressure scanning electron microscope chamber with uncoated sample
- Argon-ion milling: Used to prepare surface of sample for analysis. Ion milling provides a polished surface suitable for backscattered electron imaging
- Backscattered electron imaging: Used to reduce/eliminate charging effects on uncoated samples

Energy dispersive analysis of X-rays (EDX)

- Qualitative analysis: Used to identify elemental composition; can also be used to identify framework grains or cements in a reservoir rock or to identify layers/bands in a scale
- Dot maps: Data from this method can be used to illustrate the distribution of elements in an SEM image

Thin section petrography

- Descriptions: A detailed description of the texture, framework grains, pore system, and authigenic minerals identified in a thin section. Useful to describe mudrock, siltstone, and some limestones such as mudstone, wackestone (a mud-supported carbonate rock that contains greater than 10% grains), and boundstone (a type of calcareous rock sediment where the original components have been bound together after deposition)
- Point count: Used in addition to a thin section description; provides compositional data based on framework grains, cements, and matrix; used to describe sandstone and some limestone such as grainstone (a grain-supported carbonate rock that contains less than 1% mud-grade material)
- Touchstone petrography: Petrography used in touchstone-based reservoir characterization models. Includes minimum 300-count point count, 100 framework grain measurements, and estimation of grain coating on a minimum of 50 framework grains

[a] The relevant test methods are listed and described in the Annual Book of Standards (ASTM International) or in the lists of other standards organizations. See also ASTM D5753, 2017; ASTM D6167, 2017; ASTM D6274, 2017; and ASTM D6727, 2017.

The former test method (ASTM D5890, 2017) is a method that allows the evaluation of swelling properties of a clay mineral in water for the estimation of its usefulness for permeability or hydraulic conductivity reduction in geosynthetic clay liners (GCL). The latter test method (ASTM E1915, 2017) is primarily intended to test materials for compliance with compositional specifications and for monitoring. At this time, samples are usually taken through the borehole to determine not only reservoir properties and gas properties to conclude if the reservoir is an economic viability.

Once the samples are acquired, reservoir mineralogy can be determined both qualitatively and quantitatively in a number of ways. The three primary quantitative methods are (i) thin section analysis, (ii) X-ray diffraction (XRD), and (iii) Fourier transform infrared spectroscopy (FTIR). Thin section analysis is the oldest of these three methods, and XRD has become the method of choice in many analytical laboratories. However, despite the prevalence of XRD in laboratories, the FTIR does offer advantages in that it is a more rapid method of analysis and can be applied at the well site. For example, after sample preparation, it may take only a matter of minutes to collect the infrared spectrum (Ballard, 2007).

Analytical programs assist in the determination of the reservoir mineralogy and gas quality, leading to the best and most economical methods for extracting gas (Burnett, 1995). Furthermore, the objective of such work is to provide guidelines for the preparation of stable, representative, oxidized, relatively unpolluted, aquatic natural-matrix bed-sediment reference test samples. The test samples should be useful for collaborative method testing, to evaluate the precision and bias of test methods, and to evaluate test method performance during their development (ASTM D5074, 2017). The methods selected will vary with the reservoir and even across a gas reservoir due to differences such as (i) reservoir thickness, (ii) reservoir extent, (iii) reservoir pressure, (iv) reservoir depth, (v) mineralogy of the reservoir, (vi) gas composition, and (vii) water content, all of which require reliable analytical data in order for the project to proceed with the potential for success (Table 2.2). Thus, as part of any comprehensive all-of-the-above reservoir development program, a strategy to ensure safe and economic development of the natural gas reservoir is a necessity. This must also include methods for the development of natural gas from gas hydrates (methane hydrates) that can occur in many reservoirs.

Moreover, the need for accurate and meaningful analytical data has expended with the technological advances in hydraulic fracturing and horizontal drilling that have led to dramatic growth in natural gas development, with tremendous economic potential for state and local economies (Speight, 2016). Although hydraulic fracturing has been employed for decades, its use has rapidly increased in the past few years, and some states are taking steps to ensure that water and air quality are adequately protected during surface and subsurface natural gas development activities (Speight, 2016).

Table 2.2 Analytical needs for safe natural gas recovery and development.

Stage of development	Analysis needed	Comment
Origin	Strata mineralogy	Identify potential for gas adsorption
Exploration	Gas quality	Identify potential for gas processing
	Gas reserves	Identify economic aspects of resource development
Reservoir	Reservoir mineralogy	Tendency of gas constituents to adsorb or react with mineral
Production	Gas quality	Identify need for wellhead processing prior to transportation
Transportation	Gas composition	Identify potential for corrosion during transportation

While natural gas from different sources may seemingly have a similar analysis, the composition of the gas is not necessarily the same. In fact, changes can occur from one step in the train to the next:

$$\text{Reservoir gas} \rightarrow \text{produced gas} - \text{wellhead gas} \rightarrow \text{transported gas} \\ \rightarrow \text{stored gas} \rightarrow \text{sales gas}$$

Since the different gas supplies enter the gas system at different locations, the exact composition at any site will vary among the different regions and over time. As an example, the heating value will depend on the composition of the gas. The anticipation of differences in the composition must also take into account the differences in the reservoir gas due to differences in the maturation processes from one locale to another (Speight, 2007, 2014a; Faramawy et al., 2016). Thus, in terms of natural gas definition, there are three basic approaches to defining the gas: (i) a qualitative description of the natural gas by origin, type, and constituents, (ii) classification by characteristics based upon testing procedures, and (iii) classification as a result of the concentration of specific constituents in preference to other constituents.

In addition, the recent increases in domestic natural gas supplies have been made possible by two technologies, (i) horizontal drilling and (ii) hydraulic fracturing, which allow energy companies to tap natural gas supplies once thought to be inaccessible (Speight, 2016). Rapid expansion of hydraulic fracturing in densely populated regions where the process is unfamiliar, however, has focused efforts on ensuring that the practice is well regulated, is transparent, and protects public health and the environment. This is achieved through the acquisition of relevant and reliable analytical data. Knowledge of local geology, reservoir mineralogy, and environmental conditions allows developers as well as regulators and lawmakers to tailor regulations to meet the environmental needs, particularly to protect drinking water.

Thus, there is an opinion from some observers that the analysis of natural gas is only specific to sales to ensure that the natural gas meets the sales specifications, but nothing could be further from the truth. The application of standard test methods and consideration of the analytical data are needed throughout the natural gas exploration–discovery–production–transportation–processing–use train.

The following sections describe natural gas from its origin in the Earth to production, thereby confirming that any successful development project needs analytical monitoring at all aspects of the development. This will assist the reader to further understand the need for analysis at each stage of the train.

2.2 Origin

In order to understand the need for analysis of natural gas as it exists in the reservoir or is produced at the wellhead, understanding the origin of the gas can offer pointers to the analyst as to the number and types of test methods that should be applied to the gas. The key is to recognize that natural gas varies in composition not only because of the reservoir in which the gas exists but also because of the relative placement of wells within a reservoir.

There are many different theories as to the origin of the various fossil fuels, specifically (in the current context) natural gas and petroleum (Speight, 2007, 2014a; Faramawy et al., 2016). The most generally accepted theory is that natural gas and petroleum are formed when organic matter or *organic debris* (such as the remains of plants or animals) is compressed under the Earth, at very high pressure for periods of geologic time that are measured in terms of millions of years (millennia). At the start of the oil- and gas-producing periods, the remains of plants and animals decayed and built up in thick layers, and, over time as sediment and mud and other debris piled on top of the organic matter, metamorphosis of the organic debris occurred. As a result, the sediment, mud, and other inorganic debris were changed to rock, causing pressure to be put on the organic matter. The increasing pressure compressed the organic matter and, combined with other subterranean effects, decomposed the individual constituents into gas and petroleum (Table 2.3).

The deep-lying source rock that contains the organic precursors is often referred to as the *kitchen* (Speight, 2014a). In theory, the deeper and hotter the kitchen, the more likelihood of gas being produced from the precursor material. The increase of heat with depth is due to the presence of a *geothermal gradient* that is generally on the order of 25–30 °C per km (0.008–0.009 °C per foot of depth or 8–9 °C per 1000 feet of depth; 0.015–0.016 °F per foot of depth or 15–16 °F per 1000 feet of depth), i.e. approximately one degree for

Table 2.3 Events that are believed to have occurred in the formation of natural gas and petroleum.

400–300 million years ago	Tiny sea plants and animals died and were buried on the ocean floor
	Over time they were covered with layers of silt and sand
300–100 million years ago	The organic debris started change by simple chemical reactions
100–50 million years ago	The organic debris was buried deeper and deeper; pressure increased and (possibly) temperature increased (but, as noted above, the level of the temperature is largely unknown and, at best, very speculative)
50–1 million years ago	The organic debris reacted, under the prevalent conditions underground, to produce methane and other hydrocarbon products that eventually because natural gas that migrated to reservoirs where it was trapped and awaiting discovery

every one hundred feet below the surface of the Earth. However, this does not mean that nonassociated gas will always be found at greater depths than crude oil. Gas and crude oil have migrated from the *kitchen*, in a sideways direction and/or in an upward direction, until crude oil and gas are trapped in a reservoir that forms part of a subsurface formation. Thus, a field may have a series of layers of crude oil/gas and gas reservoirs over a thousand meters subsurface.

The methane (natural gas) produced in this manner is referred to as *thermogenic* methane (thermogenic natural gas) because it is presumed that the pressure effects and the increased depth of the precursors include the influence of heat to convert the organic matter to natural gas and petroleum includes the effects of heat. However, the actual temperature is not known and, like the remainder of the theory, is at best speculative. Experimentalists will note that laboratory studies at specified temperatures on the order of 300–350 °C (570–650 °F) will convert organic materials to methane and other hydrocarbon gases. Be that as it may, the temperatures used in the laboratory are often high on the basis that the increase in temperature makes up for the lack of geological time in the laboratory.

In addition, many experimentalists often ignore that higher temperatures change the chemistry of the process and the findings may not be completely true. The fact remains that the high temperature involved in the thermogenic origin of natural gas and other fossil fuels is not conclusively proven. This theory answers many questions, namely, the fact that natural gas and crude oil are only found under a small portion of the Earth. While there are other theories, this one seems to be the most widely accepted. However, any verbal or written differences of opinion about the formation of crude oil and natural gas should not affect the work of the analytical chemist, but she/he must be aware of the prevalent theories in order to select the most relevant test methods for application to the sample to be investigated.

Methane (natural gas) can also be formed through the transformation of organic matter by tiny microorganisms. This type of methane is referred to as *biogenic* methane (*biogenic* natural gas). In this process, the methane-producing microorganisms (*methanogens*) chemically break down organic matter to produce methane. The *methanogen* microorganisms are commonly found in oxygen-deficient areas near the surface of the Earth. Thus, the formation of methane in this manner usually takes place close to the surface of the Earth, and the methane produced is usually lost into the atmosphere, which can be seen as an environmental problem. In certain circumstances, however, this methane can be trapped underground, recoverable as natural gas. An example of biogenic methane is landfill gas. Waste-containing landfills produce a relatively large amount of natural gas, from the decomposition of the waste materials that they contain. New and evolving technologies are allowing this gas to be collected to add to the supply of natural gas, especially the production of gas from biomass sources (John and Singh, 2011; Ramroop Singh, 2011; Singh and Sastry, 2011).

A third way in which methane (and natural gas) may be formed is through *abiogenic* processes. Extremely deep under the Earth's crust, there exist hydrogen-rich gases, and as these gases rise toward the surface of the Earth, they may interact through the catalytic activity of minerals in the absence of oxygen (Speight, 2013, 2017b). This interaction may result in a reaction to form products that are found in the atmosphere (including nitrogen, oxygen, carbon dioxide, argon, and water). If these gases are under very high pressure as they move toward the surface of the Earth, they are likely to form methane (*abiogenic methane*), similar to thermogenic methane. Whatever the mode of formation of methane, the standard test methods remain the same. Methane is methane! However, there may be some minor differences in applying some of the test methods and interpreting the data when methane is not the only constituent of the gas to be tested, but other constituents are present with the methane.

Because of the manner in which it is formed, and the diverse nature of the precursors, natural gas contains constituents other than methane, which affect the composition and the properties of the gas (Table 2.4), which can raise concerns in the selection of the most appropriate analytical test method. While the major constituent of natural gas is indeed methane, there are components such as saturated hydrocarbon derivatives (C_nH_{2n+2}), low-boiling aromatic hydrocarbon derivatives (derivatives of benzene), nitrogen, carbon dioxide (CO_2), hydrogen sulfide (H_2S), and mercaptan derivatives (RSH, also called thiol derivative), as well as trace amounts of other constituents such as the

Table 2.4 General properties of unrefined natural gas (left-hand number) and refined natural gas (right-hand number).

Relative molar mass	20–16
Carbon content (% w/w)	73–75
Hydrogen content (% w/w)	27–25
Oxygen content (weight % w/w)	0.4–0
Hydrogen-to-hydrogen atomic ratio	3.5–4.0
Density relative to air at 15 °C	1.5–0.6
Boiling temperature (°C/1 atmosphere)	−162
Autoignition temperature (°C)	540–560
Octane number	120–130
Methane number	69–99
Vapor flammability limits (% v/v)	5–15
Flammability limits	0.7–2.1
Lower heating/calorific value (Btu lb−1)	900
Methane concentration (% v/v)	100–80
Ethane concentration (% v/v)	5–0
Nitrogen concentration (% v/v)	15–0
Carbon dioxide concentration (% v/v)	5–0
Sulfur concentration (ppm, w/w)	5–0

always valuable helium and argon. In cases where the natural gas does not contain hydrogen sulfide, there may also be a relative lack of carbon dioxide, which is not true for the gases produced from coal or biomass.

2.3 Reservoirs

A reservoir is a subsurface rock structure, with sufficient size and closure that contains a three-dimensional network of interconnected void (pore) space and is overlain by a fine-grained water-saturated rock. Its subsurface porous permeable rock body or formation has the capability to store and transmit fluids (crude oil, water, and natural gas) and has been created by the sequential steps of deposition, conversion, migration, and entrapment. Thus, porosity and permeability are the key features of the pore structure of the reservoir and are important factors in the production of crude oil and natural gas from the reservoir. Furthermore, a reservoir is characteristically large and extensive in volume and capacity – requirement that makes the reservoir economically attractive for development.

Deposition occurs when organic-rich sediment occurring in areas of high primary productivity is deposited and settles due to low energy currents. Such environments include rivers, deltas, estuaries, and continental shelves. Further deposition of sediment results in buildup of layers of sediment, for example, silts or clays atop the previously organic-rich sediment reduce oxygen from effectively getting into the system, effectively preserving organic matter. Further burial of the organic matter occurs due to more deposition, and pressures and temperatures of the layers increase due to increasing depth and the geothermal gradient 25–30 °C per km or 15 °F per 1000 ft, resulting in the conversion of organic matter to petroleum. *Conversion* is a breakdown of the organic matter that involved three main stages – diagenesis, catagenesis, and metagenesis – to form kerogen (Speight, 2014a), which is the high molecular weight organic residue found in many types of sediment, particularly in oil shale. *Migration* relates to the movement of oil or gas from its source rock where it was formed to a reservoir rock. Porosity and permeability of the rocks control the migration of the hydrocarbon. *Entrapment* occurs when the hydrocarbon migrates into the formation that is effectively sealed due to a change in lithology of the formation. For example, shale is commonly porous, but because of its fine grain size, it has very high capillary forces that prevent fluid flow.

Natural gas and crude oil (conventional crude oil or heavy crude oil) cannot be retained as an accumulation unless there is a trap, and this requires that the boundary between the cap rock or other sealing agent and the reservoir rock generally be convex upward, but the exact form of the boundary varies widely. The simplest forms are the flat-lying convex lens, the anticline, and the dome, each of which has a convex upper surface. Many crude oil and natural gas accumulations are trapped in anticlinal structures or in dome-type structures, which are generally more easily detected than some other types of traps (Speight, 2014a).

The most common reservoir rocks are sandstone (SiO_2), limestone ($CaCO_3$), and dolomite (a mixed calcium magnesium carbonate mineral ($CaCO_3 \cdot MgCO_3$)). The five basic elements of a reservoir system include (i) a source rock, which is the rock containing the organic material that converted into petroleum reservoir fluid; (ii) a migratory pathway, which is a route for the partially formed or completely formed crude oil and natural gas to migrate from the source rock to the reservoir; (iii) a reservoir rock, which is the rock of suitable porosity that can store and suitable permeability that can allow the fluid to move through the reservoir to a production well; and (iv) a seal, which is an impermeable cap rock that prevents the upward escape of the crude oil or natural gas to the surface of the Earth. In essence, the reservoir is a subsurface trap where the basement rock and the cap rock are arranged to trap petroleum. Thus, the reservoir is a geological arrangement of subsurface formations that prevents migration of any crude oil and natural gas.

The reservoir rock is a porous medium made up of pores conventionally envisaged as irregularly shaped holes in rock some 1–100 μm in length and diameter connected to maybe six other pores – there can be some 10^6 pores in 1 cm^3 of rock and possibly over 10^{22} pores in a typical reservoir. However, the variables that are typically sought to describe a reservoir are those that influence the amount, the position, the accessibility, and the flow of fluids through the reservoir.

2.3.1 Natural Gas Reservoirs

Since natural gas has a low density, once formed it will rise toward the surface of the Earth through loose, shale-type rock and other material. Most of this methane will simply rise to the surface and dissipate into the air. However, a great deal of this methane will rise up into geological formations that trap the gas under the ground. These formations are made up of layers of porous sedimentary rock with a denser, impermeable layer of rock on top. This impermeable rock traps the natural gas under the ground. If these formations are large enough, they can trap a great deal of natural gas underground, in what is known as a reservoir.

There are a number of different types of these formations, but the most common is created when the impermeable sedimentary rock appears as *dome*-shaped formation that traps all of the natural gas (Speight, 2014a). There are a number of ways that this sort of dome may be formed. For instance, faults are a common location for oil and natural gas deposits to exist. A fault occurs when the normal sedimentary layers sort of split vertically, so that impermeable rock shifts down to trap natural gas in the more permeable limestone or sandstone layers. Essentially, the geological formation that layers impermeable rock over more porous, oil- and gas-rich sediment has the potential to form a reservoir.

Once a potential natural gas deposit has been located by a team of exploration geologists and geophysicists, it is up to a team of drilling experts to actually dig down to where the natural gas is thought to exist. Although the process of digging deep into the Earth's crust to find deposits of natural gas that may or may not actually exist seems daunting, a number of innovations and techniques have been developed that increase the efficiency of drilling for natural gas, with a concurrent decrease in cost. The advance of technology has also contributed greatly to the increased efficiency and success rate for drilling natural gas wells. However, the search for natural gas always carries some drawbacks, and there is always the inherent risk that no natural gas will be found.

However, on the positive side, if a new well, once drilled, does in fact come in contact with a natural gas reservoir, it is developed to allow for the extraction of this natural gas and is termed a *development well* or *productive well*. At this point, with the well drilled and hydrocarbons present, the well may be completed to facilitate its production of natural gas. However, if the exploration team was incorrect in its estimation of the existence of marketable quantity of natural gas at a well site, the well is termed a *dry well*, and production does not proceed. Then, in order to successfully bring natural gas to the surface, a hole must be drilled through the impermeable rock to release the natural gas that is under pressure. Thus, natural gas that is trapped under in a subsurface formation can be recovered by drilling through the impermeable rock. Gas in these reservoirs is typically under pressure, allowing it to escape from the reservoir on its own. The decision of whether or not to drill a well into a reservoir depends on several factors, not the least of which is the economic characteristics of the potential natural gas reservoir.

The exact placement of the drill site depends on a variety of factors, including the nature of the potential formation to be drilled, the characteristics of the subsurface geology, and the depth and size of the target deposit. After the geophysical team identified the optimal location of a well (ASTM D6276, 2017, ASTM D6431, 2017), it is necessary for the drilling company to ensure that they complete all the necessary steps to ensure that they can legally drill in that area. This usually involves securing permits for the drilling operations, establishment of a legal arrangement to allow the natural gas company to extract and sell the resources under a given area of land, and a design for gathering lines that will connect the well to the pipeline. There are a variety of potential owners of the land and mineral rights of a given area.

In order to combat the presence of sulfur compounds in natural gas (many pipeline owners specify maximum sulfur content), scrubbers are installed, usually at or near the wellhead. The scrubbers serve primarily to remove sand and other large-particle impurities such as hydrogen sulfide. Heaters also ensure that the temperature of the gas does not drop too low. With natural gas that contains even low quantities of water, natural gas hydrates have a tendency to form when temperatures drop. These hydrates are solid or semisolid compounds, resembling icelike crystals. If the hydrates accumulate, they can impede the passage of natural gas through valves and gathering systems. To reduce the occurrence of hydrates, small natural gas-fired heating units are typically installed along the gathering pipe wherever it is likely that hydrates may form.

2.3.2 Crude Oil Reservoirs

In order to process and transport associated dissolved natural gas, it must be separated from the oil in which it is dissolved. This separation of natural gas from petroleum is most often done using equipment installed at or near the wellhead. The actual process used for water–crude oil–natural gas separation starts at the wellhead, and the equipment used can vary considerably. Although dry pipeline quality natural gas is virtually identical across different geographic areas, raw natural gas from different regions will vary in composition (Table 2.2), and therefore separation requirements may emphasize or de-emphasize any one of several optional separation processes. Typically, it is expected that a separator vessel will cause the water–oil–gas to separate into the three individual phases in the separator. The most basic type of separator is known as a conventional separator and consists of a simple closed tank, where the force of gravity serves to separate the heavier liquids, like oil, and the lighter gases, like natural gas.

However, in many instances, natural gas is dissolved in oil underground primarily due to the formation pressure and may give rise to what is known as foamy oil. In some cases, when natural gas and crude oil are produced, it is possible that the natural gas will readily separate from the crude oil. In these cases, separation of oil and gas is relatively easy, and the two hydrocarbons are sent separate ways for further processing. In many instances, however, specialized equipment is necessary to separate oil and natural gas. An example of this type of equipment is the low-temperature separator, and this type of separator is most often used for wells producing high-pressure gas along with light crude oil or condensate. These separators use pressure differentials to cool the wet natural gas and separate the oil and condensate.

In the separation process, the natural gas enters the separator, being cooled slightly by a heat exchanger. The gas then travels through a high-pressure liquid *knockout pot,* which serves to remove any liquids into a low-temperature separator. The gas then flows into this low-temperature separator through a choke mechanism, which expands the gas as it enters the separator. This rapid expansion of the gas allows for the lowering of the temperature in the separator. After liquid removal, the dry gas then travels back through the heat exchanger and is warmed by the incoming wet gas. By varying the pressure of the gas in various sections of the separator, it is possible to vary the temperature, which causes the oil and some water to be condensed out of the wet gas stream. This basic pressure–temperature (PT) relationship can work in reverse as well, to extract gas from a liquid oil stream. In addition, the potential for corrosion of the equipment and pipelines due to the presence of water and acid gases (hydrogen sulfide, H_2S, and carbon dioxide, CO_2) is high (Speight, 2014b) – hence the need to know the composition of the natural gas mixture that is produced at the wellhead.

2.3.3 Gas Condensate Reservoirs

Natural gas condensate is a low-density mixture of hydrocarbon liquids that are present as gaseous components in the raw natural gas produced from many natural gas fields. Some gas species within the raw natural gas will condense to a liquid state if the temperature is reduced to below the hydrocarbon dew-point temperature at a definitive pressure.

A gas condensate reservoir (also called a *dew-point reservoir*) is a reservoir in which condensation causes a liquid to leave the gas phase. The condensed liquid remains immobile at low concentrations. Thus, the gas produced at the surface will have a lower liquid content, and the producing gas–oil ratio therefore rises. This process of *retrograde* condensation continues until a point of maximum liquid volume is reached. The term *retrograde* is used because generally vaporization, rather than condensation, occurs during isothermal expansion. After the dew point is reached, because the composition of the produced fluid changes, the composition of the remaining reservoir fluid also changes.

Typically, a gas condensate reservoir will have a reservoir temperature located between the critical point and the cricondentherm on the reservoir fluid PT diagram. This is one way of identifying a gas condensate reservoir – any other definition such as condensate–gas ratio or molecular weight of the C^{7+} fraction or the API gravity of the C_{7+} fraction may leave gaps in knowledge of the behavior of the reservoir and the condensate (Thomas et al., 2009).

Each condensate has its own unique gas condensate composition. In general, gas condensate has a specific gravity ranging from 0.5 to 0.8 and is composed of low-boiling hydrocarbons such as up to an including octane (C_8) derivatives and, in some, cases up to an including dodecane (C_{12}). The C_8 to C_{12} hydrocarbon derivatives include the higher-boiling hydrocarbon compounds such as cyclohexane derivatives and aromatic derivative such as benzene, toluene ($C_6H_5CH_3$), xylene isomers (*ortho-*, *meta-*, and *para-*$CH_3C_6H_4CH_3$), and ethyl benzene ($C_6H_5C_2H_5$). Additionally, condensate may contain additional impurities such as (i) hydrogen sulfide; (ii) thiol derivative, also called mercaptans, which are denoted as RSH, where R is an organic group; and (iii) carbon dioxide.

Drip gas, so named because it can be drawn off the bottom of small chambers (called *drip chambers* or *drips*) sometimes installed in pipelines from gas wells, is another name for natural gas condensate, a naturally occurring form of gasoline obtained as a by-product of natural gas extraction. Drip gas is defined in the US Code of Federal Regulations as consisting of butane, pentane, and hexane hydrocarbons. Within set ranges of distillation, drip gas may be extracted and used to as a cleaner and solvent as well as a lantern and stove fuel. Accordingly, each type of condensate (including drip gas, natural gasoline, and casinghead gas) requires careful analysis to show the potential methods of refining and purification at the wellhead facilities prior to transportation through a pipeline to a gas processing plant (Chapter 4) or in a refinery (Speight, 2014a, 2017a).

Because condensate is typically liquid in ambient conditions and also has very low viscosity, condensate is often used to dilute highly viscous heavy crude oil that cannot otherwise be efficiently transported by means of a pipeline.

In particular, condensate is frequently mixed with bitumen from tar sand (called oil sand in Canada) to create the blend known as *dilbit*. The increased use of condensate as diluent has significantly increased its price in certain regions. However, caution is required when condensate having an unidentified (nonanalyzed) composition is blended with heavy oil and/or bitumen since the potential for incompatibility of the blended material can become a reality. This is especially true if the condensate is composed predominantly of *n*-alkane hydrocarbons. These hydrocarbons are used in laboratory deasphalting and in commercial deasphalting units in which the asphaltene fraction is produced as a solid insoluble as a solid product from the heavy oil or bitumen feedstock (Speight, 2014a).

2.4 Reservoir Fluids

Reservoir fluids are the fluids (including gases and solids) that exist in a reservoir, and each fluid requires application of one or more of the relevant standard test methods (Table 2.5) to estimate behavior of the fluid both in and out of the reservoir. The fluid type must be determined very early in the life of a reservoir (often before sampling or initial production) because fluid type is the critical factor in many of the decisions that must be made about producing the fluid from the reservoir.

Reservoirs contain complex mixtures of fluids and the behavior of these fluids is strongly dependent on chemical properties and physical properties. Heavy crude oil is a fluid that is also a multicomponent mixture composed of nonhydrocarbon derivatives and a variety of hydrocarbons – especially of the alkane series of hydrocarbon derivatives. Typical hydrocarbons encountered in heavy oil are the higher-boiling hydrocarbons – the amount depending on the original source materials and the maturation pathways; volatile hydrocarbon derivatives boiling lower than C_{12} are not present in a ready abundance. Thus, the properties of reservoir fluids can play a key role in the design and optimization of injection/production strategies and surface facilities for efficient reservoir management. Inaccurate fluid characterization often leads to high uncertainties in in-place volume estimates and recovery predictions and hence affects asset value. Indeed, when evaluating reservoir fluids the application of standard test methods to property determination is an important aspect of prospect evaluation, development planning, and reservoir management. Although many different types of fluid exist, the composition of the reservoir fluids provides important information that can influence the recovery process.

In order to produce a wellhead product from a reservoir, there has to be flow of the fluids to the wellbores through the heterogeneous porous media of the reservoir. Fluid movements within the reservoir are governed by the local fluid

Table 2.5 Analytical data required for the identification of reservoir fluids, including natural gas and gas condensate.

Formation volume factor (FVF)	The ratio of a phase volume (water, oil, gas, or gas plus oil) at reservoir conditions, relative to the volume of a surface phase (water, oil, or gas) at standard conditions resulting when the reservoir material is brought to the surface. Denoted mathematically as B_w (bbl/STB), B_0 (bbl/STB), B_g (ft^3/SCF), and B_t (bbl/STB)
Solution gas–oil ratio (GOR)	The amount of surface gas that can be dissolved in a stock tank oil when brought to a specific pressure and temperature. Denoted mathematically as R_s (SCF/STB)
Solution oil–gas ratio (OGR)	The amount of surface condensate that can be vaporized in a surface gas at a specific pressure and temperature; sometimes referred to as liquid content. Denoted mathematically as r_s (STB/MMSCF)
Liquid specific gravity	The ratio of density of any liquid measured at standard conditions (usually 14.7 psia and 60 °F) to the density of pure water at the same standard conditions. Denoted mathematically as γ_o (where water = 1)
API gravity	A common measure of the specific gravity of crude oil, defined by $\gamma_{API} = (141.5/\gamma_o) - 131.5$, with units in °API
Gas specific gravity	The ratio of density of any gas at standard conditions (14.7 psia and 60 °F) to the density of air at standard conditions; based on the ideal gas law ($pV = nRT$), gas gravity is also equal to the gas molecular weight divided by air molecular weight ($M_{air} = 28.97$). Denoted mathematically as γ_g (where air = 1)
Bubblepoint pressure	At a given temperature, this condition occurs when an oil releases an infinitesimal bubble of gas from solution when pressure drops below the bubblepoint
Retrograde pressure	At a given temperature, this condition occurs when a gas condenses a drop of oil from solution when pressure drops below the dew point; also called "retrograde dew-point pressure"
Saturation pressure	An oil at its bubblepoint pressure or a gas at its dew-point pressure
Critical point	The pressure and temperature of a reservoir fluid where the bubblepoint pressure curve meets the retrograde dew-point pressure curve; a unique state where all properties of the bubblepoint oil are identical to the dew-point gas
Composition	Quantifies the amount of each component in a reservoir mixture, usually reported in mole fraction. Typical components in petroleum reservoir mixtures include the nonhydrocarbons N_2, CO_2, and H_2S and the hydrocarbons C_1, C_2, C_3, iC_4, nC_4, iC_5, nC_5, C_6, and C_{7+} (heptanes plus)
Saturated condition	A condition where an oil and gas are in thermodynamic equilibrium, that is, the chemical force exerted by each component in the oil phase is equal to the chemical force exerted by the same component in the gas phase, thereby eliminating mass transfer of components from one phase to the other
Undersaturated condition	A condition when crude oil or natural gas is in a single phase but not at its saturation point (bubblepoint or dew point), that is, the mixture is at a pressure greater than its saturation pressure

potential gradients and reservoir effective permeability, the injection and production points, and the fluid viscosities. However, fluid flow is also governed by the character of the fluids in the reservoir.

The term *fluid* as used by reservoir engineers can mean that the contents of the reservoir that are in the gaseous state or in the liquid state and the fluids exist either as (i) a solid phase, of which wax is an example; (ii) a liquid phase, of which crude oil is an example; or (iii) a gaseous phase, of which natural gas is an example. Gas and liquids coexist in a given reservoir. More specifically in the current context, a reservoir typically contains three main *fluids*: (i) natural gas, (ii) crude oil, and (iii) water with minor constituents being acid gases (carbon dioxide and hydrogen sulfide). These components will vary greatly in combination and proportion within each reservoir, and, in the case of heavy crude oil, the amount of gas will be substantially less than would be found in a conventional oil reservoir. Reservoir fluids vary greatly in composition and chemical properties. Typically, heavy oil has an API gravity between 10 and 20° (Chapter 1) and is more viscous than conventional petroleum and has the commonality of being, or ability to be, in the liquid state and, therefore, has mobility in the reservoir. As a result of this mobility in the reservoir, heavy crude oil can be recovered from a reservoir by the use of conventional (including enhanced) oil recovery techniques (Speight, 2009).

In reservoirs that contain crude oil and natural gas, it is often assumed that natural gas, being the least dense, is found at the top of the reservoir, with the crude oil beneath the natural gas, typically followed by water at the base of the reservoir. However, rather than discrete zones of natural gas, crude oil, and water, it is more likely that a reservoir will consist of several boundary zones, such as (i) an oil-in-water zone, (ii) a water-in-oil zone, (iii) a gas-in-oil zone, and (iv) a gas-in-water zone, as well as various zones of mixed composition. It is safe to assume that in the reservoir, zones are not always ordered according to density. Thus, the distribution of the fluids in a reservoir rock is not always dependent on the density of the reservoir fluids as well as on the properties of the rock.

However, if the pores of the reservoir rock are of uniform size and evenly distributed, there is (i) an upper zone where the pores are filled mainly by gas (the gas cap), (ii) a middle zone in which the pores are occupied principally by oil with gas in solution, and (iii) a lower zone with its pores filled by water. A certain amount of water (approximately 10–30%) occurs along with the oil in the middle zone. There is a transition zone from the pores occupied entirely by water to pores occupied mainly by oil in the reservoir rock, and the thickness of this zone depends on the densities and interfacial tension of the oil and water as well as on the sizes of the pores. Similarly, there is some water in the pores in the upper gas zone that has at its base a transition zone from pores occupied largely by gas to pores filled mainly by oil.

The water found in the oil and gas zones (*interstitial water*, the analysis of which can often provide information about the mineralogy of the reservoir from the presence of water-soluble minerals) usually occurs as collars around grain contacts, as a filling of pores with unusually small throats connecting with adjacent pores, or, to a much smaller extent, as wetting films on the surface of the mineral grains when the rock is preferentially wet by water. The three-dimensional reservoir network allows continuity to exist for the hydrocarbons by means of connections on every side of the sand grains. The so-called gas–oil and oil–water contacts are generally horizontal but have been known to exist as a very gentle incline. On occasion, part of an accumulation of the oil or gas has its lower boundary marked, not by the water-bearing zone of the reservoir rock but by an adjacent sealing rock that has characteristics similar to those of the cap rock. When the pressure and temperature conditions are suitable in relation to the proportions and the nature of the gas and oil, there may be no gas cap but only oil, with dissolved gas overlying the water.

The water (brine) produced with oil has been trapped with the oil and is brought to the surface along with oil. Because the water has been in contact with the oil, it contains some of the chemical characteristics of the formation and the oil itself. Oil and gas wells produce more water than oil (7 bbl/bbl of crude oil in some fields). The composition (salt content) of coproduced water determines the need for antiscaling additives. There are strict regulations to limit disposal and beneficial use options as well as environmental impacts that pertain to oil field waters.

Reservoir temperatures may vary up to 90 °C (194 °F) or even higher, while surface conditions are around 20 °C (68 °F). Pressure can vary from its atmospheric value (or lower in the case of vacuum distillation) to a number in the hundred million pascals (Pa). Within such an ample range of conditions, hydrocarbon fluids undergo severe transformations and exist as a single phase (gas, liquid, or solid) or coexist in several forms (liquid plus gas, solid plus liquid, vapor plus solid, or even in liquid-plus-liquid combinations). Understanding the methods by which the hydrocarbon fluids interact with and react to their thermodynamic surroundings is essential to adequately analyze systems of interest, and one of the most useful phase-behavior visualizations is the PT diagram or PT envelope. Each envelope represents a thermodynamic boundary separating the two-phase conditions (inside the envelope) from the single-phase region (outside). The correct identification of the type of hydrocarbon fluid is critical for the proper design and development of the correct production strategy and the analytical program for the field under consideration. This identification is critical for proper hydrocarbon reservoir modeling as well. When a fluid is assumed to behave as heavy crude oil, it should be assumed that the behavior of the heavy crude oil will be complex and selecting an appropriate model for heavy crude oil behavior becomes a limiting factor.

Reservoir fluids are brought to the surface as a mixture of crude oil, natural gas, and water, which is sent to a surface production facility before they can be disposed or sold to an industrial consumer (e.g. a refinery). A surface production facility is the system in charge of the separation of the well stream fluids into its three single-phase components – oil, gas, and water – and of their transport and processing into marketable products and/or their disposal in an environmentally acceptable manner. Once separated, the oil, natural gas, and water follow different paths. Water is typically reinjected for reservoir pressure maintenance operations. The oil usually goes through a process of dehydration, which removes basic sediments, and hydrocarbon fluids are assumed to comprise two components – stock tank oil and surface gas.

2.5 Production

Production is the process of extracting the hydrocarbons and separating the mixture of liquid hydrocarbons, gas, water, and solids, removing the constituents that are nonsaleable and selling the liquid hydrocarbons and gas (Figure 2.1). Production sites often handle crude oil from more than one well. Crude oil is typically processed at a refinery, whereas natural gas may be processed to remove impurities either in the field (at the wellhead) or at a nearby natural gas processing plant.

The recovery methods that are required to produce natural gas, specifically from shale formations, have attracted considerable attention in the last decade. Safe and reliable extraction, transport, and delivery of natural gas to consumers remain the priority for natural gas industry participants, and this can only be planned and accomplished through the application of analytical programs that produce data to define the quality of the natural gas. In the development of a natural gas resource, analysis of the gas streams can play an important role in

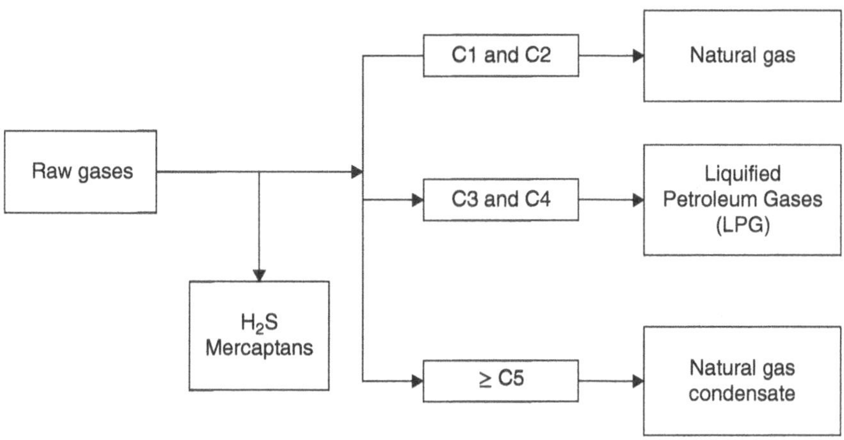

Figure 2.1 Schematic representation of natural gas processing.

developing and maintaining process economics as well as providing gas with the necessary properties required by the consumer.

During the many years when natural gas was produced as a by-product of oil productions, its value was largely ignored. In many cases large volumes of natural gas were flared with no effort being made to conserve this valuable material. Only a slight improvement was realized when many of the first sales contracts were written for ridiculously low prices. Until recently, large volumes of natural gas were sold for less than the cost of compression to the sales pressures. Such contracts completely ignored the intrinsic value of this high-quality fuel. This poor situation was made worse when these distress prices were recognized by a federal regulatory body as proper price levels for natural gas moving in interstate commerce.

In fact, natural gas reservoirs, like crude oil reservoirs, exist in many forms such as the dome (syncline–anticline) structure, with water below, or a dome of gas with a crude oil rim and water below the oil. When the water is in direct contact with the gas, pressure effects may dictate that a considerable portion of the gas (20% v/v or more) is dissolved in the crude oil as well as in the water. As the gas is produced, the reservoir pressure will decline, allowing the gas to enter the gas phase. In addition, and because of the variability of reservoir structure, gas does not always flow equally throughout the length, breadth, and depth of the reservoir, and at equal pressure, recovery wells must be distributed throughout the reservoir to recover as much of the gas as possible.

As the gas pressure in the reservoir declines, the reservoir energy declines and the natural gas requires stimulation for continued production. Furthermore, reduction in the gas pressure may allow compaction of the reservoir rock by the weight of rock above, eventually resulting in subsidence of the surface above the reservoir. This can be gradual process or a sudden catastrophic process depending on the structures of the above geological features.

A reservoir containing *wet* gas with a large amount of valuable natural gas liquids (ethane, propane, and higher molecular weight hydrocarbons) and even light crude oil and condensate has to be treated carefully. When the reservoir pressure drops below the critical point for the mixture, the liquids may condense out and remain in the reservoir. Thus, it is necessary to implement a *cycling* process in which the wet gas is produced to the surface and the natural gas liquids are condensed as a separate stream and the gas is compressed and injected back into the reservoir to maintain the pressure.

The production of natural gas from a reservoir is inherently more efficient than the production of crude oil from a reservoir. In volumetric reservoirs, at moderate depths with high formation permeabilities, recoveries of 90% v/v of the gas originally in place are common. However, in many cases, high recoveries are not attained. In a reservoir with a low permeability, the producing rates may fall below the economic limit, while large portions of the original gas are still in the reservoir. The minimum economic gas production rate is raised by the production of liquids or by any other difficulties, which tend to increase operating problems and expenses. In reservoirs with strong water drives and unequal water invasion, large amounts of gas may be lost in the residual gas saturation or by premature watering out of the producing wells. In such cases, the recovery of the gas originally in place may be as low as 50% of the volume in place.

Thus, the suitability of a natural gas reservoir for the production of gas is impacted by evaluation and implementation of sound reservoir management and investment decisions during different stages of field life (delineation, primary depletion, waterflooding, infill primary depletion, infill drilling, and flooding). Reservoir characterization has a big impact on field development decisions and process and facilities design. The two most challenging issues in reservoir characterization are (i) estimation and description of key reservoir properties (porosity, permeability, transmissivity, geometry, saturations and permeability, pressures, and rock–fluid interactions) at locations not pressures, and rock–fluid interactions) at locations not sampled by well data and (ii) representation of these properties as effective properties for reservoir simulation properties and prediction of reservoir performance.

Although the precise level of detail from the application of analytical methods is not always well defined, the detail to which a reservoir description is needed depends on its intended use (Chopra et al., 1990; Damsleth, 1994). Nevertheless, the necessary properties are evident when the reservoir is first prepared for gas production through following categories: (i) gas wells; (ii) well completion; (iii) wellhead, which must be suitable for preliminary treatment of the gas when necessary; (iv) well treatment; and (v) natural gas production. Each of these necessary issues is presented in the following sections.

2.5.1 Gas Wells

The well development is an essential part of the production process and is instituted immediately after exploration has located a reservoir (or *field*) that can economically produce natural gas. This involves the construction of one or more wells from the beginning (often referred to as *spudding*) to either abandonment if no hydrocarbons are found or to well completion if hydrocarbons are found in sufficient quantities. The well development program depends upon the quality of the natural gas and the reservoir pressure.

Gas wells are technically similar to crude oil wells with casing, tubing, and a wellhead and controls at the top. Conventional wells use casings *telescoped* inside each other and given pressure resistance by cement. However, with *expanded tubulars* each tubular casing is expanded against the previous one by pumping a tool called a *mandrel* down the casing. Thus, the well can either be thinner or cheaper for a given size of final tubing, or alternatively, using the conventional 20 inch diameter external casing, there is space to install a wider and thus higher capacity tubing.

When gas is contained in deep, relatively tight, low-permeability reservoirs, it is often economic to drill large-diameter wells with horizontal sections through the reservoirs to collect more gas. In addition, gas production can be increased by fracturing the reservoir rock using overpressure and by pumping in ceramic beads that maintain the integrity of the fractures. It is also attractive to install *smart wells* that have measurement gauges permanently installed downhole. This eliminates the need for lowering gauges down the wells during production. With the high gas flow rates, sand can often be pulled into the well with the gas, and therefore wearing parts are hardened against sand erosion.

2.5.2 Well Completion

Once a natural gas or oil well is drilled, and it has been verified that commercially viable quantities of natural gas are present for extraction, the well must be completed to allow for the flow of petroleum or natural gas out of the formation and up to the surface. This process includes strengthening the wellhole with casing, evaluating the pressure and temperature of the formation, and then installing the proper equipment to ensure an efficient flow of natural gas out of the well.

Since oil is commonly associated with natural gas deposits, a certain amount of natural gas may be obtained from wells that were drilled primarily for oil production. In some cases, this *associated natural gas* is used to help in the production of oil, by providing pressure in the formation for the oil extraction. The associated natural gas may also exist in large enough quantities to allow its extraction along with the oil.

Condensate wells are wells that contain natural gas, as well as a liquid condensate that is predominantly hydrocarbon derivative. Thus, the gas condensate is actually a mixture of hydrocarbon derivatives that are liquid at ambient temperature and pressure mixture that is often separated from the natural gas either at the wellhead or during the processing of the natural gas. Depending on the type of well that is being drilled, completion may differ slightly. It is important to remember that natural gas, being lighter than air and given the appropriate conditions, will naturally rise to the surface of a well. Because of this, in many natural gas and condensate wells, lifting equipment and well treatment are not necessary.

Completing a well consists of a number of steps: installing the well casing, completing the well, installing the wellhead, and installing lifting equipment or treating the formation should that be required. The *well casing* consists of a series of metal tubes installed in the freshly drilled hole, and installation of the casing is an important part of the drilling and completion process. Casing serves to strengthen the sides of the wellhole, to ensure that no oil or natural gas seeps out of the wellhole as it is brought to the surface, and to keep other fluids or gases from seeping into the formation through the well. In addition to strengthening the production well, the casing also provides a conduit to allow hydrocarbons to be extracted without intermingling with other fluids and formations found underground. It is also instrumental in preventing blowouts, allowing the formation to be sealed from the top should dangerous pressure levels be reached. The types of casing used depend on the subsurface characteristics of the well, including the diameter of the well (which is dependent on the size of the drill bit used) and the pressures and temperatures experienced throughout the well. In most wells, the diameter of the wellhole decreases the deeper it is drilled, leading to a type of conical shape that must be taken into account when installing casing. There are five different types of well casing, which include (i) conductor casing, (ii) surface casing, (iii) intermediate casing, (iv) liner string, and (v) production casing.

Conductor casing is installed first, usually prior to the arrival of the drilling rig. The hole for conductor casing is often drilled with a small auger drill, mounted on the back of a truck. Conductor casing, which is usually no more than 20–50 ft long, is installed to prevent the top of the well from caving in and to help in the process of circulating the drilling fluid up from the bottom of the well. Onshore, this casing is usually 16–20 inches in diameter, while offshore casing usually measures 30–42 inches. The conductor casing is cemented into place before drilling begins.

Surface casing is the next type of casing to be installed and can be anywhere from a few hundred to 2000 ft long and is smaller in diameter than the conductor casing. When installed, the surface casing fits inside the top of the conductor casing. The primary purpose of surface casing is to protect freshwater deposits near the surface of the well from being contaminated by leaking hydrocarbons or salt water from deeper underground. It also serves as a conduit for drilling mud returning to the surface and helps protect the drill hole from being damaged during drilling. Surface casing, like conductor casing, is also cemented into place. Regulations often dictate the thickness of the cement to be used to ensure that there is little possibility of freshwater contamination.

Intermediate casing is usually the longest section of casing found in a well. The primary purpose of intermediate casing is to minimize the hazards that come along with subsurface formations that may affect the well. These include abnormal underground pressure zones, underground shale formations, and formations that might otherwise contaminate the well, such as underground saltwater deposits. In many instances, even though there may be no evidence of an unusual underground formation, intermediate casing is run as insurance against the possibility of such a formation affecting the well. These intermediate casing areas may also be cemented into place for added protection. *Liner strings* are sometimes used instead of intermediate casing and are commonly run from the bottom of another type of casing to the open well area. However, liner strings are usually just attached to the previous casing with *hangers*, instead of being cemented into place. This type of casing is thus less permanent than intermediate casing.

Production casing (alternatively called the *oil string* or *long string*) is installed last and is the deepest section of casing in a well. This casing provides a conduit from the surface of the well to the petroleum-producing formation. The size of the production casing depends on a number of considerations, including the lifting equipment to be used, the number of completions required, and the possibility of deepening the well at a later time. For example, if it is expected that the well will be deepened at a later date, the production casing must be wide enough to allow the passage of a drill bit when required.

Once the casing has been set, and in most cases cemented into place, proper lifting equipment is installed to bring the hydrocarbons from the formation to the surface. Once the casing is installed, tubing is inserted inside the casing, from the opening well at the top to the formation at the bottom. The hydrocarbons that are extracted run up this tubing to the surface. This tubing may also be attached to pumping systems for more efficient extraction, should that be necessary.

Well completion commonly refers to the process of finishing a well so that it is ready to produce oil or natural gas. In essence, completion consists of deciding on the characteristics of the intake portion of the well in the targeted hydrocarbon formation. There are a number of types of completions, including (i) open hole completion, (ii) conventional perforated completion, (iii) sand exclusion completion, (iv) permanent completion, (v) multiple zone completion, and (vi) drain hole completion. The use of any type of completion depends on the characteristics and location of the hydrocarbon formation to be mined.

Open hole completion is the most basic type and is only used in very competent formations, which are unlikely to cave in. An open hole completion consists of simply running the casing directly down into the formation, leaving the end of the piping open, without any other protective filter. Very often, this type of completion is used on formations that have been treated with hydraulic of acid fracturing. *Conventional perforated completion* consists of production casing being run through the formation. The sides of this casing are perforated, with tiny holes along the sides facing the formation, which allows for the flow of hydrocarbons into the wellhole, but still provides a suitable amount of support and protection for the wellhole. The process of actually perforating the casing involves the use of specialized equipment designed to make tiny holes through the casing, cementing, and any other barrier between the formation and the open well. In the past, *bullet perforators* were used, which were essentially small guns lowered into the well. The guns, when fired from the surface, sent off small bullets that penetrated the casing and cement. Currently, *jet perforating* is preferred and consists of small, electrically ignited charges, lowered into the well that, when ignited, poke tiny holes through to the formation, in the same manner as bullet perforating.

Sand exclusion completion is designed for production in an area that contains a large amount of loose sand. These completions are designed to allow for the flow of natural gas and oil into the well but at the same time prevent sand from entering the well. Sand inside the wellhole can cause many complications, including erosion of casing and other equipment. The most common method of keeping sand out of the wellhole is screening, or filtering systems. This includes analyzing the sand experienced in the formation and installing a screen or filter to keep sand particles out. This filter may either be a type of screen hung inside the casing or adding a layer of specially sized gravel outside the casing to filter out the sand. Both of these types of sand barriers can be used in open hole and perforated completions.

Permanent completion is the type of completion in which the completion, and wellhead, is assembled and installed only once. Installing the casing, cementing, perforating, and other completion work are done with small-diameter tools to ensure the permanent nature of the completion. Completing a well in this manner can lead to significant cost savings compared with other types.

Multiple zone completion is the practice of completing a well such that hydrocarbons from two or more formations may be produced simultaneously, without mixing with each other. For example, a well may be drilled that passes through a number of formations on its way deeper underground, or alternately, it may be efficient in a horizontal well to add multiple completions to drain the formation most effectively. Although it is common to separate multiple completions so that the fluids from the different formations do not intermingle, the complexity of achieving complete separation is often a barrier. In some instances, the different formations being drilled are close enough in nature to allow fluids to intermingle in the wellhole. When it is necessary to separate different completions, hard rubber packing instruments are used to maintain separation.

Drain hole completion is a form of horizontal or slant drilling and consists of drilling out horizontally into the formation from a vertical well, essentially providing a "drain" for the hydrocarbons to run down into the well. In certain formations, drilling a drain hole completion may allow for more efficient and balanced extraction of the natural gas. This type of completion is more commonly associated with oil wells than with natural gas wells.

2.5.3 The Wellhead

The wellhead consists of an assemblage of necessary equipment that is mounted at the opening of the well to regulate and monitor the extraction of hydrocarbons from the underground formation. The equipment also prevents leaking of crude oil or natural gas out of the well and prevents blowouts due to high-pressure formations. Also, the wellhead is the place where the first analysis of the gas brought to the surface will be required. A formation that is under high pressure typically requires wellheads that can withstand a great deal of upward pressure from the escaping gases and liquids. These wellheads must be able to withstand pressures of up to 20 000 psi. The wellhead consists of three components: (i) the casing head, (ii) the tubing head, and (iii) the *Christmas tree*.

The *casing head* consists of heavy fittings that provide a seal between the casing and the surface. The casing head also serves to support the entire length of casing that is run all the way down the well. This piece of equipment typically contains a gripping mechanism that ensures a tight seal between the head and the casing itself. The tubing head is much like the casing head and provides a seal between the tubing, which is run inside the casing, and the surface. Like the casing head, the tubing head is designed to support the entire length of the casing, as well as provide connections at the surface, which allow the flow of fluids out of the well to be controlled.

The *Christmas tree* (so named because of its many branches that make it appear somewhat like a Christmas tree) is the piece of equipment that fits atop the casing and tubing heads and contains tubes and valves that serve to control the flow of hydrocarbons and other fluids out of the well. The *Christmas tree* is the most visible part of a producing well and allows for the surface monitoring and regulation of the production of hydrocarbons from a producing well.

2.5.4 Well Treatment

Well treatment is another method of ensuring the efficient flow of hydrocarbons out of a formation. Essentially, this type of well stimulation consists of injecting acid, water, or gases into the well to open up the formation and allow the petroleum to flow through the formation more easily. Acidizing a well consists of injecting acid (usually hydrochloric acid) into the well. In limestone or carbonate formations, the acid dissolves portions of the rock in the formation, opening up existing spaces to allow for the flow of petroleum. Fracturing consists of injecting a fluid into the well, the pressure of which "cracks" or opens up fractures already present in the formation.

In addition to the fluid being injected, *propping agents* are also used and consist of sand, glass beads, epoxy, or silica sand and serve to prop open the newly widened fissures in the formation. Hydraulic fracturing involves the injection of water into the formation, while carbon dioxide fracturing uses gaseous carbon dioxide. Fracturing, acidizing, and lifting equipment may all be used on the same well to increase permeability. For example, carbon dioxide–sand fracturing involves using a mixture of sand and liquid carbon dioxide to fracture formations, creating and enlarging cracks through which oil and natural gas may flow more freely. The carbon dioxide then vaporizes, leaving only sand in the formation, holding the newly enlarged cracks open. Because there are no other substances used in this type of fracturing, there are no *leftovers* from the fracturing process that must be removed. This type of fracturing effectively opens the formation and allows for increased recovery of natural gas and (i) does not damage the deposit, (ii) generates no belowground wastes, and (iii) protects groundwater resources.

Because it is a low-density gas under pressure, the completion of natural gas wells usually requires little more than the installation of casing, tubing, and the wellhead. Unlike crude oil, natural gas is much easier to extract from an underground formation. However, as deeper and less conventional natural gas wells are drilled, it is becoming more common to use stimulation techniques on gas wells.

2.5.5 Natural Gas Production

Once the well is completed, it may begin to produce natural gas. In some instances, the hydrocarbons that exist in pressurized formations will naturally rise up through the well to the surface. This is most commonly the case with natural gas. Since natural gas is lighter than air, once a conduit to the surface is opened, the pressurized gas will rise to the surface with little or no interference. This is most common for formations containing natural gas alone or with a light condensate. In these scenarios, once the *Christmas tree* is installed, the natural gas will flow to the surface on its own.

In order to more fully understand the nature of the well, a potential test is typically run in the early days of production. This test allows well engineers to determine the maximum amount of natural gas that the well can produce in a 24 h period. From this, and other knowledge of the formation, the engineer may make an estimation of the *most efficient recovery rate* (MER, the rate at which the greatest amount of natural gas may be extracted without harming the formation itself).

Another important aspect of producing wells is the *decline rate*. When a well is first drilled, the formation is under pressure and produces natural gas at a very high rate. As more natural gas is extracted from the formation, the production rate of the well decreases (the *decline rate*). Certain techniques, including lifting equipment and well stimulation, can increase the production rate of a well. In some natural gas wells, and oil wells that have associated natural gas, it is more difficult to ensure an efficient flow of hydrocarbons up the well. The underground formation may be very *tight* (i.e. low permeability such as is found in shale formations), making the movement of crude oil and natural gas through the formation and up the well a very slow and inefficient process. In these cases, lifting equipment or well treatment is required.

Lifting equipment consists of a variety of specialized equipment used to help lift petroleum out of a formation. The most common lifting method is known as *rod pumping* in which the pumping mechanism is powered by a surface pump that moves a cable and rod up and down in the well, providing the lifting pressure required to bring the oil to the surface. The most common type of cable rod lifting equipment is the *horse head* (*balanced conventional beam, sucker rod*) pump. These pumps are recognizable by the distinctive shape of the cable feeding fixture, which resembles a horse's head.

2.6 Environmental Aspects

Natural gas production is an industrial activity with the potential for many environmental consequences, most of which can be identified by application of a variety of standard test methods (Table 2.5). Effects can range from water contamination related to drilling and disposal of drilling fluids, air quality degradation from internal combustion engines on drill rigs and trucks, excess dust from equipment transportation, impacts to solitude and night skies from noise and lighting, and safety concerns associated with the large number of vehicles needed to support drilling operations. While the horizontal drilling and hydraulic fracturing practices (Speight, 2016) expected to be used in developing a natural gas resource may have the potential to exert negative environmental effects on the surrounding area, when compared with development of conventional oil and gas resources, this development method could result in fewer impacts than conventional vertical wells due to greater flexibility in the location of the well(s).

The analyst can (and should) play an important role in the environmental aspects of natural gas production, which are an important part in defining the characteristics of natural gas insofar as it (i) exist in a gaseous state at room temperature and (ii) contains predominantly hydrocarbon compounds with one to four carbon atoms, which are the dominant hazard in the substance and (iii) which, after additional fractionation, are sold on the open market in the United States as fungible products. The physical and chemical characteristics require that they be identified through the application of standard test methods that provide reliable data.

Significant volumes of natural gas can also be produced from tight (low permeability) sandstone reservoirs and coal seams, both unconventional reservoir rocks (Chapter 1). Through the application of analytical programs, a better understanding of the underlying physics of natural gas production from these rocks and the relationship between well completion practices and productivity will be achieved. Using the various analytical methods, it will be possible to find ways to maximize recovery of natural gas from these rock formations, which, in the mid to late twentieth century, were considered to be unproductive. Emphasis (when resources of shale gas are to be developed) is typically (and rightly so) often placed on water management during resource development, including the enhancement of water treatment technologies. Resource development based on nothing more than emotional issues rather than on reliable analytical data is to be treated with extreme caution.

Some of the nonhydrocarbon constituents of natural gas would not be expected to biologically degrade as these substances do not contain the chemical linkages necessary for microbial metabolism. For this reason, hydrogen, nitrogen, and carbon dioxide would not be susceptible to biodegradation. Furthermore, carbon dioxide is the final product in the biological mineralization of organic compounds. Although volatilization largely determines environmental distribution of hydrocarbon gases, some constituents have sufficient aqueous solubility that they might be present in aqueous environments at sufficient levels and/or sufficient times to make them potentially available for microbial metabolism. Gaseous hydrocarbons are widespread in nature, and their use as a carbon source for cell growth has been reported to be common among autotrophic organisms. Some microbes have evolved that are capable of utilizing these substances as their sole energy source. Higher chain length hydrocarbons typical of naphtha streams (such as the C_5 and C_6 hydrocarbon derivatives) have been shown to be inherently biodegradable in the environment. Not surprisingly, environmental and natural resource-related permits are required for oil and gas operations. A state may handle compliance under a general environmental impact statement (EIS). Therefore, further environmental analysis may not need to be undertaken on individual oil and gas operations that obtain permits to operate.

In terms of the standard test methods that can be applied, it is of primary importance to recognize that while the major constituent of natural gas is methane, there are components such as carbon dioxide (CO), hydrogen sulfide (H_2S), and mercaptans (thiols; RSH), as well as trace amounts of sundry other emissions. The fact that methane has a foreseen and valuable end use makes it a desirable product, but in several other situations it is considered a pollutant, having been identified as one of several greenhouse gases.

Carbon dioxide (ASTM D1137, 2017; ASTM D1945, 2017; ASTM D4984, 2017) in excess of 3% v/v is normally removed for reasons of corrosion prevention (ASTM D1838, 2017). Hydrogen sulfide (ASTM D2420, 2017; ASTM D2725, 2017; ASTM D4084, 2017; ASTM D4810, 2017; IP 103, IP 272) is also removed, and the odor of the gas must not be objectionable (ASTM D6273, 2017) so mercaptan content (ASTM D1988, 2017) is important. A simple lead acetate test (ASTM D2420, 2017; ASTM D4084, 2017) is available for detecting the presence of hydrogen sulfide and is an additional safeguard that hydrogen sulfide not be present (ASTM D1835, 2017). The odor of the gases must not be objectionable. Methyl mercaptan, if present, produces a transitory yellow stain on the lead acetate paper that fades completely in less than 5 min. Other sulfur compounds (ASTM D5504, 2017; ASTM D6228, 2017) present in liquefied petroleum gas do not interfere. These test methods offer techniques to determine individual sulfur species in gaseous fuel and the total sulfur content by calculation. Gas chromatography is used commonly and extensively to determine other components in gaseous fuels including fixed gas and organic components. For example, one test method (ASTM D1945, 2017) describes guidelines for the determination of the chemical composition of natural gases and similar gaseous mixtures within a range of composition through the use of a specific gas chromatographic technique with one of the more common detectors for measurement. In addition, the test method may be abbreviated for the analysis of lean natural gases containing negligible amounts of hexanes and higher hydrocarbons or for the determination of one or more components, as required. The method is also of significance for providing data for calculating physical properties of the sample, such as heating value and relative density, or for monitoring the concentrations of one or more of the components in a mixture. Liquefied petroleum gases and their products of combustion must not be unduly corrosive to the materials with which they come in contact. The potential personnel exposure hazards of hydrogen sulfide also make the detection and measurement of hydrogen sulfide important, even in low concentrations, and the odor of the gases should not be at a level that is objectionable.

In the lead acetate test (ASTM D2420, 2017), the vaporized gas is passed over moist lead acetate paper under controlled conditions. Hydrogen sulfide reacts with lead acetate [$Pb(OCOCH_3)_2$] to form lead sulfide (PbS), resulting in a stain on the paper varying in colors from yellow to black, depending on the amount of hydrogen sulfide present. Other pollutants can be determined by gas chromatography (ASTM D5504, 2017; ASTM D6228, 2017) (Tipler, 2011; Esteves et al., 2016). The total sulfur content (ASTM D1072, 2017) is normally acceptably low and frequently so low that it needs augmenting by means of alkyl sulfides, mercaptans, or thiophenes in order to maintain an acceptable safe level of odor.

The hydrocarbon dew point is reduced to such a level that retrograde condensation, i.e. condensation resulting from pressure drop, cannot occur under the worst conditions likely to be experienced in the gas transmission system. Similarly, the water dew point is reduced to a level sufficient to preclude formation of C_1 to C_4 hydrates in the system. The natural gas after appropriate treatment for acid gas reduction, odorization, and hydrocarbon and moisture dew-point adjustment would then be sold within prescribed limits of pressure, calorific value and, on occasion when necessary, the Wobbe index or Wobbe number (cv/(sp. gr.)) (Chapter 6).

Some gaseous fuels contain vapors of hydrocarbons or other components that easily condense into liquid and sometimes interfere with or mask the water dew point. For example, excess water vapor can cause corrosive conditions that often lead to corrosive degradation of pipelines and equipment. The water vapor can also condense and freeze or form methane hydrates causing blockages. The water vapor content also affects the heating value of natural gas, thus influencing the quality of the gas. One particular test method (ASTM D1142, 2017) permits the determination of water content of natural gas. In the test method, it is sometimes very helpful to supplement the apparatus described in the test method (ASTM D1142, 2017) with an optical attachment that uniformly illuminates the dew-point mirror and also magnifies the condensate on the mirror. With this attachment it is possible, in many cases, to observe separate condensation points of water vapor, hydrocarbons, and glycolamine derivatives as well as the ice point. However, if the dew point of the condensable hydrocarbons is higher than the dew point of the water vapor, when such hydrocarbons are present in large amounts, they may flood the mirror and obscure or wash off the water dew point.

Finally, it is necessary to address the issue of particulate matter, which occurs as a complex emission that is classified as either *suspended particulate matter*, *total suspended particulate matter*, or simply *particulate matter*. Particulate matter is not typically analyzed as a constituent of natural gas but may be a conspicuous constituent of process gas, especially when the process gas is an effluent from a catalytic reactor. Through time, attrition of the catalyst can result in the formation of fine (micron-size) particle that will exit the reactor as part of the gas flow. In fact, process gas emissions associated with petroleum refining are more extensive than methane and carbon dioxide and typically include process gases, petrochemical gases, volatile organic compounds (VOCs), carbon monoxide (CO), sulfur oxides (SO_x), nitrogen oxides (NO_x), particulates, ammonia (NH_3), and hydrogen sulfide (H_2S) that must be removed (Chapter 4). These effluents may be discharged as air emissions and must be treated. Particulate matter emissions are typically determined through spectroscopic analysis of the captured particulate matter that has been collected on a sampling filter (ASTM D3685, 2017; ASTM D7035, 2017).

The measurement of particulate matter and collected residue emission rates is an important test widely used in the practice of air pollution control (ASTM D3685, 2017). Particulate matter measurements after control devices are necessary to determine total emission rates to the atmosphere. These measurements, when approved by federal and state agencies, are often required for the purpose of determining compliance with regulations and statutes. The measurements made before and after control devices are often necessary as a means of demonstrating conformance with contractual performance specifications. The collected residue obtained with these test methods is also important in characterizing stack emissions. However, the utility of these data is limited unless a chemical analysis of the collected residue is performed.

Another test method (ASTM D7035, 2017) specifies a procedure for collection, sample preparation, and analysis of airborne particulate matter for the content of metals and metalloids using inductively coupled plasma atomic emission spectrometry (ICP-AES). The method is applicable to personal sampling of the inhalable or respirable fraction of airborne particles and to area sampling. This test method specifies a number of alternative methods for preparing test solutions from samples of airborne particulate matter. One of the specified sample preparation methods is applicable to the measurement of soluble metal or metalloid compounds. Other specified methods are applicable to the measurement of total metals and metalloids. However, this test method is not applicable to the sampling of elemental mercury or to inorganic compounds of metals and metalloids that are present in the gaseous or vapor state. The method also contains notes that are explanatory and are not part of the mandatory requirements of this test method.

References

Archer, J.S. (1985). Reservoir volumetrics and recovery factors. In: *Developments in Petroleum Engineering: 1* (ed. R. Dawe and D. Wilson). New York: Elsevier Applied Science Publishers.

ASTM (2017). *Annual Book of Standards*. West Conshohocken, Pennsylvania: ASTM International.

ASTM D1072 (2017). *Standard Test Method for Total Sulfur in Fuel Gases by Combustion and Barium Chloride Titration*, Annual Book of Standards. West Conshohocken, Pennsylvania: ASTM International.

ASTM D1137 (2017). *Method for Analysis of Natural Gases and Related Types of Gaseous Mixtures by the Mass Spectrometer*, Annual Book of Standards. West Conshohocken, Pennsylvania: ASTM International.

ASTM D1142 (2017). *Standard Test Method for Water Vapor Content of Gaseous Fuels by Measurement of Dew-Point Temperature*, Annual Book of Standards. West Conshohocken, Pennsylvania: ASTM International.

ASTM D1835 (2017). *Standard Specification for Liquefied Petroleum (LP) Gases*, Annual Book of Standards. West Conshohocken, Pennsylvania: ASTM International.

ASTM D1838 (2017). *Standard Test Method for Copper Strip Corrosion by Liquefied Petroleum (LP) Gases*, Annual Book of Standards. West Conshohocken, Pennsylvania: ASTM International.

ASTM D1945 (2017). *Standard Test Method for Analysis of Natural Gas by Gas Chromatography*, Annual Book of Standards. West Conshohocken, Pennsylvania: ASTM International.

ASTM D1988 (2017). *Standard Test Method for Mercaptans in Natural Gas Using Length-of-Stain Detector Tubes*, Annual Book of Standards. West Conshohocken, Pennsylvania: ASTM International.

ASTM D2113 (2017). *Standard Practice for Rock Core Drilling and Sampling of Rock for Site Exploration*, Annual Book of Standards. West Conshohocken, Pennsylvania: ASTM International.

ASTM D2420 (2017). *Standard Test Method for Hydrogen Sulfide in Liquefied Petroleum (LP) Gases (Lead Acetate Method)*, Annual Book of Standards. West Conshohocken, Pennsylvania: ASTM International.
ASTM D2725 (2017). *Test Method for Hydrogen Sulfide in Natural Gas (Methylene Blue Method)*, Annual Book of Standards. West Conshohocken, Pennsylvania: ASTM International.
ASTM D3685 (2017). *Standard Test Methods for Sampling and Determination of Particulate Matter in Stack Gases*, Annual Book of Standards. West Conshohocken, Pennsylvania: ASTM International.
ASTM D4084 (2017). *Test Method for Analysis of Hydrogen Sulfide in Gaseous Fuels (Lead Acetate Reaction Rate Method)*, Annual Book of Standards. West Conshohocken, Pennsylvania: ASTM International.
ASTM D4810 (2017). *Standard Test Method for Hydrogen Sulfide in Natural Gas Using Length-of-Stain Detector Tubes*, Annual Book of Standards. West Conshohocken, Pennsylvania: ASTM International.
ASTM D4984 (2017). *Standard Test Method for Carbon Dioxide in Natural Gas Using Length-of-Stain Detector Tubes*, Annual Book of Standards. West Conshohocken, Pennsylvania: ASTM International.
ASTM D5074 (2017). *Standard Practice for Preparation of Natural-Matrix Sediment Reference Samples for Major and Trace Inorganic Constituents by Partial Extraction Procedures*, Annual Book of Standards. West Conshohocken, Pennsylvania: ASTM International.
ASTM D5504 (2017). *Standard Test Method for Determination of Sulfur Compounds in Natural Gas and Gaseous Fuels by Gas Chromatography and Chemiluminescence*, Annual Book of Standards. West Conshohocken, Pennsylvania: ASTM International.
ASTM D5753 (2017). *Standard Guide for Planning and Conducting Borehole Geophysical Logging*, Annual Book of Standards. West Conshohocken, Pennsylvania: ASTM International.
ASTM D5890 (2017). *Standard Test Method for Swell Index of Clay Mineral Component of Geosynthetic Clay Liners*, Annual Book of Standards. West Conshohocken, Pennsylvania: ASTM International.
ASTM D6167 (2017). *Standard Guide for Conducting Borehole Geophysical Logging – Mechanical Caliper*, Annual Book of Standards. West Conshohocken, Pennsylvania: ASTM International.
ASTM D6228 (2017). *Standard Test Method for Determination of Sulfur Compounds in Natural Gas and Gaseous Fuels by Gas Chromatography and Flame Photometric Detection*, Annual Book of Standards. West Conshohocken, Pennsylvania: ASTM International.
ASTM D6273 (2017). *Standard Test Methods for Natural Gas Odor Intensity*, Annual Book of Standards. West Conshohocken, Pennsylvania: ASTM International.

ASTM D6274 (2017). *Standard Guide for Conducting Borehole Geophysical Logging – Gamma*, Annual Book of Standards. West Conshohocken, Pennsylvania: ASTM International.

ASTM D6276 (2017). *Standard Guide for Conducting Borehole Geophysical Logging – Electromagnetic Induction*, Annual Book of Standards. West Conshohocken, Pennsylvania: ASTM International.

ASTM D6431 (2017). *Standard Guide for Using the Direct Current Resistivity Method for Subsurface Investigation*, Annual Book of Standards. West Conshohocken, Pennsylvania: ASTM International.

ASTM D6727 (2017). *Standard Guide for Conducting Borehole Geophysical Logging – Neutron*, Annual Book of Standards. West Conshohocken, Pennsylvania: ASTM International.

ASTM D7035 (2017). *Standard Test Method for Determination of Metals and Metalloids in Airborne Particulate Matter by Inductively Coupled Plasma Atomic Emission Spectrometry (ICP-AES)*, Annual Book of Standards. West Conshohocken, Pennsylvania: ASTM international.

ASTM E1915 (2017). *Standard Test Methods for Analysis of Metal Bearing Ores and Related Materials for Carbon, Sulfur, and Acid-Base Characteristics*, Annual Book of Standards. West Conshohocken, Pennsylvania: ASTM international.

Ballard, B.D. (2007). Quantitative Mineralogy of Reservoir Rocks Using Fourier Transform Infrared Spectroscopy. Paper No. SPE-113023-STU. Proceedings. SPE Annual Technical Conference and Exhibition, Anaheim, California. November 11–14. Society of Petroleum Engineers, Richardson, Texas.

Burnett, A. (1995). A quantitative X-ray diffraction technique for analyzing sedimentary rocks and soils. *Journal of Testing and Evaluation* 23 (2): 111–118.

Chopra, A.K., Severson, C.D., and Carhart, S.R. (1990). Evaluation of Geostatistical Techniques for Reservoir Characterization. Paper No. SPE-20734-MS. Proceedings. SPE Annual Technical Conference and Exhibition, 23–26 September, New Orleans, Louisiana. Society of Petroleum Engineers, Richardson, Texas.

Damsleth, E. (1994). Mixed Reservoir Characterization Methods. Paper No. SPE-27969-MS. Proceedings. University of Tulsa Centennial Petroleum Engineering Symposium, Tulsa, Oklahoma. August 29–31. Society of Petroleum Engineers, Richardson, Texas.

Esteves, I.A.A.C., Sousa, G.M.R.P.L., Silva, R.J.S., Ribeiro, R.P.P.L., Eusébio, M.F.J., and Mota, J.P.B. (2016). A sensitive method approach for chromatographic analysis of gas streams in separation processes based on columns packed with an adsorbent material. *Advances in Materials Science and Engineering*. 2016, Article ID 3216267. Hindawi Publishing Corporation: http://dx.doi.org/10.1155/2016/3216267; accessed December 1, 2017)

Faramawy, S., Zaki, T., and Sakr, A.A.-E. (2016). Natural gas origin, composition, and processing: a review. *Journal of Natural Gas Science and Engineering* 34: 34–54.

Hurst, A. and Archer, J.S. (1986). Sandstone reservoir description: an overview of the role of geology and mineralogy. *Clay Minerals* 21: 791–809.

John, E. and Singh, K. (2011). Production and properties of fuels from domestic and industrial waste. In: *The Biofuels Handbook* (ed. J.G. Speight), 333–376. London, UK: Royal Society of Chemistry.

Mokhatab, S., Poe, W.A., and Speight, J.G. (2006). *Handbook of Natural Gas Transmission and Processing*. Amsterdam: Elsevier.

Ramroop Singh, N. (2011). Biofuels. In: *The Biofuels Handbook* (ed. J.G. Speight), 160–198. London, UK: Royal Society of Chemistry.

Saha, R., Uppaluri, R.V.S., and Tiwari, P. (2017). Effect of mineralogy on the adsorption characteristics of surfactant-reservoir rock system. *Colloids and Surfaces A: Physiochemical and Engineering Aspects* 531: 121–132.

Singh, K. and Sastry, M.K.S. (2011). Production of fuels from landfills. In: *The Biofuels Handbook* (ed. J.G. Speight), 408–453. London, UK: Royal Society of Chemistry.

Speight, J.G. (2007). *Natural Gas: A Basic Handbook*. Houston, Texas: GPC Books, Gulf Publishing Company.

Speight, J.G. (2009). *Enhanced Recovery Methods for Heavy Oil and Tar Sands*. Houston, Texas: Gulf Publishing Company.

Speight, J.G. (2013). *Shale Gas Production Processes*. Oxford, UK: Gulf Professional Publishing, Elsevier.

Speight, J.G. (2014a). *The Chemistry and Technology of Petroleum*, 5e. Boca Raton, Florida: CRC Press, Taylor & Francis Group.

Speight, J.G. (2014b). *Oil and Gas Corrosion Prevention*. Oxford, UK: Gulf Professional Publishing, Elsevier.

Speight, J.G. (2016). *Handbook of Hydraulic Fracturing*. Hoboken, New Jersey: Wiley.

Speight, J.G. (2017a). *Handbook of Petroleum Refining*. Boca Raton, Florida: CRC Press, Taylor & Francis Group.

Speight, J.G. (2017b). *Deep Shale Oil and Gas*. Oxford, UK: Gulf Professional Publishing, Elsevier.

Thomas, F.B., Bennion, D.B., and Andersen, G. (2009). Gas condensate reservoir performance. *Journal of Canadian Petroleum Technology* 48 (7): 18–24.

Tipler, A. (2011). The Determination of C1 to C5 Hydrocarbons in Gas Streams Using the PerkinElmer Swafer Technology. https://www.perkinelmer.com/lab-solutions/resources/docs/APP_HydrocarbonsinGasStreamsSwafer.pdf (accessed December 1, 2017).

3

Storage and Transportation

3.1 Introduction

Natural gas exists in nature under pressure in rock reservoirs in the crust of the Earth, either in conjunction with and dissolved in crude oil (*associated gas*) or without crude oil (*nonassociated gas*). Reservoirs containing natural gas vary in size and can be anything from a few hundred yards to miles across in plan and tens to hundreds of yards thick, with the gas trapped against an impermeable layer similar to crude oil traps (Speight, 2014a). Natural gas is a member of the class of fuels known as gaseous fuels (ASTM D4150, 2017). This standard test method defines the terms used in standards that are used in (i) the sampling of gaseous fuels, (ii) the analysis of gaseous fuels for composition and various other physical properties, and (iii) other practices related to the processing, transmission, and distribution of gaseous fuels.

Typically, the gas is produced from the reservoir in a manner similar to crude oil (when it occurs in conjunction with crude oil). However, extracting natural gas from deposits deep underground is dependent upon several factors in the underground environment including (i) the pressure of the gas in the reservoir, (ii) the composition of the gas, (iii) the porosity of the reservoir rock, and (iv) the permeability of the reservoir rock. Primary recovery relies on underground pressure to drive the gas through the production well to the surface. When the pressure decreases due to production of the gas, the remaining gas is brought to the surface using artificial lift technologies, such as the horsehead pump at the surface or a downhole pump at the bottom of the production well.

The use of natural gas in the past six decades has grown steadily – even replacing coal gas in many markets (Speight, 2013) – and, currently, extensively in residential as well as in commercial and industrial applications. While natural gas supplies approximately one quarter of the energy used worldwide and makes up nearly a quarter of electricity generation, as well as playing a

Handbook of Natural Gas Analysis, First Edition. James G. Speight.

crucial role as a feedstock for the industry, it is, in fact, a predominant source of energy in North America, where slightly more than 50% of homes are heated by natural gas. Natural gas power-generating facilities are replacing those generating power from coal and crude oil. This trend is expected to continue well into the foreseeable future. In addition, whether produced via conventional methods or renewable methods, this clean-burning alternative fuel is also being used as fuel for various vehicles, but for such use, the gas must be compressed or liquefied.

Environmentally, natural gas (which is predominantly methane even in the raw unrefined form; Table 3.1) (Burrus and Ryder, 2003, 2014) is the cleanest fossil fuel when burned as it produces (by combustion) primarily carbon dioxide, water vapor, and small amounts of nitrogen oxides.

$$CH_4 + O_2 \rightarrow CO_2 + H_2O$$

Thus, natural gas is a versatile fuel, and its growth is linked in part to its environmental benefits relative to other fossil fuels, particularly for air quality as well as greenhouse gas emissions.

As a brief introduction, natural gas that is scheduled to be transported and stored must meet specific quality measures so that the pipeline network (or grid) can provide uniform quality gas. Wellhead natural gas contains other hydrocarbons, inert gases, and contaminants that must be removed before the natural gas can be safely delivered to the high-pressure, long-distance pipelines that transport natural gas to consumers. Natural gas processing can be complex and usually involves several processes, or stages, to remove oil, water,

Table 3.1 Constituents of natural gas.

Name	Formula	Vol.%
Methane	CH_4	>85
Ethane	C_2H_6	3–8
Propane	C_3H_8	1–5
Butane	C_4H_{10}	1–2
Pentane$^+$	C_5H_{12}	1–5
Carbon dioxide	CO_2	1–2
Hydrogen sulfide	H_2S	1–2
Nitrogen	N_2	1–5
Helium	He	<0.5

Pentane$^+$: pentane and higher molecular weight hydrocarbons up to octane as well as benzene and toluene.

hydrocarbon gas liquids, and other impurities such as sulfur, helium, nitrogen, hydrogen sulfide, and carbon dioxide. The composition of the wellhead natural gas determines the number of stages and the process required to produce pipeline quality dry natural gas. These processes or stages may be integrated into one unit or operation, be performed in a different order or at alternative locations (lease/plant), or not be required at all.

The stages of natural gas processing/treatment are (i) gas–oil–water separators, in which pressure relief causes a natural separation of the liquids from the gases in the natural gas; (ii) condensate separator, in which gas condensate and/or natural gasoline is removed from the natural gas stream at the wellhead with separators much like gas–oil–water separators; (iii) dehydrator, in which water is removed to reduce the potential for corrosion in the pipeline as well as the formation of undesirable hydrates and water condensation in pipelines; (iv) contaminant removal units, in which nonhydrocarbon gases such as hydrogen sulfide, carbon dioxide, water vapor, helium, nitrogen, and oxygen are removed from the natural gas stream; (v) nitrogen rejection unit, in which nitrogen is removed and the gas stream is further dehydrated using molecular sieve beds; (vi) methane separator, in which methane is removed from the gas stream can occur as a separate operation by cryogenic processing and absorption methods; and (vii) fractionator, in which the gas stream is separated into component hydrocarbons (ethane, propane, butane) using the varying boiling points of the individual hydrocarbons. Each of these unit operations is described in more detail elsewhere in this text (Chapter 4). Also, there is a need to determine the purity of the gas streams as well as the efficiency of each process unit by means of gas stream analysis, usually through online or offline monitoring of the product streams, for example, gas chromatography (Table 3.2).

3.2 Storage

Gas storage facilities largely contribute to the reliability of gas supplies to consumers and enable the gas providers to level off daily gas consumption fluctuations and meet the peak demand in winter. Natural gas in storage also serves as insurance against any unforeseen accidents, natural disasters, or other occurrences that may affect its production or delivery.

Traditionally, the demand for natural gas varies according to the season insofar as it is typically usually higher during the winter, partly because it is used for heating in residential and commercial settings. As a result, storage of natural gas plays an extremely important role in ensuring that any excess supply delivered during the summer months is available to meet the increased demand of the winter months. However, with the recent trend toward natural gas-fired

Table 3.2 Standard test method used to provide information about the composition and properties of domestic and industrial fuel gases.

ASTM D1265 Practice for Sampling Liquefied Petroleum (LP) Gases, Manual Method
ASTM D1267 Test Method for Gage Vapor Pressure of Liquefied Petroleum (LP) Gases (LP-Gas Method)
ASTM D1657 Test Method for Density or Relative Density of Light Hydrocarbons by Pressure Hydrometer
ASTM D1837 Test Method for Volatility of Liquefied Petroleum (LP) Gases
ASTM D1838 Test Method for Copper Strip Corrosion by Liquefied Petroleum (LP) Gases
ASTM D2158 Test Method for Residues in Liquefied Petroleum (LP) Gases
ASTM D2163 Test Method for Determination of Hydrocarbons in Liquefied Petroleum (LP) Gases and Propane/Propene Mixtures by Gas Chromatography
ASTM D2420 Test Method for Hydrogen Sulfide in Liquefied Petroleum (LP) Gases (Lead Acetate Method)
ASTM D2598 Practice for Calculation of Certain Physical Properties of Liquefied Petroleum (LP) Gases from Compositional Analysis
ASTM D2713 Test Method for Dryness of Propane (Valve Freeze Method)
ASTM D2784 Standard Test Method for Sulfur in Liquefied Petroleum Gases (Oxy-Hydrogen Burner or Lamp)
ASTM D3700 Practice for Obtaining LPG Samples Using a Floating Piston Cylinder
ASTM D5504 Test Method for Determination of Sulfur Compounds in Natural Gas and Gaseous Fuels by Gas Chromatography and Chemiluminescence
ASTM D5623 Test Method for Sulfur Compounds in Light Petroleum Liquids by Gas Chromatography and Sulfur Selective Detection
ASTM D6667 Test Method for Determination of Total Volatile Sulfur in Gaseous Hydrocarbons and Liquefied Petroleum Gases by Ultraviolet Fluorescence
ASTM D6897 Test Method for Vapor Pressure of Liquefied Petroleum Gases (LPG) (Expansion Method)
ASTM D7756 Test Method for Residues in Liquefied Petroleum (LP) Gases by Gas Chromatography with Liquid, On-Column Injection
ASTM D7828 Test Method for Determination of Residue Composition in Liquefied Petroleum Gas (LPG) Using Automated Thermal Desorption/Gas Chromatography (ATD/GC)

power generation, the demand for natural gas during the summer months is now increasing (due to the demand for electricity to power equipment such as air conditioners).

Thus, natural gas storage plays a vital role in maintaining the reliability of supply needed to meet the demands of consumers. Historically, when natural gas was a regulated commodity, storage was part of the bundled product sold by the pipelines to distribution utilities. This all changed in 1992 with the introduction of Order 636 of the Federal Energy Regulatory Commission (FERC), which opened up the natural gas market to deregulation. Essentially, by the institution of Order 636, it meant that where natural gas storage was required prior to the Order for the operational requirements of the pipelines in meeting the needs of the utilities, natural gas became available to anyone seeking storage for commercial purposes or operational requirements. Storage used to serve only as a buffer between transportation and distribution to ensure adequate supplies of natural gas was in place for seasonal demand shifts and unexpected demand surges.

The increased use of natural gas and the need to provide a plentiful supply of this fuel is causing an expansion of the need for natural gas storage facilities so that this domestically produced gaseous fuel will be readily available through the utility infrastructure. Storage is the main link between the three segments of the natural gas industry that are involved in delivering natural gas from the wellhead to the consumer: (i) the production company explores, drills, and extracts the natural gas from the ground, (ii) the transportation company operates the pipelines that link the gas fields to major consuming areas, and (iii) the distribution company – in the form of local utility companies – delivers natural gas to the customer.

As simple as it may seem to many observers, the storage of natural gas is not only a matter of putting the gas into a container without the potential for any adverse consequences. However, there are several analytical measures that must be taken to ensure that the gas is not only stored safely but also recoverable safely and in a usable form. Furthermore, because of the geographic and geologic diversity in gas resources and storage operations in the United States, there is no single approach that can be taken without the application of analytical test methods and consideration of the data relating to (i) the potential storage facility, such as a depleted reservoir, (ii) the composition and properties of the gas to be stored, and (iii) whether the gas is adequate for each and every storage container; in this case, the word container includes natural container such as underground natural storage facilities as described below.

For example, in the case of underground storage facilities, the operators check for weak points and leakage and investigate suspicious indications through a variety of downhole logging techniques, including formation evaluation tools (such as neutron logging), fluid flow and movement indicators (noise and temperature surveys), casing inspections (magnetic flux leakage and ultrasonic methods), mechanical calipers, downhole cameras, and cathodic protection profile surveys. Multiple methods to make decisions on the mitigation of any adverse effects require continuous maintenance work that must ensure the integrity of the storage facility. Of necessity, storage facility operators are using risk-based assessments by collection of analytical data, data analysis, and recognition of the indicator of the storage area's integrity, to drive the underground storage management programs. In addition to tool-based assessment methods, the owner of a storage facility industry uses pressure tests and pressure monitoring to evaluate the integrity of the facility. Common approaches used by operators include (i) shut-in pressure monitoring at each facility, (ii) annulus pressure or flow monitoring, and (iii) mechanical integrity tests.

Facility owners may also employ a risk-based approach that considers risks and threats specific to each facility when selecting well-integrity practices. Commonly assessed risks or threats include, but are not limited to, the following: (i) casing physical properties, such as diameter, weight, and grade, (ii) the presence of atmospheric or external corrosion at or near the surface, (iii) known metal loss indications from casing inspection surveys, (iv) the presence of annulus pressure or flow, (v) water production in the well, (vi) the presence of corrosive hydrogen sulfide and bacteria, (vii) naturally corrosive zones, (viii) the flow potential of the well, and (ix) the work history of the well and its reliability in providing natural gas service. Each category requires the production of data by application of the relevant test methods. In addition, there are several volumetric measures that are used to quantify the fundamental characteristics of an underground storage facility and the gas contained within it. Indeed, it is important to distinguish between the characteristic of a facility, such as the *capacity* of the facility, and the characteristic of the natural gas within the facility, such as the actual *inventory level*.

By way of definition, the injection capacity (or injection rate) is the complement of the deliverability or withdrawal rate and is the amount of natural gas that can be injected into a storage facility on a daily basis (or unit-time basis). As with deliverability, injection capacity is usually expressed in millions of standard cubic feet per day (MMcf day^{-1}), although dekatherms/day is also used. A further definition that is necessary at this point, a dekatherm (dth) is a unit of energy used primarily to measure natural gas and was developed in about 1972 by the Texas Eastern Transmission Corporation – a natural gas pipeline company. The dekatherm is equal to 10 therms or 1 000 000 British thermal units (MMBtu) or, using the SI system, 1.055 gigajoules (GJ). The dekatherm is also approximately equal to one thousand cubic feet (Mcf, Mft^3) of natural gas or exactly one thousand cubic feet of natural gas with a heating value of 1000 Btu ft^{-3}.

The total natural gas storage capacity of an underground storage facility is the maximum volume of natural gas that can be stored in the facility in accordance with the design of the facility, which comprises (i) the physical characteristics of the facility, such as if it is a depleted reservoir, (ii) the installed equipment, and (iii) the operating procedures that are particular to the site. Following from this, the total gas in storage is the volume of natural gas in the underground facility at a particular time. Thus, the injection capacity of a storage facility is also variable and is dependent on factors comparable with those that determine the deliverability of the natural gas. In contrast, or as can be anticipated, the injection rate varies inversely with the total amount of gas in storage: the injection rate is at its lowest when the facility is full, but increases as gas is withdrawn.

The base gas (or cushion gas) is the volume of natural gas intended as permanent inventory in a storage facility to maintain adequate pressure and deliverability rates throughout the withdrawal season. On the other hand, the working gas is the volume of gas in the storage facility above the level of base gas and is, simply, the natural gas that is available for withdrawal and sales. In relation to the definition of working gas, the working gas capacity is the total gas storage capacity minus the base gas. The deliverability of the gas from the storage facility (also referred to as the deliverability rate, withdrawal rate, or withdrawal capacity) is often expressed as a measure of the amount of gas that can be delivered (withdrawn) from a storage facility on a daily basis.

Gas deliverability is usually expressed in terms of million cubic feet per day (MMcf day^{-1}). Occasionally, deliverability is expressed in terms of equivalent heat content of the gas withdrawn from the facility, most often in dekatherms per day (one therm is equal 100 000 Btu, which is roughly equivalent to 100 ft^{-3} of natural gas; a dekatherm is the equivalent of about one thousand cubic feet, Mcf). The deliverability of a given storage facility is variable, and it depends on factors such as, but not limited to, (i) the amount of natural gas in the storage facility, (ii) the pressure within the storage facility, (iii) the compression capability available to the storage facility, and (iv) the configuration and capabilities of surface facilities associated with the underground storage facility. In general, the deliverability rate of a storage facility will vary directly with the total amount of natural gas in the facility. Thus, gas deliverability of the gas is at its highest when the facility is full and declines as working gas is withdrawn.

In addition, the mineralogy of the underground storage facility is important since the presence of certain minerals may enhance the adsorption (and nonrecovery) of certain constituents of the gas that can result in changes in the composition of the gas. Also, other mineralogical constituents of the storage facility may cause adverse chemical changes to the gas. Thus, in addition to the use of geological, geophysical, and core analysis methods, the presence of discontinuities in the cap rock (e.g. because of the presence of faults or fractures) is often tested using long water pump tests. In these tests, water is either withdrawn or injected into the target formation, and pressure (or water level) is monitored in the formation (aquifer) above the cap rock (Katz and Coats, 1968; Crow et al., 2008).

Because of the potential variability in gas composition (Burrus and Ryder, 2003; Mokhatab et al., 2006; Burruss and Ryder, 2014; Speight, 2014a; Speight, 2017) and mineralogical characteristics of the storage facility, none of the aforementioned measures for any given storage facility are fixed or absolute. The injection rate and the withdrawal rate can (will) change as the level of natural gas varies within the facility. In practice, a storage facility may be able to exceed certificated total capacity in some circumstances by exceeding certain operational parameters. The total capacity of the storage facility can also vary, temporarily or permanently, as its defining parameters vary. Measures of base gas, working gas, and working gas capacity also can also change from time to time. These changes occur, for example, when a storage operator reclassifies one category of natural gas to the other, often as a result of new wells, equipment, or operating practices (such a change generally requires approval by the appropriate regulatory authority and is subject to data derived by chemical and physical analysis of the gas). Finally, storage facilities can withdraw base gas for supply to the market during times of particularly heavy demand, although, by definition, this gas may not be intended for that use.

3.3 Storage Facilities

Natural gas may be stored in a number of different ways. In the modern world, natural gas is most commonly held in an inventory underground under pressure in three types of facilities. These underground facilities are (i) depleted reservoirs in oil and/or natural gas fields, (ii) aquifers, and (iii) salt cavern formations. Natural gas is also stored in liquid or gaseous form in aboveground storage tanks, which were the conventional method of storing coal gas in the early part of the twentieth century (Speight, 2013).

More pertinent to underground storage facilities, each type of facility has its own physical characteristics (porosity, permeability, retention capability) and economics (site preparation and maintenance costs, deliverability rates, and cycling capability), which govern its suitability for storage applications. In relation to the type of facility, two important characteristics of an underground storage facility are (i) the capacity to hold natural gas for future use and (ii) the rate at which gas inventory can be withdrawn (the deliverability rate).

3.3.1 Depleted Reservoirs

Most existing natural gas storage in the United States is in depleted natural gas reservoirs or depleted crude oil reservoirs that are close to consumption centers. A benefit of the use of such reservoirs is that conversion of a gas field or crude oil field from production to storage duty takes advantage of existing wells, gathering systems, and pipeline connections. Because of their wide availability, depleted crude oil reservoirs and depleted natural gas reservoirs are the most commonly used underground of all storage facilities.

By definition, depleted reservoirs are those formations that have already been tapped of all their recoverable natural gas and/or crude oil. This leaves an underground formation with porosity and permeability that is capable of holding natural gas. In addition, using an already developed reservoir for storage purposes allows the use of the extraction and distribution equipment left from when the field was productive. Having this extraction network in place reduces the cost of converting a depleted reservoir into a storage facility. The factors that determine whether or not a depleted reservoir will make a suitable storage facility are both geographic and geologic. Geographically, depleted reservoirs must be relatively close to consuming regions, and they must also be close to transportation infrastructure, including pipelines, and distribution systems.

To be suitable for storage of natural gas, a depleted reservoir formation must have high porosity and high permeability. The porosity of the formation determines the amount of natural gas that it may hold, while the permeability of the formation determines the rate at which natural gas flows through the formation, which in turn determines the rate of injection and withdrawal of working gas. However, a depleted reservoir where withdrawal of the crude oil or natural gas has caused a decrease in the porosity and/or a decrease in the permeability may not be acceptable for natural gas storage. In certain instances, the formation may be stimulated to increase permeability (Speight, 2016).

In some cases, and in order to maintain pressure in depleted reservoirs, approximately 50% v/v of the natural gas in the formation must be kept as cushion gas. However, a depleted reservoir that has already been used as a source of filled with natural gas may not require the injection of what will become physically unrecoverable gas (cushion gas) since that gas already exists in the formation.

In regions without depleted reservoirs, like the upper Midwest of the United States, one of the other two storage options is required – either an aquifer or a salt cavern.

3.3.2 Aquifers

An aquifer is an underground highly porous and permeable rock formation that can act as natural water reservoirs. In some cases, an aquifer may be reconditioned and used as a facility for natural gas storage. Typically, an aquifer is costlier to develop than a depleted reservoir, and, consequently, these aquifer-derived storage facilities are used only in areas where there are no nearby (and cheaper to develop) depleted reservoirs. As might be anticipated since aquifers have not always been explored to the same extent as crude oil or natural gas reservoirs, seismic testing must be performed, much the same as that for the exploration of potential natural gas formations, and application of text method to determine the mineralogy of the aquifer would also be a benefit, perhaps even a necessity. The area of the formation, the composition and porosity of the formation itself, and the existing formation pressure as well as the capacity for gas storage must all be determined prior to development of the aquifer for gas storage. Typically, an aquifer is suitable for gas storage if the water-bearing sedimentary rock formation is overlaid with an impermeable cap rock. Although the geology of aquifers is similar to depleted production formations, their use for natural gas storage usually requires more base gas (cushion gas) and allows less flexibility in gas injection and withdrawal.

Furthermore, in order to develop a natural aquifer into an effective natural gas storage facility, all of the associated infrastructure must also be developed. This includes installation of wells, extraction equipment, pipelines, dehydration facilities, and possibly compression equipment. Since aquifers are naturally full of water, in some instances powerful injection equipment must be used to allow sufficient injection pressure to push down the resident water and replace it with natural gas.

While natural gas being stored in aquifers has already undergone all of its processing, upon extraction from a water-bearing aquifer formation, the gas typically requires further dehydration prior to transportation, which would, more likely, require the installation of specialized equipment at or near to the wellhead. In addition, the aquifer may not have the same natural gas retention capabilities like that of a depleted crude oil reservoir or a natural gas reservoir and a part of the injected natural gas that may escape from the formation and must be gathered and extracted by *collector wells* that are specifically designed to collect any gas that has escaped from the primary aquifer formation.

In addition, an aquifer formation may typically require cushion gas that is required in a depleted crude oil reservoir or a depleted natural gas reservoir. Since there is no naturally occurring gas in the aquifer formation to begin with, a portion of the natural gas that is injected will ultimately prove to be physically unrecoverable. While it is possible to extract cushion gas from depleted reservoirs, extracting cushion gas from an aquifer formation could have negative effects, among which formation damage might be extensive. As such, most of the cushion gas that is injected into an aquifer formation may remain unrecoverable, even after the storage facility is shut down and abandoned.

Nevertheless, in spite of these issues, in some areas of the United States, most notably the Midwestern United States, natural aquifers have been converted to storage reservoirs for natural gas. Deliverability rates may be enhanced by the presence of an active water drive, which supports the reservoir pressure through the injection and production cycles.

3.3.3 Salt Caverns

Essentially, salt caverns are formed out of existing salt deposits. These underground salt deposits may exist in two possible forms: salt domes and salt beds. Salt domes are thick formations created from natural salt deposits that, over time, leach up through overlying sedimentary layers to form large dome-type structures. They can be as large as a mile in diameter and 30 000 ft in height. Typically, salt domes used for natural gas storage are between 6000 and 1500 ft beneath the surface, although in certain circumstances they can come much closer to the surface. Salt beds are shallower, thinner formations. These formations are usually no more than 1000 ft in height. Because salt beds are wide, thin formations, once a salt cavern is introduced, they are more prone to deterioration and may also be more expensive to develop than salt domes.

Once a suitable salt dome or salt bed deposit is discovered, and deemed suitable for natural gas storage, it is necessary to develop a "salt cavern" within the formation. Essentially, this consists of using water to dissolve and extract a certain amount of salt from the deposit, leaving a large empty space in the formation. This is done by drilling a well down into the formation and cycling large amounts of water through the completed well. This water will dissolve some of the salt in the deposit and be cycled back up the well, leaving a large empty space that the salt used to occupy – this process is known as "salt cavern leaching." Some of the salt is dissolved leaving a void, and the water, now saline, is pumped back to the surface. The process continues until the cavern is the desired size. Once created, a salt cavern offers an underground natural gas storage vessel with very high deliverability.

Salt cavern leaching is used to create caverns in both types of salt deposits and can be quite expensive. However, once created, a salt cavern offers an underground natural gas storage vessel with very high deliverability. In addition, cushion gas requirements are the lowest of all three storage types, with salt caverns only requiring about 33% of total gas capacity to be used as cushion gas.

These underground salt caverns offer another option for natural gas storage. These formations are well suited to natural gas storage in that salt caverns, once formed, allow little injected natural gas to escape from the formation unless specifically extracted. The walls of a salt cavern also have the structural strength of steel, which makes it very resilient against degradation over the life of the storage facility. Salt caverns allow very little of the injected natural gas to escape from storage unless specifically extracted. The walls of a salt cavern are strong and impervious to gas over the life-span of the storage facility.

Salt cavern storage facilities are primarily located along the Gulf Coast, as well as in the Northern United States, and are best suited for peak load storage. Salt caverns are typically much smaller than depleted gas reservoirs and aquifers; in fact underground salt caverns usually take up only one one-hundredth of the acreage taken up by a depleted gas reservoir. As such, salt caverns cannot hold the volume of gas necessary to meet base load storage requirements. However, deliverability from salt caverns is typically much higher than for either aquifers or depleted reservoirs. Therefore, natural gas stored in a salt cavern may be more readily (and quickly) withdrawn, and caverns may be replenished with natural gas more quickly than in either of the other types of storage facilities. Moreover, salt caverns can readily begin flowing gas on as little as one hour's notice, which is useful in emergency situations or during unexpected short-term demand surges. Salt caverns may also be replenished more quickly than other types of underground storage facilities.

Typically, a salt cavern will provide a high withdrawal rate and a high injection rate relative to the working gas capacity of the cavern. Most salt cavern storage facilities have been developed in salt dome formations located in the

Gulf Coast of the United States. Salt caverns have also been formed (by a leaching process) in bedded salt formations in Northeastern, Midwestern, and Southwestern United States.

3.3.4 Gas Holders

A gas holder (also called a gasometer) is a large container in which natural gas (or town gas) can be stored at or near atmospheric pressure and at ambient temperature. Gas holders were initially developed in the early part of the twentieth century for the storage of coal gas (Speight, 2013). The volume of a gas holder follows the quantity of stored gas, with pressure coming from the weight of a movable cap. Typical volumes for large gas holders are approximately 1 800 000 ft^3, with 200 ft diameter structures. Gas holders tend to be used for balancing purposes (making sure gas pipes can be operated within a safe range of pressures) rather than for actually storing gas for later use.

A gas holder can provide on-site storage for purified gas and can act as a buffer by removing the need for continuous gas processing. The weight of the gas holder lift (cap) controlled the pressure of the gas in the mains and provided back pressure for the gas-making plant. A water-sealed gas holder consisted of two parts: a deep tank of water that was used to provide a seal and a vessel that rose above the water as the gas volume increased. In the well-known telescoping type of gas holder, the tank floats in a circular or annular water reservoir, held up by the roughly constant pressure of a varying volume of gas, the pressure determined by the weight of the structure, and the water providing the seal for the gas within the moving walls.

The rigid waterless gas holder is a design that neither expands nor contracts. The modern version of the waterless gas holder is the dry-seal type (membrane type) of gas holder that consists of a static cylindrical shell, within which a piston rises and falls. As it moves, a grease seal, tar/oil seal, or a sealing membrane that is rolled out and in from the piston keeps the gas from escaping.

The benefit of using a gas holder (although possibly limited in storage capacity) is that the gas holder can store gas at district pressure and can provide extra on-site gas very quickly and at peak times. Furthermore, the gas holder is the only storage method that can maintain the gas at the required pressure, which is the pressure required in local gas lines, and thus the gas holder may hold a large advantage over other methods of storage.

3.4 Transportation

After extraction from the reservoir, natural gas must be transported to different places to be processed, stored, and then finally delivered to the end consumer, and this can occur by means of a pipeline or ship. The efficient and effective movement of natural gas from the production site to the consumers requires an extensive and elaborate transportation system. In many instances, natural gas produced from the wellhead will have to travel a considerable distance to reach its point of use. The typical transportation system for natural gas consists of a complex network of pipelines, designed to quickly and efficiently transport natural gas from its origin to areas of high natural gas demand. Transportation of natural gas is closely linked to storage insofar as if the natural gas being transported is not required when it reaches the end of the pipeline (or transportation system); it can be put into storage facilities to be released at a time when it is needed.

Natural gas, as it is used by consumers, is much different from the natural gas that is brought from underground up to the wellhead. In some cases, constituents of the gas may remain in the reservoir because of various property constraints or because of interactions with the reservoir rock. Furthermore, although the processing (treating, refining) of natural gas is, in many respects, less complicated than the processing and refining of crude oil (Chapter 4), it is equally as necessary before its use by end users (Parkash, 2003; Mokhatab et al., 2006; Gary et al., 2007; Speight, 2007; Riazi et al., 2013; Speight, 2014a; Faramawy et al., 2016; Hsu and Robinson, 2017; Speight, 2017). The natural gas used by consumers is composed almost entirely of methane.

However, natural gas produced at the wellhead, although still composed primarily of methane, is by no means as pure and includes several varieties of impurities such as higher molecular weight hydrocarbons, carbon dioxide, and hydrogen sulfide (including mercaptans). Raw natural gas comes from three types of wells: oil wells, gas wells, and condensate wells. Natural gas that comes from oil wells is typically termed *associated gas* that can exist separate from oil in the formation (free gas) that occur with crude oil in the reservoir or gas that dissolved in the crude oil (dissolved gas). Natural gas from gas and condensate wells, in which there is little or no crude oil, is termed *nonassociated gas* (Chapter 1).

Whatever the source of the natural gas (Chapter 1), it commonly exists in mixtures with other hydrocarbons, principally ethane, propane, butane, and pentanes. In addition, raw natural gas contains water vapor, hydrogen sulfide (H_2S), carbon dioxide, helium, nitrogen, and other compounds. Natural gas processing (Chapter 4) consists of separating all of the various hydrocarbons and fluids from the pure natural gas to produce pipeline quality dry natural gas. Major transportation pipelines usually impose restrictions on the makeup of the natural gas that is allowed into the pipeline. That means that before the natural gas can be transported, it must be purified. The associated hydrocarbons, which include ethane, propane, butane, isobutane, and natural gasoline and are often referred to as *natural gas liquids* (NGLs), can be very valuable by-products of natural gas processing. These NGLs are sold separately and have a variety of different uses, including enhancing oil recovery in oil wells, providing raw materials for oil refineries or petrochemical plants, and as sources of energy.

While some of the needed processing can be accomplished at or near the wellhead (field processing), the complete processing of natural gas takes place at a processing plant (Chapter 4). The extracted natural gas is transported to these processing plants through a network of gathering pipelines, which are small-diameter, low-pressure pipes. In addition to processing performed at the wellhead and at centralized processing plants, some final processing is also sometimes accomplished at *straddle extraction plants*, which are located on major pipeline systems. Although the natural gas that arrives at these straddle extraction plants is already of pipeline quality, in certain instances there still exist small quantities of NGLs that are extracted at the straddle plants.

Major transportation pipelines usually impose restrictions on the makeup of the natural gas that is allowed into the pipeline. Natural gas processing consists of separating all of the various hydrocarbons and fluids from the pure natural gas to produce what is known as "pipeline quality" dry natural gas.

3.4.1 Pipelines

There are essentially three major types of pipelines along the transportation route: (i) the gathering pipeline system, (ii) the transmission pipeline system, sometimes called the interstate pipeline system, and (iii) the distribution system.

Gathering systems, primarily made up of small-diameter, low-pressure pipelines, move raw natural gas from the wellhead to a natural gas processing plant or to an interconnection with a larger mainline pipeline. Transmission pipelines are typically wide-diameter, high-pressure transmission pipelines that transport natural gas from the producing and processing areas to storage facilities and distribution centers. Compressor stations (or pumping stations) on the pipeline network keep the natural gas flowing forward through the pipeline system. Local distribution companies deliver natural gas to consumers through small-diameter, lower-pressure service lines. Whatever the pipeline system, there is the need to determine if the natural gas received at the wellhead has a high sulfur content and a high carbon dioxide content (sour gas); a specialized sour gas-gathering pipe must be installed. Sour gas is extremely corrosive and dangerous; thus its transportation from the wellhead to the sweetening plant must be done carefully (Speight, 2014b).

Thus, an issue that always arises at the wellhead when natural gas is transported by pipeline is the degree of processing at the wellhead to remove potential corrosive contaminants that would seriously affect the integrity (corrosivity) of the pipeline. While carbon dioxide and hydrogen sulfide are often considered to be noncorrosive in the dry state, the presence of water in the natural gas can render these two gases extremely corrosive (Speight, 2014b). This emphasizes the need for compositional analysis of the natural gas as it exits the production well. As a result of the compositional analysis, the pipeline operators can make the decision about the extent of the wellhead treating such as separation of (i) hydrocarbon gas liquids, (ii) nonhydrocarbon gases, and (iii) water from the natural gas before the (treated) gas is delivered into a mainline transmission system.

Major transportation pipelines usually impose restrictions on the makeup of the natural gas that is allowed into the pipeline, and natural gas must be processed to produce "pipeline quality" dry natural gas. Some field processing can be accomplished at or near the wellhead; however, the complete processing of natural gas takes place at a processing plant, usually located in a natural gas-producing region. Thus, from the wellhead, natural gas is transported to processing plants through a network of small-diameter, low-pressure gathering pipelines that may consist of a complex gathering system and can consist of thousands of miles of pipes, interconnecting the processing plant to upward of 100 wells in the area.

Thus, it is essential that the composition of the gas be known so that natural gas processing can begin at the wellhead. The composition of the raw natural gas extracted from producing wells depends on the type, depth, and location of the underground deposit and the geology of the area. A natural gas processing plant is a facility designed to clean raw natural gas by separating impurities and various nonmethane hydrocarbons and fluids to produce what is known as *pipeline quality* dry natural gas. A gas processing plant is also used to recover NGLs (condensate, natural gasoline, and liquefied petroleum gas [LPG]) and sometimes other substances such as sulfur-containing constituents and should be (at least) checked for sulfur content (ASTM D1988, 2017; ASTM D2420, 2017; ASTM D6667, 2017; ASTM D7751, 2017) and for residues (ASTM D7756, 2017; ASTM D7828, 2017) to ensure that the LPG meets the specification (ASTM D1835, 2017). The test method that presents the specifications of gaseous fuels (ASTM D1835, 2017) describes those products commonly referred to as liquefied petroleum gases, consisting of propane, propylene (propene), butane, and mixtures of these materials. Four basic types of LPGs are provided to cover the common use applications, particularly LPGs consisting of propane, propene (propylene), butane, and mixtures of these materials that are intended for use as domestic, commercial, and industrial heating and engine fuels. However, care must be taken in sampling of the liquefied gases to ensure that the sample is representative; otherwise the test results may not be significant (Chapter 5). All four types of LPGs covered by this specification should conform to the specified requirements for vapor pressure, volatile residue, residue matter, relative density, and corrosion. There is also another series of standard test methods (Table 3.2) that can be used to provide information about the composition and properties of domestic and industrial fuel gases.

The actual practice of processing natural gas to pipeline dry gas quality levels can be quite complex but usually involves four main processes to remove the various impurities: (i) oil and condensate removal, (ii) water removal, (iii) separation of NGLs, (iv) hydrogen sulfide removal, and (v) carbon dioxide removal (Chapter 4) (Mokhatab et al., 2006; Speight, 2007, 2014a). If mercury is present in the gas, typically as trace amounts, opinions differ whether or not the mercury should be removed at the wellhead. The presence of mercury can cause corrosion of aluminum heat exchangers and environmental pollution. If needed, there are two forms of removal processes: (i) regenerative processes and (ii) nonregenerative processes. The former uses sulfur-activated carbon or alumina, while the latter uses silver on a molecular sieve (Mokhatab et al., 2006; Speight, 2007, 2014a).

In addition, hydrogen sulfide and carbon dioxide can be removed by olamine scrubbing (Mokhatab et al., 2006; Speight, 2007; Kidnay et al., 2011; Speight, 2014a), and heaters and scrubbers are installed, usually at or near the wellhead. The scrubbers serve primarily to remove sand and other large particle impurities. The heaters ensure that the temperature of the gas does not drop too low. With natural gas that contains even low quantities of water, natural gas hydrates (NGH) tend to form when temperatures drop. These hydrates are solid or semi-solid compounds, resembling icelike crystals, and when gas hydrates accumulate, they can impede the passage of natural gas through valves and gathering systems. After processing, the pipeline quality natural gas is injected into gas transmission pipelines and transported to the end users. This often involves transportation of the gas over hundreds of miles, as the location of gas production is generally not the location where the gas is used.

Nevertheless, whatever of the extent of the gas processing operations (at the wellhead or in a processing facility), it is essential that the composition of the gas be known through analysis in order to ensure a (relatively) pure product for transportation and that the gas meets sales specification goals.

3.4.2 Liquefied Natural Gas

Liquefied natural gas (LNG) is the liquid form of natural gas and is used principally for transporting natural gas to markets, where it is regasified and distributed as pipeline natural gas. Contrary to some loose and inaccurate definitions, LNG is not the same as LPG (often referred to as *propane*), which is determined by a standard test method (ASTM D7940, 2017).

This test method (ASTM D7940, 2017) is particularly applicable to determining the composition of liquefied gaseous fuels (LNG, LPG), which is important for custody transfer and production. In addition, compositional data are used to calculate the heating value, which is important to ensure regulatory compliance, and to determine the efficiency of liquefied hydrocarbon gas production and ensure the quality of the processed fluids. The standard test method (ASTM D7940, 2017) can be applied to both online and laboratory instrument-based determination of composition for LNG using Raman spectroscopy. The basic methodology can also be applied to other low-boiling mixtures of hydrocarbon derivatives in either the liquid phase or the gaseous phase, if the needs of the application are met, although the rest of this practice refers specifically to liquids. From the composition, gas properties such as heating value and the Wobbe index may be calculated. The components commonly determined according to this test method are methane (CH_4), ethane (C_2H_6), C_3H_8, isobutane (i-C_4H_{10}), n-butane (n-C_4H_{10}), isopentane (i-C_5H_{12}), n-pentane (n-C_5H_{12}), neopentane (neo-C_5H_{12}), nitrogen (N_2), and oxygen (O_2). The applicable range of this standard is 200 ppm v/v to 100 mol%. Components having higher molecular weights than pentane (C_5) derivatives are not measured as part of this test method. The advantage of this test method over other standard test methods is that Raman spectroscopy can determine composition by directly application to LNG. Unlike chromatography, no vaporization step is necessary. Since incorrect operation of online vaporizers can lead to poor precision and accuracy, elimination of the vaporization step offers a significant improvement in the analysis of LNG. However, Raman spectroscopy cannot be used to directly quantify the component percentages of noble gases, but the concentration of any so-called inert constituents (typically the noble gases) in the gas sample can be estimated indirectly by subtracting the sum of the other species from 100%.

Alternatives to compositional measurement using Raman spectroscopy are (i) ASTM D1945 (2017) (Standard Test Method for Analysis of Natural Gas by Gas Chromatography), which describes the determination of the chemical composition of natural gases and similar gaseous mixtures and is used to analyze lean natural gases containing negligible amounts of hexanes and higher hydrocarbons or to determine one or more components, as required; (ii) ASTM D1946 (2017) (Standard Practice for Analysis of Reformed Gas by Gas Chromatography), which describes the determination of the chemical composition of reformed gases and similar gaseous mixtures containing hydrogen, oxygen, nitrogen, carbon monoxide, carbon dioxide, methane, ethane, and ethylene; and (iii) ASTM D7833 (2017) (Standard Test Method for Determination of Hydrocarbons and Non-Hydrocarbon Gases in Gaseous Mixtures by Gas Chromatography), which is used to quantitatively determine the noncondensed hydrocarbon gases with carbon numbers from C_1 to C_{5+} and nonhydrocarbon gases, such as hydrogen, carbon dioxide, carbon monoxide, oxygen, and nitrogen, in gaseous samples. ASTM D7833 (2017) differs from the other test methods (ASTM D1945, 2017; ASTM D1946, 2017) by incorporating the use of capillary columns instead of packed columns for the analysis. Hydrogen sulfide can be detected by this method but may not be accurate by due to loss in sample containers or sample lines and possible reactions unless special precautions are taken.

LPG is predominantly a mixture of propane and butane in a liquid state at room temperatures when under moderate pressures of less than 200 psig (pounds per square inch gauge (psig) is a common measure of pressure). The common interchanging of the terms liquefied petroleum gas and liquefied propane is explained by the fact that in the United States and Canada, the LPG consists primarily of propane, whereas in many European countries, the propane content in LPG can be lower than 50% v/v. In addition, LNG is not the same as compressed natural gas (CNG), which is natural gas that is pressurized and stored in welding bottle-like tanks at pressures up to 3600 psig. Typically, CNG has the same composition as pipeline quality natural gas, i.e. the gas has been dehydrated (water removed) and all other elements reduced. LNG is also not synonymous with gas to liquids (GTL), which refer to the conversion of natural gas to products like methanol, dimethyl ether (DME), middle distillates (diesel and jet fuel), and specialty chemicals and waxes.

LNG can be used in natural gas vehicles, although it is more common to design vehicles to use CNG. The relatively high cost of production of LNG and the need to store the liquid in expensive cryogenic tanks have prevented its widespread use in commercial applications. Prior to and during transportation, CNG is stored on the vehicle in high-pressure tanks, usually on the order of 3000–3600 psi.

The production of LNG process involves removal of certain components, such as dust, acid gases (such as hydrogen sulfide and carbon dioxide), helium, water vapor, and higher molecular weight hydrocarbon derivatives, which could cause difficulty downstream of the wellhead. The natural gas is then condensed into a liquid at close to atmospheric pressure by cooling it to approximately −162 °C (−260 °F); maximum transport pressure is set at around 4 psi. One volume of this liquid takes up approximately 1/600th of the volume of natural gas. LNG weighs less than one-half that of water, and it is colorless, odorless, and noncorrosive. When vaporized, LNG burns only in concentrations of 5–15% v/v when mixed with air. Since LNG takes less volume and weight than that of natural gas, liquefaction enhances ease of storing and transporting. Hazards include flammability after vaporization into a gaseous state, freezing, and asphyxia.

Facilities for liquefying natural gas require complex machinery with moving parts and special refrigerated ships for transporting the LNG to market. The costs of building LNG plant have lowered over the past 25 years because of greatly improved thermodynamic efficiencies, resulting in LNG becoming a major gas export method worldwide and many plants being extended or new ones built.

LNG is natural gas stored as a supercooled (cryogenic) liquid; the temperature required to condense natural gas depends on its precise composition, but it is typically between −120 and −170 °C (−184 and −274 °F). Large cryogenic tanks are needed to store the LNG, which typically may be 230 ft in diameter and 145 ft high and hold over 26 400 000 gallons of LNG. At the consumer end of the transportation process, an infrastructure for handling the reprocessing of vast quantities of natural gas from LNG is required, which is also expensive and vulnerable to sabotage.

The advantage of LNG is that it offers an energy density comparable with that of petrol and diesel fuels, extending range and reducing the frequency of refueling. The disadvantage, however, is the high cost of cryogenic storage on vehicles and the major infrastructure requirement of dispensing station for the LNG, production, plants, and transportation facilities. For transportation, the LNG is loaded onto double-hulled ships that are used for both safety and insulating purposes. Once the ship arrives at the receiving port, the LNG is typically off-loaded into well-insulated storage tanks. Regasification is used to convert the LNG back into its gas form, which enters the domestic pipeline distribution system and is ultimately delivered to the end user. The current largest specially built refrigerated tankers can carry 135 000 m^3 LNG, which is approximately 4 767 480 ft^3 of gas.

3.4.3 Liquefied Petroleum Gas

Liquefied petroleum gas is the term applied to certain specific hydrocarbons and their mixtures, which exist in the gaseous state under atmospheric ambient conditions but can be converted to the liquid state under conditions of moderate pressure at ambient temperature. LPG is a hydrocarbon mixture containing propane ($CH_3{\cdot}CH_2{\cdot}CH_3$) and butane ($CH_3{\cdot}CH_2{\cdot}CH_2{\cdot}CH_3$). To a lesser extent, *iso*-butane [$CH_3{\cdot}CH(CH_3){\cdot}CH_3$] may also be present. The most common commercial products are propane, butane, or a specific of the two gases (Table 3.3) and are generally extracted from natural gas or crude petroleum. For analysis, the LPG must be sampled by the recommended procedure (ASTM D1265, 2017) and stored in the recommended manner correctly until tested (ASTM D6849, 2017).

The former test method (ASTM D1265, 2017) describes equipment and procedures for obtaining a representative sample of specification of LPG, such as specified in another test method (ASTM D1835, 2017) as well as in comparable international standards. This test method may also be used for other NGL products that are normally single phase (such as a NGL mix and field butane), defined in other industry specifications or contractual agreements. It is not intended for nonspecification products that contain significant quantities of undissolved gases (such as nitrogen and carbon dioxide), free water, or other separated phases, such as raw or unprocessed gas/liquids mixtures and related materials. The same equipment can be used for these purposes, but additional precautions are generally needed to obtain representative samples of multiphase products (ASTM D3700, 2017). The method is suitable for obtaining representative samples for all routine tests for LPGs

Table 3.3 Properties of liquefied petroleum gas.

Constituent	Propane	Butane
Formula	C_3H_8	C_4H_{10}
Boiling point, °F	−44°	32°
Specific gravity of the gas (air = 1.00)	1.53	2.00
Specific gravity of the liquid (water = 1.00)	0.51	0.58
lbs gal^{-1}: liquid at 60 °F	4.24	4.81
Btu gal^{-1}: gas at 60 °F	91 690	102 032
Btu lb^{-1}: gas	21 591	21 221
Btu ft^{-3}: gas at 60 °F	2 516	3 280
ft^3 of vapor at 60 °F/gal of liquid, 60 °F	36.39	31.26
ft^3 of vapor at 60 °F/lb of liquid, 60 °F	8.547	6.506
Latent heat of vaporization at boiling point Btu gal^{-1}	785.0	808.0
Combustion data:		
Flash point, °F	−155	−76
Autoignition temperature, °F	878	761
Maximum flame temperature in air, °F	3 595	3 615
Flammability limits, % v/v of gas in air mixture:		
Lower limit, %	2.4	1.9
Upper limit, %	9.6	8.6
Octane number (isooctane = 100)	100+	92

required by ASTM D1835 (2017). In addition, if this is a question about sample integrity when sampling for testing against the necessary requirements (ASTM D1835, 2017), another test method (ASTM D3700, 2017) can be used as the referee sampling procedure.

The latter test method (ASTM D6849, 2017) recognizes that samples of LPG can change composition during storage and use from preferential vaporization of lower molecular weight (lower-boiling) hydrocarbon components, dissolved inert gases (such as nitrogen, argon, and helium), and other dissolved gases/liquids (such as ammonia, carbon dioxide, hydrogen sulfide, and water). Thus, careful selection of cylinder type, cylinder volume, and use of inert gas for pressurizing cylinders is required to ensure that composition changes are small enough to maintain the integrity of the LPG when used as a reference material for various LPG test methods and to ensure quality control in accordance with the standard practice (ASTM D6299, 2017). The test method provides information for the storage and use of LPG samples in standard cylinders of the type used in the manual sampling method (ASTM D1265, 2017) and when floating piston cylinders are used for acquiring the samples (ASTM D3700, 2017). The method is especially applicable when the sample of the LPG sample is used as a quality control reference material for test methods that use gas chromatographic analysis (ASTM D2163, 2017) or vapor pressure (ASTM D6897, 2017) that use only a few milliliters per test. However, test methods that require a large amount of sample per test, for example, the manual vapor pressure test method (ASTM D1267, 2017), require volumes in excess of 1000l if stored in standard Department of Transportation cylinders or in vessels defined by the American Society of Mechanical Engineers (ASME).

As already noted, the compositions of natural, manufactured, and mixed gases can vary so widely; no single set of specifications could cover all situations. The requirements are usually based on performances in burners and equipment, on minimum heat content, and on maximum sulfur content. Gas utilities in most states come under the supervision of state commissions or regulatory bodies, and the utilities must provide a gas that is acceptable to all types of consumers and that will give satisfactory performance in all kinds of consuming equipment. However, there are specifications for LPG (ASTM D1835, 2017) that depend upon the required volatility.

The different methods for gas analysis (that may also be applied to natural gas itself and other gases) include absorption, distillation, combustion, mass spectroscopy, infrared spectroscopy, and gas chromatography (ASTM D2163, 2017; ASTM D2650, 2017; ASTM D4424, 2017). Absorption methods involve absorbing individual constituents one at a time in suitable solvents and recording of contraction in volume measured. Distillation methods depend on the separation of constituents by fractional distillation and measurement of the volumes distilled. In combustion methods, certain combustible elements are caused to burn to carbon dioxide and water, and the volume changes are used to calculate composition. Infrared spectroscopy is useful in particular applications. For the most accurate analyses, mass spectroscopy and gas chromatography are the preferred methods.

The specific gravity of product gases, including LPG, may be determined conveniently by a number of methods and a variety of instruments (ASTM D1070, 2017). This test method covers the determination of relative density of gaseous fuels, including LPGs, in the gaseous state at normal temperatures and pressures. The methods specified as subsections of the test method are sufficiently varied in nature so that one or more may be used for laboratory, control, reference (quality control), and gas measurement or, in fact, for any purpose in which it is desired to know the relative density of gas or gases as compared with the density of dry air at the same temperature and pressure.

The *heat value* of a fuel gas is generally determined at constant pressure in a flow calorimeter in which the heat released by the combustion of a definite quantity of gas is absorbed by a measured quantity of water or air. Test method

ASTM D4891 (2017) provides an accurate and reliable procedure to measure the total heating value of a fuel gas, on a continuous basis, which is used for regulatory compliance, custody transfer, and process control. Some instruments that conform to the requirements set forth in this test method can have response times on the order of 1 min or less and can be used for online measurement and control. The method is sensitive to the presence of oxygen and nonparaffin fuels. For the components that fall outside of the range of composition not listed in the test method, modifications to the method and changes to the calibration gas being used may be required to obtain reliable results.

There is also a standard test method that uses a continuous recording calorimeter that is available for measuring heat values of natural gases (ASTM D1826, 2017). This method describes the determination of the total calorific (heating) value of fuel gas produced or sold in the natural gas ranging from 900 to 1200 Btu standard^{-1} ft^{-3}. The method provides an accurate and reliable method to measure the total calorific value of a fuel gas, on a continuous basis, which is used for regulatory compliance, custody transfer, and process control.

The lower and upper limits of *flammability* indicate the percentage of combustible gas in air below which and above which flame will not propagate. When flame is initiated in mixtures having compositions within these limits, it will propagate, and therefore the mixtures are flammable. A knowledge of flammable limits and their use in establishing safe practices in handling gaseous fuels is important, such as when purging equipment is used (i) in gas service, (ii) in controlling the atmospheric composition in a factory or in a mine, or (iii) in handling liquefied gases.

Many factors enter into the experimental determination of flammable limits of gas mixtures, including the diameter and length of the tube or vessel used for the test, the temperature and pressure of the gases, and the direction of flame propagation – upward or downward. For these and other reasons, great care must be taken in the application of the data. In monitoring closed spaces where small amounts of gases enter the atmosphere, often the maximum concentration of the combustible gas is limited to one-fifth of the concentration of the gas at the lower limit of flammability of the gas–air mixture.

LPG can be transported in a number of ways, including by ship, rail, tanker trucks, intermodal tanks, cylinder trucks, pipelines, and local gas reticulation systems. However, a challenge with LPG is that it can vary widely in composition, leading to variable engine performance and cold-starting performance. At normal temperatures and pressures, LPG will evaporate, and, because of this tendency, LPG is stored in pressurized steel bottles. Unlike natural gas, LPG is heavier than air and thus will flow along floors and tend to settle in low spots, such as basements or, if outside, in depressions in the earth. Such accumulations can cause explosion hazards and are the reason that LPG-fueled vehicles are prohibited from indoor parking area in some jurisdictions.

3.4.4 Compressed Natural Gas

CNG (methane stored at high pressure) is a fuel that can be used in place of gasoline, diesel fuel, and LPG. The composition and properties of CNG are dependent on the properties and composition of the original natural gas feedstock, and test methods for application to the CNG should be designed based on the test methods applied to the gaseous (noncompressed) feedstock.

CNG is often confused with LNG – the main difference between the two (CNG and LNG) is that CNG is stored at ambient temperature and high pressure, while LNG is stored at low temperature and approximately ambient pressure. In their respective storage conditions, LNG is a liquid, while CNG is a supercritical fluid (a fluid at a temperature and pressure above the critical point where distinct liquid and gas phases do not exist). Also, CNG does not require an expensive cooling process and cryogenic tanks, but CNG does require a much larger volume to store the energy equivalent of gasoline and the use of very high pressures (3000–4000 psi). As a consequence of this, LNG is often used for transporting natural gas over large distances, in ships, trains, or pipelines, where the gas is converted into CNG before distribution to the consumers.

The combustion of CNG produces fewer undesirable gases than many other fuels. Also, CNG is safer than other fuels in the event of a spill because natural gas is lighter than air and disperses quickly when released. CNG may be found above oil deposits or may be collected from landfills or wastewater treatment plants where it is known as biogas. CNG is produced by compressing natural gas to less than 1% of the volume it occupies at standard atmospheric pressure. It is stored and distributed in hard containers at a pressure on the order of 2900–3600 psi, usually in cylindrical-shaped or spherical-shaped containers. Also, gas can be transported in containers at high pressures, typically 1800 psig for a rich gas (significant amounts of ethane, propane, etc.) to roughly 3600 psig for a lean gas (mainly methane). Gas at these pressures is termed *compressed natural gas*.

CNG has been used as a transportation fuel, mostly in public transit as an alternative to conventional fuels (gasoline or diesel). CNG, which is compressed at over 3000 psi to 1% of the volume the gas would occupy at normal atmospheric pressure, can be burned in an internal combustion engine that has been appropriately modified. Compared with gasoline, CNG vehicles emit far less carbon monoxide, nitrogen oxides (NOx), and particulates. The main disadvantage of CNG is its low energy density compared with liquid fuels. A gallon of CNG has only a quarter of the energy in a gallon of gasoline. CNG vehicles therefore require big, bulky fuel tanks, making CNG practical mainly for large vehicles such as buses and trucks. The filling stations can be supplied by pipeline gas, but the compressors needed to get the gas to 3000 psig can be expensive to purchase, maintain, and operate.

An alternative approach has dedicated transport ships carrying straight, long, large-diameter pipes in an insulated cold storage cargo package. The gas has to be dried, compressed, and chilled for storage onboard. By careful control of temperature, more gas should be transported in any ship of a given payload capacity, subject to volume limitation and amount and weight of material of the pipe (pressure and safety considerations). Suitable compressors and chillers are needed but would be much less expensive than a natural gas liquefier, and would be standard, so that costs could be further minimized. According to the proposers, the terminal facilities would also be simple and hence would be of low cost.

3.4.5 Gas to Solid

Gas can be transported as a solid, with the solid being the gas hydrate (Børrehaug and Gudmundsson, 1996; Gudmundsson, 1996; Gudmundsson, and Børrehaug, 1996; Gudmundsson et al., 1997, 1998). NGH is the product of mixing natural gas with liquid water to form a stable water crystalline icelike substance (Chapters 1 and 9). The transport of NGH is (with due consideration of the safety issues) a viable alternative to LNG or pipelines for the transportation of natural gas from source to demand. Consideration of the various aspects leads to the conclusion that transporting natural gas as the solid gas hydrates can be done at a higher temperature and at a lower pressure than the transportation of LNG and the risk of ignition in transport is much lower remembering that gas hydrates may decompose explosively with simultaneous ignition.

The gas-to-solid (GTS) process involves three stages: (i) production, (ii) transportation, and (iii) regasification. NGH are created when certain small molecules, particularly methane, ethane, and propane, stabilize the hydrogen bonds within water to form a three-dimensional cage-like structure with the gas molecule trapped within the cages. A cage is made up of several water molecules held together by hydrogen bonds, and the solid gas hydrate has a snow-like appearance. Hydrates are formed from natural gas in the presence of liquid water, provided that the pressure is above and the temperature is below the equilibrium line of the phase diagram of the gas and liquid.

For the most part, in the crude oil and natural gas industries, gas hydrates are a pipeline nuisance and safety hazard and require considerable care by the operators to ensure that they do not form as they can block pipelines if precautions, such as methanol injection, are not taken. On the other hand, vast quantities of gas hydrate have been found in permafrost and at the seabed in depths below 1500 ft (500 m) and if properly exploited could become the major energy source in the next 30 years.

For gas transport, NGH can be deliberately formed by mixing natural gas and water at approximately 1175–1500 psi and 2–10 °C (35–50 °F). If the slurry is refrigerated to approximately −15 °C (5 °F), it decomposes very slowly at atmospheric pressure, so that the hydrate can be transported by ship to market in simple insulated under near adiabatic conditions – conditions that occur without transfer of heat or matter between a thermodynamic system and its surroundings.

At the market, the slurry is melted back to gas and water by controlled warming for use after appropriate drying in electricity power generation stations or other requirements. The hydrate mixture yields up to 5600 ft^3 (approximately 160 m^3) of natural gas per ton of hydrate, depending on the manufacturing process. The manufacture of the hydrate could be carried out using mobile equipment for onshore and ship for offshore using a floating production, storage, and offloading vessel with minimal gas processing (cleaning) prior to hydrate formation, which is attractive commercially.

The water can be used at the destination if there is water shortage or returned as ballast to the hydrate generator and, since it is saturated with gas, will not take more gas into solution. Process operability of continuous production of hydrate in a large-scale reactor, long-term hydrate storage, and controlled regeneration of gas from storage has all been demonstrated.

The hydrate mixture can be stored at normal temperatures (0–10 °C; 32–14 °F) and pressures (10–1 atm) where 1 m^3 of hydrate should contain about 160 m^3 gas per m^3 of water. This concentration of gas is attractive as it is easier to produce and safer and cheaper to store compared with the 200 m^3 per 1 m^3 of compressed gas (high pressure ca. 3000 psig) or the 637 m^3 gas per 1 m^3 of LNG (low temperatures of −162 °C, −260 °F).

Gas storage in hydrate form becomes especially efficient at relatively low pressures where substantially more gas per unit volume is contained in the hydrate than in the free state or in the compressed state when the pressure has dropped. When compared with the transportation of natural gas by pipeline or as LNG, the hydrate concept has lower capital and operating costs for the movement of quantities of natural gas over adverse conditions.

Thus, gas hydrate is very effective for gas storage and transport as it eliminates low temperatures and the necessity of compressing the gas to high pressures. To recall, dry hydrate pellets yield about 160 m^3 of gas under standard conditions from 1 m^3 of hydrate compared with the 637 m^3 per 1 m^3 of LNG. This is a considerable volume penalty (and hence transport cost) if considered in isolation, and with the cheaper ships for hydrate transport, the process could be economic.

3.4.6 Gas to Power

Electric power can be an intermediate product, such as in the case of mineral refining in which electricity is used to refine bauxite into aluminum; or it can be an end product that is distributed into a large utility power grid. Thus, the concept of gas to power (GTP) (sometimes referred to as *gas to grid* [GTG]) is not a new thought, but is certainly worthy of consideration in this section.

Currently, much of the transported gas destination is fuel for electricity generation. Electricity generation at or near the storage facility and transportation by cable to the destination(s) (GTP) is possible. Thus, for instance, offshore or isolated gas could be used to fuel an offshore power plant (may be sited in less hostile waters), which would generate electricity for sale onshore or to other offshore customers. Unfortunately, installing high-power lines to reach the shoreline appears to be almost as expensive as pipelines; GTP could be viewed as defeating the purpose of an alternative cheaper solution for transporting gas. There is significant energy loss from the cables along the long-distance transmission lines, more so if the power is AC rather than DC; additionally, losses also occur when the power is converted to DC from AC and when it converted from the high voltages used in the transmission to the lower values needed by the consumers.

Some observers consider having the energy as gas at consumers' end gives greater flexibility and better thermal efficiencies, because the waste heat can be used for local heating and desalination. This view is strengthened by the economics as power generation uses approximately one million scf per day of gas for every 10 MW of power generated, so that even large generation capacity would not consume much of the gas from larger fields and thus not generate large revenues for the gas producers. Nevertheless, GTP has been an option much considered in the United States for getting energy from the Alaskan gas and oil fields to the populated areas.

There are other practical considerations to note such as if the gas is associated gas, and if there is a generator shutdown and no other gas outlet, the whole oil production facility might also have to be shut down or the gas released to flare. Also, if there are operational problems within the generation plant, the generators must be able to shut down quickly (on the order of sixty seconds or less) to keep a small incident from escalating to a major incident. Additionally, the shutdown system itself must be safe so that any plant that has complicated processes that require a purge cycle or a cooldown cycle before it can shut down is clearly unsuitable (Ballard, 1965). Finally, if the plant cannot shut down easily and/or be able to start up again quickly (perhaps in an hour), operators will be hesitant to ever shut down the process, for fear of financial retribution from the power distributors.

3.4.7 Gas to Liquids

In GTL transport, the natural gas is converted to a liquid, such as syncrude methanol, and transported as such (Knott, 1997; Skrebowski, 1998; Thomas, 1998; Gaffney Cline and Associates, 2001). In the process, methane is first mixed with steam and converted to syngas (mixtures of carbon monoxide and hydrogen, $CO + H_2$) by one of a number of routes using suitable new catalyst technology.

In the steam-methane reforming process, methane reacts with steam under pressure (45–360 psi) in the presence of a catalyst to produce hydrogen, carbon monoxide, and a relatively small amount of carbon dioxide. The process is endothermic insofar as heat must be supplied to the process for the reaction to proceed. Subsequently, in what is referred to as the *water–gas shift reaction*, the carbon monoxide and steam are reacted using a catalyst to produce carbon dioxide and more hydrogen. In a final process (pressure swing adsorption step), carbon dioxide and other impurities are removed from the gas stream, leaving essentially pure hydrogen. Steam reforming can also be used to produce hydrogen from other fuels, such as ethanol, propane, or even gasoline.

$$CH_4 + H_2O \rightarrow CO + 3H_2 \text{ (steam-methane reforming reaction)}$$

$$CO + H_2O \rightarrow CO_2 + H_2 \text{ (+small amount of heat)(water – gas shift reaction)}$$

$$2H_2 + CO \rightarrow CH_3OH \text{ (methanol synthesis)}$$

Also:

$$2CH_3OH \rightarrow CH_3OCH_3 + H_2O$$

$$CH_3OCH_3 \rightarrow C_2H_4 + H_2O$$

The ethylene is polymerized and hydrogenated to give gasoline with hydrocarbon constituents having five or more carbon atoms that constitute approximately making up 80% w/w of the fuel.

In the partial oxidation process, methane and other hydrocarbons in natural gas react with a limited amount of oxygen (typically from air) that is not enough to completely oxidize the hydrocarbons to carbon dioxide and water. With less than the stoichiometric amount of oxygen available, the reaction products contain primarily hydrogen and carbon monoxide (and nitrogen, if the reaction is carried out with air rather than pure oxygen) and a relatively small amount of carbon dioxide and other compounds. Subsequently, in a water–gas shift reaction, the carbon monoxide reacts with water to form carbon dioxide and more hydrogen.

Partial oxidation is an exothermic process insofar as heat is evolved during the process. It is, typically, a much more rapid reaction than steam reforming and requires a smaller reactor vessel. As can be seen in chemical reactions of partial oxidation, this process initially produces less hydrogen per unit of the input fuel than is obtained by steam reforming of the same fuel.

$$CH_4 + \tfrac{1}{2}O_2 \rightarrow CO + 2H_2 \text{ (partial oxidation of methane)}$$

$$CO + H_2O \rightarrow CO_2 + H_2 \text{ (water – gas shift reaction)}$$

The synthesis gas (syngas) is then converted into a liquid using a Fischer–Tropsch process (in the presence of a catalyst) or an oxygenation method (mixing synthesis gas with oxygen in the presence of a suitable catalyst).

$$CO + nH_2 \rightarrow H(-CH_2-)_x H + H_2O$$

The produced liquid can be a fuel, usually a clean-burning motor fuel (syncrude) or lubricant, ammonia or methanol, or some precursor for plastics manufacture (e.g. urea and DME, which is also used as a transportation fuel, LPG substitute, or power generation fuel as well as a chemical feedstock).

Methanol is a GTL option that has been in commission since the mid-1940s. While methanol produced from gas was originally a relatively inefficient conversion process, optimized technology has improved the efficiency. Methanol can be used in internal combustion engines as a fuel, but the current market for methanol as a fuel is limited, although the development of fuel cells for motor vehicles may change this. Methanol is best used as a basic chemical feedstock for the manufacture of plastics.

Other GTL processes are being developed to produce clean fuels, e.g. syncrude, diesel, or many other products including lubricants and waxes, from gas but require complex (expensive) chemical plant with novel catalyst technology.

3.4.8 Gas to Commodity

The gas-to-commodity (GTC) concept involves the use of the components in natural gas, methane, ethane, propane, *n*- and isobutanes, and pentanes that useful in their own right. The higher paraffins are particularly valuable for a wealth of chemicals and polymer precursors such as acetic acid, formaldehyde, olefins, polyethylene, polypropylene, acrylonitrile, ethylene glycol, etc., as well as portable premium fuels, such as propane. In addition, methane can be converted via syngas to methanol, ammonia, syncrude, lubricant, or some precursor for chemicals manufacture, e.g. DME, urea, etc., and then used to make chemicals for export.

In the GTC concept, the gas is converted to thermal or electrical power, which is then used in the production of the commodity, which is then sold on the open market. It is the energy from the gas, heat via electricity or direct combustion, and not the components of the GTL concept that is used. The gas energy is, in essence, transported via the commodity.

In some case, the gas feedstock may be suitable for hydrogen production that can be obtained through various thermochemical methods utilizing methane (natural gas), LPG, coal gasification, or biomass (biomass gasification), from electrolysis of water, or by a thermolysis process. Hydrogen or H_2 gas is highly flammable and will burn at concentrations as low as 4% H_2 in air. For automotive applications, hydrogen is generally used in two forms: internal combustion or fuel cell conversion. In combustion, it is essentially burned as conventional gaseous fuels are, whereas a fuel cell uses the hydrogen to generate electricity that in turn is used to power electric motors on the vehicle. Hydrogen gas must be produced and is therefore an energy storage medium, not an energy source. The energy used to produce it usually comes from a more conventional source. Hydrogen holds the promise of very low vehicle emissions and flexible energy storage; however, the technical challenges required to realize these benefits may delay the widespread implementation for several decades. In fact, on occasions when the use of hydrogen as a fuel is proposed, the memory of the Hindenburg disaster (which occurred on 6 May 1937) always arises. The German passenger airship LZ 129 Hindenburg caught fire and was destroyed during its attempt to dock with its mooring mast at the Lakehurst Naval Air Station in Manchester Township, New Jersey, United States.

References

ASTM D1070 (2017). *Standard Test Methods for Relative Density of Gaseous Fuels*, Annual Book of Standards. West Conshohocken, Pennsylvania: ASTM International.

ASTM D1265 (2017). *Standard Practice for Sampling Liquefied Petroleum (LP) Gases, Manuel Method*, Annual Book of Standards. West Conshohocken, Pennsylvania: ASTM International.

ASTM D1267 (2017). *Standard Test Method for Gage Vapor Pressure of Liquefied Petroleum (LP) Gases (LP-Gas Method)*, Annual Book of Standards. West Conshohocken, Pennsylvania: ASTM International.

ASTM D1826 (2017). *Standard Test Method for Calorific (Heating) Value of Gases in Natural Gas Range by Continuous Recording Calorimeter*, Annual Book of Standards. West Conshohocken, Pennsylvania: ASTM International.

ASTM D1835 (2017). *Standard Specification for Liquefied Petroleum (LP) Gases*, Annual Book of Standards. West Conshohocken, Pennsylvania: ASTM International.

ASTM D1945 (2017). *Standard Test Method for Analysis of Natural Gas by Gas Chromatography*, Annual Book of Standards. West Conshohocken, Pennsylvania: ASTM International.

ASTM D1946 (2017). *Standard Practice for Analysis of Reformed Gas by Gas Chromatography*, ASTM International. West Conshohocken, Pennsylvania: Annual Book of Standards.

ASTM D1988 (2017). *Standard Test Method for Mercaptans in Natural Gas Using Length-of-Stain Detector Tubes*, Annual Book of Standards. West Conshohocken, Pennsylvania: ASTM International.

ASTM D2163 (2017). *Standard Test Method for Determination of Hydrocarbons in Liquefied Petroleum (LP) Gases and Propane/Propene Mixtures by Gas Chromatography*, Annual Book of Standards. West Conshohocken, Pennsylvania: ASTM International.

ASTM D2420 (2017). *Standard Test Method for Hydrogen Sulfide in Liquefied Petroleum (LP) Gases (Lead Acetate Method)*, Annual Book of Standards. West Conshohocken, Pennsylvania: ASTM International.

ASTM D2650 (2017). *Standard Test Method for Chemical Composition of Gases by Mass Spectrometry*, Annual Book of Standards. West Conshohocken, Pennsylvania: ASTM International.

ASTM D3700 (2017). *Standard Practice for Obtaining LPG Samples Using a Floating Piston Cylinder*, Annual Book of Standards. West Conshohocken, Pennsylvania: ASTM International.

ASTM D4150 (2017). *Standard Terminology relating to Gaseous Fuels*, Annual Book of Standards. West Conshohocken, Pennsylvania: ASTM International.

ASTM D4424 (2017). *Standard Test Method for Butylene Analysis by Gas Chromatography*, Annual Book of Standards. West Conshohocken, Pennsylvania: ASTM International.

ASTM D4891 (2017). *Standard Test Method for Heating Value of Gases in Natural Gas and Flare Gases Range by Stoichiometric Combustion*, Annual Book of Standards. West Conshohocken, Pennsylvania: ASTM International.

ASTM D6299 (2017). *Standard Practice for Applying Statistical Quality Assurance and Control Charting Techniques to Evaluate Analytical Measurement System Performance*, Annual Book of Standards. West Conshohocken, Pennsylvania: ASTM International.

ASTM D6667 (2017). *Standard Test Method for Determination of Total Volatile Sulfur in Liquefied Petroleum Gases by Ultraviolet Fluorescence*, Annual Book of Standards. West Conshohocken, Pennsylvania: ASTM International.

ASTM D6849 (2017). *Standard Practice for Storage and Use of Liquefied Petroleum Gases in Sample Cylinders for LPG Test Methods*, Annual Book of Standards. West Conshohocken, Pennsylvania: ASTM International.

ASTM D6897 (2017). *Standard Test Method for Vapor Pressure of Liquefied Petroleum Gases (LPG) (Expansion Method)*, Annual Book of Standards. West Conshohocken, Pennsylvania: ASTM International.

ASTM D7751 (2017). *Standard Test Method for Determination of Total Volatile Sulfur in Gaseous Hydrocarbons and Liquefied Petroleum Gas and Natural Gas by Ultraviolet Fluorescence*, Annual Book of Standards. West Conshohocken, Pennsylvania: ASTM International.

ASTM D7756 (2017). *Standard Test Method for Residues in Liquefied Petroleum (LP) Gases by Gas Chromatography with Liquid On-Column Injection*, Annual Book of Standards. West Conshohocken, Pennsylvania: ASTM International.

ASTM D7828 (2017). *Standard Test Method for Determination of Residue Composition in Liquefied Petroleum Gas (LPG) Using Automated Thermal Desorption/Gas Chromatography*, Annual Book of Standards. West Conshohocken, Pennsylvania: ASTM International.

ASTM D7833 (2017). *Standard Test Method for Determination of Hydrocarbons and Non-Hydrocarbon Gases in Gaseous Mixtures by Gas Chromatography*, Annual Book of Standards. West Conshohocken, Pennsylvania: ASTM International.

ASTM D7940 (2017). *Standard Practice for Analysis of Liquefied Natural Gas (LNG) by Fiber-Coupled Raman Spectroscopy*, Annual Book of Standards. West Conshohocken, Pennsylvania: ASTM International.

Ballard, D. (1965). How to Operate Quick-Cycle Plants. *Hydrocarbon Processing and Petroleum Refiner* 44 (4): 131.

Børrehaug, A. and Gudmundsson, J.S. (1996). *Gas Transportation in Hydrate Form, EUROGAS 96* (pp. 35–41), Trondheim (3–5 June).

Burruss, R.C. and Ryder, R.T. (2003). Composition of Crude Oil and Natural Gas Produced from 14 Wells in the Lower Silurian "Clinton" Sandstone and Medina Group, Northeastern Ohio and Northwestern Pennsylvania. *Open-File Report 03-409*. Reston, Virginia: United States Geological Survey.

Burruss, R.C. and Ryder, R.T. (2014). Composition of natural gas and crude oil produced from 10 wells in the Lower Silurian "Clinton" Sandstone, Trumbull County, Ohio. In: *Coal and Petroleum Resources in The Appalachian Basin; Distribution, Geologic Framework, And Geochemical Character*, Professional Paper 1708 (ed. L.F. Ruppert and R.T. Ryder). Reston, Virginia: United States Geological Survey.

Crow, W., Williams, B., Carey, J.W., et al. (2008). Wellbore integrity analysis of a natural CO_2 producer. *Proceedings. 9th International Conference on Greenhouse Gas Control Technologies*, Washington, DC (16–20 November 2008). New York: Elsevier.

Faramawy, S., Zaki, T., and Sakr, A.A.-E. (2016). Natural gas origin, composition, and processing: a review. *Journal of Natural Gas Science and Engineering* 34: 34–54.

Riazi M., Eser, S., Agrawal, S., and Peña Díez, J. 2013. *Petroleum Refining and Natural Gas Processing. Manual 58 MNL58*. ASTM International, West Conshohocken, Pennsylvania.

Gaffney Cline and Associates (2001). Prepared for the Gas to Liquids Taskforce. GTL Discussion Paper. Australian Dept. of Industry, Science & Resources, Commonwealth of Australia, June.

Gary, J.G., Handwerk, G.E., and Kaiser, M.J. (2007). *Petroleum Refining: Technology and Economics*, 5e. Boca Raton, Florida: CRC Press, Taylor & Francis Group.

Gudmundsson, J.S. (1996). Method for production of gas hydrate for transportation and storage. US Patent No. 5,536,893, filed 7 January 1994 and issued 16 July 1996.

Gudmundsson, J.S. and Børrehaug, A. 1996. Frozen hydrate for transport of natural gas. *Proceedings of the 2nd International Conference Natural Gas Hydrates* (pp. 415–422), Toulouse (2–6 June).

Gudmundsson, J.S., Andersson, V. and Levik, O.I. 1997. Gas storage and transport using hydrates. *Offshore Mediterranean Conference*, Ravenna (19–21 March—).

Gudmundsson, J.S., Andersson, V., Levik, O.I., and Parlaktuna, M. (1998). Hydrate concept for capturing associated gas. *Proceedings. SPE European Petroleum Conference*, The Hague, the Netherlands (20–22 October 1998).

Hsu, C.S. and Robinson, P.R. ed. (2017). *Handbook of Petroleum Technology, C.S.* Cham, Switzerland: Springer International Publishing AG.

Katz, D.L. and Coats, K.H. (1968). *Underground Gas Storage of Fluids*. Ann Arbor, Michigan: Ulrich Books Inc.

Kidnay, A., McCartney, D., and Parrish, W. (2011). *Fundamentals of Natural Gas Processing*. Boca Raton, Florida: CRC Press, Taylor & Francis Group.

Knott, D. (1997). Gas-to-Liquids Projects Gaining Momentum as Process List Grows. *Oil & Gas Journal* 95: 16–21.

Mokhatab, S., Poe, W.A., and Speight, J.G. (2006). *Handbook of Natural Gas Transmission and Processing*. Amsterdam: Elsevier.

Parkash, S. (2003). *Refining Processes Handbook*. Amsterdam: Gulf Professional Publishing, Elsevier.

Skrebowski, C. (1998). Gas-to-liquids or LNG? *Petroleum Review* January: 38–39.

Speight, J.G. (2007). *Natural Gas: A Basic Handbook*. Houston, Texas: GPC Books, Gulf Publishing Company.

Speight, J.G. (2013). *The Chemistry and Technology of Coal*, 5e. Boca Raton, Florida: CRC Press, Taylor & Francis Group.

Speight, J.G. (2014a). *The Chemistry and Technology of Petroleum*, 5e. Boca Raton, Florida: CRC Press, Taylor & Francis Group.

Speight, J.G. (2014b). *Oil and Gas Corrosion Prevention*. Oxford, UK: Gulf Professional Publishing, Elsevier.

Speight, J.G. (2016). *Handbook of Hydraulic Fracturing*. Hoboken, New Jersey: Wiley.

Speight, J.G. (2017). *Handbook of Petroleum Refining*. Boca Raton, Florida: CRC Press, Taylor & Francis Group.

Thomas, M. (1998). Water into wine: gas-to-liquids technology the key to unlocking future reserves, Euroil (17–21 May).

4

Gas Cleaning

4.1 Introduction

The use of natural gas as a fuel and as a source of petrochemical feedstocks will, like the use of crude oil, continue at least for the next three to five decades (Speight, 2011). Moreover, it is also predictable that use of natural gas in the transportation sector will increase as increases in travel offset increased efficiency. Consequently, for the generation of contaminant-free products, there is the need to prepare natural gas for use by a range of domestic and industrial consumers.

Gas processing cleaning (also called *gas cleaning* and *gas refining*) is a necessary procedure to produce the product that meet the various specifications. Gas processing is a necessary part of the natural gas value chain and is, in fact, an integrated system of unit processes that are used to remove objectionable products such as acid gases (e.g. carbon dioxide and hydrogen sulfide) and to separate natural gas into other useful gas streams. Thus, gas processing is instrumental in ensuring that the natural gas intended for use is as clean and pure as possible, making it the clean-burning and also an environmentally sound energy choice.

Natural gas, as it is used by consumers, is composed almost entirely of methane and is much different from the natural gas that exists in the reservoir and that is brought up to the wellhead. Although the processing of natural gas is in many respects less complicated than the processing and refining of crude oil, it is equally as necessary before its use by end users. The natural gas used by consumers is composed almost entirely of methane, but the natural gas produced at the wellhead, although still composed primarily of methane, is by no means as pure. Raw (untreated) natural gas is produced from three types of wells: (i) wells that produce crude oil and natural gas, (ii) wells that produce natural gas only, and (iii) wells that produce gas condensate and natural gas.

Natural gas that is produced with crude oil is typically termed *associated gas*, which can exist as a separate phase from the crude oil (free gas) or dissolved in

Handbook of Natural Gas Analysis, First Edition. James G. Speight.

the crude oil (dissolved gas). Natural gas from gas-only wells and condensate wells, in which there is little or no crude oil, is termed *nonassociated gas*. Gas wells typically produce raw natural gas by itself, while condensate wells produce free natural gas along with a semiliquid hydrocarbon condensate. Whatever the source of the natural gas, once separated from crude oil (if present), it commonly exists in mixtures with other hydrocarbons: principally ethane, propane, butane, and pentanes. In addition, raw natural gas contains water vapor, hydrogen sulfide (H_2S), carbon dioxide (CO_2), helium (He), nitrogen (N_2), and a variety of other compounds, often in trace amounts (typically less than 1% v/v in total) (Table 4.1).

Major transportation pipelines usually impose restrictions on the makeup of the natural gas that is allowed into the pipeline. That means that before the natural gas can be transported, it must be purified. This can commence at the wellhead where efforts are made to remove (at least) water, carbon dioxide, and hydrogen sulfur to prevent damage (corrosion) to the pipeline and associated equipment (Speight, 2014b). The goal of this first stage of natural gas processing consists of separating (some or all of) the various hydrocarbon derivatives (C_{2+} as well as natural gasoline) that, depending upon the dew point of the hydrocarbon and the conditions in the pipeline, could separate in the pipeline and cause liquid blockage.

These associated hydrocarbons, known as *natural gas liquids* (NGLs), can be very valuable by-products of natural gas processing. The NGLs may undergo further separation at a gas processing plant to be sold separately and have different uses, including providing raw materials for crude oil refineries or petrochemical plants and being sources of energy. Also at the wellhead, contaminants

Table 4.1 Constituents of natural gas.

Name	Formula	% v/v
Methane	CH_4	>85
Ethane	C_2H_6	3–8
Propane	C_3H_8	1–5
Butane	C_4H_{10}	1–2
Pentane$^+$	$C_5H_{12}{}^+$	1–5
Carbon dioxide	CO_2	1–2
Hydrogen sulfide	H_2S	1–2
Nitrogen	N_2	1–5
Helium	He	<0.5

Pentane$^+$: pentane and higher molecular weight hydrocarbons, including benzene and toluene.

such as carbon dioxide and hydrogen sulfide are also separated from the gas steam to produce *pipeline quality* dry natural gas. While ethane, propane, butane, and pentanes must be removed from natural gas, this does not mean that they are waste products, but they are valuable by-products (Speight, 2014a, 2017).

Briefly, gas processing (Mokhatab et al., 2006; Speight, 2007, 2014a) consists of separating all of the various hydrocarbons and any other impurities from the methane (Figure 4.1). While the necessary preprocessing of natural gas can be accomplished at or near the wellhead (field processing), the complete processing of natural gas takes place at a processing plant, usually located in a natural gas producing region. The extracted natural gas is transported to these processing plants through a network of gathering pipelines, which are small-diameter, low-pressure pipes.

A complex gathering system can consist of thousands of miles of pipes, interconnecting the processing plant to upward of 100 wells in the area. In addition to processing done at the wellhead and at centralized processing plants, some final processing is also sometimes accomplished at *straddle extraction plants*.

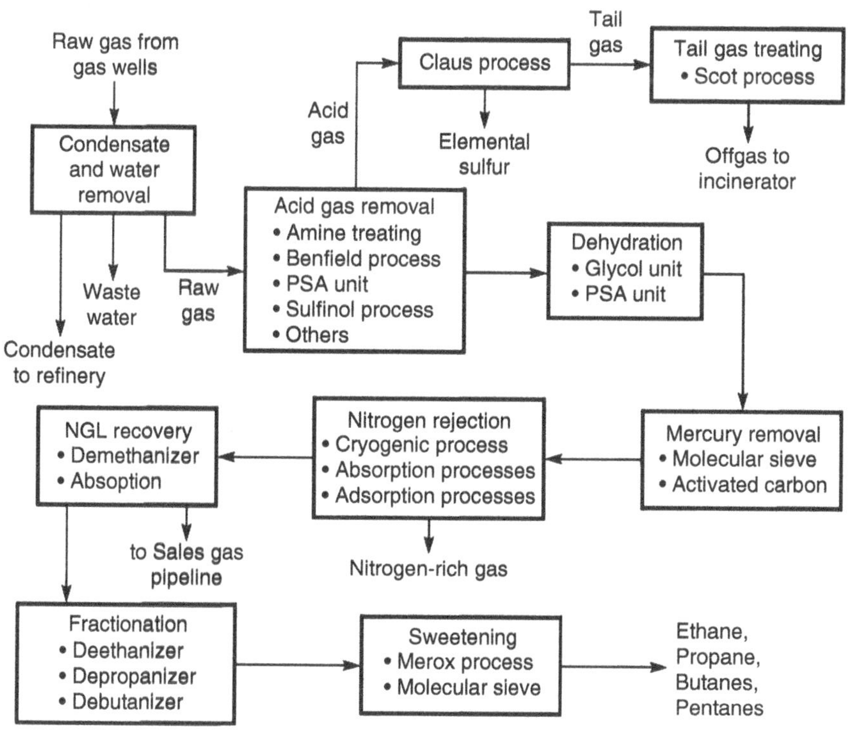

Figure 4.1 Schematic of the integrated unit processes used for gas processing.

These plants are located on major pipeline systems, and although the natural gas that arrives at these straddle extraction plants is already of pipeline quality, in certain instances there still exist small quantities of NGLs, which are extracted at the straddle plants.

Sulfur and a variety of sulfur-containing compounds (such as hydrogen sulfide, H_2S, mercaptans, RSH – also called thiols – and carbonyl sulfide, COS) occur in gases. These compounds are detrimental (especially in the presence of water) by causing corrosion to the pipeline and can cause deactivation of catalysts used in gas-treating systems. The sulfur levels in the natural gas (or, for that matter, in any gas stream) are an important parameter for gas treatment. Therefore, it is essential that sulfur-containing compounds such as hydrogen sulfide (as well as acid gases such as carbon dioxide) be removed from a gas stream and converted to more useful products such as sulfur or sulfuric acid (Bartoo, 1985; Mokhatab et al., 2006; Speight, 2007, 2014a).

To satisfy the chemical principles for the combustion of natural gas, knowledge of the concentration of the various constituents is essential. Natural gas is available in different forms (Chapter 1), and knowledge of gas composition (ASTM D1945, 2017) and heating value (ASTM D4891, 2017) are important parameters to be met for optimum and economical processing to separate the gas into the individual constituents and ensuring that each constituent meets the necessary sales specifications. In this context, the term concentration describes the amount of a substance, expressed as mass or volume in a unit mass or volume of the gas. Also, the heating value can be presented as Btu lb^{-1} (the commonly used units in the United States) or in units that are consistent with the SI system of units.

NGLs include ethane, propane, butane, *iso*-butane, and natural gasoline that are sold separately and have a variety of different uses, including enhancing oil recovery in oil wells, providing raw materials for oil refineries or petrochemical plants, and as sources of energy.

Finally, it is also worthy of note that, because of the variation of gas volume with changes in temperature and changes in pressure, it is necessary to use one of the following alternatives for describing a concentration value: (i) additional specification of gas temperature and pressure values existing during measurement or (ii) conversion of the measured concentration value into the corresponding value at standard zero conditions.

In terms of gas processing (Figure 4.1), the analysis of natural gas and its constituents (Chapter 10) includes the detection of gases and vapors in connection with control of chemical and metallurgical processes and control of environment and in the field of safety control. A whole class of analytical instruments, which use physical or physicochemical detecting methods, is available for gas analysis. Although analyzers are considered to be electrical instruments, some knowledge of chemistry is required to understand their operating principle and to obtain correct and accurate measuring results.

Process analysis, in contrast to laboratory analysis, includes all continuous measuring methods for real-time determination of physical and chemical properties and concentrations in process streams. Measuring points with continuously operating sampling devices, including the analyzers, are located at selected locations spots in a gas processing plant. The difference to laboratory analysis is where the analyzers are installed in a laboratory under the best operating conditions and where the samples are taken discontinuously from the process and brought to the laboratory. Laboratory analyzers are normally designed for sophisticated measurements and require trained operators.

The results of analysis in gas processing at the wellhead and in gas processing plants (Figure 4.1) are used for (i) process control by, for example, controlling/monitoring the properties of the raw feedstock and the products by optimizing the process steps, (ii) safety issues such as the protection of personnel and plant equipment by controlling the plant atmosphere for toxic gas mixtures or for explosive gas mixtures, (iii) product quality control by monitoring the production process steps and controlling the final product specification, and (iv) environmental protection by monitoring the exhaust gases for compliance with the specified limits of the types of pollutants.

Knowledge of those constituents that can have a detrimental effect on the combustion process or, for that matter, on the sales specification must be known and given attention. It is for this reason that gas processing (also called gas processing or gas refining) is an important aspect of natural gas technology along with the relevant analytical test methods for natural gas feedstocks as well as the properties of the separated products.

Using methane as the example, when natural gas is used as a combustible fuel gas, the presence of noncombustible (inert) constituents reduces the gross and net calorific value of the gas and increases contamination of the furnace walls. Increasing water content raises the water dew point and consumes energy to evaporate water in the flue gas. The sulfur contained in the gas is burnt (oxidized) to sulfur dioxide (SO_2) and sulfur trioxide (SO_3), which, at temperatures below the dew point, may lead to the formation of aggressive sulfurous acid and sulfuric acid:

$$SO_2 + H_2O \rightarrow H_2SO_3 \text{ Sulfurous acid}$$

$$SO_3 + H_2O \rightarrow H_2SO_4 \text{ Sulfuric acid}$$

$$2SO_2 + 2O_2 + 2H_2O \rightarrow 2H_2SO_4 \text{ Sulfuric acid}$$

After production of the natural gas, the analysis starts with transportation (to detect and remove corrosive contaminants) and in the gas processing units where the gas is cleaned of all possible contaminants and prepared for use by domestic and industrial consumers.

In addition, the oxygen required for a natural gas combustion process is supplied as part of the combustion air that consists of nitrogen (N_2), oxygen (O_2), varying small amounts of carbon dioxide and rare gases, and a variable content of water vapor. In some processes pure oxygen or an air/oxygen mixture may be used for the combustion. The minimal amount of oxygen required to burn all combustible components completely depends on the fuel composition. The actual volume of air required for the combustion of natural gas can be calculated from (i) the air volume needed for the ideal combustion, depending on the composition of the gas, (ii) the desired excess oxygen value, and (iii) the relative oxygen content of the used air or the air/oxygen mixture. If, however, the ambient air is not dry, the water content must also be considered for correct air volume calculation.

4.2 Gas Streams

Natural gas streams are variable in composition (Chapter 1), and gas processing consists of separating all of the various hydrocarbons and fluids from the pure natural gas to produce what is known as "pipeline quality" dry natural gas.

The actual practice of processing the various gas streams to either pipeline dry gas at the wellhead or specification-quality gas in a gas processing plant levels can be quite complex but usually involves four main processes to remove the impurities. Gas streams produced from a gas reservoir or from a crude oil reservoir, while ostensibly being hydrocarbon in nature, may contain large amounts of acid gases such as hydrogen sulfide and carbon dioxide. Most commercial plants employ hydrogenation to convert organic sulfur compounds into hydrogen sulfide. Hydrogenation is achieved by means of recycled hydrogen-containing gases or external hydrogen over a nickel molybdate or cobalt molybdate catalyst (Parkash, 2003; Gary et al., 2007; Speight, 2014a; Hsu and Robinson, 2017; Speight, 2017).

In summary, refinery process gas, in addition to hydrocarbons, may contain other contaminants, such as carbon oxides (CO_x, where $x = 1$ and/or 2), sulfur oxides (SO_x, where $x = 2$ and/or 3), as well as ammonia (NH_3), mercaptans (R–SH), and carbonyl sulfide (COS) as well as olefins (Chapter 2). On the other hand, raw biogas produced from digestion is roughly 60% v/v methane and 29% v/v carbon dioxide with trace elements of hydrogen sulfide. Landfill gas is produced by wet organic waste decomposing under anaerobic conditions in a digester and may contain the same constituents as biogas but, like biogas, in different proportions depending upon the composition of the source material and the process parameters under which the gas was produced (Chapter 1), and there are processes for converting biogas to biomethane (Ryckebosch et al., 2011).

Landfill gas must be treated to remove impurities, condensates, and particulates, hence the need for analysis to determine the composition of the gas. However, the treatment system depends on the end use: (i) minimal treatment is needed for the direct use of gas in boiler, furnaces, or kilns and (ii) using the gas in electricity generation typically requires more in-depth treatment. Thus, treatment systems are divided into primary and secondary treatment processing. Primary processing systems remove moisture and particulates. Gas cooling and compression are common in primary processing. Secondary treatment systems employ multiple cleanup processes, physical and chemical, depending on the specifications of the end use. Two constituents that may need to be removed are siloxane derivatives and sulfur compounds, which are damaging to equipment and significantly increase maintenance cost. Adsorption and absorption are the most common technologies used in secondary treatment processing, such as for the removal of siloxane derivatives (Yao et al., 2016).

The presence of these impurities may eliminate some of the sweetening processes, since some processes remove large amounts of acid gas but not to a sufficiently low concentration. On the other hand, there are those processes not designed to remove (or incapable of removing) large amounts of acid gases, whereas they are capable of removing the acid gas impurities to very low levels when the acid gases are present only in low to medium concentration in the gas (Katz, 1959; Mokhatab et al., 2006; Speight, 2007, 2014a). The processes that have been developed to accomplish gas purification vary from a simple once-through wash operation to complex multistep recycling systems (Mokhatab et al., 2006; Speight, 2007, 2014a). In many cases, the process complexities arise because of the need for recovery of the materials used to remove the contaminants or even recovery of the contaminants in the original, or altered, form (Katz, 1959; Kohl and Riesenfeld, 1985; Newman, 1985; Mokhatab et al., 2006; Speight, 2007, 2014a).

Because of the variable number of constituents and the variable composition of the gas, there are many variables in treating refinery gas or natural gas. The precise area of application, of a given process, is difficult to define. Several factors (not necessarily in order of importance) must be considered: (i) the types and (ii) concentration of contaminants in the gas; (iii) the degree of contaminant removal desired, (iv) the selectivity of acid gas removal required; (v) the temperature, (vi) pressure, (vii) volume, and (viii) composition of the gas to be processed; (ix) the carbon dioxide–hydrogen sulfide ratio in the gas; and (x) the desirability of sulfur recovery due to process economics or environmental issues.

In addition to hydrogen sulfide (H_2S) and carbon dioxide (CO_2), natural gas (and other gases: processes gases, biogases, and landfill gases) may contain other contaminants, such as mercaptans (RSH) and carbonyl sulfide (COS). Higher amounts of these contaminants may occur in biogas and landfill, depending upon the source of these gases and the process parameters that were operative when they were produced.

Process selectivity indicates the preference with which the process removes one acid gas component relative to (or in preference to) another. For example, some processes remove both hydrogen sulfide and carbon dioxide; other processes are designed to remove hydrogen sulfide only. It is important to consider the process selectivity for, say, hydrogen sulfide removal compared with carbon dioxide removal that ensures minimal concentrations of these components in the product, thus the need for consideration of the carbon dioxide to hydrogen sulfide in the gas stream.

Gas processing involves the use of several different types of processes, but there is always an overlap between the individual processing concepts. In addition, the terminology used for gas processing can often be confusing and/or misleading because of the overlap (Nonhebel, 1964; Curry, 1981; Maddox, 1982; Mokhatab et al., 2006; Speight, 2007, 2014a).

At power plants, flue gas (Chapter 8) is often treated with a series of chemical processes and scrubbers, which remove pollutants. Most flue gas desulfurization systems employ two stages: (i) one for fly ash removal and (ii) the other for sulfur dioxide removal. For example, as environmental regulations regarding sulfur dioxide emissions have been enacted and enforced in many countries, sulfur dioxide is removed from flue gases by a variety of methods. Common methods use (i) wet scrubbing, which uses a slurry of alkaline sorbent, usually limestone or lime, or seawater to scrub the gases; (ii) spray-dry scrubbing, which uses similar sorbent slurries as described in the first category; (iii) a wet sulfuric acid process, which allows the recovery of sulfur in the form of commercial quality sulfuric acid; (iv) a process commonly referred to as a SNOX flue gas desulfurization process, which removes sulfur dioxide, nitrogen oxides, and particulate matter from flue gases; and (v) a dry sorbent injection process, which introduces powdered hydrated lime or other sorbent material into the exhaust ducts to eliminate sulfur dioxide and sulfur trioxide from the process emissions.

In wet scrubbing systems, the flue gas normally passes first through a fly ash removal device, either an electrostatic precipitator or a baghouse, and then into an absorber that removes sulfur dioxide. However, in dry injection or spray drying operations, the sulfur dioxide is first reacted with the lime, and then the flue gas passes through a particulate control device. Another important design consideration associated with wet flue gas desulfurization systems is that the flue gas exiting the absorber is saturated with water and still contains some sulfur dioxide. These gases are highly corrosive to any downstream equipment such as fans, ducts, and stacks, and two methods that may minimize corrosion are (i) reheating the gases to above the dew point or (ii) using materials of construction and designs that allow equipment to withstand the corrosive conditions.

Electrostatic precipitators or fabric filters remove particulate matter, and flue gas desulfurization units capture the sulfur dioxide produced by burning fossil fuels, particularly coal. Nitrogen oxides are treated either by modifications to the combustion process to prevent their formation or by high-temperature or catalytic reaction with ammonia (NH_3) or urea (H_2NCONH_2). In either case, the aim is to produce nitrogen gas, rather than nitrogen oxides. In the United States, there is a rapid deployment of technologies to remove mercury from flue gas. This is typically accomplished by absorption on sorbents or by capture in inert solids as part of the flue gas desulfurization process. Such scrubbing can lead to meaningful recovery of sulfur for further industrial use (Mokhatab et al., 2006; Speight, 2007, 2014a).

Technologies based on regenerative capture by amines (usually called *olamine processes*) for the removal of carbon dioxide from flue gas have been deployed to provide high-purity carbon dioxide for further industrial use as well as for enhanced oil recovery (Speight, 2014a).

Nevertheless, there are four general processes used for emission control (often referred to in another more specific context as flue gas desulfurization): (i) adsorption, (ii) absorption, (iii) catalytic oxidation, and (iv) thermal oxidation (Soud and Takeshita, 1994; Mokhatab et al., 2006; Speight, 2007, 2014a).

4.2.1 Gas Streams from Natural Gas

Natural gas is also capable of producing emissions that are detrimental to the environment. While the major constituent of natural gas is methane, there are components such as carbon dioxide (CO), hydrogen sulfide (H_2S), and mercaptans (thiols; R—SH), as well as trace amounts of sundry other emissions such as carbonyl sulfide (COS). The fact that methane has a foreseen and valuable end use makes it a desirable product, but in several other situations, it is considered a pollutant, having been identified as a greenhouse gas.

A sulfur removal process must be very precise, since natural gas contains only a small quantity of sulfur-containing compounds that must be reduced in several orders of magnitude. Most consumers of natural gas require less than 4 ppm in the gas.

A characteristic feature of natural gas that contains hydrogen sulfide is the presence of carbon dioxide (generally in the range of 1–4% by volume). In cases where the natural gas does not contain hydrogen sulfide, there may also be a relative lack of carbon dioxide.

In practice, heaters and scrubbers are installed, usually at or near the wellhead. The scrubbers serve primarily to remove sand and other large-particle impurities, and the heaters ensure that the temperature of the gas does not drop too low. With natural gas that contains even low quantities of water, natural gas hydrates tend to form when temperatures drop. These hydrates are solid or semisolid compounds, resembling icelike crystals. If the hydrates accumulate, they can impede the passage of natural gas through valves and gathering systems (Zhang et al., 2007). To reduce the occurrence of hydrates, small natural gas-fired heating units are typically installed along the gathering pipe wherever it is likely that hydrates may form.

While natural gas hydrates are usually considered as possible nuisances in the development of oil and gas fields, mainly in deep-water drilling operations, hydrates can be used for the safe and economic storage of natural gas, mainly in cold countries. In remote offshore areas, the use of hydrates for natural gas transportation is also presently considered as an economic alternative to the processes based either on liquefaction or on compression (Lachet and Béhar, 2000).

4.2.2 Gas Streams from Crude Oil

In order to process and transport associated dissolved natural gas, it must be separated from the oil in which it is dissolved. This separation of natural gas from oil is most often done using equipment installed at or near the wellhead. The actual process used to separate oil from natural gas, as well as the equipment that is used, can vary widely. Although dry pipeline quality natural gas is virtually identical across different geographic areas, raw natural gas from different regions will vary in composition (Chapter 1) and therefore separation requirements may emphasize or deemphasize the optional separation processes.

In many instances, natural gas is dissolved in oil underground primarily due to the formation pressure. When this natural gas and oil is produced, it is possible that it will separate on its own, simply due to decreased pressure; much like opening a can of soda pop allows the release of dissolved carbon dioxide. In these cases, separation of oil and gas is relatively easy, and the two hydrocarbons are sent separate ways for further processing. The most basic type of separator is known as a conventional separator. It consists of a simple closed tank, where the force of gravity serves to separate the heavier liquids like oil, and the lighter gases, like natural gas.

In certain instances, however, specialized equipment is necessary to separate oil and natural gas. An example of this type of equipment is the low-temperature separator. This is most often used for wells producing high-pressure gas along with light crude oil or condensate. These separators use pressure differentials to cool the wet natural gas and separate the oil and condensate. Wet gas enters the separator, being cooled slightly by a heat exchanger. The gas then travels through a high-pressure liquid vessel (often referred to as a *knockout pot*) that serves to remove any liquids into a low-temperature separator. The gas then flows into this low-temperature separator through a choke mechanism, which expands the gas as it enters the separator. This rapid expansion of the gas allows for the lowering of the temperature in the separator. After liquid removal, the dry gas then travels back through the heat exchanger and is warmed by the incoming wet gas. By varying the pressure of the gas in various sections of the separator, it is possible to vary the temperature, which causes the oil and some water to be condensed out of the wet gas stream. This basic pressure–temperature relationship can work in reverse as well to extract gas from a liquid oil stream.

On the other hand, crude oil refining produces gas streams that contain substantial amounts of acid gases such as hydrogen sulfide and carbon dioxide. These gas streams are produced during initial distillation of the crude oil and during the various conversion processes. Of particular interest is the hydrogen sulfide (H_2S) that arises from the hydrodesulfurization of feedstocks that contain organic sulfur:

$$[S]_{feedstock} + H_2 \rightarrow H_2S + \text{hydrocarbons}$$

The terms *refinery gas* and *process gas* are also often used to include all of the gaseous products and by-products that emanate from a variety of refinery processes (Table 4.2). There are also components of the gaseous products that must be removed prior to release of the gases to the atmosphere or prior to use of the gas in another part of the refinery, i.e. as a fuel gas or as a process feedstock.

Crude oil refining involves, with the exception of heavy crude oil, *primary distillation* (Speight, 2014a, 2017) that results in separation into fractions differing in carbon number, volatility, specific gravity, and other characteristics.

Table 4.2 Low-boiling products of a crude oil refinery (excluding hydrogen sulfide, sulfur oxides, and carbon dioxide).

Hydrocarbon	Carbon atoms	Mol. wt.	Boiling range 0 °C	Boiling range °F	Uses
Methane	1	16	−182	−296	Fuel gas
Ethane	2	30	−89	−128	Fuel gas
Ethylene	2	28	−104	−155	Fuel gas, petrochemicals
Propane	3	44	−42	−44	Fuel gas, LPG
Propylene	3	42	−48	−54	Fuel gas, petrochemicals, polymer gasoline
Isobutane	4	58	−12	11	Alkylate, motor gasoline
n-Butane	4	58	−1	31	Automotive gasoline
Isobutylene	4	56	−7	20	Synthetic rubber and chemicals, polymer gasoline, alkylate, motor gasoline
Butylene-1[a]	4	56	−6	21	Synthetic rubber and chemicals
Butylene-2[a]	4	56	1	34	Alkylate, polymer gasoline, motor gasoline

[a]Numbers refer to the positions of the double bond.

The most volatile fraction, which contains most of the gases that are generally dissolved in the crude, is referred to as *pipestill gas* or *pipestill light ends* and consists essentially of hydrocarbon gases ranging from methane to butane(s), or sometimes pentane(s).

The gas varies in composition and volume, depending on crude origin and on any additions to the crude made at the loading point. It is not uncommon to re-inject low-boiling hydrocarbons such as propane and butane into the crude oil before transportation by tanker or by pipeline. This results in a higher vapor pressure of the crude oil that is measurable (ASTM D5191, 2017, ASTM D5482, 2017, ASTM D6377, 2017, ASTM D7975, 2017) (Rey, 2008), but it allows a definitive increase in the quantity of low molecular weight products obtained at the refinery. Since light ends (hydrocarbon gases) in most crude oil markets command a premium, while in the oil field itself propane and butane may have to be re-injected or flared, *spiking* crude oil with liquefied crude oil gas is a common practice.

Vapor pressure is a very important physical property of volatile liquids and is typically regulated by government agencies in various countries. The specifications for volatile crude oil products generally include vapor pressure limits to ensure products of suitable volatility performance. Thus, the determination of the vapor pressure of natural gas and other gaseous fuels (ASTM D4953, 2017, ASTM D5191, 2017) is an important aspect of gas analysis as is used to determine the correct sequence of units in gas processing.

The former test method (ASTM D4953, 2017) is particularly applicable to gasoline that has a vapor pressure range from 5 to 15 psi. This method, a modification of the Reid vapor pressure test method (ASTM D323, 2017), provides two procedures to determine the vapor pressure of gasoline. The method is also applicable to low-boiling naphtha fractions, gas condensate, and natural gasoline (Chapter 10). However, because the external atmospheric pressure is counteracted by the atmospheric pressure initially present in the air chamber, this vapor pressure is an absolute pressure at 37.8 °C (100 °F) in pounds-force per square inch. This vapor pressure differs from the true vapor pressure of the sample due to some small vaporization of the sample and air in the confined space.

The alternate test method (ASTM D5191, 2017) is more precise (than ASTM D4953, 2017) and requires a small sample size (1–10 ml) with about 7 min to complete the test. This method covers the use of automated vapor pressure instruments to determine the total vapor pressure exerted in vacuum by air-containing, volatile, liquid crude oil products, including automotive spark-ignition fuels. Furthermore, the method is suitable for testing samples with boiling points above 0 °C (32 °F) that exert a vapor pressure between 7 and 130 kPa (1.0 and 18.6 psi) at 37.8 °C (100 °F) at a vapor-to-liquid ratio of 4 : 1. Measurements are made on liquid sample sizes in the range from 1 to 10 ml, but, it must be noted, no account is made for dissolved water in the sample. As mentioned earlier, the method is also applicable to low-boiling naphtha fractions, gas condensate, and natural gasoline (Chapter 10).

In addition to the gases obtained by distillation of crude oil, more highly volatile products result from the subsequent processing of naphtha and middle distillate to produce gasoline. Hydrogen sulfide is produced in the desulfurization processes involving hydrogen treatment of naphtha, distillate, and residual fuel and from the coking or similar thermal treatments of vacuum gas oils and residua (Speight, 2014a, 2017). The most common processing step in the production of gasoline is the catalytic reforming of hydrocarbon fractions in the heptane (C_7) to decane (C_{10}) range.

Additional gases are produced in *thermal cracking processes*, such as the coking or visbreaking processes for the processing of heavy feedstocks (Parkash, 2003; Gary et al., 2007; Speight, 2014a; Hsu and Robinson, 2017; Speight, 2017). In the visbreaking process, fuel oil is passed through externally fired tubes and undergoes liquid phase cracking reactions, which result in the formation of lighter fuel oil components. Oil viscosity is thereby reduced, and some gases, mainly hydrogen, methane, and ethane, are formed. Substantial quantities of both gas and carbon are also formed in coking (both fluid coking and delayed coking) in addition to the middle distillate and naphtha. When coking a residual fuel oil or heavy gas oil, the feedstock is preheated and contacted with hot carbon (coke) that causes extensive cracking of the feedstock constituents of higher molecular weight to produce lower molecular weight products ranging from methane, via liquefied crude oil gas(es) and naphtha, to gas oil and heating oil. Products from coking processes tend to be unsaturated, and olefin components predominate in the tail gases from coking processes.

Another group of refining operations that contributes to gas production is that of the *catalytic cracking processes* (Parkash, 2003; Gary et al., 2007; Speight, 2014a; Hsu and Robinson, 2017; Speight, 2017). These consists of fluid-bed catalytic cracking, and there are many process variants in which heavy feedstocks are converted into cracked gas, liquefied crude oil gas, catalytic naphtha, fuel oil, and coke by contacting the heavy hydrocarbon with the hot catalyst. Both catalytic and thermal cracking processes, the latter being now largely used for the production of chemical raw materials, result in the formation of not only unsaturated hydrocarbons, particularly ethylene ($CH_2{=}CH_2$), but also propylene (propene, $CH_3CH{=}CH_2$), *iso*-butylene [*iso*-butene, $(CH_3)_2C{=}CH_2$], and *n*-butenes ($CH_3CH_2CH{=}CH_2$, and $CH_3CH{=}CHCH_3$) in addition to hydrogen (H_2), methane (CH_4) and smaller quantities of ethane (CH_3CH_3), propane ($CH_3CH_2CH_3$), and butanes [$CH_3CH_2CH_2CH_3$, $(CH_3)_3CH$]. Diolefins such as butadiene ($CH_2{=}CHCH{=}CH_2$) are also present.

Another source of refinery gas is *hydrocracking*, a catalytic high-pressure pyrolysis process in the presence of fresh and recycled hydrogen (Parkash, 2003; Gary et al., 2007; Speight, 2014a; Hsu and Robinson, 2017; Speight, 2017). The feedstock is again heavy gas oil or residual fuel oil, and the process is mainly directed at the production of additional middle distillates and gasoline. Since hydrogen is to be recycled, the gases produced in this process again have to be separated into lighter and heavier streams; any surplus recycle gas and the liquefied crude oil gas from the hydrocracking process are both saturated.

In a series of *reforming processes* (Parkash, 2003; Gary et al., 2007; Speight, 2014a; Hsu and Robinson, 2017; Speight, 2017), commercialized under names such as *platforming*, paraffin, and naphthene (cyclic nonaromatic) hydrocarbons are converted in the presence of hydrogen and a catalyst are converted into aromatics or isomerized to more highly branched hydrocarbons. Catalytic reforming processes thus not only result in the formation of a liquid product of higher octane number but also produce substantial quantities of gases. The latter not only are rich in hydrogen, but also contain hydrocarbons from methane to butanes, with a preponderance of propane ($CH_3CH_2CH_3$), *n*-butane ($CH_3CH_2CH_2CH_3$), and *iso*-butane [$(CH_3)_3CH$].

The composition of the process gas varies in accordance with reforming severity and reformer feedstock. All catalytic reforming processes require substantial recycling of a hydrogen stream. Therefore, it is normal to separate reformer gas into a propane ($CH_3CH_2CH_3$) and/or a butane stream [$CH_3CH_2CH_2CH_3$ plus $(CH_3)_3CH$], which becomes part of the refinery liquefied crude oil gas production, and a lighter gas fraction, part of which is recycled. In view of the excess of hydrogen in the gas, all products of catalytic reforming are saturated, and there are usually no olefin gases present in either gas stream.

Both hydrocracker gases and catalytic reformer gases are commonly used in catalytic desulfurization processes. In the latter, feedstocks ranging from light to vacuum gas oils are passed at pressures of 500–1000 psi with hydrogen over a hydrofining catalyst. This results mainly in the conversion of organic sulfur compounds to hydrogen sulfide,

$$[S]_{feedstock} + H_2 \rightarrow H_2S + \text{hydrocarbons}$$

This process also produces some light hydrocarbons by hydrocracking.

The presence of impurities in gas streams may eliminate some of the sweetening processes, since some processes remove large amounts of acid gas but not to a sufficiently low concentration. On the other hand, there are those processes not designed to remove (or incapable of removing) large amounts of acid gases, whereas they are capable of removing the acid gas impurities to very low levels when the acid gases are present only in low to medium concentration in the gas.

From an environmental viewpoint, it is not the means by which these gases can be utilized, but it is the effects of these gases on the environment when they are introduced into the atmosphere. Thus, in addition to the corrosion of equipment of acid gases, the escape into the atmosphere of sulfur-containing gases can eventually lead to the formation of the constituents of acid rain, i.e.

the oxides of sulfur (SO_2 and SO_3). Similarly, the nitrogen-containing gases can also lead to nitrous and nitric acids (through the formation of the oxides NO_x, where $x = 1$ or 2), which are the other major contributors to acid rain. The release of carbon dioxide and hydrocarbons as constituents of refinery effluents can also influence the behavior and integrity of the ozone layer.

Finally, another acid gas, hydrogen chloride (HCl), although not usually considered to be a major emission, is produced from mineral matter and the brine that often accompany crude oil during production and is gaining increasing recognition as a contributor to acid rain. However, hydrogen chloride may exert severe local effects because it does not need to participate in any further chemical reaction to become an acid. Under atmospheric conditions that favor a buildup of stack emissions in the areas where hydrogen chloride is produced, the amount of hydrochloric acid in rain water could be quite high.

From an environmental viewpoint, crude oil processing can result in a variety of gaseous emissions. It is a question of degree insofar as the composition of the gaseous emissions may vary from process to process, but the constituents are, in the majority of cases, the same.

Thus, before any gas is subjected to the various aspects of gas processing, it is essential to know the properties of the gas, such as (i) the density, (ii) the vapor pressure, and (iii) the potential for a residue as well as other relevant properties, such as vapor pressure (ASTM D1267, 2017), density or relative density (ASTM D1657, 2017), a residue in the liquefied gas (ASTM D2158, 2017), composition including soluble gases in the liquefied gas (ASTM D2163, 2017), and the means of interconversion of gas volume to liquid volume and vice versa (ASTM D2421, 2017).

4.3 Water Removal

Water is a common impurity in gas streams, and removal of water is necessary to prevent condensation of the water and the formation of ice or gas hydrates ($C_nH_{2n+2}{\cdot}xH_2O$). Water in the liquid phase causes corrosion or erosion problems in pipelines and equipment, particularly when carbon dioxide and hydrogen sulfide are present in the gas. The simplest method of water removal (dehydration) is to cool the gas (refrigeration or cryogenic separation) to a temperature at least equal to or (preferentially) below the dew point.

In addition to separating crude oil and some condensate from the wet gas stream, it is necessary to remove most of the associated water. Most of the liquid, free water associated with extracted natural gas is removed by simple separation methods at or near the wellhead. However, the removal of the water vapor that exists in solution in natural gas requires a more complex treatment. This treatment consists of dehydrating the natural gas, which usually involves one of two processes: either absorption or adsorption. Absorption occurs when the water vapor is taken out by a dehydrating agent, whereas adsorption occurs when the water vapor is condensed from the gas stream and collected on the surface.

In a majority of cases, cooling alone is insufficient and, for the most part, impractical for use in field operations. Other more convenient water removal options use (i) *hygroscopic* liquids (e.g. diethylene glycol [DEG] or triethylene glycol [TEG]) and (ii) solid adsorbents or desiccants (e.g. alumina, silica gel, and molecular sieves). Ethylene glycol can be directly injected into the gas stream in refrigeration plants.

4.3.1 Absorption

Absorption differs from *adsorption*, in that it is not a physical–chemical surface phenomenon, but an approach in which the absorbed gas is ultimately distributed throughout the absorbent (liquid). The process depends only on physical solubility and may include chemical reactions in the liquid phase (*chemisorption*). Common absorbing media used are water, aqueous amine solutions, caustic, sodium carbonate, and nonvolatile hydrocarbon oils, depending on the type of gas to be absorbed. Usually, the gas–liquid contactor designs that are employed are plate columns or packed beds (Mokhatab et al., 2006; Speight, 2007, 2014a). Thus, absorption is achieved by dissolution (a physical phenomenon) or by reaction (a chemical phenomenon) (Barbouteau and Dalaud, 1972; Ward, 1972; Mokhatab et al., 2006; Speight, 2007, 2014a). Chemical adsorption processes adsorb sulfur dioxide onto a carbon surface where it is oxidized (by oxygen in the flue gas) and absorbs moisture to give sulfuric acid impregnated into and on the adsorbent.

Liquid absorption processes (which usually employ temperatures below 50 °C (120 °F) are classified either as *physical solvent processes* or *chemical solvent processes*. The former processes employ an organic solvent, and absorption is enhanced by low temperatures, or high pressure, or both. Regeneration of the solvent is often accomplished readily (Staton et al., 1985; Mokhatab et al., 2006; Speight, 2007, 2014a). In chemical solvent processes, absorption of the acid gases is achieved mainly by use of alkaline solutions such as amines or carbonates (Kohl and Riesenfeld, 1985). Regeneration (desorption) can be brought about by use of reduced pressures and/or high temperatures, whereby the acid gases are stripped from the solvent.

Solvents used for emission control processes should have (i) a high capacity for acid gas, (ii) a low tendency to dissolve hydrogen, (iii) a low tendency to dissolve low molecular weight hydrocarbons, (iv) a low vapor pressure at operating temperatures to minimize solvent losses, (v) a low viscosity, (vi) a low thermal stability, (vii) an absence of reactivity toward gas components, (viii) a low tendency for fouling, and (ix) a low tendency for corrosion and should be (x) economically acceptable (Mokhatab et al., 2006; Speight, 2007, 2014a).

Amine washing of gas emissions involves chemical reaction of the amine with any acid gases with the liberation of an appreciable amount of heat and it is necessary to compensate for the absorption of heat. Amine derivatives such as ethanolamine (monoethanolamine [MEA], diethanolamine (DEA), triethanolamine (TEA), methyldiethanolamine (MDEA)), diisopropanolamine (DIPA), and diglycolamine (DGA) have been used in commercial applications (Katz, 1959; Jou et al., 1985; Kohl and Riesenfeld, 1985; Maddox et al., 1985; Polasek and Bullin, 1985;Pitsinigos and Lygeros, 1989; Mokhatab et al., 2006; Speight, 2007, 2014a).

The chemistry can be represented by simple equations for low partial pressures of the acid gases:

$$2RNH_2 + H_2S \rightarrow (RNH_3)_2 S$$

$$2RHN_2 + CO_2 + H_2O \rightarrow (RNH_3)_2 CO_3$$

At high acid gas partial pressure, the reactions will lead to the formation of other products:

$$(RNH_3)_2 S + H_2S \rightarrow 2RNH_3HS$$

$$(RNH_3)_2 CO_3 + H_2O \rightarrow 2RNH_3HCO_3$$

The reaction is extremely fast, the absorption of hydrogen sulfide being limited only to mass transfer; this is not so for carbon dioxide. Regeneration of the solution leads to near-complete desorption of carbon dioxide and hydrogen sulfide. A comparison between MEA, DEA, and DIPA shows that MEA is the cheapest of the three but shows the highest heat of reaction and corrosion; the reverse is true for DIPA.

An example of absorption dehydration is known as *glycol dehydration*, and DEG ($HOCH_2CH_2CH_2CH_2OH$), the principal agent in this process, has a chemical affinity for water and removes water from the gas stream. In this process, a liquid desiccant dehydrator serves to absorb water vapor from the gas stream. In the process, glycol dehydration involves using a glycol solution, usually either DEG or TEG ($HOCH2CH2CH2CH_2CH_2CH_2OH$), which is brought into contact with the wet gas stream in a *contactor*. The glycol solution absorbs the water from the wet gas and, once absorbed, the glycol–water mix increases in density (becomes heavier) and sinks to the bottom of the contactor where it (the mixture) is removed. The natural gas, having been stripped of most of its water content, is then transported out of the dehydrator. The glycol solution, bearing all of the water stripped from the natural gas, is put through a specialized boiler designed to vaporize only the water out of the solution. The boiling point differential between water (100 °C, 212 °F) and glycol (204 °C, 400 °F) makes it relatively easy to remove water from the glycol solution, allowing it to be reused in the dehydration process.

As well as absorbing water from the wet gas stream, the glycol solution occasionally carries with it small amounts of methane and other compounds found in the wet gas. In the past, this methane was simply vented out of the boiler. In addition to losing a portion of the natural gas that was extracted, this venting contributes to air pollution and the greenhouse effect. In order to decrease the amount of methane and other compounds that are lost, flash tank separator–condensers work to remove these compounds before the glycol solution reaches the boiler. Essentially, a flash tank separator consists of a device that reduces the pressure of the glycol solution stream, allowing the methane and other hydrocarbons to vaporize (*flash*).

After dehydration, the glycol solution then travels to the boiler, which may also be fitted with air or water-cooled condensers, which serve to capture any remaining organic compounds that may remain in the glycol solution. The regeneration (stripping) of the glycol is limited to temperature: DEG and TEG decompose at or before their respective boiling points. Such techniques as stripping of hot TEG with dry gas or vacuum distillation are recommended. In practice, absorption systems recover 90–99% v/v of methane that would otherwise be flared into the atmosphere.

4.3.2 Adsorption

Adsorption is a physical–chemical phenomenon in which the gas is concentrated on the surface of a solid or liquid to remove impurities. Removal of water by adsorption of the water onto a solid adsorbent (often referred to as solid-desiccant dehydration) is another option for the dehydration of gas streams.

In the process, it is typically carbon that is the adsorbing medium – the carbon can be regenerated by *desorption* of the adsorbed material (Fulker, 1972; Mokhatab et al., 2006; Speight, 2007, 2014a). The quantity of material adsorbed is proportional to the surface area of the solid, and, consequently, adsorbents are usually granular solids with a large surface area per unit mass. Subsequently, the captured gas can be desorbed with hot air or steam either for recovery or for thermal destruction.

Adsorber units are widely used to increase a low gas concentration prior to incineration unless the gas concentration is very high in the inlet air stream. Adsorption also is employed to reduce problem odors from gases. There are several limitations to the use of adsorption systems, but it is generally felt that the major one is the requirement for minimization of particulate matter and/or condensation of liquids (e.g. water vapor) that could mask the adsorption surface and drastically reduce its efficiency (Table 4.1). Thus, in any gas processing plant, it is necessary to know the composition of the gas not only entering the plant but also entering (and exiting) each of the unit processes that constitute the gas processing plant and methods involving gas chromatographic analysis that are particularly useful (ASTM D1945, 2017, ASTM D1946, 2017,

ASTM D2597, 2017). This type of analytical data can prevent overloading the unit process and (through the removal of corrosive constituents) mitigate the potential for equipment corrosion.

Solid-adsorbent or solid-desiccant dehydration is the primary form of dehydrating natural gas using adsorption and usually consists of two or more adsorption towers, which are filled with a solid desiccant. Typical desiccants include activated alumina or a granular silica gel material. Wet natural gas is passed through these towers, from top to bottom. As the wet gas passes around the particles of desiccant material, water is retained on the surface of these desiccant particles. Passing through the entire desiccant bed, almost all of the water is adsorbed onto the desiccant material, leaving the dry gas to exit the bottom of the tower.

Solid-**desiccant** dehydrators are typically more effective than glycol dehydrators and are usually installed as a type of straddle system along natural gas pipelines. These types of dehydration systems are best suited for large volumes of gas under very high pressure and are thus usually located on a pipeline downstream of a compressor station. Two or more towers are required due to the fact that after a certain period of use, the desiccant in a particular tower becomes saturated with water. To "regenerate" the desiccant, a high-temperature heater is used to heat gas to a very high temperature. Passing this heated gas through a saturated desiccant bed vaporizes the water in the desiccant tower, leaving it dry and allowing for further natural gas dehydration.

Molecular sieves – a class of aluminosilicates that produce the lowest water dew point and that can be used to simultaneously sweeten dry gases and liquids (Maple and Williams, 2008) – are commonly used in dehydrators ahead of plants designed to recover ethane and other NGLs. These plants operate at very cold temperatures and require very dry feed gas to prevent formation of hydrates. Dehydration to −100 °C (−148 °F) dew point is possible with molecular sieves. Water dew points less than −100 °C (−148 °F) can be accomplished with special design and definitive operating parameters (Mokhatab et al., 2006; Speight, 2007, 2014a).

Although two-bed adsorbent treaters have become more common (while one bed is removing water from the gas, the other undergoes alternate heating and cooling), on occasion, a three-bed system is used: one bed adsorbs, one is being heated, and one is being cooled. An additional advantage of the three-bed system is the facile conversion of a two-bed system so that the third bed can be maintained or replaced, thereby ensuring continuity of the operations and reducing the risk of a costly plant shutdown.

Silica gel (SiO_2) and alumina (Al_2O_3) have good capacities for water adsorption (up to 8% by weight). Bauxite (crude alumina, Al_2O_3) adsorbs up to 6% by weight water, and molecular sieves adsorb up to 15% by weight water. Silica is usually selected for dehydration of sour gas because of its high tolerance to hydrogen sulfide and to protect molecular sieve beds from plugging by sulfur.

Alumina *guard beds* (which serve as protectors by the act of attrition and may be referred to as an *attrition catalyst*) (Speight, 2000) may be placed ahead of the molecular sieves to remove the sulfur compounds. Downflow reactors are commonly used for adsorption processes, with an upward flow regeneration of the adsorbent and cooling in the same direction as adsorption.

Solid-desiccant units generally cost more to buy and operate than glycol units. Therefore, their use is typically limited to applications such as gases having a high hydrogen sulfide content, very low water dew point requirements, simultaneous control of water, and hydrocarbon dew points. In processes where cryogenic temperatures are encountered, solid-desiccant dehydration is usually preferred over conventional methanol injection to prevent hydrate and ice formation (Kindlay and Parrish, 2006).

4.3.3 Chemical Treatment

In *chemical conversion processes*, contaminants in gas emissions are converted to compounds that are not objectionable or that can be removed from the stream with greater ease than the original constituents. For example, a number of processes have been developed that remove hydrogen sulfide and sulfur dioxide from gas streams by absorption in an alkaline solution.

Carbonate washing is a mild alkali process for emission control by the removal of acid gases (such as carbon dioxide and hydrogen sulfide) from gas streams (Mokhatab et al., 2006; Speight, 2007, 2014a) and uses the principle that the rate of absorption of carbon dioxide by potassium carbonate increases with temperature. It has been demonstrated that the process works best near the temperature of reversibility of the reactions:

$$K_2CO_3 + CO_2 + H_2O \rightarrow 2KHCO_3$$

$$K_2CO_3 + H_2S \rightarrow KHS + KHCO_3$$

Water washing, in terms of the outcome, is analogous to washing with potassium carbonate (Kohl and Riesenfeld, 1985; Mokhatab et al., 2006; Speight, 2007, 2014a), and it is also possible to carry out the desorption step by pressure reduction. The absorption is purely physical, and there is also a relatively high absorption of hydrocarbons, which are liberated at the same time as the acid gases.

Catalytic oxidation is a chemical conversion process that is used predominantly for the destruction of volatile organic compounds and carbon monoxide. These systems operate in a temperature regime of 205–595 °C (400–1100 °F) in the presence of a catalyst. Without the catalyst, the system would require higher temperatures. Typically, the catalysts used are a combination of noble metals deposited on a ceramic base in a variety of configurations (e.g. honeycomb-shaped) to enhance good surface contact.

Catalytic systems are usually classified on the basis of bed types such as *fixed bed* (or *packed bed*) and *fluid bed* (*fluidized bed*). These systems generally have very high destruction efficiencies for most volatile organic compounds, resulting in the formation of carbon dioxide, water, and varying amounts of hydrogen chloride (from halogenated hydrocarbons). The presence in emissions of chemicals such as heavy metals, phosphorus, sulfur, chlorine, and most halogens in the incoming air stream acts as poison to the system and can foul up the catalyst.

Thermal oxidation systems, without the use of catalysts, also involve chemical conversion (more correctly, chemical destruction) and operate at temperatures in excess of 815 °C (1500 °F) or 220–610 °C (395–1100 °F) higher than catalytic systems.

4.3.4 Filters and Scrubbers

Historically, *particulate matter control* (*dust control*) (Mody and Jakhete, 1988) has been one of the primary concerns of industries, since the emission of particulate matter is readily observed through the deposition of fly ash and soot as well as in impairment of visibility. Differing ranges of control can be achieved by use of various types of equipment. Upon proper characterization of the particulate matter emitted by a specific process, the appropriate piece of equipment can be selected, sized, installed, and performance tested. The general classes of control devices for particulate matter are as follows:

Cyclone collectors are the most common of the inertial collector class. Cyclones are effective in removing coarser fractions of particulate matter. The particle-laden gas stream enters an upper cylindrical section tangentially and proceeds downward through a conical section. Particles migrate by centrifugal force generated by providing a path for the carrier gas to be subjected to a vortex-like spin. The particles are forced to the wall and are removed through a seal at the apex of the inverted cone. A reverse-direction vortex moves upward through the cyclone and discharges through a top center opening. Cyclones are often used as primary collectors because of their relatively low efficiency (50–90% is usual). Some small-diameter high-efficiency cyclones are utilized. The equipment can be arranged either in parallel or in series to both increase efficiency and decrease pressure drop. These units for particulate matter operate by contacting the particles in the gas stream with a liquid. In principle the particles are incorporated in a liquid bath or in liquid particles that are much larger and therefore more easily collected.

Fabric filters are typically designed with nondisposable filter bags. As the dusty emissions flow through the filter media (typically cotton, polypropylene, Teflon, or fiberglass), particulate matter is collected on the bag surface as a dust cake. Fabric filters are generally classified on the basis of the filter bag processing mechanism employed. Fabric filters operate with collection efficiencies up to 99.9% although other advantages are evident.

Wet scrubbers are devices in which a countercurrent spray liquid is used to remove particles from an air stream. Device configurations include plate scrubbers, packed beds, orifice scrubbers, venturi scrubbers, and spray towers, individually or in different combinations. Other methods include use of high-energy input *venturi scrubbers* or electrostatic scrubbers where particles or water droplets are charged and flux force/condensation scrubbers where a hot humid gas is contacted with cooled liquid or where steam is injected into saturated gas. In the latter scrubber the movement of water vapor toward the cold-water surface carries the particles with it (*diffusiophoresis*), while the condensation of water vapor on the particles causes the particle size to increase, thus facilitating collection of fine particles. The *foam scrubber* is a modification of a wet scrubber in which the particle-laden gas is passed through a foam generator, where the gas and particles are enclosed by small bubbles of foam.

Electrostatic precipitators operate on the principle of imparting an electric charge to particles in the incoming air stream, which are then collected on an oppositely charged plate across a high-voltage field. Particles of high resistivity create the most difficulty in collection. Conditioning agents such as sulfur trioxide (SO_3) have been used to lower resistivity. Important parameters include design of electrodes, spacing of collection plates, minimization of air channeling, and collection-electrode rapping techniques (used to dislodge particles). Techniques under study include the use of high-voltage pulse energy to enhance particle charging, electron-beam ionization, and wide plate spacing. Electrical precipitators are capable of efficiencies greater than 99% under optimum conditions, but performance is still difficult to predict in new situations.

4.3.5 Membrane Processes

Membrane separation processes are very versatile and are designed to process a wide range of feedstocks and offer a simple solution for removal and recovery of higher boiling hydrocarbons (NGLs) from natural gas (Foglietta, 2004) as well as cleaning other gases such as biogas (Schweigkofler and Niessner, 2001; Popat and Deshusses, 2008; Deng and Hagg, 2010; Matsui and Imamura, 2010). Synthetic membranes are made from a variety of polymers including polyethylene, cellulose acetate, polysulfone, and polydimethylsiloxane (Table 4.3) (Isalski, 1989; Robeson, 1991). The material from which the membrane is manufactured plays an important role in the ability of the membrane to provide the desired performance characteristics. For process optimization, the membrane should have a high permeability as well as sufficient selectivity. It is also important to match the membrane properties to that of the system operating conditions (for example, pressures and gas composition).

The separation process is based on high-flux membranes that selectively permeate higher boiling hydrocarbons (compared with methane) and are recovered as a liquid after recompression and condensation. The residue

Table 4.3 Examples of polymer structures used for the production of membranes.

Polyethylene:

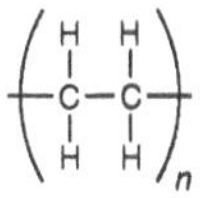

Cellulose acetate:

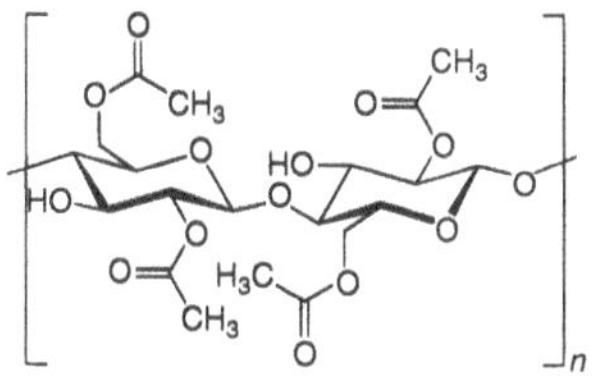

Polysulfone:

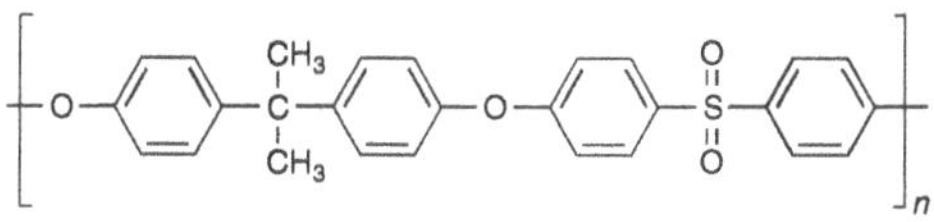

Polydimethylsiloxane:

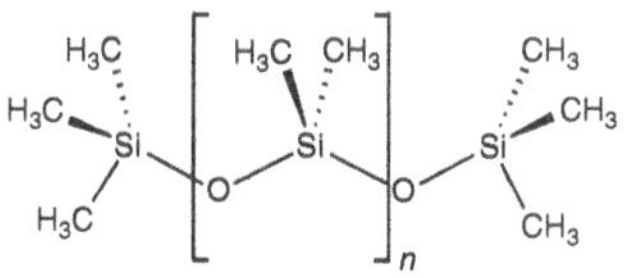

stream from the membrane is partially depleted of higher boiling hydrocarbons and is then sent to sales gas stream. Gas permeation membranes are usually made with vitreous polymers that exhibit good selectivity, but, to be effective, the membrane must be very permeable with respect to the separation process.

Polymeric membranes are a common option for the separation of carbon dioxide from flue gas because of the maturity of the technology in a variety of industries, namely, the petrochemical industry. The ideal polymer membrane has both high selectivity and permeability. Polymer membranes are examples of systems that are dominated by the solution–diffusion mechanism. The membrane is considered to have holes that the gas can dissolve (solubility) and the molecules can move from one cavity to the other (diffusion).

Silica membranes can be produced with high uniformity (the same structure throughout the membrane), and the high porosity of these membranes is accompanied by high permeability. Synthesized membranes have smooth surfaces and can be modified on the surface to drastically improve selectivity. For example, silica membrane surfaces that have been functionalized by inclusion of an amine (on the surface) allow the membranes to separate carbon dioxide from flue gas streams more effectively (Jang et al., 2011).

Zeolites (crystalline aluminosilicate minerals) with a regular repeating structure of molecular-sized pores also can be used to produce workable membranes. These membranes selectively separate molecules based on pore size and polarity and are thus highly tunable to specific gas separation processes. In general, smaller molecules and those with stronger zeolite-adsorption properties are adsorbed onto zeolite membranes with larger selectivity. The capacity to discriminate based on both molecular size and adsorption affinity makes zeolite membranes an attractive candidate for carbon dioxide separation from natural gas.

4.4 Liquids Removal

Natural gas coming directly from a well contains many NGLs that are commonly removed. In most instances, NGLs have higher value as separate added-value products, and it is thus economical to remove them from the gas stream. The removal of NGLs usually takes place in a relatively centralized processing plant and uses techniques similar to those used to dehydrate natural gas. Recovery of the liquid hydrocarbons can be justified either because it is necessary to make the gas salable or because economics dictate this course of action. The justification for building a liquid recovery (or a liquid removal) plant depends on the price differential between the enriched gas (containing the higher molecular weight hydrocarbons) and lean gas with the added value of the extracted liquid.

4.4.1 Extraction

There are two basic steps to the treatment of NGLs in the natural gas stream: (i) the liquids must be extracted from the natural gas, and (ii) the NGLs must be separated themselves, down to their base components. These two processes account for approximately 90% v/v of the total production of NGLs.

There are two principle techniques for removing NGLs from the natural gas stream: (i) the absorption method and (ii) the cryogenic expander process.

4.4.1.1 Absorption Process

The absorption method of extraction is very similar to using absorption for dehydration. The main difference is that, in the absorption of NGLs, absorbing oil is used as opposed to glycol. This absorbing oil has an affinity for NGLs in much the same manner as glycol has an affinity for water. Before the oil has picked up any NGLs, it is termed *lean* absorption oil. As the natural gas is passed through an absorption tower, it is brought into contact with the absorption oil that soaks up a high proportion of the NGLs. The rich absorption oil, now containing NGLs, exits the absorption tower through the bottom. It is now a mixture of absorption oil, propane, butanes, pentanes, and other heavier hydrocarbons. The rich oil is fed into lean oil stills, where the mixture is heated to a temperature above the boiling point of the NGLs, but below that of the oil. This process allows for the recovery of around 75% v/v butane derivatives and 85–90% v/v pentane derivatives as well as higher molecular weight hydrocarbons from the natural gas stream.

4.4.1.2 Cryogenic Process

Cryogenic processes are also used to extract NGLs from natural gas. While absorption methods can extract almost all of the higher molecular weight NGLs, the lighter hydrocarbons, such as ethane, are often more difficult to recover from the natural gas stream. In certain instances, it is economic to simply leave the lower molecular weight NGLs in the natural gas stream. However, if it is economic to extract ethane and other lighter hydrocarbons, cryogenic processes are required for high recovery rates. Essentially, cryogenic processes consist of dropping the temperature of the gas stream to approximately −85 °C (−120 °F). The rapid temperature drop condenses ethane and other hydrocarbon derivatives in the gas stream, but methane remains in gaseous form.

The process allows for the recovery of about 90–95% v/v of the ethane originally in the gas stream. The extraction of NGLs from the natural gas stream produces both cleaner, purer natural gas, as well as the valuable hydrocarbons that are the NGLs themselves.

4.4.2 Fractionation of Natural Gas Liquids

After the NGLs have been removed from the natural gas stream, they must be broken down into their base components to be useful. That is, the mixed stream of different NGLs must be separated. The process used to accomplish this task is called fractionation that is based on the different boiling points of the different hydrocarbons in the NGLs stream. Essentially, fractionation occurs in stages consisting of the boiling off of hydrocarbons one by one.

The fractionation process occurs in stages consisting of the boiling off of hydrocarbons one by one. The name of a particular fractionator gives an idea as to its purpose, as it is conventionally named for the hydrocarbon that is boiled off. The entire fractionation process is broken down into steps, starting with the removal of the lighter NGLs from the stream. The fractionators are used in the following order: (i) the deethanizer, which separates the ethane from the NGLs stream; (ii) the depropanizer, which separates propane from the deethanized stream; and (iii) the debutanizer, which removes the butane isomers from the deethanized–depropanized stream and leaves the pentane derivatives and higher molecular weight hydrocarbon derivatives in what reasons of the NGLs stream. There is also a butane splitter (deisobutanizer) that separates the iso and normal butanes.

4.5 Nitrogen Removal

Nitrogen may often occur in sufficient quantities in natural gas and, consequently, lower the heating value of the gas. Thus, several plants for *nitrogen removal* from natural gas have been built, but it must be recognized that nitrogen removal requires liquefaction and fractionation of the entire gas stream, which may affect process economics. In many cases the nitrogen-containing natural gas is blended with a gas having a higher heating value and sold at a reduced price depending upon the thermal value (Btu ft^{-3}).

4.6 Acid Gas Removal

In addition to water and natural gas liquids removal, one of the most important parts of gas processing involves the removal of hydrogen sulfide and carbon dioxide. Natural gas from some wells contains significant amounts of hydrogen sulfide and carbon dioxide and is usually referred to as *sour gas.* Sour gas is undesirable because the sulfur compounds it contains can be extremely harmful, even lethal, to breathe and the gas can also be extremely corrosive. The process for removing hydrogen sulfide from sour gas is commonly referred to as *sweetening* the gas.

4.6.1 Hydrogen Sulfide and Carbon Dioxide Removal

The primary process for sweetening sour natural gas is quite similar to the processes of glycol dehydration and removal of NGLs by absorption. In this case, however, amine (*olamine*) solutions are used to remove the hydrogen sulfide (the *amine process*). The sour gas is run through a tower, which contains the olamine solution. There are two principle amine solutions used, MEA and DEA. Either of these compounds, in liquid form, will absorb sulfur compounds from natural gas as it passes through. The effluent gas is virtually free of sulfur compounds and thus loses its sour gas status. Like the process for the extraction of NGLs and glycol dehydration, the amine solution used can be regenerated for reuse. Although most sour gas sweetening involves the amine absorption process, it is also possible to use solid desiccants like iron sponge (*q.v.*) to remove hydrogen sulfide and carbon dioxide.

Treatment of gas to remove the acid gas constituents (hydrogen sulfide and carbon dioxide) is most often accomplished by contact of the natural gas with an alkaline solution. The most commonly used treating solutions are aqueous solutions of the ethanolamine or alkali carbonates, although a considerable number of other treating agents have been developed in recent years (Mokhatab et al., 2006; Speight, 2007, 2014a). Most of these newer treating agents rely upon physical absorption and chemical reaction. When only carbon dioxide is to be removed in large quantities or when only partial removal is necessary, a hot carbonate solution or one of the physical solvents is the most economical selection.

The most well-known hydrogen sulfide removal process is based on the reaction of hydrogen sulfide with iron oxide (often also called the iron sponge process or the dry box method) in which the gas is passed through a bed of wood chips impregnated with iron oxide. The iron oxide process is the oldest and still the most widely used batch process for sweetening natural gas and NGLs (Zapffe, 1963; Duckworth and Geddes, 1965; and Anerousis and Whitman, 1984). The process was implemented during the nineteenth century. In the process the sour gas is passed down through the bed. In the case where continuous regeneration is to be utilized, a small concentration of air is added to the sour gas before it is processed. This air serves to continuously regenerate the iron oxide, which has reacted with hydrogen sulfide, which serves to extend the on-stream life of a given tower but probably serves to decrease the total amount of sulfur that a given weight of bed will remove.

The process is usually best applied to gases containing low to medium concentrations (300 ppm) of hydrogen sulfide or mercaptans. This process tends to be highly selective and does not normally remove significant quantities of carbon dioxide. As a result, the hydrogen sulfide stream from the process is usually high purity. The use of iron sponge process for sweetening sour gas is based on adsorption of the acid gases on the surface of the solid sweetening agent followed by chemical reaction of ferric oxide (Fe_2O_3) with hydrogen sulfide:

$$2Fe_2O_3 + 6H_2S \rightarrow 2Fe_2S_3 + 6H_2O$$

The reaction requires the presence of slightly alkaline water and a temperature below 43 °C (110 °F) and bed alkalinity (pH + 8–10) should be checked regularly, usually on a daily basis. The pH level should be maintained through the injection of caustic soda with the water. If the gas does not contain sufficient water vapor, water may need to be injected into the inlet gas stream.

The ferric sulfide produced by the reaction of hydrogen sulfide with ferric oxide can be oxidized with air to produce sulfur and regenerate the ferric oxide:

$$2Fe_2S_3 + 3O_2 \rightarrow 2Fe_2O_3 + 6S$$

$$S_2 + 2O_2 \rightarrow 2SO_2$$

The regeneration step is exothermic, and, as a result, air must be introduced slowly so the heat of reaction can be dissipated. If air is introduced quickly the heat of reaction may ignite the bed. Some of the elemental sulfur produced in the regeneration step remains in the bed. After several cycles this sulfur will form a cake on the surface of the ferric oxide, decreasing the reactivity of the bed. Typically, after 10 cycles the bed must be removed, and a new bed introduced into the vessel.

The iron oxide process is one of several metal oxide-based processes that scavenge hydrogen sulfide and organic sulfur compounds (mercaptans) from gas streams through reactions with the solid-based chemical adsorbent (Kohl and Riesenfeld, 1985; Mokhatab et al., 2006; Speight, 2007, 2014a). They are typically nonregenerable, although some are partially regenerable, losing activity upon each regeneration cycle. Most of the processes are governed by the reaction of a metal oxide with hydrogen sulfide to form the metal sulfide. For regeneration, the metal oxide is reacted with oxygen to produce elemental sulfur and the regenerated metal oxide. In addition, to iron oxide, the primary metal oxide used for dry sorption processes is zinc oxide.

Molecular sieves are highly selective for the removal of hydrogen sulfide (as well as other sulfur compounds) from gas streams and over continuously high absorption efficiency. They are also an effective means of water removal and thus offer a process for the simultaneous dehydration and desulfurization of gas. However, gas that has excessively high water content may require upstream dehydration (Mokhatab et al., 2006; Speight, 2007, 2014a).

The *molecular sieve process* is similar to the iron oxide process. Regeneration of the bed is achieved by passing heated clean gas over the bed. As the temperature of the bed increases, it releases the adsorbed hydrogen sulfide into the regeneration gas stream. The sour effluent regeneration gas is sent to a flare stack and up to 2% v/v of the gas seated can be lost in the regeneration process. A portion of the natural gas may also be lost by the adsorption of hydrocarbon components by the sieve (Mokhatab et al., 2006; Speight, 2007, 2014a).

In this process, unsaturated hydrocarbon components, such as olefins and aromatics, tend to be strongly adsorbed by the molecular sieve. Molecular sieves are susceptible to poisoning by such chemicals as glycols and require thorough gas processing methods before the adsorption step. Alternatively, the sieve can be offered some degree of protection by the use of *guard beds* in which a less expensive catalyst is placed in the gas stream before contact of the gas with the sieve, thereby protecting the catalyst from poisoning. This concept is analogous to the use of guard beds or attrition catalysts in the crude oil industry (Speight, 2000).

4.6.2 Claus Process

The disposition of hydrogen sulfide, a toxic gas that originates in crude oils and is also produced in the coking, catalytic cracking, hydrotreating, and hydrocracking processes, is an issue with many refiners. Burning hydrogen sulfide as a fuel gas component or as a flare gas component is precluded by safety and environmental considerations since one of the combustion products is the highly toxic sulfur dioxide (SO_2), which is also toxic. As described above, hydrogen sulfide is typically removed from the refinery light ends gas streams through an olamine process after which application of heat regenerates the olamine and forms an acid gas stream. Following this, the acid gas stream is treated to convert the hydrogen sulfide elemental sulfur and water. The conversion process utilized in most modern refineries is the Claus process or a variant thereof.

The Claus process involves combustion of approximately one-third of the hydrogen sulfide to sulfur dioxide and then reaction of the sulfur dioxide with the remaining hydrogen sulfide in the presence of a fixed bed of activated alumina, cobalt molybdenum catalyst, resulting in the formation of elemental sulfur:

$$2H_2S + 3O_2 \rightarrow 2SO_2 + 2H_2O$$

$$2H_2S + SO_2 \rightarrow 3S + 2H_2O$$

Different process flow configurations are in use to achieve the correct hydrogen sulfide/sulfur dioxide ratio in the conversion reactors.

Overall, conversion of 96–97% of the hydrogen sulfide to elemental sulfur is achievable in a Claus process. If this is insufficient to meet air quality regulations, a Claus process tail gas treater is utilized to remove essentially the entire remaining hydrogen sulfide in the tail gas from the Claus unit. The tail gas treater may employ a proprietary solution to absorb the hydrogen sulfide followed by conversion to elemental sulfur.

4.7 Enrichment

The purpose of *enrichment* is to produce natural gas for sale and enriched tank oil. The tank oil contains more light hydrocarbon liquids than natural crude oil, and the residue gas is drier (leaner, i.e. has lesser amounts of the higher molecular weight hydrocarbons). Therefore, the process concept is essentially the separation of hydrocarbon liquids from the methane to produce a lean, dry gas.

Crude oil enrichment is used when there is no separate market for light hydrocarbon liquids or when the increase in API gravity of the crude provides a substantial increase in the price per unit volume as well as volume of the stock tank oil. A very convenient method of enrichment involves manipulation of the number and operating pressures of the gas–oil separators (traps). However, it must be recognized that alteration or manipulation of the separator pressure affects the gas compression operation as well as influences other processing steps.

One method of removing light ends involves the use of a pressure reduction (vacuum) system. Generally, stripping of light ends is achieved at low pressure, after which the pressure of the stripped crude oil is elevated so that the oil acts as an absorbent. The crude oil, which becomes enriched by this procedure, is then reduced to atmospheric pressure in stages or using fractionation (rectification).

4.8 Epilog

There is a wide variety of fuels currently in widespread use of which the simplest in composition is natural gas that consists primarily of methane but includes several other constituents from ethane to butane (collectively known as NGLs) as well as higher boiling hydrocarbons from pentane to octane (C_8H_{18}) or, in some cases, to dodecane ($C_{12}H_{26}$) decane (collectively known as gas condensate and generally referred to as the C_{5+} fraction). In addition, process gas, biogas, and landfill gas are also gases of interest in this present content.

Information on the composition of natural gas (including the condensate fraction) is generally much more available than the information for liquid fuels. Rarely is the molecular composition of liquid fuels known with any degree of certainty (for example, there are many suppositions that go into describing the molecular composition of, say, fuel oil) because of the complexity of the liquid fuels contain a large (sometime undefinable) number of hydrocarbon species. The most commonly reported composition data are derived from the *ultimate analysis*, which consists of measurements of the elemental composition of the fuel, generally presented as mass fractions (% w/w) of carbon, hydrogen, sulfur, oxygen, nitrogen, and ash, where appropriate as well as the potential for low-boiling (even gaseous) hydrocarbons (ASTM D2427, 2017, ASTM D8110, 2017) (Nadkarni, 2005) that, with suitable choice of the column and instrumental parameters, can also include determination of olefin constituents (ASTM D2163, 2017).

For example, test method ASTM D2427 is a method appropriate for the determination of C_2 through C_5 hydrocarbons in gasoline (and by inference low-boiling naphtha, gas condensate, and natural gasoline) by gas chromatography (GC). However, this test method does not cover the determination of cyclic olefin derivatives, diolefin derivatives, or acetylene derivatives, which are typically minor components in finished gasolines although the presence of acetylene derivatives in low-boiling naphtha fractions (as well as in gas condensate and natural gasoline) is not considered likely. The concentration of the various constituents as determined by this test method is typically reported as:

Ethylene plus ethane
Propane
Propylene
iso-Butane
n-Butane
Butene-1 plus isobutylene
trans-Butene-2
cis-Butene-2
iso-Pentane
3-Methylbutene-1
n-Pentane
Pentene-1
2-Methylbutene-1
trans-Pentene-2
cis-Pentene-2
2-Methylbutene-2

However, as extensive as the method is for the analysis of the relevant sample, it is not suitable for the analysis of samples that contain significant amounts of material boiling lower than ethylene.

In addition to test method ASTM D2427, various elemental analytical techniques such as atomic absorption spectrometry (AAS) (ASTM D3605, 2017; ASTM D4628, 2017), inductively coupled plasma atomic emission spectrometry (ICP-AES) (ASTM D7111, 2017; ASTM D4951, 2017; ASTM D5185, 2017), X-ray fluorescence (XRF) (ASTM D7343, 2017; ASTM D7220, 2017; ASTM D4927, 2017; ASTM D6443, 2017), or graphite furnace atomic absorption spectrometry (GFAAS) (ASTM D6732, 2017) are used for this purpose.

Another test method (ASTM D8110, 2017) can be used for elemental analysis of crude oil products, especially the presence of trace elements in the sample. The method covers the rapid determination of seven elements in distillate crude oil products and describes the procedure for the determination of trace elements in light and middle distillate crude oil products using inductively coupled plasma mass spectrometry (ICP-MS). Test times approximate a few minutes per test specimen, and quantification for most elements is in the low to sub $ng\,g^{-1}$ (ppb mass) range. High analysis sensitivity can be achieved for some elements that are difficult to determine by other techniques.

The heating value (also called the *heat content* or *calorific value*) is a measure of the heat release during complete combustion and is also reported *with* the ultimate analysis. In addition, the properties that influence the handling and use of a particular fuel are frequently measured – for liquid fuels, specific gravity or API gravity, viscosity (possibly at several temperatures), flash point (a measure of the temperature at which the fuel is sufficiently volatile to ignite readily), and distillation profiles (fraction vaporized as a function of temperature) may be reported. From these, data predictability of behavior in use can be estimated.

On the other hand, natural gas, which is predominantly methane, offers more informative data from which behavior can be predicted with a higher degree of certainty that in the case of the liquid fuels. It is often for this reason and the often-erroneous assumption that natural gas will behave in the same manner as methane, but errors (because of erroneous suppositions) have been encountered during the use of the gas and when the gas is released to the environment. For example, the density of any gas compared with the density of air is the *vapor density* and is a very important characteristic of the constituents of natural gas and natural gas constituents (Chapter 8). The C_{3+} constituents (propane and higher molecular weight gaseous constituents) of raw natural gas have a vapor density that shows that these constituents are heavier than air. Thus, when raw natural gas is inadvertently released into the atmosphere, these higher molecular weight constituents will collect in low spots (depressions) in the surface of the Earth and are liable to explosive ignition.

Thus, a statement is often made that *natural gas is lighter than air* (because of the continued insistence by some engineers and scientists that the properties of a mixture are determined by the mathematical average of the properties of the individual constituents of the mixture) is incorrect. Such mathematical bravado and inconsistency of thought is *detrimental to safety* and needs to be qualified by consideration of the available analytical data. If such data are not available, they should be!

In fact, at each stage of natural gas production, wellhead treating, transportation, and processing analysis of the gas by standard test methods is an essential part of the chemistry and technology of natural gas. Use of analytical methods offers vital information about the behavior of natural gas during recovery, wellhead processing, transportation, gas processing, and use. However, the various purposes to be served by the analytical data are listed, and the necessary accuracy with which each constituent of each type of gas must be known in order to serve each specific purpose is then estimated. These estimates afford the criteria by means of which the suitability of analytical methods and apparatus may be judged, but they are subject to revision when more is known about the limiting attainable accuracies of the analytical methods.

Thus, because of the worldwide use of natural gas and the varying composition of gas from different wells, there is a need to apply a series of standard test to the gas in order that it can meet sales specifications (Chapter 9). Natural gas testing includes the analysis of conventional and shale gas, liquefied natural gas, and other hydrocarbon condensates and components. Furthermore, regulation of the environmental practices related to natural gas development can more easily address the regional and state-specific character of the activities, compared with a one-size-fits-all management by level of government or industry. Some of these factors include geology, hydrology, climate, topography, industry characteristics, development history, state legal structures, population density, and local economics. Thus, the regulation of natural gas production is a detailed monitoring of each stage of the development (production, wellhead processing, transportation, gas processing facilities, and sales) through the many controls at the state level.

Each aspect of gas technology requires detailed analysis (and thence knowledge) of the properties of the gas, and to understand the applicability of the analytical methods, it is helpful for the reader to understand (i) the origin of natural gas, (ii) the production of natural gas, (iii) the transportation of natural gas, (iv) the refining of natural gas, including gas cleaning, and (v) the influence of natural gas on the environment. Consequently, an introduction to the essentials of natural gas formation, production, wellhead processing, transportation, gas processing, and use is presented below.

When biogas is sent to processing for impurity removal, the cleaned gas has characteristics of the biogas that are close to the characteristics of a natural gas. In this instance the producer of the biogas can utilize the local gas distribution networks. The gas must be very clean to reach pipeline quality. Water (H_2O), hydrogen sulfide (H_2S), and particulates are removed if present at high levels or if the gas is to be completely cleaned. Carbon dioxide is less frequently removed, but it must also be separated to achieve pipeline quality gas. If the gas is to be used without extensively cleaning, it is sometimes cofired with natural gas to improve combustion. Biogas cleaned up to pipeline quality is called renewable natural gas. In this form the gas can be used in any application in which natural gas is used. The heating value of the raw biogas is on the order of 600 Btu ft^{-3} (5340 cal m^{-3}). The raw gas will burn in an engine, but corrosion can occur unless proper materials of construction are selected.

The previous chapters will serve to introduce the reader to the various aspects of natural gas recovery and processing that are involved in making the gas ready for use by the consumer. The awareness of the various processes involved will assist analysts to design test programs that will produce the needed data for the client. There have also been citations to ASTM standard test methods that produce data by use of highly selective sensitive analytical equipment, such as the so-called piggyback technique of gas chromatography/mass spectrometry (GC/MS) combined. As a quality control step, it may be necessary (on the case of processes gas, biogas, and landfill gas) to derivatize (chemically modify) nonvolatile and heat degradable target chemicals prior to analysis by the two most commonly used instrumental methods in gas analysis, GC or GC/MS.

Gas processing is a necessary segment of the natural gas value chain insofar as the series of unit processes that ensure that the natural gas intended for use is as clean and pure as possible and meets the specifications of being a clean-burning and environmentally sound energy choice. Once the natural gas has been fully processed, and is ready to be consumed, it must be transported from those areas that produce natural gas, to those areas that require it.

Just as gas processing is a necessary segment of the gas value chain, chemical analysis is used to determine the qualitative or quantitative content of a given sample with the help of chemical and physical methods. In fact, the analysis of the chemical and physical properties of natural gas is an essential part of the chemistry and technology of the use of natural gas. The data from the of analytical methods (Part II) offers vital information about the behavior of natural gas during recovery, wellhead processing, transportation, gas processing, and use.

Typically, natural gas is a mixture of light hydrocarbon gases rich in methane – the methane content of natural gases is usually above 75% with C_{5+} fraction (i.e. the fraction commonly known as *gas condensate*) less than 1% v/v. If the mole fraction of hydrogen in a natural gas is less than 4 ppm v/v, it is called "sweet" gas. Dry gases contain no C5+ and have more than 90 mol% methane. The main difference between natural gas and other reservoir fluids is that the amount of C_{5+} or even C_{5+} in the mixture is quite low and the main components are light paraffinic hydrocarbons.

Another type of reservoir fluid that is in gaseous phase under reservoir conditions is gas condensate systems (the C_{5+} fraction). The C_{5+} fraction of a mixture should be treated as an undefined fraction, and its properties may be determined by a series of standard test methods that are different to the test methods used for various gases as well as the liquid constituents of gas streams (Part II). The C_{5+} content is more than that of natural gases and it is about few percent, while its methane content is less than that of natural gas. However, these reservoir fluids generally contain components such as carbon dioxide (CO_2), hydrogen sulfide (H_2S), or nitrogen (N_2). Presence of such compounds affects the properties of the gas mixture.

In the description of the composition of natural gas, the C_{5+} (condensate) fraction is a relatively narrow boiling range fraction, and characterization methods are available that can be used to determine the various properties of this fraction. However, contrary to the C_{5+} fraction of natural gas, the standard test methods applied to this condensate are not the typical standard test methods applied to crude oil liquids, but should be selected with discretion from the test methods to be applied to low-boiling naphtha that, even when selectively chosen, may require some modification for application to the condensate.

Gases at atmospheric pressure condition have much larger free space between molecules than do liquids. As a result, the interaction between various like and unlike molecules in a gaseous state is less than the molecular interactions in similar liquid mixtures. Therefore, the role of natural gas composition on properties of gas mixtures is not as evident as in the case of liquids. However, the effect of composition on properties of gas mixtures increases as pressure increases and free space between molecules decreases. The role of composition on properties of dense gases cannot be ignored. Under low-pressure conditions where most gases behave like ideal gases, all gas mixtures regardless of their composition have the same molar density at the same temperature and pressure.

Generally, volume and mole fractions are used interchangeably for all types of gas mixtures. Composition of gas mixtures is rarely expressed in terms of weight fraction and this type of composition has very limited application for gas systems. Whenever composition in a gas mixture is expressed only in percentage, it should be considered as mole % or volume % (i.e. % v/v).

In all aspects of natural gas analysis (as well as the analysis of other gaseous products and fuels) technology, including the calculation of physical properties, accurate characterization of the gas mixture requires (i) assurance that the correct sampling methods have been used (Chapter 5) (ASTM F307, 2017 Standard Practice for Sampling Pressurized Gas for Gas Analysis), (ii) the appropriate type of equipment (ASTM E355, 2017 Standard Practice for Gas Chromatography Terms and Relationships) leading to a complete compositional analysis of the mixture with known mole fraction and carbon number of the gas (ASTM D7833, 2017 Standard Test Method for Determination of Hydrocarbons and Non-Hydrocarbon Gases in Gaseous Mixtures by Gas Chromatography), and method for calculation of the physical properties (for example, ASTM D3588 Standard Practice for Calculating Heat Value, Compressibility Factor, and Relative Density of Gaseous Fuels) to assist in the design of the most appropriate gas processing sequence (Zhu et al., 2014; El-Rahman et al., 2017).

The analysis of natural gas to determine not only the presence but also the relative amount of the different constituents is a widely used in natural gas technology. When natural gas is extracted from the sources, the gas is treated in gas treating plants before use as an energy source in industries or households. The hydrogen sulfide, carbon dioxide, and other constituents of the gas are removed leave methane as the so-called *residue gas*. The importance of the removal of constituents other than methane can be found in a number of reasons such as (i) the production of added-value products, (ii) the toxicity of hydrogen sulfide, (iii) the penchant for hydrogen sulfide and carbon dioxide to enhance metals corrosion in the presence of water, and (iv) the potential for environmental pollution when the gas is use.

The methods for the measurement of the non-methane constituents are numerous, but the choice of the analytical method is dependent on the range and composition of the natural gas as well as on the analytical technology available at the time of choice. Part II presents information that will be useful in method selection. Thus, by applying the appropriate standard test methods to the gas (as presented in Part II), it is possible to determine the distribution of the constituents from methane to the constituents of the condensate from which the behavior of the reservoir, including the gas composition from different wells within a reservoir, can be assessed.

References

Anerousis, J.P., and Whitman, S.K. (1984). An updated examination of gas sweetening by the iron sponge process. Paper No. SPE 13280. SPE Annual Technical Conference and Exhibition, Houston, Texas (16–19 September).

ASTM D1267 (2017). *Test Method for Gage Vapor Pressure of Liquefied Crude Oil (LP) Gases (LP-Gas Method)*, Annual Book of Standards. West Conshohocken, Pennsylvania: ASTM International.

ASTM D1657 (2017). *Test Method for Density or Relative Density of Light Hydrocarbons by Pressure Hydrometer*, Annual Book of Standards. West Conshohocken, Pennsylvania: ASTM International.

ASTM D1945 (2017). *Standard Test Method for Analysis of Natural Gas by Gas Chromatography*, Annual Book of Standards. West Conshohocken, Pennsylvania: ASTM International.

ASTM D1946 (2017). *Standard Practice for Analysis of Reformed Gas by Gas Chromatography*, Annual Book of Standards. West Conshohocken, Pennsylvania: ASTM International.

ASTM D2158 (2017). *Test Method for Residues in Liquefied Crude Oil (LP) Gases*, Annual Book of Standards. West Conshohocken, Pennsylvania: ASTM International.

ASTM D2163 (2017). *Standard Test Method for Determination of Hydrocarbons in Liquefied Crude Oil (LP) Gases and Propane/Propene Mixtures by Gas Chromatography*, Annual Book of Standards. West Conshohocken, Pennsylvania: ASTM International.

ASTM D2421 (2017). *Practice for Interconversion of Analysis of C5 and Lighter Hydrocarbons to Gas-Volume, Liquid-Volume, or Mass Basis*, Annual Book of Standards. West Conshohocken, Pennsylvania: ASTM International.

ASTM D2427 (2017). *Standard Test Method for Determination of C2 through C5 Hydrocarbons in Gasolines by Gas Chromatography*, Annual Book of Standards. West Conshohocken, Pennsylvania: ASTM International.

ASTM D2597 (2017). *Test Method for Analysis of Demethanized Hydrocarbon Liquid Mixtures Containing Nitrogen and Carbon Dioxide by Gas Chromatography*, Annual Book of Standards. West Conshohocken, Pennsylvania: ASTM International.

ASTM D323 (2017). *Standard Test Method for Vapor Pressure of Petroleum Products (Reid Method)*, Annual Book of Standards. West Conshohocken, Pennsylvania: ASTM International.

ASTM D3588 (2017). *Standard Practice for Calculating Heat Value, Compressibility Factor, and Relative Density of Gaseous Fuels*, Annual Book of Standards. West Conshohocken, Pennsylvania: ASTM International.

ASTM D3605 (2017). *Standard Test Method for Trace Metals in Gas Turbine Fuels by Atomic Absorption and Flame Emission Spectroscopy*, Annual Book of Standards. West Conshohocken, Pennsylvania: ASTM International.

ASTM D4628 (2017). *Standard Test Method for Analysis of Barium, Calcium, Magnesium, and Zinc in Unused Lubricating Oils by Atomic Absorption Spectrometry*, Annual Book of Standards. West Conshohocken, Pennsylvania: ASTM International.

ASTM D4891 (2017). *Standard Test Method for Heating Value of Gases in Natural Gas and Flare Gases Range by Stoichiometric Combustion*, Annual Book of Standards. West Conshohocken, Pennsylvania: ASTM International.

ASTM D4927 (2017). *Standard Test Methods for Elemental Analysis of Lubricant and Additive Components—Barium, Calcium, Phosphorus, Sulfur, and Zinc by Wavelength-Dispersive X-Ray Fluorescence Spectroscopy*, Annual Book of Standards. West Conshohocken, Pennsylvania: ASTM International.

ASTM D4951 (2017). *Standard Test Method for Determination of Additive Elements in Lubricating Oils by Inductively Coupled Plasma Atomic Emission Spectrometry*, Annual Book of Standards. West Conshohocken, Pennsylvania: ASTM International.

ASTM D4953 (2017). *Standard Test Method for Vapor Pressure of Gasoline and Gasoline-Oxygenate Blends (Dry Method)*, Annual Book of Standards. West Conshohocken, Pennsylvania: ASTM International.

ASTM D5191 (2017). *Standard Test Method for Vapor Pressure Measurement of Crude Oil Products (Mini Method)*, Annual Book of Standards. West Conshohocken, Pennsylvania: ASTM International.

ASTM D5185 (2017). *Standard Test Method for Multielement Determination of Used and Unused Lubricating Oils and Base Oils by Inductively Coupled Plasma Atomic Emission Spectrometry (ICP-AES)*, Annual Book of Standards. West Conshohocken, Pennsylvania: ASTM International.

ASTM D5482 (2017). *Standard Test Method for Vapor Pressure Measurement of Crude Oil Products (Mini Method – Atmospheric)*, Annual Book of Standards. West Conshohocken, Pennsylvania: ASTM International.

ASTM D6377 (2017). *Standard Test Method for Determination of Vapor Pressure of Crude Oil: VPCRx (Expansion Method)*, Annual Book of Standards. West Conshohocken, Pennsylvania: ASTM International.

ASTM D6443 (2017). *Standard Test Method for Determination of Calcium, Chlorine, Copper, Magnesium, Phosphorus, Sulfur, and Zinc in Unused Lubricating Oils and Additives by Wavelength Dispersive X-ray Fluorescence Spectrometry (Mathematical Correction Procedure)*, Annual Book of Standards. West Conshohocken, Pennsylvania: ASTM International.

ASTM D6732 (2017). *Standard Test Method for Determination of Copper in Jet Fuels by Graphite Furnace Atomic Absorption Spectrometry*, Annual Book of Standards. West Conshohocken, Pennsylvania: ASTM International.

ASTM D7111 (2017). *Standard Test Method for Determination of Trace Elements in Middle Distillate Fuels by Inductively Coupled Plasma Atomic Emission Spectrometry (ICP-AES)*, Annual Book of Standards. West Conshohocken, Pennsylvania: ASTM International.

ASTM D7220 (2017). *Standard Test Method for Sulfur in Automotive, Heating, and Jet Fuels by Monochromatic Energy Dispersive X-ray Fluorescence Spectrometry*, Annual Book of Standards. West Conshohocken, Pennsylvania: ASTM International.

ASTM D7343 (2017). *Standard Practice for Optimization, Sample Handling, Calibration, and Validation of X-ray Fluorescence Spectrometry Methods for Elemental Analysis of Petroleum Products and Lubricants*, Annual Book of Standards. West Conshohocken, Pennsylvania: ASTM International.

ASTM D7833 (2017). *Standard Test Method for Determination of Hydrocarbons and Non-Hydrocarbon Gases in Gaseous Mixtures by Gas Chromatography*, Annual Book of Standards. West Conshohocken, Pennsylvania: ASTM International.

ASTM D7975 (2017). *Standard Test Method for Determination of Vapor Pressure of Crude Oil: VPCRx – F(Tm °C) (Manual Expansion Field Method)*, Annual Book of Standards. West Conshohocken, Pennsylvania: ASTM International.

ASTM D8110 (2017). *Standard Test Method for Elemental Analysis of Distillate Products by Inductively Coupled Plasma Mass Spectrometry (ICP-MS)*, Annual Book of Standards. West Conshohocken, Pennsylvania: ASTM International.

ASTM E355 (2017). *Practice for Gas Chromatography Terms and Relationships*, Annual Book of Standards. West Conshohocken, Pennsylvania: ASTM International.

ASTM F307 (2017). *Practice for Sampling Pressurized Gas for Gas Analysis*, Annual Book of Standards. West Conshohocken, Pennsylvania: ASTM International.

Barbouteau, L. and Dalaud, R. (1972). *In Gas Purification Processes for Air Pollution Control* (ed. G. Nonhebel). London: Butterworth and Co. Chapter 7.

Bartoo, R.K. (1985). *In Acid and Sour Gas Treating Processes* (ed. S.A. Newman). Houston, Texas: Gulf Publishing.

Curry, R.N. (1981). *Fundamentals of Natural Gas Conditioning*. Tulsa, Oklahoma: PennWell Publishing Co.

Deng, L. and Hagg, M.B. (2010). Techno-economic evaluation of biogas upgrading process using CO2 facilitated transport membrane. *International Journal of Greenhouse Gas Control* 4 (4): 638–646.

Duckworth, G.L. and Geddes, J.H. (1965). Natural gas desulfurization by the iron sponge process. *Oil & Gas Journal* 63 (37): 94–96.

El-Rahman Sayed, A., Ashour, I., and Gadalla, M. (2017). Integrated process development for an optimum gas processing plant. *Chemical Engineering Research and Design* 124: 114–123.

Foglietta, J.H. (2004). Dew point turboexpander process: a solution for high pressure fields. *Proceedings. IAPG Gas Conditioning Conference*, Neuquen, Argentina (18 October). Houston, TX: IAPG.

Fulker, R.D. (1972). *In Gas Purification Processes for Air Pollution Control* (ed. G. Nonhebel). London: Butterworth and Co. Chapter 9.

Gary, J.G., Handwerk, G.E., and Kaiser, M.J. (2007). *Crude oil Refining: Technology and Economics*, 5e. Boca Raton, Florida: CRC Press, Taylor & Francis Group.

Hsu, C.S. and Robinson, P.R. ed. (2017). *Handbook of Crude Oil Technology.* Cham: Springer International Publishing AG.

Isalski, W.H. (1989). *Separation of Gases. Monograph on Cryogenics No. 5*, 228–233. Oxford: Oxford University Press.

Jang, K.-S., Kim, H.-J., Johnson, J.R. et al. (2011). Modified mesoporous silica gas separation membranes on polymeric hollow fibers. *Chemistry of Materials* 23 (12): 3025–3028.

Jou, F.Y., Otto, F.D., and Mather, A.E. (1985). *In Acid and Sour Gas Treating Processes* (ed. S.A. Newman). Houston, Texas: Gulf Publishing Company Chapter 10.

Katz, D.K. (1959). *Handbook of Natural Gas Engineering.* New York: McGraw-Hill Book Company.

Kindlay, A.J. and Parrish, W.R. (2006). *Fundamentals of Natural Gas Processing.* Boca Raton, Florida: CRC Press, Taylor & Francis Group.

Kohl, A.L. and Riesenfeld, F.C. (1985). *Gas Purification*, 4e. Houston, Texas: Gulf Publishing Company.

Lachet, V. and Béhar, E. (2000). Industrial perspective on natural gas hydrates. *Oil & Gas Science and Technology* 55: 611–616.

Maddox, R.N. (1982). *Gas Conditioning and Processing. Volume 4. Gas and Liquid Sweetening.* Norman, Oklahoma: Campbell Publishing Co.

Maddox, R.N., Bhairi, A., Mains, G.J., and Shariat, A. (1985). *In Acid and Sour Gas Treating Processes* (ed. S.A. Newman). Houston, Texas: Gulf Publishing Company Chapter 8.

Maple, M.J. and Williams, C.D. (2008). Separating nitrogen/methane on zeolite-like molecular sieves. *Microporous Mesoporous Mater.* 111: 627–631.

Matsui, T. and Imamura, S. (2010). Removal of siloxane from digestion gas of sewage sludge. *Bio resource Technology* 101 (1 Suppl): S29–S32.

Mody, V. and Jakhete, R. (1988). *Dust Control Handbook.* Park Ridge, New Jersey: Noyes Data Corp.

Mokhatab, S., Poe, W.A., and Speight, J.G. (2006). *Handbook of Natural Gas Transmission and Processing.* Amsterdam: Elsevier.

Nadkarni, R.A.K. (2005). *Elemental Analysis of Fuels and Lubricants: Recent Advances and Future Prospects*, Publication No. STP 1468. West Conshohocken, Pennsylvania: ASTM International.

Newman, S.A. (1985). *Acid and Sour Gas Treating Processes*. Houston, Texas: Gulf Publishing.

Nonhebel, G. (1964). *Gas Purification Processes*. London: George Newnes Ltd.

Parkash, S. (2003). *Refining Processes Handbook*. Amsterdam: Gulf Professional Publishing, Elsevier.

Pitsinigos, V.D. and Lygeros, A.I. (1989). Predicting H_2S-MEA equilibria. *Hydrocarbon Processing* 58 (4): 43–44.

Polasek, J. and Bullin, J. (1985). *In Acid and Sour Gas Treating Processes* (ed. S.A. Newman). Houston, Texas: Gulf Publishing Company Chapter 7.

Popat, S.C. and Deshusses, M.A. (2008). Biological removal of siloxanes from landfill and digester gases: opportunities and challenges. *Environmental Science and Technology* 42 (22): 8510–8515.

Rey, M. (2008). *Distillation and Vapor Pressure Measurement in Crude Oil Products*, Manual No. 51 MNL51. West Conshohocken, Pennsylvania: ASTM International.

Robeson, L.M. (1991). Correlation of separation factor versus permeability for polymeric membranes. *Journal of Membrane Science* 62: 165.

Ryckebosch, E., Drouillon, M., and Vervaeren, H. (2011). Techniques for transformation of biogas to biomethane. *Biomass and Bioenergy* 35 (5): 1633–1645.

Schweigkofler, M. and Niessner, R. (2001). Removal of siloxanes in biogases. *Journal of Hazardous Materials* 83 (3): 183–196.

Soud, H. and Takeshita, M. (1994). *FGD Handbook*, No. IEACR/65. London: International Energy Agency Coal Research.

Speight, J.G. (2000). *The Desulfurization of Heavy Oils and Residua*, 2e. New York: Marcel Dekker Inc.

Speight, J.G. (2007). *Natural Gas: A Basic Handbook*. Houston, Texas: GPC Books, Gulf Publishing Company.

Speight, J.G. (2011). *An Introduction to Crude Oil Technology, Economics, and Politics*. Salem, Massachusetts: Scrivener Publishing.

Speight, J.G. (2014a). *The Chemistry and Technology of Crude Oil*, 5e. Boca Raton, Florida: CRC Press, Taylor & Francis Group.

Speight, J.G. (2014b). *Oil and Gas Corrosion Prevention*. Oxford: Gulf Professional Publishing, Elsevier.

Speight, J.G. (2017). *Handbook of Crude Oil Refining*. Boca Raton, Florida: CRC Press, Taylor & Francis Group.

Staton, J.S., Rousseau, R.W., and Ferrell, J.K. (1985). *In Acid and Sour Gas Treating Processes* (ed. S.A. Newman). Houston, Texas: Gulf Publishing Company Chapter 5.

Ward, E.R. (1972). *In n Gas Purification Processes for Air Pollution Control* (ed. G. Nonhebel). London: Butterworth and Co. Chapter 8.

Yao, P., Boardman, G.D., and Li, E.T. (2016). Research progress for removing siloxane from biogas by adsorption. *Chemical Industry and Engineering Progress* 35 (2): 604–611.

Zapffe, F. (1963). Iron sponge process removes mercaptans. *Oil & Gas Journal* 61 (33): 103–104.
Zhang, L.Q., Shi, L.B., and Zhou, Y. (2007). Formation prediction and prevention technology of natural gas hydrate. *Natural Gas Technology* 1 (6): 67–69.
Zhu, L., Li, L., Zhu, J. et al. (2014). Analytical methods to calculate water content in natural gas. *Chemical Engineering Research and Design* 93: 148–162.

Part II

Analysis

5

Sampling and Measurement

5.1 Introduction

Natural gas exhibits a wide range of composition that influences the various chemical and physical properties (Table 5.1) (Speight, 2007, 2014a; Mokhatab et al., 2006). The chemical and physical properties of natural gas and the yields and properties of products or factions produced, therefrom vary considerably and are dependent on the concentration of the hydrocarbon constituents and nonhydrocarbon constituents present in the raw gas as it emanates from the well. Some types of natural gas have pronounced advantages as sources of fuels and petrochemical feedstocks because the gas requires less specialized processing than that needed for production of the same products from other types of natural gas. On the other hand, natural gas from other sources may contain unusually low concentrations of components that are desirable for use as fuel or as petrochemical feedstocks, and the production of these products may require extensive treating at the wellhead and at the gas processing plant. One of the main points of treating at the wellhead is to remove carbon dioxide and hydrogen, which, in themselves, are not necessarily corrosive but, in the presence of water, can cause considerable corrosion to a pipeline (Speight, 2014b). The same rationale applies to coal gas, biogas, and landfill gas (Chapter 1) where the composition of the gas is dependent upon the source material and the process by which the gas was produced.

Thus, the evaluation of natural gas for use as a fuel or as a petrochemical feedstock typically involves an examination of one or more of its physical properties (Mokhatab et al., 2006; Speight, 2007, 2014a). By this means, a set of basic characteristics can be obtained that can be correlated with utility (Poling et al., 2004). To satisfy specific needs regarding the type of processing required for the gas, as well as to the nature of the products, various standards organizations, such as the American Society for Testing and Materials in

Handbook of Natural Gas Analysis, First Edition. James G. Speight.

Table 5.1 Constituents of natural gas.

Name	Formula	Vol.%
Methane	CH_4	>85
Ethane	C_2H_6	3–8
Propane	C_3H_8	1–5
Butane	C_4H_{10}	1–2
Pentane$^+$	C_5H_{12}	1–5
Carbon dioxide	CO_2	1–2
Hydrogen sulfide	H_2S	1–2
Nitrogen	N_2	1–5
Helium	He	<0.5

Pentane$^+$: pentane and higher molecular weight hydrocarbons up to octane as well as benzene and toluene.

North America, have devoted considerable time and effort to the correlation and standardization of methods for the inspection and evaluation of natural gas and the various products, and, moreover, these same test methods can also be applied with little or no adaptation to coal gas, biogas, and landfill gas (ASTM, 2017). Many other countries (Chapter 1) also have similar organizations for developing standard test methods or industrial products.

As applied in the context of this book, a standard test method is a method for a test in science or engineering to determine the properties and behavior of gas streams that have been published by international, regional, or national standard-writing bodies, by reputable technical organizations (Table 5.2). The method may be a test to determine a chemical property or a physical property and is, put simply, a definitive procedure that produces a test result, which can be qualitative or quantitative and may be from a personal observation or an output of a precision measuring instrument. The test result may also be the measured response based on the conditions of the test (a dependent variable) or an independent variable. Some tests, however, may involve changing the independent variable to determine the level at which a certain response occurs: in this case, the test result is the independent variable.

Nonstandard methods are those methods not taken from authoritative, validated sources. These may not have undergone validation such as a collaborative study or process to evaluate their performance capabilities. Nonstandard methods are selected for use when a customer's request cannot be addressed with the use of a standard method. Such methods are subject to agreement with the customer, and a clear specification of the customer's work requests, including the purpose of the test, is made. Nonstandard and

Table 5.2 Examples of standards organizations in various countries (listed alphabetically).

Organization	Acronym	Country
American National Standards Institute	ANSI	US
Badan Standardisasi Nasional	BSN	Indonesia
Brazilian National Standards Organization	ABNT	Brazil
British Standards Institution	BSI	UK
Bureau of Indian Standards	BIS	India
Bureau of Standards of Jamaica	BSJ	Jamaica
Colombian Institute of Technical Standards and Certification	ICONTEC	Colombia
Deutsches Institut für Normung	DIN	Germany
Dirección General de Normas	DGN	Mexico
Ente Nazionale Italiano di Unificazione	UNI	Italy
Estonian Centre for Standardization	EVS	Estonia
Euro-Asian Council for Standardization, Metrology, and Certification	GOST	Russia
Finnish Standards Association	SFS	Finland
French Association for Standardization	AFNOR	France
Instituto Argentino de Normalización y Certificación	IRAM	Argentina
Instituto Português da Qualidade	IPQ	Portugal
Japanese Industrial Standards Committee	JISC	Japan
Jabatan Standard Malaysia	DSM	Malaysia
Korean Agency for Technology and Standards	KATS	Korea (Republic)
Luxembourg Institute for Standardization	ILNAS	Luxembourg
Nederlandse Norm	NEN	Netherlands
Romanian Standards Association	ASRO	Romania
South African Bureau of Standards	SABS	South Africa
Spanish Association for Standardization and Certification	AENOR	Spain
Standardization Administration of China	SAC	China
Standards Australia	SAI	Australia
Standards Council of Canada	SCC	Canada
Standards New Zealand	SNZ	New Zealand
Standards Norway	SN	Norway
Standards Organisation of Nigeria	SON	Nigeria
Swedish Standards Institute	SIS	Sweden
Swiss Association for Standardization	SNV	Switzerland

laboratory-developed methods need method validation. This activity is planned and assigned to qualified laboratory personnel. The performance characteristics of the method are based on its intended use. For example, if the method will be used for qualitative analysis, there is no need to test and validate the linearity of the method over the full dynamic range of the equipment employed.

In gas technology, it is vital for all producers and customers to understand and agree upon standard test methods of obtaining relevant data. It is common for a physical property to be influence by the precise method of testing, and it is essential for the test to be fully documented in order to provide the needed data for specifications that will satisfy the customer that the gas has the stated properties and behavior.

Using a standard test method, as published by a recognized standards organization (Table 5.2), is a part of analytical practice. A well-written test method is important. However, even more important is choosing a method of measuring the correct property or characteristic. Not all tests and measurements are equally useful: usually a test result is used to predict or imply suitability for a certain purpose. For example, as is the case of natural gas and other process gases and several constituents, the apple test method may have various levels of connections: (i) the test results of a raw material should connect with tests of a component made from that material, (ii) the test results of a component should connect with performance testing, and (iii) the results of laboratory performance testing should connect with field performance. These connections or correlations may be based on published literature, scientific studies, or engineering studies, and validation of the suitability of the test method is often required. There are always those situations where it may be necessary to modify an existing test method or to develop a new test method. Again, documentation of the new or modified data relating to test method as well as the reproducibility and repeatability of the data is very necessary. Test methods, whether an original published method or a modified method or a new method, must be scrutinized for their validity, applicability, and accuracy. It is essential that the scope of the test method is clearly defined, and any aspect included in the scope is shown to be accurate (precise) and repeatable (especially when a test method is new or has been modified considerably) through validation through interlaboratory testing of the method (ASTM E691, 2017; ASTM E1169, 2017; ASTM E1301, 2017; ASTM E1488, 2017; ASTM E2655, 2017).

For example, test method ASTM E691 (2017) describes the need for precision statements in all test methods in terms of repeatability and reproducibility. This method can be used to obtain the needed information as simple as possible, which can then be used to prepare a precision statement (ASTM E177). A statement of precision allows potential users of a test method to assess in general terms the test method's usefulness with respect to variability in proposed applications. A statement of precision is not intended to exhibit values that can be exactly duplicated in every user's laboratory. Instead, the statement provides guidelines as to the magnitude of variability that can be expected between test results when the method is used in one or in two or more reasonably competent laboratories. Furthermore, a definition or statement of the precision of the test method is useful in commerce and in technical work when comparing test results against standard values (such as specification limits) or between data sources, such as between different laboratories, is involved in producing and comparing data or when different instruments are employed as part of the method.

As another example, test method (ASTM E2655, 2017) describes the standard guide for reporting uncertainty of test results and use of the term measurement uncertainty in ASTM test methods. The concepts necessary for understanding the term "uncertainty" when applied to a quantitative test result since several measures of uncertainty can be applied to a given measurement result. As a result of applying this method, the interpretation of some of the common forms can be described and express the uncertainty of the test result, which relates to standard statistical methodology. In addition, relationships between uncertainty and concepts of precision and bias are described.

When a test method is applied to a large number of portions of a material that are as nearly alike as possible, the test results obtained will not all have the same value. A measure of the degree of agreement among these test results describes the precision of the test method for that material. Numerical measures of the variability between such test results provide inverse measures of the precision of the test method. Greater variability implies smaller (that is, poorer) precision and larger imprecision. The precision of the data is reported as a standard deviation, coefficient of variation (relative standard deviation), variance, or precision limit (a data range indicating no statistically significant difference between test results). Thus, the procedures presented in this practice consist of three basic steps: (i) planning an interlaboratory study, (ii) guiding the testing phase of the study, and (iii) analyzing the test result data.

In North America, the standard test methods are prime assets for the evaluation and assessment of the physical, mechanical, rheological, thermal, and chemical properties of natural gas, fuel gases, hydrocarbon streams, petrochemical feedstocks, and other naturally occurring energy resources that are used for various commercial and domestic applications. For the purposes of this text, it is appropriate that in any discussion of the physical properties of petroleum and petroleum products, reference should be made to the corresponding standard test methods (designated by the letters ASTM along with a number) and, accordingly, the various test numbers included in the text.

It is the purpose of the next section to discuss how sampling of natural gas can be accomplished. The scope of any sampling program should be well defined. Technicians responsible for installing the sampling system, gathering the samples, and conducting the analysis should be well versed in their tasks.

A lapse in any step of the process could skew the results. If any doubt exists regarding the proper use of composite sampling systems and their ancillary equipment, most manufacturers offer operational training and technical assistance (GPA, 1986; API, 2001; Kimmel, 2003). Furthermore, where the term "natural gas" is used in this text, it is intended to convey that the statement or method that is made relevant to natural gas can also, within the constraints of the gas composition, be applied to coal gas, biogas, and landfill gas.

5.2 Sampling

Natural gas sampling is performed for a variety of reasons. Sampling is performed to determine total gas composition, hydrocarbon dew point, specific gravity, and most importantly, the value of the gas. For any seller or purchaser of natural gas (or any gaseous fuel), it is important to have knowledge about the gas quality to be prepared to handle operating problems such as the formation of environmentally hazardous products or corrosion by the constituents of the gas or the combustion products. Requirements should be listed for all relevant fuel parameters for each individual gas producing plant to ensure the necessary gas quality and the purchase price. Quality control of the supplied gas consignment constitutes important follow-up. Knowing the quality of the gas allows the seller to document this to the customers, and by performing one or more standard test methods to the gas, the seller and purchaser will be certain of the quality of the gas.

Three techniques are normally used to obtain gas samples: continuous composite sampling, continuous online sampling, or spot sampling. Improper methods utilized during the collection and sample conditioning of products are among the largest sources of error and uncertainty in the overall process of hydrocarbon compositional measurement. The accurate determination of product quality and value is totally dependent upon representative, reliable, and technically defensible sample collection, conditioning, and analytical processes. Ensuring that each of these processes is accomplished with optimal integrity is absolutely essential.

In the current context, samples of gases for analysis are typically taken by means of a tube and an aspirator bulb being either drawn directly into the analyzer or stored in glass sampling tubes for analysis at a later date. Sample tubes must be made of a gastight material. Many laboratories follow standard procedure to capture samples of natural gas (and the other gases mentioned in Chapter 1), but some laboratories do have in-house methods that have been developed through experience and have proven to be reliable (the data are producible and repeatable).

Examples of gaseous fuels in common use are (i) natural gas, (ii) liquefied petroleum gas (LPG), refinery gas, coal gas, and biogas (such as digester gas and landfill gas). These gases can vary considerably in composition because of the variability of the feedstock from which they are produced (Chapter 1). In addition, coal gas and biogas produced by pyrolytic processes may contain substantial amount of particulate matter (Speight, 2011, 2013). Therefore, careful analysis (Wallis, 1986) is necessary to assure that buyer that the purchased gas meets the sales specifications. More particularly, when using natural gas (or any gas) as a fuel gas, the composition and energy content of the gas can fluctuate heavily. When burning fuel gas in a furnace, control over the energy flow is key to optimize the product quality, throughput, and efficiency.

More particularly, natural gas is sampled to determine quality for custody transfer applications, balance a plant process, or gathering system. Natural gas, as a fuel gas, is purchased based on the amount of energy delivered. The quantity of energy delivered is calculated by multiplying the gas volume per unit time by the energy value (Btu or BTU) per unit volume. In order to identify individual components of the sample and their quantity as well as the overall quality of the gas, it is conventional to use a gas chromatograph to determine the quality of the sample taken. The production, wellhead, transport, and process analysis of the sample is crucial to the accuracy required for custody transfer of natural gas. In addition, accurate gas sampling can only be accomplished if the flowing stream is in a gas-only phase. Therefore, it is critical that the sample be taken at a point in the system where single-phase flow is present. This may appear to be a simple requirement, but there are a number of factors that conspire to make a single-phase flow difficult to maintain, especially in cold temperature climates.

Since gases consist of various compounds, their composition is of interest and important from the viewpoint of (i) safety, (ii) efficiency, and (iii) computation of specific values, which might affect the continuity of a process. In combustion process the common personnel hazard is carbon monoxide, which can result from incomplete combustion. An operational hazard can be the formation of an explosive mixture. This hazard can be present wherever combustible gases or vapors are in contact with a substance, which will support combustion, and the properties are in the explosive range. Gas analysis can be used to detect and to confirm and locate suspected condition of this nature. As an example, there are many uses for LPG, the major ones being as (i) petrochemical, synthetic rubber, and motor gasoline feedstocks and as (ii) commercial, domestic, and industrial fuel.

The reliability of a sampling method is the degree of perfection with which the identical composition and properties of the entire whole petroleum or petroleum product are obtained (preserved completely without alteration) in the sample – such a sample is referred to as a *representative sample*. The reliability of the storage procedure is the degree to which the sample remains unchanged, thereby guaranteeing the accuracy and usefulness of the analytical data. At this point, a review of the sampling methods applied to petroleum and petroleum products is worth of inclusion.

One of the more critical aspects for the analysis of light hydrocarbons is the question of sampling. Sampling of gaseous and liquefied materials is addressed in a variety of specific sampling methods, and many of the test methods themselves contain additional sampling requirements (for example, ASTM D1265, 2017; ASTM D5287, 2017; ASTM D3700, 2017; ASTM D4057, 2017; ASTM D4177, 2017; ASTM D5842, 2017). While sampling of C_1 and C_2 hydrocarbons is typically performed using stainless steel cylinders, either lined or unlined, other containers are employed dependent upon particular situations, for example, glass cylinder containers or sampling bags. The preferred method for sampling C_3 and C_4 hydrocarbons is by the use of piston cylinders (ASTM D3700, 2017), although sampling these materials as gases is also acceptable in many cases. The sampling of C_3 and higher hydrocarbons is dependent upon the vapor pressure of the sample. Piston cylinders or pressurized steel cylinders are advisable for high vapor pressure samples (containing significant amounts of light gases), while atmospheric sampling may be used for low vapor pressure samples.

The use of piston cylinders allows the collection of a representative sample of LPG that may contain trace volatile dissolved components such as methane, ethane, and nitrogen. Sampling using ASTM D1265 (2017) can result in a small, but predictable, loss of the lower-boiling constituents, and while this method (ASTM D1265, 2017) is suitable for collecting samples for routine specification testing, the small loss of lower-boiling constituents is not significant under the specification requirements. Nevertheless, the more stringent test procedure (ASTM D3700, 2017) is recommended whenever the highly accurate determination of the lower-boiling constituents is required. For example, the composition of the gas, as determined on samples collected stringently (ASTM D3700, 2017), may be used to establish the product value of mixtures' natural gas liquids.

Since the value of any product is judged by the characteristics of the sample as determined by laboratory tests, the ability to collect and preserve a sample that is representative of the site is a critically important step. Furthermore, obtaining a representative sample of petroleum or a petroleum product is always a challenge due to the heterogeneity of the materials. Additional difficulties are encountered due to the ranges in volatility, solubility, and adsorption potential of individual constituents. And the procedures used for sample collection and preparation must be analytically reproducible with a high degree of accuracy (Weisman, 1998; Dean, 2003; Patnaik, 2004; Speight, 2014a, 2015).

Furthermore, it is essential that the sample to be used for the test methods must be representative of the bulk material, or data will be produced that are not a true representation of the material and will be, to be blunt, incorrect no matter how accurate or precise the test method and no matter how much *fine-tuning* is applied by the submitter of the sample. In addition, the type and cleanliness of sample containers are important if the container is contaminated or is made of material that either reacts with the product or is a catalyst; the test results may be wrong.

The following may be accepted as a general guide for the common uses for the four fuel types covered by these specifications: (i) commercial propane is the preferred fuel type for domestic, commercial, and industrial fuels (it is also a suitable fuel for low severity internal combustion engines); (ii) commercial butane is used principally as feedstock for petrochemicals, synthetic rubber, and blending stocks or feedstocks in the manufacture of motor gasoline (use as a fuel is generally limited to industrial applications where vaporization problems are not encountered; however, small quantities are used as domestic fuel); (iii) commercial butane–propane mixtures cover a broad range of mixtures, which permits the tailoring of fuels or feedstocks to specific needs; and (iv) propane is less variable in composition and combustion characteristics than other products covered by these specifications as it is also suitable as a fuel for internal combustion engines operating at moderate to high engine severity.

The uses of gas analysis to promote efficiency are varied and many. Efficiency and economy do not necessarily imply the complete oxidation in the least amount of oxygen or air. It is the reason by which the goals for good utilization can be reached: (i) for gas appliances to operate safely without liberating injurious quantities of toxic gases and (ii) for gas appliances to operate near optimum efficiency.

5.2.1 General Protocols

Most gases, especially the hydrocarbon gases, are amenable to analytical techniques, and there has been the tendency, and it remains, for the determination of both major constituents and trace constituents that is the case with the heavier hydrocarbons. The complexity of the mixtures, which is evident as the boiling point of petroleum fractions and petroleum products increases, makes identification of many of the individual constituents difficult, if not impossible. Samples of petroleum and petroleum products are obtained for many reasons, including the determination of chemical and physical properties. These properties may be used for calculating standard volumes, establishing product value, and often safety and regulatory reporting. In addition, methods have been developed for the determination of physical characteristics such as calorific value, specific gravity, and enthalpy from the analyses of mixed hydrocarbon gases, but the accuracy does suffer when compared with the data produced by methods for the direct determination of these properties.

Bulk physical property tests, such as density and heating value, as well as some compositional tests, such as the Orsat analysis (the measurement of the quantitative amounts by selective absorption of carbon dioxide, oxygen, and carbon monoxide) and the mercuric nitrate method for the determination of unsaturation, are still used. However, the choice of a particular test rests upon the decision of the analyst that then depends upon the nature of the gas under study. For example, judgment by the analyst is necessary whether or not a test that is applied to LPG is suitable for natural gas insofar as inference from the nonhydrocarbon constituents will be minimal.

Gas will be transferred from the process line to the sample chamber via tubing or piping. Often, the terms tubing and piping are used interchangeably. However, it is important that tubing utilized to transfer the gas from the process to the collection chamber is as short in length and small in diameter as possible. It will reduce the amount of time between the sample point and the collection chamber, thereby maintaining the sample's integrity.

Isolating different parts of the sample system is important for obvious reasons during different times in the sample procedure, installation, and maintenance of equipment. An important factor to consider when positioning valves is the phenomenon of condensation. When gas experiences a restriction (partially closed valve), a pressure drop ensues causing condensation, which causes inaccurate sampling. Valves should also be checked for leaks during installation and normal operation. Leaks at the sample cylinder are very important because these will cause the lighter components of the gas to escape, leaving the higher molecular weight components (such as the C_3 and C_4) overrepresented in the sample and therefore providing a nonrepresentative sample. Larger orifices in valves will reduce fractionalization of the sample.

In fact, there are inherent limitations when performing any type of sampling; any one of which may affect the representative nature of the sample. As examples, a spot sample provides a sample from only one particular point in the tank, vessel compartment, or pipeline. In the case of running or all-level samples, the sample only represents the column of material from which it was taken. Based on the product and testing to be performed, one test method (ASTM D4057, 2017) provides guidance on sampling equipment, container preparation, and manual sampling procedures for petroleum and petroleum products of a liquid, semiliquid, or solid state from the storage tanks, flowlines, pipelines, marine vessels, process vessels, drums, cans, tubes, bags, kettles, and open discharge streams into the primary sample container. Thus, all procedure used for sampling gas streams should conform to standard specification (ASTM D1265, 2017; ASTM D4057, 2017). Proper selection of filter flow capacity and particle size is of significant consideration due to filters with insufficient drip pot capacity for gases that have entrained water, liquid, or condensate. Filters that have insufficient flow capacity and drip pot capacity will directly impact accuracy.

To prevent condensation of the sample, heating elements maintain a constant temperature above the hydrocarbon dew point preventing any part of the sample from condensing. Knowledge of ambient temperatures and the dew points of the various constituents will determine the necessity of including heating elements in specific sampling applications (ASTM D3700, 2017).

As the part of the sampling system that is directly installed into the process, certain considerations should be taken in regard to the probe assembly. Probes should also be placed in an area of the process pipe in which turbulence is at a minimum. Turbulence causes contaminants at the bottom of the pipe to be stirred up and could potentially cause the contaminants to be included in the sample, thereby providing a sample that is not representative of the gas being sampled. Turbulence is caused by bends, valves, headers, and any other items that restrict, change, or impact the flow of gas. Installation of the probe is recommended on the top of the pipe with the probe extending into the middle 1/3 of the pipe from which the sample is being gathered. For large diameter pipes, the probe should extend at least 200 mm or 8 inches into the pipe if the middle 1/3 of the pipe is greater than 8 inches.

Until the gas is released, the protocols for sampling methane hydrate (or any gas hydrate for that matter) offer challenges to the analyst. The analysis of methane hydrate has mostly focused on (i) documenting the geologic parameters that control the occurrence and stability of methane hydrates in nature, (ii) assessing the volume of natural gas stored within various methane hydrate accumulations, (iii) analyzing the production response and characteristics of methane hydrates, (iv) identifying and predicting natural and induced environmental and climate impacts of natural methane hydrates, (v) analyzing the methane hydrate role as a geohazard, (vi) establishing the means to detect and characterize methane hydrate accumulations using geologic and geophysical data, and (vii) establishing the thermodynamic phase equilibrium properties of methane hydrates as a function of temperature, pressure, and gas composition. The method must be quantitative, and the sample must be representative of the total volume of methane present in the hydrate; there are means for estimating the amount of methane that should be released by the depressurization approach.

Once methane gas is generated from dissociated methane hydrate, then the above approach may be used to collect it using a similar principle, methods, equipment, and facilities as employed for the development of natural gas. To dissociate methane hydrate that is stable at low temperature and under high pressure, it is necessary to (i) increase the temperature – the heating method – or (ii) decrease the pressure – the depressurization method.

However, the difficulties and limitations of the depressurization method are due to the fact that gas hydrate dissociation is an endothermic process, so gas production and recovery rates are highly governed by how much thermal energy can be supplied from natural heat sources. In a simple depressurization, the thermal energy needed for hydrate dissociation is expected to be supplied from sediment grains, pore fluids, and the hydrate crystals themselves. In addition, heat transferred from surrounding formations to the reduced temperature zone may support the heat supply. The portion of the hydrate origin gas produced by the depressurization method can be regarded as a primary recovery of methane hydrate in which gas can be extracted from the solid material by natural heat, not by artificial thermal energy.

Moreover, the mechanical properties of the hydrate-containing sediments, such as stiffness, strength, and stress conditions, are important for safe and stable gas production, since they affect the stability of both the borehole and sediments after production technologies have been applied. These mechanical properties impact operational risks, including borehole instabilities, well integrity, formation damage due to consolidation, water breakthrough from aquifers and the seabed through faults and fractures that may be created or reactivated, and large-scale seafloor instabilities. These properties also introduce the potential for errors in the compositional analysis of the hydrates as they exist in the sediment(s).

When examining methane hydrate – existing as a solid substance in layers – in the laboratory, it is necessary to dissociate methane hydrate to methane gas and water and collect only the methane gas. As mentioned earlier, once methane gas is generated from dissociated methane hydrate, then the above approach may be used to collect it using a similar principle, methods, equipment, and facilities as employed for the development of natural gas. The operation of the increasing temperature or decreasing pressure of layers bearing methane hydrate is the actual way of the production of methane from the hydrate.

In order to scientifically conclude that the methane release method is effective, it is necessary to know in detail the physical properties of methane hydrate itself and the state and physical properties of methane hydrate-bearing layers. Since methane hydrate is a substance that is stable at low temperatures and high pressure, investigating its state and measuring its physical properties may be necessary to fabricate a special experimental system capable of observing and analyzing targets with variations in a low-temperature and high-pressure environment. In terms of analysis, it is also necessary to ensure that the methane hydrate decomposes in a controlled manner and that all of the emitted methane is collected.

Degassing of hydrate-bearing pressure cores and measurement of natural gas quantity are used to calculate gas hydrate saturations. Pressure cores trap a volume of sediment along with any gas hydrate, dissolved gas, or free gas, enabling a mass balance approach to be applied to this sealed system, and thermodynamic equilibrium is assumed in these calculations. The degassing of pressure cores is the only hydrate quantification method that can prove that gas hydrate does not exist in a sample of sediment (by showing that the pore-water is undersaturated in natural gases) and is considered the "gold standard" for hydrate quantification, as it is used to ground-truth other hydrate quantification techniques. One major drawback of the destructive depressurization of pressure cores is that the valuable pressure core is no longer available for other studies. If a nondestructive method of testing could be used for hydrate quantification in pressure cores, this would allow further analyses to take place on the core. In addition, as in any analytical method, there should also be close attention paid to the mass balance.

However, a fast depressurization rate can result in a fast hydrate decomposition rate, therefore, a rapid gas production rate. In addition, the radial extension behavior of the decomposition front confirms that ambient heat transfer is a critical factor driving hydrate decomposition into free gas and liquid water. Obvious hydrate reformation and ice generation phenomenon, as seen in some of the sudden depressurization experiments, can be effectively avoided using piecewise and continuous depressurization methods. Thus, the production pressure affects the gas production behavior from hydrate deposits and provides the potential for errors in the analytical data when seeking to establish optimal production techniques for utilizing hydrate resources in the field.

Thus, an accurate sampling procedure of gas is vital to determining components and quality of the gas. Spot sampling is not always representative of gas quality and components over a given time or volume of gas. Testing methods, techniques, and equipment can all affect the quality of the sample taken, and great care should be taken when obtaining samples in order to ensure that the most accurate sample is obtained.

5.2.2 Sampling Methods

Natural gas sampling can be accomplished using several different methods. However, the most important aspect of taking a sample is where and how the sample is taken. Both aspects can influence the quality of the sample (i.e. a representative sample or a nonrepresentative sample), and the analyst should be made aware of the method of sample – if not he or she should ask the sample taker about the protocols that were used to acquire the sample. A sample can be taken as a spot, a composite, or a continuous sample connected to a chromatograph. Thus, continuous sampling is exactly like it sounds, continuous.

The most common method, spot sampling (GPA, 1986), involves manually filling a sample cylinder connected to the center of the pipeline by a probe and separator. The entire volume of gas that has flowed between sample periods is evaluated for its quality with that one sample, which is an indication of the composition and/or the properties of the gas at a particular moment in time. From this statement, it should be obvious to the sampler and the analysts that this method of sampling should only be used for gas streams that are constant in terms of composition and volumetric flow rate.

In locations where high volumes of gas are transferred, it is common to use online chromatographs to determine the real-time composition of the flowing gas stream. In such cases, temperature compensating regulators are necessary to supply a constant representative sample from the sample location in the pipeline to the gas chromatograph. However, many other applications have volumetric flow rates, which require the same accuracy for custody transfer and require the use of an automatic composite sampling system. These systems take periodic samples based on time or volumetric flow rate to create a composite sample, which represents the true real-time composition of the gas stream during the sampling period. Once a sample is taken, the gas should be stored in a sample cylinder and transported to a laboratory where the composition and quality of the gas are determined by gas chromatography.

In the laboratory, one of the more critical aspects for the analysis of low-boiling hydrocarbons relates to the issue of volumetric measurement (ASTM D1071, 2017) and sampling (ASTM D1265, 2017). Sampling of liquefied gases poses issues that must be addressed by the sample, and the analyst should be aware of the sampling method and the material (gas or liquid) from which the sample was taken (Table 5.3). Using LPG as an example of sample acquisition

Table 5.3 Examples of items that should be recorded in a sampling log.

i)	The precise (geographic or other) location (or site or refinery or process) from which the sample was obtained.
ii)	The identification of the location (or site or refinery or process) by name.
iii)	The character of the bulk material (solid, liquid, or gas) at the time of the sampling.
iv)	The means by which the sample was obtained.
v)	The protocols that were used to obtain the sample.
vi)	The date and the amount of sample that were originally placed into storage.
vii)	Any chemical analyses (elemental analyses, fractionation by adsorbents or by liquids, compositional analyses) that have been determined to date.
viii)	Any physical analyses (API gravity, viscosity, distillation profile) that have been determined to date.
ix)	The date of any such analyses included in items (vii) and (viii).
x)	The methods used for analyses that were employed in (vii) and (viii).
xi)	The analysts who carried out the work in (vii) and (viii).
xii)	A log sheet showing the names of the persons who removed the samples from storage.
xiii)	The date and the reason for the removal of an aliquot.
xiv)	The amount of each sample (aliquot) that was removed for testing.
xv)	The test methods that were applied to the aliquot.
xvi)	The location of the results of the analysis of the aliquot.

from a liquid storage system and obtaining a representative sample, the sampling is complicated by existence of two phases (gas and liquid), and the composition of the supernatant vapor phase will, most probably, differ from the composition of the liquid phase. Furthermore, the composition of both the gas phase and the liquid phase will vary as a sample is (or samples are) removed from one phase or from both phases. An accurate check of composition can only be made if samples are taken when the tank is filled with the sample or are taken from a fully charged tank.

In general, the sampling of gaseous constituents and of liquefied gases is the subject of a variety of sampling methods. Standard Test Method for Determination of Water Vapor (Moisture Concentration) in Natural Gas by Tunable Diode Laser Spectroscopy, TDLAS), such as the manual method (ASTM D1265, 2017; ASTM D4057, 2017), the floating piston cylinder method (ASTM D3700, 2017), and the automatic sampling method (ASTM D4177, 2017; ASTM D5287, 2017). Methods for the preparation of gaseous and liquid blends are also available (ASTM D4051, 2017; ASTM D4307, 2017) including the sampling and handling of gaseous fuels and low-boiling fuels for volatility measurements (ASTM D5842, 2017). This method (ASTM D5842, 2017) describes several subprocedure and equipment for obtaining, mixing, and handling representative samples of volatile fuels for the purpose of testing for compliance with the standards set forth for volatility-related measurements applicable to gaseous fuels and low-boiling fuels. Thus, the method is applicable to hydrocarbon fuels that have a dry vapor pressure equivalent range of 2–16 psi. However, the method is not applicable to crude oil for which other test methods should be applied (Speight, 2015; ASTM D4057, 2017; ASTM D4177, 2017).

Sampling methane (CH_4) and ethane (C_2H_6) hydrocarbons is usually achieved using either lined or unlined stainless steel cylinders. However, other containers may also be employed dependent upon the temperature and pressure conditions. For example, glass cylinder containers or polymer-based sampling bags may also be used but cannot be subjected to temperatures or pressures that are far in excess of ambient temperature and pressure

Clean sample cylinders are of the utmost importance. Any residual particles present in the sample cylinder will cause inaccurate samples. Material selection – any materials utilized in the equipment used for obtaining samples – should be compatible with the gas being sampled. Normally, 316 stainless steel and Viton elastomeric components are compatible in the clear majority of sampling situations. However, the end user is ultimately responsible for determining chemical compatibility in regard to their specific gas composition.

Thus, the importance of the correct sampling of any product destined for analysis should always be overemphasized and must be maintained in a form that can be justified in court, if ever a legal claim is involved (Table 5.3). Incorrect sampling protocols can lead to erroneous analytical data from which performance of the product in service cannot be accurately deduced. For example, properties such as specific gravity, distillation yield, vapor pressure, hydrogen sulfide content, and octane number of the gasoline are affected by the content of low-boiling hydrocarbons so that suitable cooling or pressure sampling methods have to be used, and care is taken during the subsequent handling of the sample in order to avoid the loss of any volatile components. In addition, adequate records of the circumstances and conditions during sampling have to be made, for example, in sampling from storage tanks, the temperatures and pressures of the separation plant and the atmospheric temperature would be noted.

At the other end of the volatility scale, products such as natural gasoline that have the potential to contain, or are composed of, high molecular weight paraffin hydrocarbons that are also in a solid state (wax, C_{17+}) because of dissolution at some point of the production process that may require judicious heating (to dissolve the wax) and agitation (homogenized, to ensure thorough mixing) before sampling. If room temperature sampling is the *modus operandi* and product cooling causes wax to precipitate, homogenization to ensure correct sampling is also necessary.

Representative samples are prerequisite for the laboratory evaluation of any type of product, and many precautions are required in obtaining and handling representative samples (ASTM D1265, 2017). This test method describes the equipment and procedures for obtaining a representative sample of specification, LPG, and is suitable for obtaining representative samples for all routine tests for liquefied petroleum gases. The procedure may also be used for other natural gas liquid products that are typically in a single phase, but the procedure is not intended for nonspecification products that contain significant quantities of undissolved gases (such as nitrogen and carbon dioxide), free water, or other separate phases, such as raw or unprocessed gas/liquid mixtures and related materials. In this latter case (i.e. raw or unprocessed gas/liquid mixtures and related materials), the same equipment can be used, but additional precautions are generally needed to obtain representative samples of multiphase products (ASTM D3700, 2017).

The precautions depend upon the sampling procedure, the characteristics (low-boiling or high-boiling constituents) of the product being sampled, and the storage tank, container, or tank carrier from which the sample is obtained. In addition, the sample container must be clean, and the type to be used depends not only on the product but also on the data to be produced.

When taking a sample for a pipeline, a sample should be taken on the longest section of the pipeline available to reduce inference from swirls (turbulent flow) caused within the pipeline. In addition, all samples should be taken through a sample probe. A sample probe is usually a valve with a piece of tubing welded to the bottom of it. There are several types of sample probes available, such as (i) the single flow probe, (ii) the dual flow probe, and (iii) the hot-tap insertion probe. Whichever probe is used, the most important thing to remember is to make sure the tip of the probe is as near as possible to the center of the pipeline. A probe that is inserted in an area that is close to the pipeline wall of the pipe can pick up liquids. When placing a probe on a multiple meter tube station, the probe should be placed on the tube that will be flowing all the time. The probe should not be used on a header – this can create swirling gases and dead gas pockets, which will throw doubt on the true representative nature of the sample and, hence, on the repeatability of the analytical data.

Arguably, the most important factor when performing any type of gas sampling is the hydrocarbon dew point. In the simplest terms, hydrocarbon dew point is the point at which the gas components begin to change phase from gaseous to liquid. When phase change occurs, certain components of the gas stream separate from the gas stream and form liquids, thereby making an accurate gas sample impossible to obtain. The hydrocarbon dew point is a function of gas composition and pressure. A hydrocarbon dew-point curve is a reference chart that determines the specific pressure and temperature at which condensation occurs. No two hydrocarbon dew-point curves are alike due to differing gas compositions.

Accurate sampling from gas streams with temperature at the hydrocarbon dew-point temperature is more difficult than sampling from streams with temperatures above the hydrocarbon dew-point temperature (API, 2001). If, at any part of the sampling process, the temperature of the sample should fall below the hydrocarbon dew point, a nonrepresentative sample is the likely result leading to nonreproducible and imprecise data. In order to avoid this problem, the sample gas temperature must remain above the gas hydrocarbon dew point during sampling. This can be accomplished on an *ad hoc* basis (a temporary basis that is suitable for one sample acquisition only and must be renewed for each sample to be taken) by heating the sample probes and by applying heat tracing lines, to the regulators and to the sample cylinders, or by employing a defined method of delivering heat to the fluid in the sampling system.

Due to the uncertainty in measuring or calculating the hydrocarbon dew point, it is recommended that the gas being sampled should be maintained at 20–50 °F (11–28 °C) above the expected hydrocarbon dew point throughout the sampling system. If ambient temperatures are above the hydrocarbon dew point, heating may not be required. When the sampling process involves a pressure reduction, sufficient heat should be provided at or prior to the point of pressure reduction to offset the Joule–Thomson effect (approximately 7 °F (3.9 °C) per 100 psi of pressure reduction).

The key is any sampling program, whether it is related to natural gas or any other product, is the acquisition of a representative sample of the gas (GPA, 1986; API, 2001; Mokhatab et al., 2006; Speight, 2007, 2015). In the contact of natural gas, a variety of sampling methods have been defined (Table 5.4).

Table 5.4 Sampling methods as defined by the Gas Processors Association (GPA, 1986).

Method	Comments
Fill and empty	Requires sample line separator, sample cylinder, extension tube, appropriate valves and gauges
	Involves purging the sampling apparatus and then repeatedly filling and emptying the apparatus
	The number of cycles is line pressure-dependent
Controlled rate method	Requires the same apparatus as the fill and empty method with the addition of a flow tube plug (orifice) in the extension tube
	Allows for a controlled flow of sample through the apparatus for a specific length of time
	The amount of time varies depending on line pressure at the sample source
Evacuated cylinder method	Requires an evacuated sample container with a pressure of 1 mm Hg or less
	Instead of an extension tube, this method requires a vent tube installed between the sample source and the sample cylinder
	After a careful purge procedure, the cylinder is filled with natural gas
Reduced pressure method	Similar to the evacuated cylinder method
	Not recommended when line pressure is below 100 psi
Helium pop method	The sampling begins with an evacuated cylinder (1 mm Hg)
	Carefully fill sample cylinder with helium (approximately 5 psi)
	Use the same sampling configuration as the evacuated cylinder method
	After purging the sampling apparatus, the cylinder is filled to line pressure
	Requires a separate determination for helium to calculate the correct BTU content
Glycol or water displacement method	A sample cylinder filled with clean glycol or water and a vessel to measure the displaced liquid are required for this method
	Carefully purge the sampling apparatus
	Open sample source and slowly open outlet valve allowing displacement of fluid

Table 5.4 (Continued)

Method	Comments
	Close outlet valve as soon as gas can be seen or heard
	The source gas must displace all fluid during sampling
Floating piston cylinder method	The sample cylinder has a piston that creates a precharge chamber and sample chamber
	Fill the precharge chamber with an inert gas (slightly above line pressure)
	Carefully purge the sampling apparatus
	Slowly open the outlet valve, allowing the sample to displace the piston and fill the cylinder
	The amount sampled should not exceed 80% of the cylinders capacity

These methods provide the potential for obtaining a representative example, provided that extreme care is taken while the samples are being obtained. In the current context, the following methods are the most common methods in use. As mentioned earlier, three techniques are normally used to obtain gas samples: (i) spot sampling. (ii) online sampling and analysis, and (iii) composite sampling.

5.2.2.1 Spot Sampling

Spot sampling is taking a sample at a given point in time. This difference is very important when considering the definition of a representative sample. Therefore, a spot sample acquired by this method may or may not necessarily be a *representative sample* of the gas, when considered as a whole. Thus, spot sampling, by definition, is different from composite sampling or constant sampling and was the primary method of acquiring a sample for analysis until the early 1970s, although in many gas fields, the method is still used. However, natural gas is an extremely fragile product, and at almost every step in the production, transportation, processing, and distribution, the gas will be exposed to potentially adverse effects on the composition and hence on the quality.

Early spot sampling was performed by simply connecting a sample cylinder to the pipeline until the pressure in the sample cylinder reached that of the pipeline. The cylinder was then removed and sent to a laboratory for analysis. Laboratories would test the sample using either a chromatograph or calorimeter. During this time the BTU value or "quality" of the gas was becoming increasingly important for producers, distributors, and end users. The sampling method at that time was suspected of introducing various contaminants and thereby providing erroneous data in regard to the actual gas composition and BTU value.

Gas from different wells, pressure changes as the gas flows to the wellhead, pressure changes in separators and storage vessels, and temperature changes in separators and in storage vessels are only a few of the parameters that can add to or subtract from the heating value of the gas as it is moved one step at a time to the consumer. Thus, it is necessary to recognize that a spot sample may not even represent the quality of the gas that is sent to the processing plant, where incorrect data can cause (i) transportation problems and (ii) processing problems before shipping to the consumer.

The dropout of liquid hydrocarbon derivatives from gas streams can cause a number of problems in gas transmission lines, including increased pressure drop, reduced line capacity, and equipment problems such as compressor damage (Dustman et al., 2006; Bullin et al., 2011). To avoid liquid dropout, most current operating specifications for gas transmission lines require that the lines be operated above the hydrocarbon dew-point temperature. For example, in a gas separator, phase equilibrium exists at the line pressure and line temperature, but if the temperature drops below the hydrocarbon dew point, liquid hydrocarbons will condense from the flowing stream. These liquids, which may appear as droplets, may be entrained in the flow stream or, more commonly, they can become attached to the pipe wall and are carried along the pipe wall in the direction of the gas flow. In this situation, two-phase flow exists, and sample distortion (really, sample unreliability) will occur if a sample is taken at this location unless steps are taken to maintain the temperature above the hydrocarbon dew point.

Various pipeline contaminants can have a large effect on the measured hydrocarbon dew-point temperature. One drop of compressor oil or glycol from upstream dehydration will raise the measured hydrocarbon dew-point temperature substantially. Furthermore, while the dew point identifies the condition at which vapor first begins to condense to liquid, it provides no information about the quantity of condensation, resulting from a small degree of cooling (Bergman et al., 1975). The condensation rate of liquids in gas transmission lines may vary widely depending on the composition, temperature, and pressure of the system (Bullin et al., 2011). The quantity of the liquid dropout can, however, be estimated from knowledge about the composition of the gas stream and the amounts of hydrocarbon constituents as well as the dew-point temperature of each constituent (Bullin et al., 2011).

The potential for cold ambient conditions coupled with the fact that many separator designs expose the gas to the ambient environment readily contributes to the potential for two-phase flow in the gas stream. When receiving such a sample, the analyst should ask for details of when, where, and how the sample was taken, as well as the temperature within the gas stream followed by an investigation (by the analyst) of the proposed contents of the gas stream and, in the case of hydrocarbon constituents, the dew points of the various hydrocarbons.

5.2.2.2 Online Sampling and Analysis

Online analysis is convenient, although it is dependent on the accuracy of the analyzer, its correct calibration, and the quality of the sample reaching it. It is also important to point out that with online units, there is no second or third chance at analysis and no second opinion option, as is the case of a sample in a sample cylinder. Most online analyzers can also provide other important information, such as archived data (hourly, daily, or monthly averages for BTU or gas compositions, etc.), and limit alarms for component concentrations that may go out of a specified range.

Online gas chromatographs have been widely used when rigid custody transfer standards are needed for natural gas trading and when gas specifications need to be monitored closely. Gas chromatography is a scientific method in which a gas sample is separated into its component parts for measurement. The gas chromatograph consists of subsystems that inject the sample, separate the sample, detect the components, integrate the peaks, and report the results. The injection, separation, and detection all occur in the heart of the gas chromatograph. An online gas chromatograph can provide almost real-time measurement of the natural gas, and because the gas chromatograph is online, it is able to run an analysis of the flowing gas every few minutes and supply the flow computer with up-to-date data.

5.2.2.3 Composite Sampling

Composite sampling (sometimes also called *grab sampling*) is the collection of the gas by direct introduction into a sample cylinder from a probe/valve combination or by means of a timed or proportional-to-flow sampler. It is the median between spot sampling and the continuous online analytical gas chromatographs. A composite gas sampler or gas sampling system consists of a probe, a sample collection pump, an instrumentation supply system, a timing system, and a collection cylinder for sample transportation. The sole objective is to collect and store a representative composite sample at line conditions, allowing it to be transported to the laboratory for repeatable analysis. Composite sampling is performed when a sample is taken over a specific amount of time or a specific amount of flow. Therefore, composite sampling is normally described as either *flow based* or *time based.*

Composite sampling is the most representative form of taking a sample. A composite sampler takes a small bite of sample from the pipeline and injects it into a sample cylinder. If the sampler is connected to a flow computer, it is possible to take samples proportional to flow. This means that if the flow rate goes up, the sampling rate will increase; if the flow goes down, the sample rate will decrease. The most important thing to remember when taking a sample from a flowing stream is that all sampler components must be above the hydrocarbon dew point (API 14.1). If any component of your sampling system is below the hydrocarbon dew point, the heavy components of the sample will liquefy and drop out usually causing a lean sample result (API, 2001).

The flow based or flow proportional sampler is electronically set to take a sample proportional to flow. This means that the sample rate is not fixed as in time-based sampling and will increase or decrease dependent upon the process flow rate. Typically, the signal to sample will come from a nearby flow computer, and there are a number of methods used to calculate when that signal should be sent. As with time-based calculations, the main variables are bite size and sample cylinder size. Using these and the average daily flow rate, the flow signal to sampler frequency can be determined.

Time-based sample control is designed to inject a selected amount of sample (bite) over a predetermined time to fill a sample cylinder to line pressure. Calculation tables are available to determine the time versus the amount of sample to be injected into a specified volume sample cylinder. The variables in the calculation would be the sample bite size and the size of the sample cylinder. The sample period may also be a variable but is typically 30 or 31 days. Samplers programmed as time based should have the capability to stop sampling when a no-flow condition is detected. Time-based sampling can be successfully used to obtain a representative average sample when the gas stream is known to have a constant composition or steady flow rate.

Using the composite sampling method, it is possible to obtain a representative sample over a predetermined period. It is the only practical method for collecting a continuous sample. The composite sampler will introduce a set volume taken in equal amounts to the collection cylinder over a set period and is the preferred method when a representative sample has to be taken over time.

Composite sampling has the advantage of providing a precise measurement over a given period when using a timer and can also take samples proportional to flow when taking a modified signal from a flowmeter. In addition, the sample is taken from the flowing stream at the system pressure and can be fed into the sampler or sample cylinder at the pressure of the gas flow. Thus, any change in composition is avoided. Another feature required of any sampler is that it should not have areas or pockets where residue of previous samples can accumulate and must take a fresh grab or bite of gas each time it samples.

The major considerations of a composite sampling system installation and its operations are (i) locating the right sample point/sample probe, (ii) choosing the appropriate sampling system configuration, (iii) components of the sampling system, (iv) controlling the sampling system, and (v) sample cylinder selection. Each of the components must receive proper attention, or the results of the sample may be compromised and provide unreliable analytical data.

5.2.3 Sampling Location

The location of the sampler is critical in that improper placement can affect the ability to obtain a representative average sample. The sample point should be on a horizontal level run of pipe, and consideration should be given to nearby turbulence factors. The ideal mounting location would be at least five pipe diameters downstream from elbows, tees, fittings, valves, orifice plates, or any other restricting devices. These devices can create swirls (recirculation eddies), which may have a composition markedly different than the flowing stream. These disturbances can also pull liquids off the pipe walls and create aerosols that will adversely affect the sampling process. Typically, the location of the sample point should be in a section of the pipeline that includes a positive velocity, minimum turbulence, and the sample tap being located on the top of the pipe.

In all sampling applications that are applied to natural gas, it is important that the sample probe be in a location that best represents the product to be sampled. For example, when natural gas sampling is the objective, the focus tends to be on the composition of the flowing stream that is in a gas phase. If the sample probe is in an inappropriate location due to the lack of objective consideration, the gas sample entering the system will more than likely not be a representative sample, and the end analysis from the analytical laboratory will be inaccurate. Thus, for pipeline sampling, the sample point should be at the top of the pipeline and in a central position when it is mounted in a straight and horizontal pipe section. In addition, it is recommended that the sampling location should be a distance equivalent to five pipeline diameters away from orifice plates, pipe bends, fittings, valves, or other restricting devices that create turbulent flow.

The most common mounting method for composite samplers is the direct mount method. The pump is close coupled with the sample probe and mounted directly to the thread-o-let on the process pipe. The direct mount system also referred to as probe mounted is generally preferred for several reasons. The primary one being that the sampler mounted directly on the probe provides the shortest distance to the sample point and, ultimately, the freshest sample. Most samplers will include a port for a hot loop, which keeps the gas flowing through the sample chamber. Single flow sample probes are typically used with the direct mount sampler, and dependent upon the type of sample probe used, the hot loop may or may not be needed.

Remote mounting is typically done when physical constraints such as obstructions or access issues prevent the pump from being direct mounted. The pump will typically be mounted on a two-inch pipe stand or other peripheral structure as close as possible to the sample point. Keeping the tubing between the pump and the probe as short as possible and using a dual flow probe are key to the success of a remote mount system. Dual flow probes are preferred on remote mounted samplers to enable positive flow through the sample chamber. If a dual flow probe is not used, the hot loop should return to a lower-pressure point downstream.

Continuous sampling is exactly like it sounds, continuous.

Automated composite gas sampling systems are designed to take small bite size samples from the flowing stream at regular intervals and store them in a sample container or cylinder. The samples taken can be either time based or proportional to flow. When the cylinder is filled, the typical sample period is 30 or 31 days; it is taken to the lab for analysis. These systems also provide an economical solution when increased sampling frequency is a necessity.

The fill and purge method was adopted and after a period of time, it was determined that this method was causing condensation, and therefore the method recommended by the Gas Processors Association was adopted. This method utilizes a manifold or a *pigtail* for filling the single cavity cylinder. This method reduced the negative effects connected to the fill only procedure. The manifold allows the gas to be trapped in the cylinder instead of *dead ended* into the cylinder. In addition, sampling situations with free liquids present in the pipeline required an additional approach, and a liquid separator was developed and added to the technique.

When liquids are present in the pipeline, initiation and continuation of turbulent flow can create droplets (often referred to as aerosols) that are a mixture of liquids and gases flowing together. The liquids that contribute to the creation of the aerosol can be present along the bottom or upper circumference of the wall of the pipeline. Areas of turbulence initiate movement of the liquids where they mix with the flowing gas to create a mist (aerosol). If the sample point is located in this region of aerosol occurrence (or close to the region of aerosol occurrence), the aerosol will also be a part of the sample that is stored in the sample cylinder. This can only lead to interference with the analytical method in the form of unreliable data and an inaccurate outcome.

5.2.4 Sample Probe

Selection of the sample probe is an integral part of the analytical sampling and analytical method. The type of sampling probe can dictate whether or not the analytical data are reliable and accurate. Thus, once the most appropriate sampling point has been selected, the next consideration is the type of probe to be used. Two types of probes are commonly used: (i) the single flow probe and (ii) the dual flow probes.

As stated above, the inlet of any probe must be located in the top center of the pipeline to ensure that the sample taken will be from a region of representative flow (nonturbulent flow) and avoid sampling any deposited liquids that may be present along the bottom or upper circumference of the pipe wall. In a composite sampling system, there should be an option for purging the volume of gas residing in the single flow probe before the sampling cycle begins. To achieve this, some sampling systems bleed gas continually either to the atmosphere or to a low-pressure source to ensure that the probe is purged before the sampling cycle begins.

5.2.5 Representative Sample

As mentioned earlier, representative samples are always a prerequisite for the laboratory evaluation of natural gas and its products (Speight, 2007, 2015), and precautions are required in (i) obtaining and (ii) handling representative samples. The commercial value of any petroleum product (as well as the parent petroleum) is assessed by the characteristics of the sample as determined by standard test methods. In summary, the sample must be representative of the lot, or a wrong evaluation will be made, no matter how accurately the sample may be tested. In addition, each type of petroleum or petroleum product (gas, liquid, or solid) will more than likely require application of a different sampling method to assure that the sample is truly a representative sample.

Whether or not a sample is a representative sample depends on the sampling procedure, the type of petroleum or petroleum product being sampled, and the storage tank, container, or tank carrier from which the sample is obtained. The sample container must be clean, and the type to be used depends not only on the product but also on the laboratory data desired. Thus, the objective of each procedure is to obtain a sample or a composite of several samples in such manner and from such location in the tank or other container that the sample or composite will be representative of the entire bulk material. The sample container – which must be clean and free of any possible source of sample contamination and should not be made of material which either reacts with the sample or is a catalyst – should be labeled immediately by the sampler, to indicate at least (i) the nature/name of the sample, (ii) time of sampling, (iii) location of the sampling point, and (iv) any other information necessary for the sample identification.

Thus, the basic objective of each sampling procedure is to obtain a truly representative sample or, more often, a composite of several samples that can be considered to be a representative sample. In some cases, because of the size of the storage tank and the lack of suitable methods of agitation, several samples are taken from large storage tanks in such manner that the samples represent the properties of the bulk material from different locations in the tank and, thus, the composite sample will be representative of the entire lot being sampled.

The composite sampling method compensates for any potential differences in the sample composition that might be due to the stratification of the bulk material due to tank size or temperature at the different levels of the storage tank.

5.2.6 Sampling Error

When a property of the sample (which exists as a large volume of material) is to be measured, there usually will be differences between the analytical date derived from application of the test methods to a *gross lot* or *gross consignment* and the data from the *sample lot.* This difference (the *sampling error*) has a frequency distribution with a mean value and a variance. *Variance* is a statistical term defined as the mean square of errors; the square root of the variance is more generally known as the *standard deviation* or the *standard error of sampling.*

Analytical test methods applied in the field – and such logic can be applied to all sampling protocols – must take into consideration the various site conditions comprising local heterogeneity as well as the changing ambient conditions. Furthermore, the analysts must take into account the potential for samples to be disturbed during sampling, storage, and pretreatment. Rather than attempt to rationalize the application of corrections because of such disturbances, it is better that steps be taken to avoid sample disturbance. If corrections must be applied, the manner in which the corrections are applied must be beyond reproach, or claims of the data will not stand up (when needed) to legal scrutiny, and falsification of the data will be the most likely result (Speight and Foote, 2011).

5.2.7 Sample Container

Choice of the most appropriate sample container is not a matter-of-fact issue; it is a necessarily important part of the sampling program since selection of an inferior (unsuitable) container can not only ruin the integrity of the sample but also lead to inaccurate and/or unreliable analytical data. The analyst needs to be aware that the incorrect choice of a sample contain can lead to delivery to the laboratory of a nonrepresentative sample that negates the test data, no matter how well the test protocol was followed.

There are two types of sample containers that are used: (i) the spun-end bottle and (ii) the constant pressure cylinder. The spun-end bottle is made of stainless steel or aluminum. If the bottle is longer than twelve inches or greater than four inches in diameter, the general requirement for transportation requires that a pressure-relieving device should be installed on the cylinder to protect the bottle from overpressuring and the potential for an explosion. When sampling into a spun-end bottle, it is recommended that the bottle be placed in the vertical position to prevent the settling of heavy gases during the purging process (GPA 2166–68). Also, if the bottle (sample container) is longer than twelve inches or greater than four inches in diameter, the US Department of Transportation (US DOT) requires that a relieving device be installed on the cylinder. This is a safety issue to protect the bottle from overpressuring.

If the BTU value (heat content) is in excess of 975 Btu lb^{-1}, it is advisable to use a constant pressure cylinder. Also, if the heat content is in excess of 975 Btu lb^{-1} and if the sample is being injected into a spun-end bottle, there may be a problem related to pressure change. For example, starting at atmospheric pressure in the cylinder and increasing to pipeline pressure, it is possible for some of the gas constituents to change state (flash) as the pressure builds, and, thus, the sample will, more than likely, not be representative of the pipeline gas, i.e. the sample taken will not be a representative sample. By using a constant pressure cylinder, the gas in the cylinder, as well as the gas being sampled, will stay at pipeline pressure, which prevents the gas from flashing or changing phase.

By way of explanation, in crude oil and natural gas production, the flash gas is a spontaneously generated vapor that is produced by the heating or depressurization of the extracted crude oil or natural gas mixture during the different phases of production. Flashing (also called *flash evaporation*) is the term that describes this process of volatile components suddenly vaporizing from their liquid state to the gaseous state. This often happens during the transportation of petroleum products (such as gases and low-boiling liquids) and natural gas products through pipelines and into vessels, such as when the stream from a common separation unit flows into an on-site atmospheric storage tank. Vessels that are used to intentionally flash a mixture of gas and saturated liquids are aptly named flash drums, and a venting apparatus is used in these vessels to prevent damage due to increasing pressure; extreme cases of this are referred to as boiling liquid expanding vapor explosion (BLEVE).

Once the sampling procedure is accomplished, the sample container should be labeled immediately, to indicate the product, time of sampling, location of the sampling point, and any other information necessary for the sample identification (Table 5.3). Moreover, if the samples were taken from different levels of the storage tank, the levels from which the samples were taken and the amounts taken and mixed into the composite should be indicated on the sample documentation.

When preparing the cylinders for drawing off the representative sample, there are some points to consider and these are:

1) When using a spun-end bottle, there is the need to heat the cylinder so that any hydrocarbon derivatives that might have liquefied will revert to the vapor state.
2) When using a constant pressure cylinder, there is the need to apply pressure to the precharged side equal to the bottle pressure so that when the sample is being drawn off, the pressure will remain at pipeline pressure.
3) Specifically, for the laboratory analyst, another consideration is the use of a calibration standard that is close to the pipeline gas. If the calibration

standard is not available, some of the properties that are present in the gas, the data obtained from the sample may not be truly representative of the properties of the pipeline gas – the term *unreliability of the data* raises its ugly head again!

While the above text has focused on the acquisitions of samples from storage tanks, sampling records for any procedure must be complete and should include a variety of items that document the history of the sample (Table 5.3). In summary, there must be a means to accurately track and identify the sample history so that each sample is tracked and defined in terms of source, activity, and the personnel involved in any of the above stages. Thus, the accuracy of the data from any subsequent procedures and tests for which the sample is used will be placed beyond a *reasonable doubt* and would stand the test of time and in court, should legal issues arise.

5.2.8 Transportation

After the sample has been obtained, it will be transported to a laboratory for analysis. All samples being transported must meet and comply with the US DOT certification. CFR – 49 covers the rules and regulations required for transporting samples under pressure. Other considerations include (i) proper tagging of cylinder for time, date, and location, (ii) pressure and temperature of source gas, (iii) name of the person who obtained the sample, (iv) the method used for obtaining the sample, (v) checks for leaks, (vi) proper protection of cylinder for transport, and (vii) ambient temperature concerns during transportation. Considerations for analysis include calibration standards utilized at different laboratories. It is necessary to ensure that the calibration standard includes elements similar to the gas being analyzed. If the calibration standard does not include all elements of the sample being tested, the BTU measurement can be compromised.

Accurate samples translate directly to the bottom line of all companies.

5.3 Volume Measurement

As part of the analytical procedure for natural gas, the gas processing operators also need to know not only the bulk properties of the material but also the amount of materials held in storage tanks – as part of, say, the natural gas and natural gas product inventory. In fact, once the natural gas or natural gas product has been sampled, there is the need to calculate the total volume of the material that remains in storage. Following from this, the amount of natural gas or natural gas product involved in commercial transactions can be calculated with a high degree of accuracy. The same volume-remaining issue is relevant for coal gas, biogas, and landfill gas.

Many types of instruments can be used to sample methane gas at landfills. Instruments will typically measure on the lower explosive limit (LEL) scale or by volume of methane. Instruments that measure on the LEL scale only typically detect up to 100% of the LEL or 5% v/v. When methane exceeds the LEL (or 5% v/v), determining the exact amount of methane can be problematic if an instrument with a LEL scale is used.

The complex nature of natural gas and natural gas products requires several procedures for calculating volume or density of liquid products in commercial transactions. In addition, temperature affects the volume and density of natural gas and its products. Measured volumes must be converted to net quantities at 15.6 °C (60 °F), which is the accepted base temperature used widely in commercial transactions. Again, obtaining a representative sample of the material that is held in storage or carrier tanks is essential for converting measured quantities to the standard volume. To assist with volume calculations, petroleum measurement tables that can also be applied to natural gas are available for use in the calculation of quantities of natural gas and natural gas products as they exist under designated conditions (ASTM D1250, 2017). The tables presented in the test method provide data that cover typical operating ranges for the reduction of gravity and volume to standard states, for calculation of weight–volume relationships, and interconversion of a wide variety of units that are of value for commercial sales of the gas or product. However, this test method provides factors for temperature and pressure volume correction for generalized crude oils and refined products. Natural gas liquids and liquefied petroleum gases are excluded from consideration.

Following from this, the issues that always face analytical laboratory personnel include the need to provide higher-quality results. In addition, environmental regulations may influence the method of choice. Nevertheless, the method of choice still depends to a large extent on the boiling range (or carbon number) of the sample to be analyzed. For example, there is a large variation in the carbon number range and boiling points (of normal paraffins) for some of the more common petroleum products and thus a variation in the methods that may be applied to these products (Speight, 2015).

Testing for suspended water and sediment is used primarily with fuel oil (which may also include the liquid fuels such as gas condensate and natural gasoline produced from natural gas), where appreciable amounts of water and sediment may cause fouling of facilities for handling the gaseous fuel (including the gas condensate or natural gasoline) and giving trouble in burner mechanisms. Three standard methods are available for this determination: (i) the centrifuge method, which gives the total water and sediment content of the sample by volume, (ii) the distillation method, which gives the volumetric amount of water, and (iii) the extraction method, which gives the solid sediment in per cent by weight.

Sediment usually consists of finely divided solids that may be dispersed in the liquid fuel or carried in water droplets. The solids may be fine particles of sand or pipeline debris picked up during the transport of the natural gas. In any event, the sediment can lead to serious plugging of the equipment, abrasive corrosion, and a lowering of fuel quality.

Water may be found in the natural gas as it emits from the production well and may occur in an emulsified form if the temperature drops below the hydrocarbon dew point; liquid hydrocarbons have condensed from the flowing stream. The quantity is generally limited by pipeline companies and by gas processing operators, and steps are normally taken at the wellhead to reduce the water content as low as possible. Prior to analyses, it is often necessary to separate the water from a natural gas sample, especially gas condensate or natural gasoline, which is usually achieved by one of the procedures described in the distillation of crude petroleum (ASTM D4006, 2017) or by the coulometric Karl Fischer titration method (ASTM D4928, 2017). The former test method (ASTM D4006, 2017) may not be suitable (without modification) for gas condensate and natural gasoline because of the lower-boiling ranges of these gas-related products.

The determination of density or specific gravity (ASTM D287, 2017; ASTM D1298, 2017) as used for the measurement and calculation of volume of gas condensate and natural gasoline is an important measurement since gravity is an index of the weight of a measured volume of the product. Two scales are in use in the petroleum industry, specific gravity and API gravity, the determination being made in each case by means of a hydrometer of constant weight displacing a variable volume of liquid hydrocarbon (Speight, 2014a, 2015). The reading obtained depends upon both the gravity and the temperature of the liquid.

Accurate determination of the gravity of petroleum and its products is necessary for the conversion of measured volumes to volumes at the standard temperature of 60 °F (15.6 °C). This procedure (ASTM D287, 2017) is most suitable for determining the API gravity of low viscosity transparent liquids. Gravity is a factor governing the quality of crude oils. However, the gravity of a petroleum-related product is an indication of its quality. Correlated with other properties, gravity can be used to give approximate hydrocarbon composition and heat of combustion. This test method covers the determination by means of a glass hydrometer in conjunction with a series of calculations of the API gravity of crude petroleum and petroleum products normally handled as liquids and having a Reid vapor pressure (ASTM D323, 2017) of 14.696 psi or less. Gravities are determined at 60 °F (15.6 °C) or converted to values at 60 °F (see ASTM D1250 (2017) Guide for Petroleum Measurement Tables).

The alternate procedure (ASTM D1298, 2017) is most suitable for determining the density, relative density (specific gravity), or API gravity of low viscosity transparent liquids. This procedure can also be used for viscous liquids by allowing sufficient time for the hydrometer to reach temperature equilibrium and for opaque liquids by employing a suitable meniscus correction. Additionally, for both transparent and opaque fluids, the readings shall be corrected for the thermal glass expansion effect and alternative calibration temperature effects before correcting to the reference temperature. Without serious modification, this method may not be suitable for application to gas condensate or natural gasoline.

5.4 Method Validation

Method validation is the confirmation by examination and the provision of objective evidence that the specifications for an intended use are fulfilled. Validation is conducted to confirm that the methods are fit for the intended use by addressing the needs of the given application or field of application, and the validation must be documented. The testing laboratory validates non-standard methods, laboratory-developed methods, and modified standard methods including use outside the intended scope and applications. The validation results include a statement as to whether the method is fit for the intended use. The needs of the customer define the intended use of the method. The attributes and data quality objectives include but are not limited to (i) accuracy, (ii) precision, (iii) specificity, (iv) detection limit, (v) limit of quantitation, and (vi) applicability of the method. If all of the data quality objectives are met as indicated by the data collected, the method is considered as validated. Furthermore, in any laboratory that applies standard test methods to the analysis of gas stream (or any other materials for that matter), there must be adherence to the written protocol for method performance verification, method validation, or method modification.

By way of definition, the verification of an analytical procedure is the demonstration that a laboratory is capable of replicating with an acceptable level of performance using a standard test method. Verification under conditions of use is demonstrated by meeting system suitability specifications established for the method as well as a demonstration of accuracy and precision or other method parameters for the type of method. Method performance in relation to the analysis of gas streams is accomplished by using performance characteristics such as (i) blanks to assess composition or the degree of contamination by nonhydrocarbon constituents, (ii) laboratory control samples to assess the accuracy of the method, (iii) the precision of the method based on the analysis of duplicate samples, (iv) the use of calibration check standards that are analyzed periodically in the analytical batch for quantitative analyses, and (v) monitoring quality control samples, usually through the use of, for example, control charts (Table 5.5).

Table 5.5 Example of materials used for method validation.

Tool	Comment
Blanks	Enables an assessment of the response of the analyte and any response attributable to other causes
Reference materials	Used to assess the accuracy of the method
Spiked materials and solutions	Recovery assessment made from spiking (fortification) with a known amount of analyte
Incurred materials	Materials in which the analyte of interest may be essentially alien
Measurement standards	Used for calibration or identification purposes
	Used to check the response of the test method to the analyte
Replicate analysis	Provides a check for changes in precision
Statistical techniques	Employed to evaluate accuracy, precision, limits of detection and quantification, and measurement uncertainty

In the simplest sense, to avoid contamination, the method is to collect just the gas to analyze either in an empty bag or in an evacuated bottle. One of the older (and cleaner and easier) ways to hold (and manipulate) a gas sample is to use an upside-down glass bottle initially full of water and partially submersed open mouth in a water bath: the gas can be easily introduced in the bottle through a flexible tubing and get trapped inside, displacing some water by buoyancy, and the hydraulic stop is most convenient (to have an airtight joint is very difficult) to secure the rubber stopper. A nonreturn valve may simplify sample collection by just opening the valve in an evacuated rigid container (or in a preloaded flexible or elastic bag).

In order to validate a modified test method where the testing procedure is modified from the standard or existing testing procedure and protocol, it must be demonstrated that the modifications do not adversely affect the precision and accuracy of the data obtained. In order to implement the modification, the standard or existing method should be first performed, and each major modification can then be verified against the original method. The results from the validation study are submitted for approval as designated by the laboratory. Statistical techniques are employed to evaluate the method performance and determine its use.

Throughout the process of natural gas analysis, the validation of the various standard test methods must be certain since the data produced by the method demonstrates the scientific soundness of the measurement or characterization. Thus, method validation is a process of demonstrating that the method meets the required performance capabilities. Most international standards, certifying organizations, and regulatory agencies require evidence that analytical methods are capable of producing valid results (ASTM E2857, 2017). This applies to laboratories using published standard test methods, modified standard test methods, and in-house test methods. The use of multiple validation techniques described in this guide increases confidence in the validity or application of the method (ASTM E1601, 2017; ASTM E1763, 2017).

The validation practice demonstrates that an analytical method measures the correct substance in the correct amount and in the appropriate range for the intended samples and allows the analyst to understand the behavior of the method and to establish the performance limits of the method. Moreover, in order to perform method validation, the laboratory should be following a written standard operating procedure that describes the process of conducting method validation as well as using qualified and calibrated instrumentation with a corresponding standard operating procedure. There should be a well-developed and well-documented standardized test method in place, and an approved protocol should be in place prior to the execution of any validation experiments.

Method validation is the process of demonstrating that an analytical method is acceptable for its intended purpose. In general, methods for product specifications and regulatory submission must include studies on specificity, linearity, detection limit, and quantitation limit as well as accuracy and precision (ASTM E2857 (2017) Standard Guide for Validating Analytical Methods). The process of method development and validation covers all aspects of the analytical procedure, and the best way to minimize method problems is to perform validation experiments during development. However, method validation cannot mitigate out all of the potential problems; the process of method development and validation should address the most common (anticipated) problems.

5.4.1 Requirements

In order to perform validation studies, the approach should be viewed with the understanding that validation requirements are continually changing and vary widely, depending on the type of product under tested and compliance with any necessary regulatory group. Furthermore, issues arising from the use of analytical methods increase as additional (new) people unfamiliar with the method, different laboratories, and equipment are used to perform the analytical investigation. For example, a small adjustment (no adjustment is really *small*) can usually be made to make the method work, but the flexibility to change or reverse the adjustment on an as-needed basis is lost once the method is transferred to other analysts or to other analytical laboratories.

The first step in the method development and validation cycle should be to set minimum requirements, which are essentially acceptance specifications for the method. A complete list of criteria should be agreed on during method development, and the end users before the method are developed so that expectations are clear. Once the validation studies are complete, the method developers should be confident in the ability of the method to provide good quantitation in their own laboratories. The remaining studies should provide greater assurance that the method will work well in other laboratories, where different operators, instruments, and reagents are involved and where it will be used over much longer periods of time.

The remaining precision studies comprise much of what is often referred to as *ruggedness*. *Intermediate precision* is the precision obtained when an assay is performed by multiple analysts using several instruments on different days in one laboratory. Intermediate precision results are used to identify which of the above factors contribute significant variability to the final result.

The last type of precision study is *reproducibility* (*q.v.*) that is determined by testing homogeneous samples in multiple laboratories, often as part of inter-laboratory crossover studies. The evaluation of reproducibility results often focuses more on measuring bias in results than on determining differences in precision alone. Statistical equivalence is often used as a measure of acceptable interlaboratory results. An alternative, more practical approach is the use of *analytical equivalence* in which a range of acceptable results is chosen prior to the study and used to judge the acceptability of the results obtained from the different laboratories.

Performing a thorough method validation can be a tedious process, but the quality of data generated with the method is directly linked to the quality of this process. Time constraints often do not allow for sufficient method validation. Many researchers have experienced the consequences of invalid methods and realized that the amount of time and resources required to solve problems discovered later exceeds what would have been expended initially if the validation studies had been performed properly. *Putting in time and effort up front* will help any analysts to find a way through the method validation maze and eliminate many of the problems common to inadequately validated analytical methods.

5.4.2 Detection Limit

The method detection limit (MDL) is the smallest quantity or concentration of a substance that the instrument can measure (Patnaik, 2004). It is related to the instrument detection limit (IDL) that depends on the type of instrument and its sensitivity and on the physical and chemical properties of the test substance.

The MDL is, in reality, a statistical concept that is applicable only in trace analysis of certain type of substances, such as organic pollutants by gas chromatographic methods. The MDL measures the minimum detection limit of the method and involves all analytical steps including sample extraction, concentration, and determination by an analytical instrument. Unlike the IDL, the MDL is not confined only to the detection limit of the instrument.

In the environmental analysis of organic pollutants, the MDL is the minimum concentration of a substance that can be measured and reported with 99% confidence that the analyte concentration is greater than zero and is determined from the analysis of a sample in a given matrix containing the analyte. For determination of the MDL, several replicate analyses are performed at the concentration level of the IDL or at a level equivalent to two to five times the background noise level. The standard deviation of the replicate tests is found. The MDL is determined by multiplying the standard deviation by the *t-factor*.

In environmental analysis, however, periodic determination of the MDL (e.g. once per year or with any change in personnel, location, or instrument) is part of the quality control requirement.

5.4.3 Accuracy

The *accuracy* of a test is a measure of how close the test result will be to the true value of the property being measured. As such the accuracy can be expressed as the *bias* between the test result and the true value. However, the *absolute accuracy* can only be established if the true value is known.

In the simplest sense, a convenient method to determine a relationship between two measured properties is to plot one against the other. Such an exercise will provide either a line fit of the points or a spread that may or may not be within the limits of experimental error. The data can then be used to determine the approximate accuracy of one or more points employed in the plot. For example, a point that lies outside the limits of experimental error (a *flyer*) will indicate an issue of accuracy with that test and the need for a repeat determination.

However, the graphical approach is not appropriate for finding the absolute accuracy between more than two properties. The well-established statistical technique of regression analysis is more pertinent to determining the accuracy of points derived from one property and any number of other properties. There are many instances in which relationships of this sort enable properties to be predicted from other measured properties with as good precision as they can be measured by a single test. It would be possible to examine in this way the relationships between all the specified properties of a product and to establish certain key properties from which the remainder could be predicted, but this would be a tedious task.

An alternative approach to that of picking out the essential tests in a specification using regression analysis is to take a look at the specification as a whole and extract the essential features (termed *principal components analysis*).

Principal components analysis involves an examination of set of data as points in *n*-dimensional space (corresponding to *n* original tests) and determines (first) the direction that accounts for the biggest variability in the data (*first principal component*). The process is repeated until *n* principal components are evaluated, but not all of these are of practical importance since some may be attributable purely to experimental error. The number of significant principal components shows the number of independent properties being measured by the tests considered.

Following from this, it is necessary to establish the number of independent properties that is necessary to predict product performance in service with the goals of rendering any specification more meaningful and allowing a high degree of predictability of product behavior. On a long-term approach, it might be possible to obtain new tests of a fundamental nature to replace, or certainly to supplement, existing tests. In the short term, selecting the best of the existing tests to define product quality is the most beneficial route to predictability.

5.4.4 Precision

The precision of a test method is the variability between test results obtained on the same material, using a specific test method. The precision of a test is usually unrelated to its accuracy. The results may be precise, but not necessarily accurate. In fact, the precision of an analytical method is the amount of scatter in the results obtained from multiple analyses of a homogeneous sample. To be meaningful, the precision study must be performed using the exact sample and standard preparation procedures that will be used in the final method. Precision is expressed as repeatability and reproducibility.

The *intra-laboratory precision* or the *within-laboratory precision* refers to the precision of a test method when the results are obtained by the same operator in the same laboratory using the same apparatus. In some cases, the precision is applied to data gathered by a different operator in the same laboratory using the same apparatus. Thus, intra-laboratory precision has an expanded meaning insofar as it can be applied to laboratory precision.

Repeatability or repeatability interval of a test (*r*) is the maximum permissible difference due to test error between two results obtained on the same material in the same laboratory.

$$r = 2.77 \times \text{standard deviation of test}$$

The repeatability interval (*r*) is, statistically, the 95% probability level, or the differences between two test results are unlikely to exceed this repeatability interval more than five times in a hundred.

The *intra-laboratory precision* or the *between-laboratory precision* is defined in terms of the variability between test results obtained on the aliquots of the same homogeneous material in different laboratories using the same test method.

The term *reproducibility* or *reproducibility interval* (R) is analogous to the term repeatability, but it is the maximum permissible difference between two results obtained on the same material, but now in different laboratories. Therefore, differences between two or more laboratories should not exceed the reproducibility interval more than five times in a hundred.

$$R = 2.77 \times \text{standard deviation of test}$$

The repeatability value and the reproducibility value have important implications for quality (Ralli et al., 2003). As the demand for clear product specifications and hence control over product consistency grows, it is meaningless to establish product specifications that are more restrictive than the reproducibility/repeatability values of the specification test methods.

Bias, on the other hand, is a systematic error that contributes to the difference between the mean of a large number of test results and an accepted reference value (ASTM E177 Standard Practice for Use of the Terms Precision and Bias in ASTM Test Methods). When included in a standard test method, this statement describes the bias and the methods utilized to provide corrected test results. It is important to remember that if an accepted reference value is not available, then the bias cannot be established. However, if the bias is unknown but the direction or bounds of the bias can be estimated, this information should be included in the bias statement. Thus, a statement of bias furnishes guidelines on the relationship between a set of typical test results produced by the test method under specific test conditions and a related set of accepted reference values.

Since precision and bias are mandatory sections of an ASTM standard test methods (in the context of analysis) (ASTM E177), the utmost care should be taken to ensure that the final data, as well as the steps that were taken to generate the data, are as precise and accurate as possible. A standard test method that is incapable of doing what it purports can be misleading if precision and bias are not taken into account. Precision and bias statements strengthen the perceived validity of the standard test method and provide the user of the document with the added confidence of knowing that the standard test method has been tested in a reliable setting.

5.5 Quality Control and Quality Assurance

An integral part of the laboratory quality system is the use of standard test methods, which are used whenever possible or unless otherwise specified by the customer. Nonstandard methods are used in cases where a standard method does not exist, and the customer has agreed to the use of the nonstandard method. However, a clear expression of the quality objectives and testing parameters or testing criteria must be discussed when a nonstandard method is employed.

Quality control (QC) and quality assurance (QA) programs are key components of all analytical protocols in all areas of analysis, including environmental, pharmaceutical, and forensic testing, among others (Patnaik, 2004). These programs mandate that the laboratories follow a set of well-defined guidelines to achieve valid analytical results to a high degree of reliability and accuracy within an acceptable range. Although such programs may vary depending on the regulatory authority, certain key features of these programs are more or less the same.

However, there is often confusion between the terms *QA* and *QC*, perhaps because there is also considerable overlap between certain aspects of QA and QC programs.

5.5.1 Quality Control

Quality controls are single procedures that are performed in conjunction with the analysis to help assess in a quantitative manner the success of the individual analysis. Examples of quality controls are blanks, calibration, calibration verification, surrogate additions, matrix spikes, laboratory control samples, performance evaluation samples, determination of detection limits, etc. The success of the QC is evaluated against an acceptance limit. The actual generation of the acceptance limit is a function of quality assurance; it would not be termed a QC.

Unlike quality assurance plans that mostly address regulatory requirements involving comprehensive documentation, QC programs are science based; the components of which may be defined statistically. The two most important components of QC are (i) determination of precision of analysis and (i) determination of accuracy of measurement.

Whereas *precision* is a measure of the reproducibility of data from replicate analyses, the *accuracy* (Section 4.4) of a test estimates how accurate are the data, that is, how close the data would fall to probable true values or how accurate is the analytical procedure to give results that may be close to true values. Both the precision and accuracy are measured on one or more samples selected at random from a given batch of samples for analysis. The precision of analysis is usually determined by running duplicate or replicate tests on one of the samples in a given batch of samples. It is expressed statistically as standard deviation, relative standard deviation (RSD), coefficient of variance (CV), standard error of the mean (M), and relative percent difference (RPD).

The standard deviation in measurements, however, can vary with the concentrations of the analytes. On the other hand, the RSD, which is expressed as the ratio of standard deviation to the arithmetic mean of replicate analyses and is given as a percent, does not have this problem and is a more rational way of expressing precision:

$$\mathrm{RSD} = \begin{pmatrix} \text{standard deviation/arithmetic} \\ \text{mean of replicate analysis} \end{pmatrix} \times 100\%$$

The standard error of the mean, M, is the ratio of the standard deviation and the square root of the number of measurements (n):

$$M = \frac{\text{Standard deviation}}{n}$$

This scale, too, will vary in the same proportion as standard deviation with the size of the analyte in the sample.

In routine testing, many repeat analyses of a sample aliquot may not be possible. Alternatively, therefore, the precision of a test may be determined from duplicate analyses of sample aliquots and expressed as the RPD:

$$\mathrm{RPD} = \frac{(a_1 - a_2)}{\left[(a_1 + a_2)/2\right]} \times 100\%$$

In another iteration,

$$\mathrm{RPD} = \frac{(a_2 - a_1)}{\left[(a_1 + a_2)/2\right]} \times 100\%$$

In this equation, a_1 and a_2 are the results of duplicate analyses of a sample. Since only two tests are performed on a selected sample, the RPD may not be as accurate as the measure of precision as the RSD, and since the RSD does not vary with sample size, it should be used whenever possible to estimate precision of analysis from replicate tests.

The accuracy of an analysis can be determined by several procedures. One common method is to analyze a "known" sample, such as a standard solution or a QC check standard solution that may be commercially available or a laboratory-prepared standard solution made from a neat compound, and compare the test results with the true values (theoretically expected values). Such samples must be subjected to all analytical steps, including sample extraction, digestion, or concentration, similar to regular samples. Alternatively, accuracy may be estimated from the recovery of a known standard solution *spiked* or added into the sample in which a known amount of the same substance that is to be tested is added to an aliquot of the sample, usually as a solution, prior to the analysis. The concentration of the analyte in the spiked solution of the sample is then measured. The percent spike recovery is then calculated.

A correction for the bias in the analytical procedure can then be made based on the percent spike recovery. However, in most routine analysis such bias correction is not required. Percent spike recovery then may be calculated as follows:

$$\text{Recovery, \%} = \frac{\text{measured concentration}}{\text{theoretical concentration}} \times 100\%$$

The percent spike recovery to measure the accuracy of analysis may also be determined by the US Environmental Protection Agency (EPA) method often used in environmental analysis:

$$\text{Recovery, \%} = \frac{\left[100\left(X_s - X_u\right)\right]}{K}$$

X_s is the measured value for the spiked sample, and X_u is the measured value for the unspiked sample adjusted for the dilution of the spike, and K is the known value of spike in the sample.

5.5.2 Quality Assurance

QA is an umbrella term that is correctly applied to everything that the laboratory does to assure product reliability. As the product of a laboratory is information, anything that is done to improve the reliability of the generated information falls under quality assurance.

QA includes all the quality controls, the generation of expectations (acceptance limits) from the quality controls, and a great number of other activities such as (i) analyst training and certification, (ii) data review and evaluation, (iii) preparation of final reports of analysis, (vi) information given to clients about tests that are needed to fulfill regulatory requirements, (v) use of the appropriate tests in the laboratory, (vi) obtainment and maintenance of laboratory certifications/accreditations, (vii) conducting internal and external audits, (viii) preparation of responses to the audit results, (ix) the receipt, storage, and tracking of samples, and (x) tracking of the acquisition of standards and reagents.

Thus, the objective of QA (usually in the form of a *QA* ***plan***) is to obtain reliable and accurate analytical results that may be stated with a high level of confidence (statistically), so that such results are legally defensible. The key features of any plan involve essentially documentation and record keeping. In short, QA program involves documentation of sample collection for testing, the receipt of samples in the laboratory, and their transfer to the individuals who perform the analyses. The information is recorded on chain-of-custody forms stating the dates and times along with the names and signatures of individuals who carry out these tasks. Also, other pertinent information is recorded, such as any preservatives added to the sample to prevent degradation of test analytes, the temperature at which the sample is stored, the temperature to which the sample is brought prior to its analysis, the nature of the container (which may affect the stability of the sample), and its holding time prior to testing.

In fact, the performance of QC is just one small aspect of the QA program. The functions of QA are embodied in the terms *analytically valid* and (in these days of perpetual litigation) *legally defensible*.

Analytically valid means that the target analyte has been (i) correctly identified and (ii) quantified using fully calibrated tests. In addition the sensitivity of the test (MDL) has been established. And the analysts have demonstrated that they are capable of performing the test. The accuracy and precision of the test on the particular sample must also have been determined, and the possibility of false positive and false negative results has been evaluated through performance of blanks and other test-specific interference procedures.

5.5.3 Specifications

A *feedstock specification* or *product specification* is the data that give adequate control of feedstock behavior in a refinery, or in a processing plant, or a measure of product quality. More accurately, the specifications are derived from the set of tests and data limits applicable to the natural gas or to a finished product in order to ensure that every batch is of satisfactory and consistent quality at release for sales. The specifications should include all critical parameters in which variations would be likely to affect the safety and in-service use of the product – they are, in fact, part of the *assay*.

Because of the differences in the composition of natural gas not only from different fields but also from different wells within a natural gas reservoir, the importance of the correct sampling of the natural gas that contains low-boiling hydrocarbons (such as ethane, propane, and butane) cannot be overestimated

Hence, the production of data focuses on (i) measurement, (ii) accuracy, (iii) precision, and (iv) method validation, all of which depend upon the sampling protocols that were used to obtain the sample. Without strict sampling protocols, variation and loss of accuracy (or precision) must be anticipated. For example, correct sampling of the product in storage or carrier tanks is important in order to obtain a representative sample for the laboratory tests, which are essential in converting measured quantities to the standard volume.

Elemental analysis of natural gas shows that it contains mainly carbon and hydrogen. Nitrogen, oxygen, and sulfur (heteroelements) are present in smaller amounts and, of the noncarbon and nonhydrogen elements, sulfur is the most important. The mixture of hydrocarbons may be much more complex than the present of only ethane, propane, and butane would indicate. Higher molecular weight paraffin derivatives, cycloparaffin derivatives, and even aromatic derivative may also be present. The complexity increases with an increase in the boiling range of the gas condensate and the natural gasoline.

Often, the specifications against which a finished product is tested before release for sale are referred to as the *batch release specifications.* These are the product data against which the finished product is tested to ensure satisfactory quality. If the product should be in service before a certain date, the purchases will be advised (by law) through a note on the product of the *expiry specifications*, and the product, if tested at any time within the functional life period, must comply with the requirements in the expiry specifications.

In terms of natural gas, a specification offers the luxury of *predictability* of gas behavior (i) in a treating operation at the wellhead or (ii) in a processing plant or (iii) *predictability* of product quality (therefore, product behavior) relative to market demand. Ultimately, feedstock behavior or product quality is judged by an assessment of performance. And it is *performance* that is the ultimate criterion of quality. It is therefore necessary to determine those properties, the values of which can be established precisely and relatively simply by inspection tests in a control laboratory, that correlate closely with the important performance properties.

The value of natural gas to the producer and to the processor depends upon the quality of the gas as it emerges at the wellhead and whether the gas can be treated/processed to economically produce a satisfactory product pattern that matches market demand (*market pull*). In the main, the processor may not be concerned with the actual chemical nature of the natural gas but in methods of analysis that would provide information sufficient to assess the potential quality of the gas, to supply preliminary engineering data, and also to indicate whether any difficulties might arise in handling, treating (at the wellhead), transporting, or processing. Such information may be obtained in one of two ways: (i) inspection data from a preliminary assay, such as determination of the composition through use of gas chromatography, and (ii) inspection data from a full assay, which involves the derivation of a series of physical properties of the gas as well as the analysis of fractions and product blends throughout the full range of the natural gas constituents, including gas condensate and natural gasoline.

References

API (2001). American Petroleum Institute, Chapter 14 section 1. In: *Collecting and Handling of Natural Gas Samples for Custody Transfer*, 5e. Washington, DC: American Petroleum Institute.

ASTM (2017). *Annual Book of ASTM Standards*. West Conshohocken, Pennsylvania: American Society for Testing and Materials.

ASTM D1250 (2017). *Standard Guide for Use of the Petroleum Measurement Tables*. West Conshohocken, Pennsylvania: ASTM International.
ASTM D1265 (2017). *Standard Practice for Sampling Liquefied Petroleum (LP) Gases*, Manual Method. Annual Book of Standards. West Conshohocken, Pennsylvania: ASTM International.
ASTM D1298 (2017). *Standard Test Method for Density, Relative Density, or API Gravity of Crude Petroleum and Liquid Petroleum Products by Hydrometer Method*, Annual Book of Standards. West Conshohocken, Pennsylvania: ASTM International.
ASTM D287 (2017). *Standard Test Method for API Gravity of Crude Petroleum and Petroleum Products (Hydrometer Method)*, Annual Book of Standards. West Conshohocken, Pennsylvania: ASTM International.
ASTM D3700 (2017). *Containing Hydrocarbon Fluid Samples Using a Floating Piston Cylinder*, Annual Book of Standards. West Conshohocken, Pennsylvania: ASTM International.
ASTM D4006 (2017). *Standard Test Method for Water in Crude Oil by Distillation*, Annual Book of Standards. West Conshohocken, Pennsylvania: ASTM International.
ASTM D4051 (2017). *Standard Practice for Preparation of Low-Pressure Gas Blends*, Annual Book of Standards. West Conshohocken, Pennsylvania: ASTM International.
ASTM D4057 (2017). *Standard Practice for Manual Sampling of Petroleum and Petroleum Products*, Annual Book of Standards. West Conshohocken, Pennsylvania: ASTM International.
ASTM D4177 (2017). *Automatic Sampling of Petroleum and Petroleum Products*, Annual Book of Standards. West Conshohocken, Pennsylvania: ASTM International.
ASTM D4307 (2017). *Standard Practice for Preparation of Liquid Blends for Use as Analytical Standards*, Annual Book of Standards. West Conshohocken, Pennsylvania: ASTM International.
ASTM D4928 (2017). *Standard Test Method for Water in Crude Oils by Coulometric Karl Fischer Titration*, Annual Book of Standards. West Conshohocken, Pennsylvania: ASTM International.
ASTM D5287 (2017). *Automatic Sampling of Gaseous Fuels*, Annual Book of Standards. West Conshohocken, Pennsylvania: ASTM International.
ASTM D5842 (2017). *Sampling and Handling of Fuels for Volatility Measurement*, Annual Book of Standards. West Conshohocken, Pennsylvania: ASTM International.
ASTM E1169 (2017). *Standard Guide for Conducting Ruggedness Tests*, Annual Book of Standards. West Conshohocken, Pennsylvania: ASTM International.
ASTM E1301 (2017). *Standard Guide for Proficiency Testing by Interlaboratory Comparisons*, Annual Book of Standards. West Conshohocken, Pennsylvania: ASTM International.

ASTM E1488 (2017). *Standard Guide for Statistical Procedures to Use in Developing and Applying Test Methods*, Annual Book of Standards. West Conshohocken, Pennsylvania: ASTM International.

ASTM E1601 (2017). *Practice for Conducting an Interlaboratory Study to Evaluate the Performance of an Analytical Method*, Annual Book of Standards. West Conshohocken, Pennsylvania: ASTM International.

ASTM E1763 (2017). *Guide for Interpretation and Use of Results from Interlaboratory Testing of Chemical Analysis Methods*, Annual Book of Standards. West Conshohocken, Pennsylvania: ASTM International.

ASTM E2655 (2017). *Standard Guide for Reporting Uncertainty of Test Results and Use of the Term Measurement Uncertainty in ASTM Test Methods*, Annual Book of Standards. West Conshohocken, Pennsylvania: ASTM International.

ASTM E2857 (2017). *Standard Guide for Validating Analytical Methods*, Annual Book of Standards. West Conshohocken, Pennsylvania: ASTM International.

ASTM E691 (2017). *Standard Practice for Conducting an Interlaboratory Study to Determine the Precision of a Test Method*, Annual Book of Standards. West Conshohocken, Pennsylvania: ASTM International.

Bergman, D.F., Tek, M.R., and Katz, D.L. (1975). *Retrograde Condensation in Natural Gas Pipelines*. Washington, DC: American Gas Association.

Bullin, J.A., Fitz, C., and Dustman, T. 2011. Practical Hydrocarbon Dew Point Specification for Natural Gas Transmission Lines. Digital Refining. March. www.digitalrefining.com/article/1000452 (accessed 12 November 2017).

ASTM D1071 (2017). *Standard Test Methods for Volumetric Measurement of Gaseous Fuel Samples*, Annual Book of Standards. West Conshohocken, Pennsylvania: ASTM International.

ASTM D323 (2017). *Standard Test Method for Vapor Pressure of Petroleum Products (Reid Method)*, Annual Book of Standards. West Conshohocken, Pennsylvania: ASTM International.

Dean, J.R. (2003). *Methods for Environmental Trace Analysis*. Hoboken, New Jersey: Wiley.

Dustman, T., Drenker, J., Bergman, D.F., and Bullin, J.A. (2006). An Analysis and Prediction of Hydrocarbon Dew Points and Liquids in Gas Transmission Lines. *Proceedings. Eighty-Fifth Gas Processors Association Annual Convention*, Dallas, TX (March). Tulsa, OK: Gas Processors Association,.

GPA (1986). Obtaining Natural Gas Samples for Analysis by Gas Chromatography. *2166-86*. Tulsa, OK: Gas Processors Association.

Kimmel, K. (2003). Proceedings. 2003 Meeting. American School of Gas Measurement Technology, Houston, TX (pp. 65–69).

Mokhatab, S., Poe, W.A., and Speight, J.G. (2006). *Handbook of Natural Gas Transmission and Processing*. Amsterdam, the Netherlands: Elsevier.

Patnaik, P. ed. (2004). *Dean's Analytical Chemistry Handbook*, 2e. New York: McGraw-Hill.

Poling, B.E., Prausnitz, J.M., and O'Connell, J.P. (2004). *The Properties of Gases and Liquids*, 5e. New York: McGraw-Hill.

Ralli, D.K., Pandey, S.C., Saxena, A.K., and Alamkhan, W.K. (2003). Impact of a Quality Management System on Product Quality at ONGC, Uran. *Journal of Scientific & Industrial Research* 62: 1001–1007.

Speight, J.G. (2007). *Natural Gas: A Basic Handbook*. Houston, Texas: GPC Books, Gulf Publishing Company.

Speight, J.G. ed. (2011). *The Biofuels Handbook*. London, United Kingdom: Royal Society of Chemistry.

Speight, J.G. (2013). *The Chemistry and Technology of Coal*, 3e. Boca Raton, Florida: CRC Press, Taylor & Francis group.

Speight, J.G. (2014a). *The Chemistry and Technology of Petroleum*, 5e. Boca Raton, Florida: CRC Press, Taylor & Francis Group.

Speight, J.G. (2014b). *Oil and Gas Corrosion Prevention*. Oxford, United Kingdom: Gulf Professional Publishing, Elsevier.

Speight, J.G. (2015). *Handbook of Petroleum Product Analysis*, 2e. Hoboken, New Jersey: Wiley.

Speight, J.G. and Foote, R. (2011). *Ethics in Science and Engineering*. Salem, Massachusetts: Scrivener Publishing.

Wallis, P.A. (1986). Natural Gas Analysis. *Trends in Analytical Chemistry*. 5 (3): 63–67.

Weisman, W. 1998. Analysis of Petroleum Hydrocarbons in Environmental Media. Total Petroleum Hydrocarbons Criteria Working Group Series. Volume 1. Amherst Scientific Publishers, Amherst, MA. (See also: Volume 2: Composition of Petroleum Mixtures, 1998; Volume 3: Selection of Representation Total Petroleum Hydrocarbons Fractions Based on Fate and Transport Considerations, 1997; Volume 4: Development of Fraction-Specific, 1997; and Volume 5: Human Health Risk-Based Evaluation of Petroleum Contaminated Sites – Implementation of the Working Group Approach, 1999).

6

Analytical Methods

6.1 Introduction

The chemical composition of natural gas is not straightforward because of the presence of hydrocarbon constituents that might be as high as C_{10}, including gas condensate and natural gasoline as well as nonhydrocarbon contaminants such as carbon dioxide, hydrogen sulfide, and sundry other compounds (Chapter 1). Thus, it is essential that the most appropriate analytical methods are selected from a comprehensive list of methods and techniques that are used for the analysis of samples (Miller, 2000; Speight, 2005, 2014a, 2015; ASTM, 2017). Furthermore, samples may be disturbed during sampling, storage, and pretreatment. Also, most laboratory experiments impose steady environmental conditions and the manner in which any corrections be applied to allow for the ambient conditions of the laboratory and must be beyond reproach or the production of unreliable data will be the most likely result (ASTM D6299, 2017; ASTM D6792, 2017).

In the early days of natural gas use, there was not a need to understand the character and behavior of the natural gas, the detail that is currently required. In fact, until the mid-twentieth century, coal gas was the predominant gaseous fuel, but knowledge of the gas composition was still required and not to be ignored.

However, with the startling demands on the gaseous fuels from crude oil and natural gas processing industries during and after World War II to produce specification-grade products and the emergence of the age of petrochemicals and plastics, the natural gas production and processing industries needed to produce acceptable quality natural gas that strictly adhered to the specifications demanded by the user. In addition, there became a necessity to find out more about natural gas so that producers and processors might be able to enjoy the luxury of predictability and plan a product composition that was adequate to the needs of domestic and industrial consumers. This was a difficult task when the character of the natural gas was variable on a well-by-well basis, and

Handbook of Natural Gas Analysis, First Edition. James G. Speight.

the idea that natural gas processing (also called *natural gas treating* and *natural gas refining*) should be a *hit-and-miss* affair (at the wellhead or in a processing plant) was not acceptable.

Thus, in order to meet sales specifications, natural gas production and processing requires not only knowledge of its chemical and physical properties but also knowledge of the chemical and physical reactivity of the various constituents, including the chemical and physical reactivity of the contaminants. Moreover, because natural gas varies markedly in its properties and composition according to the source of the gas, it also varies in its chemical and physical reactivity. Thus, knowledge of natural gas reactivity is required for optimization of existing processes and for the development and design of new processes.

Using gas condensate and natural gasoline as examples, valuable information can be obtained from the true boiling point (TBP) curve, which is a function of percent weight distilled and temperature, that is, a boiling point distribution (Speight, 2014a, 2015). However, the boiling range does not convey much detail about chemical reactivity of either the condensate or the natural gasoline. In addition to the boiling point distribution, it is possible to measure bulk physical properties, such as specific gravity (SG) and viscosity, which have assisted in the establishment of certain empirical relationships for processing and use of the TBP curve. Many of these relationships include assumptions that are based on experience with a range of feedstocks. However, the chemical behavior of feedstocks that contain different proportions of chemical species emphasizes the need for more definitive data that would enable more realistic predictions to be made of behavior in processing operations.

Thus, this chapter presents some of the methods that are generally applied to study the makeup of natural gas in terms of chemical structures as well as methods that might be preferred for wellhead processing and more formal processing in a plant. There are, of course, many analytical methods that can be applied to the analysis of natural gas and the various constituents, but they vary with sample condition and composition. More specifically, this chapter deals with the more common methods used to define chemical and physical properties of a sample. Any of the methods described herein might also be applied to the analysis of sample for environmental purposes. However, methods for analytical purpose are not intended to be the focus of this text and are described elsewhere (Speight, 2005).

6.2 Chemical and Physical Analyses

The importance of chemical and physical properties is dependent upon the purity of the natural gas stream. In the strictest sense, natural gas streams and gas condensate (natural gasoline) are complex chemical mixtures. These mixtures contain hydrocarbons and nonhydrocarbon compounds that confer properties on the mixture that may not be reflected in the composition. Therefore, it is necessary to apply various test methods (Table 6.1) to natural gas streams and to associated streams to determine if the material is suitable for processing and (in the case of the products) for sale with a designated use in mind. The most common tests are introduced in the following subsections – these tests are presented in alphabetical order with no preference given to any particular test method.

6.2.1 Boiling Point Distribution

The distillation tests give an indication of the types of products and the quality of the products that can be obtained from the crude oil and natural gas (especially gas condensate and natural gasoline), and the tests are used to compare different liquid fraction through the yield and quality of the various fractions.

The basic method of distillation (ASTM D86, 2017) is one of the oldest methods in use since the distillation characteristics of hydrocarbons have an important effect on safety and performance, especially in the case of liquid fuels and naphtha solvents. The boiling range gives information on the composition, the properties, and the behavior of crude oil and derived products during storage and use. Volatility is the major determinant of the tendency of a hydrocarbon mixture to produce potentially explosive vapors. Several methods are available to define the distillation characteristics of crude oil and its various crude oil products. In addition to these physical methods, other test methods based on gas chromatography (GC) are also used to derive the boiling point distribution of a sample (ASTM D2887, 2017; ASTM D3710, 2017; ASTM D6352, 2017).

In the preliminary assay of natural gas and crude oil, the method of distillation is often used to give a rough indication of the boiling range of the raw feedstock (ASTM D2892, 2017). The test method is used for the distillation of stabilized (gases removed) crude oil to a final cut temperature of up to 400 °C (750 °F) *atmospheric equivalent temperature* (AET). The crude oil is heated and separated by the distillation column into lower-boiling products such as naphtha and kerosene. The distillate and the residuum can be further examined by tests such as SG (ASTM D1298, 2017), sulfur content (ASTM D129, 2017), and viscosity (ASTM D445, 2017). In fact, using a method (ASTM D2569, 2017) developed for the determining the distillation characteristics of nonvolatile residues allows further detailed examination of residua.

In addition to the whole crude oil tests performed as part of the inspection assay, a comprehensive or full assay requires that the crude be fractionally distilled and the fractions characterized by the relevant tests. Fractionation of the crude oil begins with a TBP distillation employing a fractionating column having an efficiency of 14–18 theoretical plates and operated at a reflux ratio of 5 : 1 (ASTM D2892, 2017). The TBP distillation may be used for all fractions

Table 6.1 Examples of ASTM standard test method for application to the analysis of gas streams.

Designation	Title
ASTM D1070	Standard Test Methods for Relative Density of Gaseous Fuels
ASTM D1071	Standard Test Methods for Volumetric Measurement of Gaseous Fuel Samples
ASTM D1072	Standard Test Method for Total Sulfur in Fuel Gases by Combustion and Barium Chloride Titration
ASTM D1142	Standard Test Method for Water Vapor Content of Gaseous Fuels by Measurement of Dew-Point Temperature
ASTM D1826	Standard Test Method for Calorific (Heating) Value of Gases in Natural Gas Range by Continuous Recording Calorimeter
ASTM D1945	Standard Test Method for Analysis of Natural Gas by Gas Chromatography
ASTM D1946	Standard Practice for Analysis of Reformed Gas by Gas Chromatography
ASTM D1988	Standard Test Method for Mercaptans in Natural Gas Using Length-of-Stain Detector Tubes
ASTM D3588	Standard Practice for Calculating Heat Value, Compressibility Factor, and Relative Density of Gaseous Fuels
ASTM D4084	Standard Test Method for Analysis of Hydrogen Sulfide in Gaseous Fuels (Lead Acetate Reaction Rate Method)
ASTM D4468	Standard Test Method for Total Sulfur in Gaseous Fuels by Hydrogenolysis and Rateometric Colorimetry
ASTM D4596	Standard Practice for Collection of Channel Samples of Coal in a Mine
ASTM D4810	Standard Test Method for Hydrogen Sulfide in Natural Gas Using Length-of-Stain Detector Tubes
ASTM D4888	Standard Test Method for Water Vapor in Natural Gas Using Length-of-Stain Detector Tubes
ASTM D4891	Standard Test Method for Heating Value of Gases in Natural Gas and Flare Gases Range by Stoichiometric Combustion
ASTM D4984	Standard Test Method for Carbon Dioxide in Natural Gas Using Length-of-Stain Detector Tubes
ASTM D5192	Standard Practice for Collection of Coal Samples from Core
ASTM D5287	Standard Practice for Automatic Sampling of Gaseous Fuels
ASTM D5454	Standard Test Method for Water Vapor Content of Gaseous Fuels Using Electronic Moisture Analyzers
ASTM D5504	Standard Test Method for Determination of Sulfur Compounds in Natural Gas and Gaseous Fuels by Gas Chromatography and Chemiluminescence

Table 6.1 (Continued)

Designation	Title
ASTM D5954	Standard Test Method for Mercury Sampling and Measurement in Natural Gas by Atomic Absorption Spectroscopy
ASTM D6228	Standard Test Method for Determination of Sulfur Compounds in Natural Gas and Gaseous Fuels by Gas Chromatography and Flame Photometric Detection
ASTM D6273	Standard Test Methods for Natural Gas Odor Intensity
ASTM D6350	Standard Test Method for Mercury Sampling and Analysis in Natural Gas by Atomic Fluorescence Spectroscopy
ASTM D6968	Standard Test Method for Simultaneous Measurement of Sulfur Compounds and Minor Hydrocarbons in Natural Gas and Gaseous Fuels by Gas Chromatography and Atomic Emission Detection
ASTM D7164	Standard Practice for On-line/At-line Heating Value Determination of Gaseous Fuels by Gas Chromatography
ASTM D7165	Standard Practice for Gas Chromatograph Based On-line/At-line Analysis for Sulfur Content of Gaseous Fuels
ASTM D7166	Standard Practice for Total Sulfur Analyzer Based On-line/At-line for Sulfur Content of Gaseous Fuels
ASTM D7314	Standard Practice for Determination of the Heating Value of Gaseous Fuels using Calorimetry and On-line/At-line Sampling
ASTM D7493	Standard Test Method for Online Measurement of Sulfur Compounds in Natural Gas and Gaseous Fuels by Gas Chromatograph and Electrochemical Detection
ASTM D7551	Standard Test Method for Determination of Total Volatile Sulfur in Gaseous Hydrocarbons and Liquefied Petroleum Gases and Natural Gas by Ultraviolet Fluorescence
ASTM D7607	Standard Test Method for Analysis of Oxygen in Gaseous Fuels (Electrochemical Sensor Method)
ASTM D7800	Standard Test Method for Determination of Elemental Sulfur in Natural Gas
ASTM D7833	Standard Test Method for Determination of Hydrocarbons and Non-Hydrocarbon Gases in Gaseous Mixtures by Gas Chromatography
ASTM D7904	Standard Test Method for Determination of Water Vapor (Moisture Concentration) in Natural Gas by Tunable Diode Laser Spectroscopy (TDLAS)
ASTM D7940	Standard Practice for Analysis of Liquefied Natural Gas (LNG) by Fiber-Coupled Raman Spectroscopy
ASTM D8000	Standard Practice for Flow Conditioning of Natural Gas and Liquids

up to a maximum cut point of about 350 °C AET, but low residence time in the still (or reduced pressure) is needed to minimize cracking.

It is often useful to extend the boiling point data to higher temperatures that are possible in the fractionating distillation method previously described, and for this purpose a vacuum distillation in a simple still, with no fractionating column (ASTM D1160, 2017), can be carried out. This distillation, which is done under fractionating conditions equivalent to one theoretical plate, allows the boiling point data to be extended to about 600 °C (1112 °F) with many crude oils. This method gives useful comparative and reproducible results that are often accurate enough for refinery purposes, provided significant cracking does not occur.

Usually seven fractions provide the basis for a reasonably thorough evaluation of the distillation properties of the feedstock: (i) gas, boiling range: less than 15.6 °C/60 °F; (ii) gasoline (light naphtha), boiling range: 15.6–150 °C/60–300 °F; (iii) kerosene (medium naphtha), boiling range: 150–230 °C/300–450 °F; (iv) gas oil, boiling range: 230–345 °C/450–650 °F; (v) light vacuum gas oil, boiling range: 345–370 °C/650–700 °F; (vi) heavy vacuum gas oil, boiling range: 370–565 °C/700–1050°; and (vii) residuum, boiling range: greater than 565 °C/1050 °F. From 5 to 50l of crude oil are necessary to complete a full assay, depending on the number of cuts to be taken and the tests to be performed on the fractions.

A more recent test method (ASTM D5236, 2017) is seeing increasingly more use and appears to be the method of choice for crude assay vacuum distillations. The method employs a vacuum pot still with a low-pressure drop entrainment separator operated under total takeoff conditions. The reduced pressure allows volatilization at a lower temperature than under atmospheric conditions, thus allowing temperatures up to 565 °C (1050 °F) for most samples and that avoids thermal decomposition (cracking) of the oil (caused by prolonged exposure to temperatures above 350 °C, 650 °F). The test method applies to the higher-boiling fractions of crude oil that are in the gas oil and lubricating oil range as well as heavy crude oils and residua. Application of this test method to gas condensate and natural gasoline is of limited use.

Wiped-wall or thin-film molecular stills can also be used to separate the higher-boiling fractions under conditions that minimize cracking. In these units, however, cut points cannot be directly selected, because vapor temperature in the distillation column cannot be measured accurately under operating conditions. Instead, the wall (film) temperature, pressure, and feed rate that will produce a fraction with a given end point are determined from in-house correlations developed by matching yields between the wiped-wall distillation and the conventional distillation (ASTM D1160, 2017; ASTM D5236, 2017). And wiped-wall stills are often used because they allow higher end points and can easily provide sufficient quantities of the fractions for characterization purposes.

In the natural gas processing industry, boiling range distribution data are used (i) to assess the gas quality before purchase, (ii) to monitor gas quality during transportation, (iii) to evaluate the natural gas stream for processing, and (iv) to provide information for the optimization of the processes and processing sequence. Traditionally, boiling range distributions are determined by distillation, which may be in order for gas condensate or natural gasoline, but the boiling range of natural gas can be determined by GC through identification of the hydrocarbon constituents. The information may be reported in the form of a gas stream assay literature, which provides information on the yield of specific constituents obtained by distillation.

To some extent in the laboratory, atmospheric and vacuum distillation techniques have largely been replaced by *simulated* distillation methods, which use low resolution GC and correlate retention times to hydrocarbon boiling points. Two test methods (ASTM D2887, 2017; ASTM D3710, 2017) use external standards composed of *n*-alkane derivatives, while a similar third test method (ASTM D5307, 2017) requires two determinations to be made with each sample, one of which uses an internal standard. However, it must be observed that their test methods are more suitable for the liquid products of gas production and processing rather than for the gaseous products. From the point of view of the analysis of the more complex crude oil products (Speight, 2014a, 2015) as well as gas condensate and natural gasoline, the boiling range distribution provides an indication of volatility and component distribution. In addition, boiling range distribution data is also useful for the development of equations for predicting loss by evaporation.

6.2.2 Low-boiling Hydrocarbons

The amount of the individual light hydrocarbons in natural gas (methane, ethane, propane, and butane) is often included as part of the preliminary assay. For safety reasons, it is essential to know the content of low-boiling hydrocarbons as well as low-boiling liquid hydrocarbons natural gas prior to handling and shipping the gas.

By definition, natural gasoline (sometimes also referred to as *gas condensate*) is a mixture of low-boiling liquid hydrocarbons isolated from crude oil and natural gas wells suitable for blending with light naphtha and/or refinery gasoline (Mokhatab et al., 2006; Speight, 2007, 2014a). Because of the presence of low-boiling hydrocarbons, light naphtha can become extremely explosive, even at relatively low ambient temperatures. Some of these gases may be burned off (flared) at the field wellhead, but others remain in the liquid products extracted from the well (Speight, 2014a).

6.2.3 Density, Specific Gravity, and API Gravity

Density is the mass per unit volume of a substance and is typically reported for oils in units of $g\,ml^{-1}$ or $g\,cm^{-3}$ and less often in units of $kg\,m^{-3}$. Density is temperature dependent. It is an important property of crude oil and crude oil products (including gaseous products) as it gives the investigator(s) indications of whether or not the contaminant(s) will dissipate in air under standard conditions of temperature and pressure (STP).

Density (ASTM D1298, 2017) is an important property of gas condensate and natural gasoline since these products are usually bought and sold on that basis or, if on volume basis, converted to mass basis via density measurements. This property is almost synonymously termed as density, relative density, gravity, and SG, all terms related to each other. Usually a hydrometer, pycnometer, or more modem digital density meter is used for the determination of density or SG (Speight, 2014a; ASTM, 2017).

In the most commonly used method (ASTM D1298, 2017), the sample is brought to the prescribed temperature and transferred to a cylinder at approximately the same temperature. The appropriate hydrometer is lowered into the sample and allowed to settle, and, after temperature equilibrium has been reached, the hydrometer scale is read and the temperature of the sample is noted.

Although there are many methods for the determination of density due to the different nature of gas condensate or natural gasoline and the different products, one test method (ASTM D5002, 2017) is used for the determination of the density or relative density of the products that can be handled in a normal fashion as liquids at test temperatures between 15 and 35 °C (59 and 95 °F). This test method applies to liquid products with high vapor pressures, provided that appropriate precautions are taken to prevent vapor loss during transfer of the sample to the density analyzer. Approximately 0.7 ml of crude oil sample is introduced into an oscillating sample tube, and the change in oscillating frequency caused by the change in mass of the tube is used in conjunction with calibration data to determine the density of the sample.

Another test determines density and SG by means of a digital densimeter (ASTM D4052, 2017). In the test, a small volume (approximately 0.7 ml) of liquid sample is introduced into an oscillating sample tube, and the change in oscillating frequency caused by the change in the mass of the tube is used in conjunction with calibration data to determine the density of the sample. The test is usually applied to crude oil, its distillates, and products that are liquids at temperatures between 15 and 35 °C (59 and 95 °F) and have vapor pressures below 600 mmHg and viscosities below about 15 000 cSt at the temperature of test. However, the method should not be applied to samples so dark in color that the absence of air bubbles in the sample cell cannot be determined with certainty.

Accurate determination of the density or SG of crude oil is necessary for the conversion of measured volumes to volumes at the standard temperature of 15.56 °C (60 °F) (ASTM D1250, 2017). The SG is also a factor reflecting the quality of crude oils.

Although one of the more conventional distillation procedures might be employed, the determination of low-boiling hydrocarbons isolated from gas streams is best carried out using a gas chromatographic method (ASTM D2427, 2017).

The accurate determination of the American Petroleum Institute (API) gravity of gas condensate and natural gasoline is necessary for the conversion of measured volumes to volumes at the standard temperature of 60°F (15.56°C). Gravity is a factor governing the quality of crude oils. However, the density or SG of gas condensate and natural gasoline is not a truly certain indication of their quality. Correlated with other properties, the SG can be used to give approximate hydrocarbon composition and heat of combustion. This is usually accomplished through the use of the API gravity that is derived from the SG:

$$\text{API gravity, °API} = (141.5 / \text{sp gr } 60/60\ °\text{F}) - 131.5$$

The API gravity is also an important measure for estimating the quality and refinability of gas condensate and natural gasoline.

API gravity or density or relative density can be determined using one of the two hydrometer methods (ASTM D287, 2017; ASTM D1298, 2017). The use of a digital analyzer (ASTM D5002, 2017) is finding increasing popularity for the measurement of density and SG.

In the method (ASTM D287, 2017), the API gravity is determined using a glass hydrometer for gas condensate and natural gasoline that are normally handled as liquids normal temperature and pressure (STP) and that have a Reid vapor pressure of 26 psi (180 kPa) or less. The API gravity is determined at 15.6°C (60°F), or converted to values at 60°F, by means of standard tables. These tables are not applicable to nonhydrocarbons or essentially pure hydrocarbons such as the aromatics.

This test method is based on the principle that the gravity of a liquid varies directly with the depth of immersion of a body floating in it. The API gravity is determined using a hydrometer by observing the freely floating API hydrometer and noting the graduation nearest to the apparent intersection of the horizontal plane surface of the liquid with the vertical scale of the hydrometer after temperature equilibrium has been reached. The temperature of the sample is determined using a standard test thermometer that is immersed in the sample or from the thermometer that is an integral part of the hydrometer (thermohydrometer).

Two density-related properties of gas condensate and natural gasoline are often used: (i) SG and (ii) the API gravity. SG (*relative density*) is the ratio, at a specified temperature, of the oil density to the density of pure water. The API gravity scale (presented as API) arbitrarily assigns an API gravity of 10° to pure water, and scale is commercially important for ranking the quality of hydrocarbon liquids. Thus,

$$\text{API} = [141.5 / (\text{specific gravity at } 15.6\ ^\circ\text{C}) - 131.5]$$

Gas condensate and natural gasoline typically have an API gravity on the order of 45 + API, even as high as 65 API. Tar sand bitumen typically has an API gravity less than 10°.

6.2.4 Evaporation Loss

Evaporation is the removal of the lower-boiling constituents from a product such as gas condensate, and natural gasoline is usually observed under ambient conditions. The rate of evaporation and evaporation loss are of importance for all volatile (liquid) constituents of natural gas streams. Standard test methods such as those designated for distillation and vapor pressure determination are often used to determine evaporation properties (ASTM, 2017). Although not necessarily applicable to crude oil and crude oil products in general, evaporation loss data can be obtained at any temperature in the range from 93 to 316 °C (200 and 600 °F).

Crude oil and its products, including gas condensate and natural gasoline, typically evaporate at a logarithmic rate with respect to time (Fingas, 1998). This is attributed to the overall logarithmic appearance of many components evaporating at different linear rates. Products with fewer constituents (such as gas condensate and natural gasoline) evaporate at a rate that is square root with respect to time, which is a result of a number of components evaporating. The evaporation process is not strictly boundary-layer regulated, which is largely a result of the high saturation concentrations of oil components in air and is associated with a high boundary-layer regulated rate. Some volatile products show some effect of boundary-layer regulation at the start of the evaporation process, but after several minutes, evaporation slows because of the loss of the more volatile components, at which point evaporation ceases to be boundary-layer regulated. It must also be recognized that as evaporation occurs, both the density and the viscosity of the residual material increase, thereby causing behavioral changes in the product.

6.2.5 Fire Point and Flash Point

The *fire point* is the lowest temperature, corrected to one atmospheric pressure (14.7 psi), at which the application of a test flame to the sample surface causes the vapor of the oil to ignite and burn for at least 5 s. At any time when handling natural gas, gas condensate, or natural gasoline, fire should always be considered an imminent hazard. Related to fire point, the *flash point* is a measure of the tendency of the product to form a flammable mixture with air under controlled laboratory conditions and is an indication of a fire hazard when handling or storing such a chemical (Speight, 2014a, 2015). There are several standard test methods for measuring flash points; ASTM D56 (2017) and ASTM D93 (2017) are among the most commonly used. The minimum flash point that can be determined (ASTM D93, 2017) is 10 °C (50 °F). One method (ASTM D56, 2017) is intended for liquids with a viscosity less than 9.5 cSt at 25 °C (77 °F). The flash point and fire point of lubricating oil are determined by a separate method (ASTM D92, 2017).

The fire point and flash point are only two of a number of properties that should be considered in assessing the overall flammability hazard of volatile materials, such as natural gas, gas condensate, or natural gasoline (ASTM D92, 2017; ASTM D93, 2017). Furthermore, it must be emphasized here that fire point or the flash point of a natural gas mixture is not the average of the fire points of the individual constituents of the mixture, but is the lowest fire point or the lowest flash point of the most ignitable constituent. The ignition temperature (sometimes called the *autoignition temperature*) is related to the fire point and is the minimum temperature at which, in the current context, a constituent of natural gas, gas condensate, or natural gasoline will ignite without a spark or flame being present (ASTM E659, 2017). Also related to fire point, the flammability limits of vapor in air are an expression of the percent concentration in air (by volume). These values give an indication of relative flammability. The limits are sometimes referred to as *lower explosive limit* (LEL) and *upper explosive limit* (UEL) (Speight, 2014a, 2015).

Before a fire or explosion can occur, three conditions must be met simultaneously. A fuel (i.e. combustible gas) and oxygen (air) must exist in certain proportions, along with an ignition source, such as a spark or flame. The ratio of fuel and required oxygen varies with each combustible gas or vapor. The minimum concentration of a particular combustible gas or vapor necessary to support its combustion in air is its LEL. Below this level, the mixture is too lean to burn. The maximum concentration of a gas or vapor that will burn in air is the UEL. Above this level, the mixture is too rich to burn. The range between the LEL and UEL is known as the flammable range for that gas or vapor. The values are valid only for the conditions under which they were determined (usually room temperature and atmospheric pressure using a 2-in. tube with spark ignition). The flammability range of most materials expands as temperature, pressure, and container diameter increase.

6.2.6 Pour Point and Viscosity

The pour point of crude oil or a crude oil product is the lowest temperature at which the oil will just flow, under standard test conditions (ASTM D97, 2017). The failure to flow at the pour point is usually attributed to the separation of waxes from the oil but can also be due to the effect of viscosity in the case of very viscous oils. Also, particularly in the case of residual fuel oils, pour points may be influenced by the thermal history of the sample, that is, the degree and duration of heating and cooling to which the sample has been exposed. While this test method may not necessarily apply to gas condensate and natural gasoline, if the sample comes from a highly paraffinic source and there are waxes in the higher-boiling product, there may be a need for the test.

The *pour point* of crude oil or a crude oil product is an index of the lowest temperature at which the crude oil will flow under specified conditions. The maximum and minimum pour point temperatures provide a temperature window where the product, depending on the thermal history of the product, might appear in the liquid and the solid states. The pour point data can be used to supplement other measurements of cold flow behavior, and are particularly useful for the screening of the effect of wax interaction modifiers on the flow behavior of the product as well as a necessity for waxy gas condensate and waxy natural gasoline.

In the original (and still widely used) test for pour point (ASTM D97, 2017), a sample is cooled at a specified rate and examined at intervals of 3 °C (5.4 °F) for flow characteristics. The lowest temperature at which the movement of the oil is observed is recorded as the pour point. A later test method (ASTM D5853, 2017) covers two procedures for the determination of the pour point of crude oils down to −36 °C. One method provides a measure of the maximum (upper) pour point temperature. The second method measures the minimum (lower) pour point temperature. In the methods, the test specimen is cooled (after preliminary heating) at a specified rate and examined at intervals of 3 °C (5.4 °F) for flow characteristics. Again, the lowest temperature at which movement of the test specimen is observed is recorded as the pour point.

In any determination of the pour point, a crude oil product that contains wax produces an irregular flow behavior when the wax begins to separate. This type of crude oil product possesses viscosity relationships that are difficult to predict when the product is transported by pipeline. In addition, some waxy crude oil products are sensitive to heat treatment that can also affect the viscosity characteristics. This complex behavior limits the value of viscosity and pour point tests on waxy products. However, laboratory pumpability tests (ASTM D3829, 2017; ASTM D7528, 2017) that give an estimate of minimum handling temperature and minimum line or storage temperature are available.

The pour point of gas condensate or natural gasoline will only be of some significance to show that these liquid products contain high-boiling hydrocarbon derivative that constitutes wax, typically *n*-hydrocarbon derivatives that contain C_{17+} isomers (for example, *n*-$C_{17}H_{36}$, melting point: 21 °C, boiling point: 302 °C).

Viscosity and pour point determinations are performed principally to ascertain the handling (flow) characteristics of crude oil products – including gas condensate and natural gasoline – at low temperatures. There are, however, some general relationships about the composition of hydrocarbon liquids that can be derived from pour point and viscosity data. Commonly, a lower pour point of a hydrocarbon liquid typically corresponds to the presence of aromatic constituents in the liquid, while a higher pour point typically corresponds to the presence of aromatic constituents in the liquid.

Viscosity is usually determined at different temperatures (e.g. 25 °C, 77 °F and 100 °C, 212 °F) by measuring the time for a volume of liquid to flow under gravity through a calibrated glass capillary viscometer (ASTM D445, 2017).

In the test, the time for a fixed volume of liquid to flow under gravity through the capillary of a calibrated viscometer under a reproducible driving head and at a closely controlled temperature is measured in seconds. The kinematic viscosity is the product of the measured flow time and the calibration constant of the viscometer. Conversion of the kinematic viscosity in centistokes (cSt) at any temperature to Saybolt universal viscosity in Saybolt universal seconds (SUS) at the same temperature and for converting kinematic viscosity in centistokes at 122 and 210 °F to Saybolt Furol viscosity in Saybolt Furol seconds (SFS) at the same temperatures (ASTM D2161, 2017).

The *viscosity index* (ASTM D2270, 2017) is a widely used measure of the variation in kinematic viscosity due to changes in the temperature of a hydrocarbon liquid between 40 and 100 °C (104 and 212 °F). For crude oils of similar kinematic viscosity, the higher the viscosity index, the smaller the effect of temperature on its kinematic viscosity. The accuracy of the calculated viscosity index is dependent only on the accuracy of the original viscosity determination.

6.2.7 Sulfur Content

The sulfur content of any crude oil or natural gas product is important for a number of reasons. Downstream processes as used in crude oil refineries or in natural gas processing plants will be adversely affected by high sulfur contents. If high sulfur liquid product oils are combusted, the products can include high levels of sulfur dioxide (SO_2) that can lead to acid deposition (acid rain) (Speight, 2005). *Acid rain* occurs when, in this context, the oxides of sulfur that are released to the atmosphere during the combustion of fossil fuels are deposited (as water-soluble acids) with rainfall, usually at some location remote from the source of the emissions:

$$SO_2 + H_2O \rightarrow H_2SO_3$$

$$2SO_2 + O_2 \rightarrow 2SO_3$$

$$SO_3 + H_2O \rightarrow H_2SO_4$$

The total sulfur content of oil can be determined by numerous standard techniques (such as ASTM D129, 2017; ASTM D4294, 2017). The former method (ASTM D129, 2017) is more applicable to crude oil products of low volatility and containing at least 0.1 mass percent sulfur, whereas the latter test method (ASTM D4294, 2017) is applicable to both volatile and nonvolatile crude oil products, such as gas condensate and natural gasoline, which have a sulfur concentration on the order of 0.05–5% w/w.

The sulfur content of natural gas and process gases is an important property and compounds containing this element are among the most undesirable constituents of natural gas and crude oil since they can give rise to plant corrosion and atmospheric pollution. Furthermore, crude oil can evolve hydrogen sulfide during distillation as well as low-boiling sulfur compounds, leading to sulfur in the process gas. In the case of crude oil refining, the sulfur compounds are concentrated in the distillation residue, and the volatile ones in the distillates are removed by processes such as hydrofining and caustic washing (Speight, 2014a). Nevertheless, the presence of sulfur-containing compounds in the residue leads to the presence of hydrogen sulfur and other low-boiling sulfur-containing constituents in process gas from cracking processes, such as visbreaking and coking (Speight, 2014a, 2017). The sulfur content of gaseous fuels obtained from crude oil residua and the atmospheric pollution arising from the use of these fuels are important factors in gas processing and utilization, so that the increasing insistence on a low sulfur content gaseous fuel has increased the value of low sulfur natural gas.

Sulfur compounds cause corrosiveness in refined products and contribute to corrosion of refinery equipment, poisoning of catalysts, and environmental pollution as a result of the combustion of fuel products. They may be present throughout the boiling range of crude oils although, as a rule, they are more abundant in the higher-boiling fractions. In some crude oils, thermally labile sulfur compounds can decompose during heating to produce hydrogen sulfide, which is corrosive and toxic.

A considerable number of tests are available to estimate the sulfur in crude oil or to study its effect on various crude oil products. Hydrogen sulfide dissolved in crude oil is normally determined by absorption of the hydrogen sulfide in a suitable solution that is subsequently analyzed chemically (doctor method) (ASTM D4952, 2017). The doctor test measures the amount of sulfur available to react with metallic surfaces at the temperature of the test. The rates of reaction depend on the metal type, temperature, and time. In the test, a sample is treated with copper powder at 150 °C (300 °F). The copper powder is filtered from the mixture. Active sulfur is calculated from the difference between the sulfur contents of the sample (ASTM D129, 2017) before and after treatment with copper.

Sulfur that is chemically combined as an organic constituent of crude is usually estimated by oxidizing a sample in a bomb and converting the sulfur compounds to barium sulfate, which is determined gravimetrically (ASTM D129, 2017). This method is applicable to any sample of sufficiently low volatility (e.g. a residuum or tar sand bitumen) that can be weighed accurately in an open sample boat and that contains at least 0.1% sulfur. In the method, the sample is oxidized by combustion in a vessel (bomb) containing oxygen under pressure. The sulfur in the sample is converted to sulfate and from the bomb washings is gravimetrically determined as barium sulfate. However, the method is not applicable to samples containing elements that give residues, other than barium sulfate, which are insoluble in dilute hydrochloric acid and would interfere in the precipitation step. In addition, the method is also subject to inaccuracies that arise from interference by the sediment inherently present in crude oil.

Until recently, one of the most widely used methods for the determination of total sulfur content has been combustion of a sample in oxygen to convert the sulfur to sulfur dioxide, which is collected and subsequently titrated iodometrically or detected by nondispersive infrared (ASTM D1552, 2017). This method is particularly applicable to heavier oil and fractions such as residua that boil above 177 °C (350 °F) and contain more than 0.06% w/w sulfur.

In the iodate detection system, the sample is bummed in a stream of oxygen at a sufficiently high temperature to convert the sulfur to sulfur dioxide. The combustion products are passed into an absorber that contains an acidic solution of potassium iodide and starch indicator. A faint blue color is developed in the absorber solution by the addition of standard potassium iodate solution, and as combustion proceeds, bleaching the blue color, more iodate is added. From the amount of standard iodate consumed during the combustion, the sulfur content of the sample is calculated.

In the infrared detection system, the sample is weighed into a special ceramic boat that is then placed into a combustion furnace at 1370 °C (2500 °F) in an oxygen atmosphere. Moisture and dust are removed using traps, and the sulfur dioxide is measured with an infrared detector.

Other methods such as the lamp combustion method (ASTM D1266, 2017) and the Wickbold combustion method are used for the determination of sulfur in crude oil. Trace quantities of total sulfur in crude oil products is related to various other methods (ASTM D2384, 2017; ASTM D2784, 2017; ASTM D4045, 2017). In the lamp method (ASTM D1266, 2017), a sample is burned in a closed system using a suitable lamp and an artificial atmosphere composed of 70% carbon dioxide and 30% oxygen to prevent the formation of nitrogen oxides. The sulfur oxides are absorbed and oxidized to sulfuric acid (H_2SO_4) through hydrogen peroxide (H_2O_2) solution, which is then flushed with air to remove dissolved carbon dioxide. Sulfur as sulfate in the absorbent is determined acidimetrically by titration with a standard sodium hydroxide (NaOH) solution. Alternatively, the sample can be burned in air, and the sulfur as sulfate in the absorbent can be determined by gravimetrically barium sulfate ($BaSO_4$) after precipitation. If the sulfur content of the sample is less than 0.01% w/w, it is necessary to determine sulfur in the absorber solution turbidimetrically as barium sulfate.

The olderE classical techniques for sulfur determination are being supplanted by two instrumental methods (ASTM D2622, 2017; ASTM D4294, 2017). In the former method (ASTM D2622, 2017), the sample is placed in an X-ray beam, and the peak intensity of the sulfur Kα line at 5.373 Å is measured. The background intensity, measured at 5.190 Å, is subtracted from the peak intensity, and the resultant net counting rate is then compared with a previously prepared calibration curve or equation to obtain the sulfur concentration in % w/w. The latter method (ASTM D4294, 2017) employs energy-dispersive X-ray fluorescence spectroscopy, has slightly better repeatability and reproducibility than the high-temperature method, and is adaptable to field applications but can be affected by some commonly present interferences such as halides. Here, the sample is placed in a beam emitted from an X-ray source. The resulting excited characteristic X radiation is measured, and the accumulated count is compared with the counts from previously prepared calibration standard to obtain the sulfur concentration. Two groups of calibration standards are required to span the concentration range; one standard ranges from 0.015 to 0.1% w/w sulfur and the other from 0.1 to 5.0% w/w sulfur.

6.2.8 Surface Tension and Interfacial Tension

Interfacial tension (ASTM D971, 2017) is the force of attraction between the molecules at the interface of two fluids. At the air/liquid interface, this force is often referred to as surface tension. The SI units for interfacial tension are milli-newtons per meter ($mN\,m^{-1}$). These are equivalent to the former units of dynes per centimeter ($dyne\,cm^{-1}$).

The surface tension of gas condensate or natural gasoline, together with the viscosity, reflects the rate at which the gas will spread on water, especially when spilled. Air/oil and oil/water interfacial tensions can be used to calculate a spreading coefficient that gives an indication of the tendency for the gas condensate or natural gasoline to spread and can be represented by the following equation:

$$\text{Spreading coefficient} = S_{WA} - S_{OA} - S_{WO}$$

S_{WA} is water/air interfacial tension, S_{OA} is oil/air interfacial tension, and S_{WO} is water/oil interfacial tension.

6.2.9 Viscosity

Viscosity is a measure of a fluid's resistance to flow; the lower the viscosity of a fluid, the more easily it flows. Like density, viscosity is affected by temperature. As temperature decreases, viscosity increases. Viscosity is a very important property of gas condensate or natural gasoline because it affects the spreading coefficient, the degree to which it will penetrate shoreline substrates, and the selection of mechanical spill countermeasures equipment.

Viscosity measurements may be absolute or relative (sometimes called "apparent"). Absolute viscosities are those measured by a standard method, with the results traceable to fundamental units, and are distinguished from relative measurements made with instruments that measure viscous drag in a fluid, without known and/or uniform applied shear rates (Schramm, 1992). An important benefit of absolute viscosity is that the test results are independent of the type or make of viscometer used.

There are standard test methods for measuring the viscosity of oils (such as ASTM D445, 2017 and ASTM D4486, 2017) that make use of glass capillary kinematic viscometers and will produce absolute measurements in units of centistokes (cSt) only for oils that exhibit Newtonian flow behavior (viscosity independent of the rate of shear).

6.2.10 Water Content and Sediment

Considerable importance is attached to the presence of water or sediment in crude oil for they lead to difficulties in the refinery, e.g. corrosion of equipment, uneven running on the distillation unit, blockages in heat exchangers, and adverse effects on product quality.

The water and sediment content of crude oil and crude oil products (such as gas condensate and natural gasoline), like salt, results from production and transportation practices. Water, with its dissolved salts, may occur as easily removable suspended droplets or as an emulsion. The sediment dispersed in crude oil may be composed of inorganic minerals from the production horizon or from drilling fluids and scale and rust from pipelines and tanks used for oil transportation and storage. Usually water is present in far greater amounts than sediment, but collectively, it is unusual for them to exceed 1% of the crude oil on a delivered basis. Like salt, water and sediment can foul heaters, stills, and exchangers and can contribute to corrosion and to deleterious product quality. Also, water and sediment are principal components of the sludge that accumulates in storage tanks and must be disposed of periodically in an environmentally acceptable manner. Knowledge of the water and sediment content is also important in accurately determining net volumes of crude oil in sales, taxation, exchanges, and custody transfers.

The sediment consists of finely divided solids that may be drilling mud or sand, or scale picked up during the transport of the oil, or may consist of chlorides derived from evaporation of brine droplets in the oil. The solids may be dispersed in the oil or carried in water droplets. Sediment in gas condensate and natural gasoline, although not always likely, can lead to serious plugging of the equipment, corrosion due to chloride decomposition, and lowering of the quality of the gas condensate and natural gasoline.

Water may be found in the crude either in an emulsified form or in large droplets and can cause flooding of distillation units and excessive accumulation of sludge in tanks. Refiners generally limit the quantity, and although steps are normally taken at the oilfield to reduce the water content as low as possible, water may be later introduced during shipment. In any form, water and sediment are highly undesirable in a refinery feedstock, and the relevant tests involving distillation (ASTM D95, 2017; ASTM D4006, 2017), centrifuging (ASTM D4007, 2017), extraction (ASTM D473, 2017), and the Karl Fischer titration (ASTM D4377, 2017; ASTM D4928, 2017) are regarded important in the determination of the quality of gas condensate and natural gasoline.

The Karl Fischer test method (ASTM D1364, 2017; ASTM D6304, 2017) covers the direct determination of water in crude oil. In the test, the sample injection in the titration vessel can be performed on a volumetric or gravimetric basis. Viscous samples can be analyzed using a water vaporizer accessory that heats the sample in the evaporation chamber, and the vaporized water is carried into the Karl Fischer titration cell by a dry inert carrier gas.

Water and sediment in gas condensate and natural gasoline can be determined simultaneously (ASTM D4007, 2017) by the centrifuge method. Known volumes of gas condensate and natural gasoline and solvent are placed in a centrifuge tube and heated to 60 °C (140 °F). After centrifugation, the volume of the sediment-and-water layer at the bottom of the tube is read. For gas condensate and natural gasoline that contain wax, a temperature of 71 °C (160 °F) or higher may be required to completely melt the wax crystals so that they are not measured as sediment. The temperature must not interfere with the properties of the gas condensate and natural gasoline under test.

Sediment is also determined by an extraction method (ASTM D473, 2017) or by membrane filtration (ASTM D4807, 2017). In the former method (ASTM D473, 2017), an oil sample contained in a refractory thimble is extracted with hot toluene until the residue reaches a constant mass. In the latter test, the sample is dissolved in hot toluene and filtered under vacuum through a 0.45 μm porosity membrane filter. The filter with residue is washed, dried, and weighed.

Natural gas, as well as gas condensate or natural gasoline, may contain substantial (or, at least, measurable) amounts of water. Because any process that would separate water from the gas condensate or natural gasoline would also change the composition of the gas condensate or natural gasoline, most properties are often determined on the gas condensate or natural gasoline as received at the laboratory. Therefore, for those samples with significant water contents (greater than 5% w/w), many of the properties measured will not represent the properties of the *dry* sample. Generally, water content is determined by the Karl Fischer titration (ASTM D6304, 2017). The procedure is adopted in many official standards as the standard test method for water determination in crude oil and crude oil products, including gas condensate and natural gasoline (Speight, 2014a, 2015).

6.2.11 Other Tests

The inspection assay tests discussed above are not exhaustive but are the ones most commonly used and provide data on the impurities present as well as a general idea of the products that may be recoverable. Other properties that are determined on an as-needed basis include, but are not limited to, the following: (i) wax content; (ii) vapor pressure – the Reid method; (iii) total acid number, which has rapidly become essential in terms of the data relating to the content of acidic species in crude oil and crude oil products (Speight, 2014a, b); (iv) chloride content; and as the occasion demands (v) the *aniline point* (or *mixed aniline point*).

On the other hand, the cloud point (ASTM D2500, 2017; ASTM D5772, 2017; ASTM D7397, 2017; ASTM D7689, 2017), which is often used to indicate the temperature at which wax deposits from oil, may be determined by cooling of a sample under prescribed conditions with stirring. The temperature at which the wax first appears is the wax appearance point.

The Reid vapor pressure test method (ASTM D323, 2017) measures the vapor pressure of volatile crude oil products, such as gas condensate and natural gasoline. The Reid vapor pressure differs from the true vapor pressure of the sample due to some small sample vaporization and the presence of water vapor and air in the confined space.

The *acid number* is the quantity of base, expressed in milligrams of potassium hydroxide per gram of sample that is required to titrate a sample in this solvent to a green/green-brown end point, using *p*-naphtholbenzein indicator solution. The *strong acid number* is the quantity of base, expressed as milligrams of potassium hydroxide per gram of sample, required to titrate a sample in the solvent from its initial meter reading to a meter reading corresponding to a freshly prepared nonaqueous acidic buffer solution or a well-defined inflection point as specified in the test method (ASTM D664, 2017).

To determine the acid number by the color-indicator method (ASTM D974, 2017), the sample is dissolved in a mixture of toluene and isopropyl alcohol containing a small amount of water, and the resulting single-phase solution is titrated at room temperature with standard alcoholic base or alcoholic acid solution, respectively, to the end point indicated by the color change of the added *p*-naphtholbenzein solution (orange in acid and green-brown in base). To determine the strong acid number, a separate portion of the sample is extracted with hot water, and the aqueous extract is titrated with potassium hydroxide solution, using methyl orange as an indicator.

To determine the acid number by the potentiometric titration method (ASTM D664, 2017), the sample is dissolved in a mixture of toluene and isopropyl alcohol containing a small amount of water and titrated potentiometrically with alcoholic potassium hydroxide using a glass indicating electrode and a calomel reference electrode. The meter readings are plotted manually or automatically against the respective volumes of titrating solution, and the end points are taken only at well-defined inflections in the resulting curve. When no definite inflections are obtained, end points are taken at meter readings corresponding to those found for freshly prepared nonaqueous acidic and basic buffer solutions.

The acid numbers obtained by this color-indicator test method (ASTM D974, 2017) may or may not be numerically the same as those obtained by potentiometric titration method (ASTM D664, 2017). In addition, the color of the crude oil sample can interfere with observation of the end point when the color indictor method is used. Determination of the acid number is more appropriate for high acid crude oils, heavy oils, and tar sand bitumen (which are often high in acid content) and various crude oil products, such as gas condensate and natural gasoline (Speight, 2014a).

The test method for the determination of the acid number by the color-indicator titration method (ASTM D3339, 2017) measures the acid number of oils obtained from laboratory oxidation test (ASTM D943, 2017) using smaller amounts of samples than those used in other acid number tests (ASTM D664, 2017; ASTM D974, 2017).

In this test, the sample is dissolved in a solvent mixture of toluene and isopropyl alcohol, and a small amount of water and the solution is titrated at room temperature under a nitrogen atmosphere with standard potassium hydroxide (KOH) in isopropyl alcohol to the stable green color of the added indicator *p*-naphtholbenzein. Dark-colored crude oils (and crude oil products) are more difficult to analyze by this method because of the difficulty in detecting color change. In such cases, the potentiometric titration method (ASTM D664, 2017) may be used if sufficient sample is available. The acid numbers will not provide the data essential to determining whether a single crude oil (or a blend of the crude oil with other crude oils) will yield the desired product slate. Such data can only be generated when a comprehensive assay of the crude oil (and its partners in the blending, including gas condensate and natural gasoline) is performed and the data from several tests are taken in relation to each other.

The chloride content of crude oil (ASTM D4929, 2017) is, like the total acid content, also a necessary piece of data – especially at the wellhead and after the desalting operation. Just as *high acid crude oils* cause corrosion in the refinery, overhead corrosion is caused by the mineral salts, magnesium, calcium, and sodium chloride that are hydrolyzed to produce volatile hydrochloric acid, causing a highly corrosive condition in the overhead exchangers (Speight, 2014c). Therefore, these salts present a significant contamination in opportunity crude oils.

Finally, the *aniline point* (or *mixed aniline point*) (ASTM D611, 2017) has been used for the characterization of crude oil although it is more applicable to pure hydrocarbons and in their mixtures and is used to estimate the aromatic content of mixtures. Aromatics exhibit the lowest aniline points, and paraffins the highest aniline points. Cycloparaffins and olefins exhibit values between these two extremes. In any hydrocarbon homologous series, the aniline point increases with increasing molecular weight.

6.3 Chromatographic Analyses

Typically, functional group (including hydrocarbon groups) analyses are conducted to determine amounts of the compound classes (e.g. saturate constituents, aromatic constituents, and polar constituents – i.e. constituents containing a nonhydrocarbon functional group) present in gas condensate and natural gasoline samples. Group-type test methods include multidimensional GC, high-performance liquid chromatography (HPLC), and thin-layer chromatography (TLC) (Miller, 2000; Patnaik, 2004; Speight, 2005, 2014a, 2015).

Test methods that analyze individual compounds (e.g. benzene–toluene–ethylbenzene–xylene mixtures and polynuclear aromatic hydrocarbons) are generally applied to detect the presence of an additive or to provide concentration data needed to estimate individual compounds in gas condensate or natural gasoline. Common constituent measurement techniques include GC with second column confirmation, GC with multiple selective detectors, and GC with mass spectrometry detection (GC-MS) (Speight, 2005).

6.3.1 Gas Chromatography

GC, which is based on the principle of a stationary phase and a mobile phase, remains a primary technique for determining hydrocarbon distribution in hydrocarbon liquids, such as gas condensate and natural gasoline products, through the identification of individual hydrocarbons. Although a measure of purity by GC is often sufficient for many purposes, it is not always sufficient for measuring absolute purity – not all possible impurities will pass through the chromatographic column, and not all those that do will be measured by the detector. Absolute purity is best measured by distillation range or freeze or solidification points. Despite this disadvantage, the technique is still a widely used technique and is the basis of many current standard test methods for the determination and measurement of hydrocarbon derivatives in crude oil products. When classes of hydrocarbons such as olefins (which occur frequently in processes gas streams) need to be measured, techniques such as the bromine index are used (ASTM D1492, 2017; ASTM D2710, 2017).

Briefly, *gas–liquid chromatography* (GLC) is a method for separating the volatile components of various mixtures (Fowlis, 1995; Grob, 1995). It is, in fact, a highly efficient fractionating technique, and it is ideally suited to the quantitative analysis of mixtures when the possible components are known, and the interest lies only in determining the amounts of each present. In this type of application, GC has taken over much of the work previously done by the other techniques; it is now the preferred technique for the analysis of hydrocarbon gases, and gas chromatographic in-line monitors are having increasing application in refinery plant control. GLC is also used extensively for individual component identification, as well as percentage composition, in the gasoline boiling range, which includes gas condensate and natural gasoline.

The mobile phase is the carrier gas, and the gas selected has a bearing on the resolution. Nitrogen has very poor resolution ability, while helium or hydrogen is a better choice with hydrogen being the best carrier gas for resolution. However, hydrogen is reactive and may not be compatible with all sets of target analytes. There is an optimum flow rate for each carrier gas to achieve maximum resolution. As the temperature of the oven increases, the flow rate of the gas changes due to thermal expansion of the gas. Most modern gas chromatographs are equipped with constant flow devices that change the gas valve settings as the temperature in the oven changes, so changing flow rates are no longer a concern. Once the flow is optimized at one temperature, it is optimized for all temperatures.

6.3.2 Gas Chromatography–Mass Spectrometry

A gas chromatography–mass spectrometry system is used to measure concentrations of target volatile and semivolatile constituents of gas condensate and natural gasoline. The advantage of the technique is the high selectivity or ability to confirm compound identity through retention time and unique spectral pattern. The method is used for identification and quantification of the constituents of crude oil fractions and crude oil products.

To reduce the possibility of false positives, the intensities of one-to-three selected ions are compared with the intensity of a unique target ion of the same spectrum. The sample ratios are compared with the ratios of a standard. If the sample ratios fall within a certain range of the standard, and the retention time matches the standard within specifications, the analyte is considered present. Quantification is performed by integrating the response of the target ion only.

Mass spectrometers are among the most selective detectors, but they are still susceptible to interferences. Isomers have identical spectra, while many other compounds have similar mass spectra.

6.4 Spectroscopic Analyses

The chemical composition of a natural gas has always been considered to be a valuable indicator of behavior of the gas mixture when treating at the wellhead or when treating in a gas processing plant. More than likely, chemical composition studies truly complement physical property and physical behavior studies, and the true picture is a combination of all of these studies.

However, the chemical composition of a natural gas stream (or gas condensate stream or a natural gasoline stream) is represented in terms of compound types and/or in terms of generic compound classes, thus allowing the analytical chemist, the process chemist, and the process engineer to determine the nature of the reactions. Hence, chemical composition can play a large part in determining the nature of the products that arise from the gas processing operations. It can also play a role in determining how a particular feedstock should be processed (Speight, 2014a, 2015). However, proper interpretation of the data resulting from the composition studies requires an understanding of chemical structures, their significance, and an open mind!

Spectroscopic studies have played an important role in the evaluation of crude oil and of crude oil products for the last six decades, and many of the methods are now used as standard methods of analysis for refinery feedstocks and products. Application of these methods to feedstocks and products is a natural consequence for the refiner.

Infrared spectroscopy can be used for the determination of specific constituents, such as benzene in gas condensate or in natural gasoline, while ultraviolet spectroscopy is employed for the evaluation of liquids (ASTM D1840, 2017; ASTM D2269, 2017). The methods include the use of *mass spectrometry* to determine the (i) hydrocarbon types in middle distillates (ASTM D2425, 2017), (ii) hydrocarbon types of gas oil saturate fractions (ASTM D2786, 2017), (iii) hydrocarbon types in low-olefin gasoline (ASTM D2789, 2017), and (iv) aromatic types of gas oil aromatic fractions (ASTM D3239, 2017). *Nuclear magnetic resonance spectroscopy* has been developed as a standard method for the determination of hydrogen types in liquid fuels (ASTM D3701, 2017). *X-ray fluorescence spectrometry* has been applied for the determination and analysis of selected elements (nitrogen, sulfur, nickel, and vanadium) and to the determination of sulfur in various crude oil products (ASTM D2622, 2017; ASTM D4294, 2017).

It is also appropriate at this point to give a brief description of other methods that are used for the identification of the constituents of natural gas, gas condensate, and natural gasoline.

6.4.1 Infrared Spectroscopy

Infrared spectroscopy is a well-established method that was developed for comparative, semiquantitative analysis, leading to the present quantitative analysis (ASTM E168, 2017; ASTM E334, 2017; ASTM E1252, 2017). Infrared spectra are displayed either as percent transmittance or as absorptivity versus frequency (cm^{-1}). Transmittance, T, defined as the ratio of transmitted light over incident light, or percent transmittance, usually shows more detail over the entire range and is generally the preferred display. Absorbance (A), on the other hand, is proportional to the concentration and is, therefore, used for quantitative measurements.

Infrared spectroscopy is a simple procedure and is one of several techniques that can provide quick information about the distribution of several structural and functional groups (Drews, 1998; Rand, 2003; Totten, 2003). In combination with nuclear magnetic resonance spectroscopy, it will provide quick yet fair detail on the distribution of carbon-hydrogen groups. It is also possible to derive structural parameters from infrared spectroscopic data, and these are (i) saturated hydrogen to saturated carbon ratio, (ii) paraffin character, (iii) cycloparaffin–naphthene character, (iv) methyl group content, and (v) paraffin chain length. In conjunction with proton magnetic resonance (PMR) (see next section), structural parameters such as the fraction of paraffinic methyl groups to aromatic methyl groups can be obtained.

Increased resolution by band narrowing techniques in conjunction with reference spectra allows the distinction of methylene (CH_2) groups next to other groups such as alkyls, aromatic rings, carbonyls, or alkoxy groups. Carbon–hydrogen (CH_n) groups and aromatic carbon can be identified, measured, and compared. The ratio of the band at 1602 cm^{-1} (aromatic C—C stretching) to that at 2920 cm^{-1} (aliphatic hydrogen–carbon stretching) in methylene groups (β or further away from the aromatic ring) is a good relative measure for the aromaticity of a sample. Fourier transform infrared spectroscopy is sufficiently sensitive that it can be used for detection in HPLC and even in GC and supercritical fluid chromatography (SFC).

Band assignments for infrared spectra of hydrocarbon constituents of crude oil were established several decades ago. Resolution enhancement by Fourier self-deconvolution (band narrowing) led to the recognition and distinction of bands that could previously not be separated. Examples of such new assignments are methyl (CH_3) groups next to aromatic rings (2948 cm^{-1}) and next to alkyl groups (2960 cm^{-1}), methylene (CH_2) groups (together with some methyl groups) next to aromatic rings (2916 cm^{-1}) and in aliphatic chains (2926 cm^{-1}), methylene groups next to an alkoxy function (2928 cm^{-1}) and next to a carbonyl group (2933 cm^{-1}), and methane (CH) groups in unspecified environments (2905 and 1897 cm^{-1}). The rocking vibration band of methylene groups in straight aliphatic chains varies slightly with their length. For *n*-pentyl groups, the band is located at 726 cm^{-1}, and with longer chains, it shifts to lower frequencies until it stabilizes at 720 cm^{-1} for chains with more than nine carbon atoms.

6.4.2 Mass Spectrometry

Mass spectrometry furnishes the molecular weight and chemical formula of compounds and their relative amounts in a mixture and offers nondestructive examination of the sample (ASTM D2650, 2017; ASTM D2650, 2017; ASTM

D2786, 2017; ASTM D2789, 2017; ASTM D3239, 2017; ASTM E1316, 2017). The technique can also provide important information about their molecular structure. The earliest and most common type of mass spectrometry, electron impact mass spectrometry (El-MS), gives a fragmentation pattern displaying both parent ion peaks and fragment ion peaks, characteristic of each molecular type. Fragmentation is frequently employed to differentiate between isomers of pure compounds and of molecules in fairly simple mixtures. However, it is usually avoided with such complex samples as high molecular weight fractions of crude oil, which are composed of such a multitude of closely related compounds that their fragmentation patterns are nondistinctive and cannot be readily interpreted.

Mass spectrometry can play a key role in the identification of the constituents of feedstocks and products either in the laboratory or online (ASTM D2425, 2017; ASTM D2786, 2017; ASTM D2789, 2017; ASTM D3239, 2017). The principal advantages of mass spectrometric methods are (i) high reproducibility of quantitative analyses, (ii) the potential for obtaining detailed data on the individual components and/or carbon number homologues in complex mixtures, and (iii) a minimal sample size is required for analysis. The ability of mass spectrometry to identify individual components in complex mixtures is unmatched by any modern analytical technique – with the exception of GC.

In fact, *gas chromatography coupled with mass spectrometry* (GC-MS) is a powerful method for crude oil distillates. However, there are disadvantages arising from the use of mass spectrometry: (i) the limitation of the method to organic materials that are volatile and stable at temperatures up to 300 °C (570 °F) and (ii) the difficulty of separating isomers for absolute identification. The sample is usually destroyed, but this is seldom a disadvantage. Nevertheless, in spite of these limitations, mass spectrometry does furnish useful information about the composition of feedstocks and products even if this information is not as exhaustive as might be required. There are structural similarities that might hinder identification of individual components. Consequently, identification by type or by homologue will be more meaningful since similar structural types may be presumed to behave similarly in processing situations. Knowledge of the individual isomeric distribution may add only a little to an understanding of the relationships between composition and processing parameters.

Mass spectrometry should be used discriminately where a maximum amount of information can be expected. The heavier nonvolatile feedstocks are for practical purposes, beyond the useful range of routine mass spectrometry. At the elevated temperatures necessary to encourage volatility, thermal decomposition will occur in the inlet, and any subsequent analysis would be biased to the low molecular weight end and to the lower molecular products produced by the thermal decomposition.

6.4.3 Nuclear Magnetic Resonance Spectroscopy

Nuclear magnetic resonance has frequently been employed for general studies of hydrogen types in crude oil and crude oil products (ASTM D4808, 2017) as well as for the structural studies of higher-boiling crude oil constituents (Speight, 1994, 2014a; ASTM E386, 2017). The technique has recently been adopted to measure the hydrogen content of low-boiling hydrocarbon fuels and other crude oil products (ASTM D3701, 2017; ASTM D4808, 2017). In fact, PMR studies (along with infrared spectroscopic studies) were, perhaps, the first studies of the modern era that allowed structural inferences to be made about the polynuclear aromatic systems that occur in the higher molecular weight constituents of crude oil and products such as high-boiling fuel oils.

For the more complex liquids, assuming that GC does not give the desired information, the proton (hydrogen) types in liquid hydrocarbon fractions can be typically subdivided into three types: (i) aromatic ring hydrogen, (ii) aliphatic hydrogen adjacent to an aromatic ring, and (iii) aliphatic hydrogen remote from an aromatic ring. In other cases, five types of hydrogen locations are identified: (i) aromatic hydrogen, (ii) substituted hydrogen next to an aromatic ring, (iii) naphthene hydrogen, (iv) methylene hydrogen, and (v) terminal methyl hydrogen remote from an aromatic ring.

It is in this regard that *carbon-13 magnetic resonance* (CMR) can play a useful role. Since carbon magnetic resonance deals with analyzing the carbon distribution types, the obvious structural parameter to be determined is the aromaticity, f_a. A direct determination from the various carbon type environments is one of the better methods for the determination of aromaticity. Thus, through a combination of proton and carbon magnetic resonance techniques, refinements can be made on the structural parameters, and, for the solid state high resolution CMR technique, additional structural parameters can be obtained.

The basic instrumentation for CMR is the same as that for PMR except that there are two radio frequency fields orthogonal (at right angles) to the main magnetic field, one for observing the carbon-13 nuclei and the other for decoupling the proton nuclei. The low abundance of carbon-13 isotopes (1.1%) and the lower gyromagnetic ratio of the carbon-13 nucleus make the signal weaker by more than two orders of magnitude, and moreover, the nuclei have longer relaxation times. The effect is that, even with Fourier transform (FT) data acquisition, CMR measurements can hours to perform.

In contrast to PMR spectroscopy, the peak areas arising from the carbon-13 nuclei in different molecular positions ordinarily are not proportional to their concentration. Quantitative measurements require that two effects must be overcome: (i) the different relaxation times of the carbon-13 nuclei in different chemical groups and (ii) the nuclear Overhauser enhancement (NOE). The latter effect refers to the rise in signal intensity when C—H coupled protons are saturated by the decoupling field. One way to do this is to add a small amount of a paramagnetic relaxation reagent, such as trisacetylacetonatochromium(III), $Cr(AcAc)_3$, which changes the dominant relaxation mechanism into one involving the interaction between unpaired electrons and '^{3}C nuclei. It also reduces the long relaxation times of some carbons.

In its simplest form, CMR can distinguish between aliphatic and aromatic carbon. In the aliphatic region of hydrocarbon liquid ^{13}C–NMR spectra, several sharp peaks stand out and are used for quantitative. The most prominent peak is usually that at 29.7 ppm. It is attributed to methylene (CH_2) carbon atoms in long alkyl chains, positioned four or more carbons (>γ) away from an aromatic ring and from terminal methyl CH_3) groups.

Normally, the absorption of 29.5–30.3 ppm is an estimation of the amount of carbon atoms in long alkyl chains (>C_5). Because this band represents methylene (CH_2) groups two or more carbons away from an aromatic ring and a terminal group, there must be four more carbons per chain that are indicated by the area under these peaks. The *number of long chains* (n_{CH_2}) can be estimated from the peaks at 14.2 ppm (ω-CH_3) and at 28.1 ppm (CH_2) next to a terminal branch point, i.e. a methane (CH) group. On the one hand, the peak at 14.2 ppm gives results that are too high for this purpose because it also indicates CH_3 groups from chain branches. Subtracting half of the 37.6 ppm peak area (CH_2 next to CH groups inside a chain) corrects for this feature. On the other hand, the peak at 14.2 ppm does not cover twin CH_3 groups (as in an isopropyl group). This is why the 28.1 peak CH next to two terminal methyl groups is needed. Thus,

$$n_{CH_2\ \text{long chains}} = A(14.2) - 1/2A(37.6) + 1/3A(28.1)$$

The number of methylene groups in long chains is the sum of the peak at 29.7 ppm, the number of methylene groups at the two ends, i.e. six times the number of long chains for each of the two ends, and the number close to branch points inside the chain, namely, those next to them and one carbon atom removed on both sides of the CH group. Thus, the total number is

$$n_{CH_2} = C(29.7) + 6n_{CH_2\ \text{long chains}} + 4C(37.6)$$

The number of methine (CH) groups in long chains can be estimated from the absorbance at 37.6 and at 39.5 ppm. The peak at 37.6 ppm represents methylene groups next to methane carbon and thus does not represent a direct measurement of the methane groups. Thus,

$$n_{CH_2\ \text{long chains}} = 1/2C(37.6) + C(39.5)$$

The methyl groups give rise to at least four peaks in carbon-13 spectra, namely, the peaks at 11.5, 14.2, 19.5, and 22.7 ppm. The peak 22.7 ppm represents twin methyl groups as in an isopropyl group, and it also has a contribution from a methylene group next to a methyl group and does not need to be represented again by a contribution to the equation from the peak at 14.2 ppm. Thus,

$$n_{CH_3 \text{ long chains}} = C(11.5) + C(19.5) + C(22.7)$$

The final estimate for carbon atoms in long chains is then the sum of the three types. The average length of long chains derives from dividing the number of carbon atoms in long chains by the number of long chains. Remember, of course, that averages derived from magnetic resonance spectra can be very misleading if the data are interpreted too literally (Speight, 2014a).

The methyl group (CH_3) gives rise to several peaks depending on its position in the molecule. A peak at 14.2 ppm signals such a group positioned at the end of an unbranched chain segment of at least two or three methylene (CH_2) groups. A methyl group next to a branch point (methine group, CH) at the end of a chain with at least two methylene groups produces a peak at 22.7 ppm. Farther away from the end of the chain, such a group gives a peak at or near 19.8 ppm. The methyl group at the end is also affected by these branch points.

Naphthenic carbon in high molecular weight crude oil fractions usually occupies so many slightly different positions that the peaks are unresolved and form a broad hump in the range 25–60 ppm under the generally well-resolved paraffinic peaks. The evaluation of the hump is the only direct method for the determination of naphthenic methylene and methane (CH_n) groups. However, its measurement may not always be reliable. In very high-boiling crude oil samples, such as asphaltene constituents, the paraffin-type resonance may be only partly resolved. Although this leads mainly to broader peaks, the overlap may in some cases add to the hump and, thus, cause erroneously high results for naphthenic carbon. The aromatic region of the ^{13}C-NMR spectrum can be evaluated by conventional integration of the peaks due to the main aromatic group types.

6.4.4 Ultraviolet Spectroscopy

The ultraviolet–visible (UV–Vis) spectrum, although not as specific for chemical group types as infrared spectroscopy and nuclear magnetic resonance spectroscopy, can distinguish between aromatic compounds with different ring numbers and configurations (e.g. ASTM D1840, 2017; ASTM D2269, 2017). The patterns are not distinct enough to recognize or distinguish these compounds in complex mixtures, but they can be useful for their identification in narrow fractions.

UV–Vis spectroscopy (ASTM E169, 2017) can be employed as a detector for the fractionation of crude oil samples, for example, for the chromatographic separation and/or identification of aromatics by ring number, especially in combination with a technique such as liquid chromatography (Speight, 1986). Thus, UV–Vis spectroscopy lends itself to studies of refining processes as an online detector.

6.4.5 X-ray Diffraction

X-ray diffraction had been used for the determination of the fraction of aromatic carbon (f_a) in crude oil constituents. This ratio can also be conveniently and precisely obtained by carbon-13 nuclear magnetic resonance spectroscopy as well as by infrared spectroscopy. The determination of the fraction of carbon that is aromatic may be in error (Ebert, 1990) since X-ray diffraction data can be very misleading (Ebert et al., 1984, 1987), especially when the data are used to translate geometric data – measurements of (aromatic) sheet diameter – into structural information. On the one hand, not all aromatic atoms contribute to the stack diameter seen by X-ray diffraction, whereas, on the other hand, nonaromatic atoms such as hydroaromatic carbons and other substituents on aromatic rings may contribute to the diffraction pattern. Thus, the interpretation of these measurements is quite arbitrary.

Extended X-ray absorption fine structure (EXAFS) and *X-ray absorption near edge structure* (XANES) spectroscopies are tools for the investigation of the immediate chemical environment of X-ray absorbing elements such as metals and sulfur.

6.5 Molecular Weight

The molecular weight (formula weight) of a compound is the sum of the atomic weights of all the atoms in a molecule and can be determined by a variety of methods. Liquid fuels, being complex mixtures of many constituents, require qualification of the molecular weight as either (i) number average molecular weight or (ii) weight average molecular weight.

The *number average molecular weight* is the ordinary arithmetic mean or average of the molecular weights of the individual constituents. It is determined by measuring the molecular weight of *n* molecules, summing the weights, and dividing by *n*. The *weight average molecular weight* is a way of describing the molecular weight of a complex mixture such as crude oil even if the molecular constituents are not of the same type and exist in different sizes. Molecular weight is often used in refineries to provide mass-average or number-average measurements. Consequently, a number of methods are available to measure molecular weight of crude oil and crude oil products.

Molecular weight may be calculated from viscosity data (ASTM D2502, 2017). The test method requires viscosity data from different temperatures, typically at 37.8 and 98.9 °C (100 and 210 °F). The method is generally applicable to a variety of hydrocarbon fractions, but the number is an average number and applicable to those fraction or products with a molecular weight in the range 250–700. Samples with unusual composition, such as aromatic-free white mineral oils, or oils with very narrow boiling range, may give atypical or questionable results. For samples with higher molecular weight (up to 3000 or more) with unusual composition or for polymers, another test method is recommended (ASTM D2503, 2017). This method uses a vapor pressure osmometer to determine the molecular weight of the sample. Low-boiling samples may not be suitable – the vapor pressure of the constituents of the sample can interfere with the method.

If the molecular weight determination involves the use of a solvent, it is recommended that to negate concentration effects and temperature effects, the molecular weight determination be carried out at three different concentrations at three different temperatures. The data for each temperature are then extrapolated to zero concentration, and the zero-concentration data at each temperature are then extrapolated to room temperature (Speight, 2014a, 2015). Furthermore, each method of molecular weight determination has proponents and opponents because of assumptions made in the use of the method or because of the mere complexity of the sample and the nature of the inter- and intramolecular interactions.

Methods for molecular weight measurement are also included in other more comprehensive standards, and there are several indirect methods that have been proposed for the estimation of molecular weight by correlation with other more readily measured physical properties. They are satisfactory when dealing with the conventional type of crude oils or their fractions and products when approximate values are desired.

6.6 Instability and Incompatibility

The problem of instability in crude oil and crude oil products may manifest itself in changes to the liquid such as (i) asphaltene separation that occurs when the paraffinic nature of the liquid medium increases, as is often the case when paraffinic liquid such as gas condensate or natural gasoline is blended with residual fuel oil; (ii) wax separation that occurs when there is a drop in temperature or the aromaticity of the liquid medium increases; and (iii) sludge/sediment formation in liquid fuels that occurs because of the interplay of several chemical and physical factors, even in low-boiling products such as gas condensate and natural gasoline (Mushrush and Speight, 1995; Speight, 2014a, 2015). In addition to the phase-separation phenomenon, there may also be a general darkening of the color of the liquid product (ASTM D1500, 2017).

Gas condensate and natural gasoline do not contain asphaltene constituents but may contribute to paraffin deposition either from the deposition of paraffin constituents when the gas condensate or the natural gasoline is blended with other crude oil-related feedstocks at the wellhead prior to transportation or in the refinery. Paraffin precipitation and deposition in crude oil pipelines and in crude oil product pipelines is an issue that decreases the cross-sectional area of the pipeline, thereby restricting operating capacities and places additional strain on pumping equipment. If such incompatibility does occur, it may be manifested by the appearance of sludge or sediment.

Sludge (or sediment) formation takes one of the following forms: (i) material dissolved in the liquid, (ii) precipitated material, and (iii) material emulsified in the liquid. Under favorable conditions, sludge or sediment will dissolve in the crude oil or product with the potential of increasing the viscosity. Sludge or sediment, which is not soluble in the hydrocarbon streams (ASTM D473, 2017; ASTM D1796, 2017; ASTM D2273, 2017; ASTM D4007, 2017; ASTM D4807, 2017), may either settle at the bottom of the storage tanks or remain in the crude oil as an emulsion. In most of the cases, the smaller part of the sludge/sediment will settle satisfactorily; the larger part will stay in the crude oil as emulsions. In any case there is a need of breaking the emulsion, whether it is a water-in-oil emulsion or whether it is the sludge itself, which has to be separated into the oily phase and the aqueous phase. The oily phase can then be processed with the crude oil, and the aqueous phase can be drained out of the system.

Compatibility in distillate products is important to commercial and industrial consumers as well as to the producer. Distillate products that are made from the refining process based on straight-run distillation show very little incompatibility problems. However, problems for refiners will continue to increase as the quality of the available crude decreases worldwide. This decrease coupled with the inevitable future use of liquid fuels from biosources, coal, and shale sources will exacerbate the present problems for the producers.

When various stocks are blended at the refinery (Gary et al., 2007; Speight, 2014a, 2017; Hsu and Robinson, 2017), incompatibility can be explained by the onset of acid–base catalyzed condensation reactions of the various organonitrogen compounds in the individual blending stock themselves. These are usually very rapid reactions with practically no observed induction time period (Mushrush and Speight, 1995).

When the product is transferred to a storage tank or some other holding tank, incompatibility can occur the free-radical hydroperoxide induced polymerization of active olefins. This is a relatively slow reaction, because the observed increase in hydroperoxide concentration is dependent on the dissolved oxygen content. Another incompatibility mechanism involves degradation when the product is stored for prolonged periods, as might occur during stockpiling of fuel for military use. This incompatibility process involves (i) the buildup of hydroperoxide moieties after the gum reactions, (ii) a free-radical reaction with the various organo-sulfur compounds present (such as mercaptan sulfur, R—SH) (ASTM D3227, 2017; ASTM D5305, 2017) that can be oxidized to sulfonic acids, and (iii) reactions such as condensations between organo-sulfur and nitrogen compounds and esterification reactions.

6.7 Use of the Data

Natural gas is a combustible gas that is a mixture of simple hydrocarbon compounds. It is a fossil fuel that contains primarily methane, along with small amounts of ethane, butane, pentane, and propane. Natural gas does not contain carbon monoxide. The by-products of burning natural gas are primarily carbon dioxide and water vapor.

Natural gas is a colorless, tasteless, odorless, and nontoxic gas. Because it is odorless, a powerful chemical called mercaptan is added to the gas, in very small amounts, to give the gas a distinctive smell of rotten eggs. This strong smell can be helpful in detecting the source of any gas leak. Natural gas (as long as it is predominantly methane since methane has a vapor density less than the vapor density of air) can dissipate into the air. Other positive attributes of natural gas are a high ignition temperature and a narrow flammability range.

Natural gas is made up of several component gases and is therefore subject to natural variation. The Wobbe index number is used to measure the energy of gas within a given volume, rather than calorific value (CV), which is subject to natural variation. This inconsistency affects the energy contained within a given volume of gas; the measure used is the CV of the gas. The Wobbe number or Wobbe index describes the way in which the gas burns and is calculated as a factored ratio of CV and SG (otherwise known as relative density).

Thus, natural gas and the other gases described in this text (Chapters 1, 7, and 8) are mixtures of hydrocarbon compounds and nonhydrocarbon compounds and generally are of little value in the natural state – processing (refining, treating) is required to produce saleable products that meet the specifications required for the designated use (Table 6.2). Thus, before natural gas can be used as a fuel, most, but not all, must be processed to remove impurities, including water, to meet the specifications of marketable natural gas. The by-products of this processing include ethane, propane butane isomers, and pentane isomers as well as higher molecular weight hydrocarbons, hydrogen sulfide (which may be converted into elemental sulfur), carbon dioxide, and, sometimes, helium and nitrogen. Furthermore, an important task in natural gas processing is the need for reliable values of the volumetric and thermodynamic properties for hydrocarbon mixtures, which are important in the design and operation of gas processing equipment.

Table 6.2 Examples of specifications for a gas stream using natural gas as the example.

Component	Composition
Methane (C_1)	Not less than 85 mol%
Ethane (C_2)	Not more than 9.2 mol%
Propane (C_3)	Not more than 3.00 mol%
Butanes (C_4) and heavier	Not more than 2.00 mol%
Pentanes (C_5) and heavier	Not more than 0.25 mol%
Nitrogen (N_2)	Not more than 1.25 mol%
Oxygen (O_2)	Not more than 0.5 mol%
Total nonhydrocarbons	Not more than 2.0 mol%
Total sulfur including H_2S	Not more than 10 ppm by weight, H_2S content less than 4 ppm v/v
Impurities	Gas shall be reasonably free from dust (max size 5 μm)
Water content	Not more than 112 kg $MMSCM^{-1}$

The application of standard test method to the analysis of natural gas is almost as old as the natural gas industry and even goes back to the older coal gas industry (Speight, 2007, 2013). Because of the increased use of gases as fuel gases, characterization and determination of physical properties has received considerable attention as a needed procedure to determine refinability. In the modern era, as a result of the expansion of computer simulators and advanced analytical tools accompanied by the availability of more accurate analytical data, the fields of natural gas analysis have expanded even further (Speight, 2015).

In this context, the end product of the analysis (or testing) of natural gas and process gases (including biogas and landfill gas) is the collection of data that allows the investigator to *specify* the character and quality of the material under investigation. Thus, a series of specifications are necessary for natural gas (and other fuel gases), and such data are derived from application of the relevant text methods to the sample(s). Specifications for many of the gas streams and their receptive products are based on bulk properties such as (but not limited to) density and boiling range (in the case of gas condensate and natural gasoline) to assure that the product can perform the intended in-service task: (i) the raw product meets the requirement in the next step of a processing sequence, and (ii) the finished product meets the requirement for sales. Furthermore, because gas streams exhibit wide variations in composition and properties, and these occur not only in natural gas from different sources, specifications are necessary to assure that the product performs in service as required. Specifications are even more extensive as increased amounts of biogas and landfill gas are introduced into gas processing plants (or refineries) either singly or as blends with other feedstocks for storage and/or transportation purposes (Chapter 3).

Providers of natural gas and their large-volume customers (particularly electric utilities and possibly other industrial) are used to defining fuel requirements in the form of specification data. In the case of natural gas, due to the nature of the product and the transportation mechanism, the quality of the delivered product is primarily determined by the pipeline. Utility generators have a need to know the expected range of quality of the fuel being delivered to them and ideally to have some control over the variability of that fuel to assure compliance with regulations, to protect their investment in generating equipment, and to be able to meet the needs of their customers in the most economic manner.

Gas transmission lines are one of the core assets of the energy infrastructure in the United States. As a result, the operation of these lines must be as trouble-free as possible. A major operational consideration for gas pipelines is hydrocarbon liquid condensation from the natural gas. The deposition of a hydrocarbon liquid in gas pipelines can cause operational issues including increased pressure drop, reduced line capacity, and equipment problems such as compressor damage. In order to avoid hydrocarbon condensation or liquid dropout in natural gas pipelines, several different control parameters have historically been monitored and assigned limits to the amount of higher molecular weight hydrocarbons including (i) C_{6+} GPM, gallons of liquid per thousand standard cubic feet of gas; (ii) mole fraction of the hexane and higher molecular weight hydrocarbons; and (iii) the hydrocarbon dew.

By way of clarification, the hydrocarbon dew point is the point at which the first droplet of hydrocarbon liquid condenses from the vapor. It can also be thought of as the minimum temperature above which no condensation of hydrocarbons occurs at a specified pressure. Hydrocarbon liquid dropout can cause a number of problems in gas transmission lines, including increased pressure drop, reduced line capacity, and equipment problems such as compressor damage. To avoid liquid dropout, most current operating specifications for gas transmission lines require that the lines be operated above the hydrocarbon dew point.

As analytical procedures evolved, and data become available, it became obvious that each piece of data not only reflected the test method applied to the sample and the exact steps followed in the analysis of a sample depend, among other factors, on the interpretation by the analyst of the written test method and the unique characteristics of the test sample – not all samples of natural gas and the various process gases (including biogas and landfill gas) are the same or even equivalent. Thus, the procedure and the resulting data will (i) reflect the background of the investigator, (ii) reflect the ability of the investigator to interpret the results in relation to the application, (iii) be dependent upon the analytical facilities available, and (iv) be shown the required levels of accuracy and precision. As a result, it has been necessary to support the objectives of *standards organizations* (such as ASTM International), which provide a forum for the exchange of information related to test methods, discussion of any issues, and cooperative development of modified and/or standard test methods.

Indeed, whatever the source, natural gas and the various process gases are complex mixtures of hydrocarbon compounds and nonhydrocarbon compounds. Furthermore, it must never be ignited that the composition of the various gas streams depends upon (i) the source of the gas that vary greatly in chemical composition and (ii) the processing sequence that is used to produce the products (Mokhatab et al., 2006; Speight, 2007, 2014a).

When the data from the various analytical test and investigations have been collected, the obvious question that comes to mind relates to the use of the data. For the most part, the data are used to determine whether or not a gas stream may produce a certain product and whether or not the product meets specifications. However, since the early days of analysis, there has been a growing tendency to use the analytical data as a means of more detailed and accurate projections of gas behavior. For example, predictive methods to determine the composition or amount of sulfur in a gaseous fuel is vital to see if a product meets specifications set by the federal government or by local authorities.

For example, using the data obtained from GC as a single test method or combined with mass spectrometry are the most important techniques for providing the hydrocarbon composition data in gas stream analysis. By gas chromatographic analysis, it is now possible to determine routinely the individual methane (CH_4) to higher molecular weight hydrocarbon derivatives as well as the individual aromatic derivatives that boil below 165 °C (330 °F) to obtain a complete picture of the hydrocarbon distribution in gas streams and in gas condensate (natural gasoline). In addition, by using a microcoulometric detector specific to sulfur, the sulfur compound distribution can be obtained throughout the distillate range. Gas chromatographic analysis can also be used to provide simulated TBP curves (for gas condensate and natural gasoline), while developments in preparative scale GLC have made possible the preparation of fractions in quantities sufficient not only for extensive spectrometric analyses but also for the normal inspection type tests to be undertaken.

The direction to clean gaseous fuels become more and more obvious over the past five decades, and now the focus divides itself into two approaches: (i) developing a theory of the chemical composition of natural gas as it influences gas processing and product yields and properties and (ii) using analytical data to construct compositional pictures of natural gas that can be used for predictability of behavior during recovery and refinery operations. A third category, namely, the ability of a product to meet specifications, will remain much the same, but it is the first two categories that help determine the methods by which the product should be produced from the raw gas stream and the *fine tuning* required to take the raw product to be a specification-grade saleable product. In addition to examining properties that contribute to product specifications, other properties may also be required for more detailed evaluation of the raw gas stream and for comparison between different raw gas streams feedstocks and the potential products slates from each gas stream or blend of gas streams.

Thus, the data derived from any one or more of the analytical techniques described in this text give not only an indication of the product characteristics but also whether or not the product is suitable (i.e. meets specification) for the proposed use. In addition, when used in conjunction with the properties of the refinery feedstock (crude oil) or the properties of a refinery (distillate) fraction, the data can be used to indicate the means by which the feedstock should be further processed to generate the desired products (Mokhatab et al., 2006; Speight, 2007, 2014a).

Thus, using the data derived from the selected standard test methods, it is possible to assess the raw gas stream and the product quality, thereby acquiring a degree of predictability of performance before processing and during product use. However, knowledge of the basic concepts of gas processing (Chapter 4) will help the analyst understand the production and, to a large extent, the anticipated properties of the product, which in turn is related to storage, sampling, and handling the various gas streams and their respective products. In addition, there are many instances in which interrelationships of the specification data enable properties to be predicted from the measured properties with as good precision as can be obtained by a single test. It would be possible to examine in this way the relationships between all the specified properties of a product and to establish certain key properties from which the remainder could be predicted.

Finally, each process chemist, process engineer, and gas processing operator has his/her own preferences, and such efforts are dependent on the raw gas stream as well as being dependent on the gas processing sequence. There is also the need to recognize that what is adequate for one raw gas stream (or blend of raw gas streams) and one processing sequence refinery will not be suitable for a different refinery with a different gas stream (or gas stream blend).

References

ASTM (2017). *Annual Book of Standards*. West Conshohocken, Pennsylvania: ASTM International.

ASTM D1160 (2017). *Standard Test Method for Distillation of Petroleum Products at Reduced Pressure*, Annual Book of Standards. West Conshohocken, Pennsylvania: ASTM International.

ASTM D1250 (2017). *Standard Guide for Petroleum Measurement Tables.* West Conshohocken, Pennsylvania: ASTM International.
ASTM D1266 (2017). *Standard Test Method for Sulfur in Petroleum Products (Lamp Method),* Annual Book of Standards. West Conshohocken, Pennsylvania: ASTM International.
ASTM D129 (2017). *Standard Test Method for Sulfur in Petroleum Products,* Annual Book of Standards. West Conshohocken, Pennsylvania: ASTM International.
ASTM D1298 (2017). *Standard Test Method for Density, Relative Density, or API Gravity of Crude Petroleum and Liquid Petroleum Products by Hydrometer Method.* West Conshohocken, Pennsylvania: ASTM International.
ASTM D1364 (2017). *Standard Test Method for Water in Volatile Solvents (Karl Fischer Reagent Titration Method),* Annual Book of Standards. West Conshohocken, Pennsylvania: ASTM International.
ASTM D1492 (2017). *Standard Test Method for Bromine Index of Aromatic Hydrocarbons by Coulometric Titration,* Annual Book of Standards. West Conshohocken, Pennsylvania: ASTM International.
ASTM D1500 (2017). *Standard Test Method for ASTM Color of Petroleum Products (ASTM Color Scale),* Annual Book of Standards. West Conshohocken, Pennsylvania: ASTM International.
ASTM D1552 (2017). *Standard Test Method for Sulfur in Petroleum Products (High-Temperature Method),* Annual Book of Standards. West Conshohocken, Pennsylvania: ASTM International.
ASTM D1796 (2017). *Standard Test Method for Water and Sediment in Fuel Oils by the Centrifuge Method (Laboratory Procedure),* Annual Book of Standards. West Conshohocken, Pennsylvania: ASTM International.
ASTM D1840 (2017). *Standard Test Method for Naphthalene Hydrocarbons in Aviation Turbine Fuels by Ultraviolet Spectrophotometry,* Annual Book of Standards. West Conshohocken, Pennsylvania: ASTM International.
ASTM D2161 (2017). *Standard Practice for Conversion of Kinematic Viscosity to Saybolt Universal Viscosity or to Saybolt Furol Viscosity,* Annual Book of Standards. West Conshohocken, Pennsylvania: ASTM International.
ASTM D2269 (2017). *Standard Test Method for Evaluation of White Mineral Oils by Ultraviolet Absorption,* Annual Book of Standards. West Conshohocken, Pennsylvania: ASTM International.
ASTM D2270 (2017). *Standard Practice for Calculating Viscosity Index from Kinematic Viscosity at 40 and 100 °C,* Annual Book of Standards. West Conshohocken, Pennsylvania: ASTM International.
ASTM D2273 (2017). *Standard Test Method for Trace Sediment in Lubricating Oils,* Annual Book of Standards. West Conshohocken, Pennsylvania: ASTM International.
ASTM D2384 (2017). *Standard Test Methods for Traces of Volatile Chlorides in Butane-Butene Mixtures,* Annual Book of Standards. West Conshohocken, Pennsylvania: ASTM International.

ASTM D2425 (2017). *Standard Test Method for Hydrocarbon Types in Middle Distillates by Mass Spectrometry*, Annual Book of Standards. West Conshohocken, Pennsylvania: ASTM International.

ASTM D2427 (2017). *Standard Test Method for Determination of C2 through C5 Hydrocarbons in Gasolines by Gas Chromatography*, Annual Book of Standards. West Conshohocken, Pennsylvania: ASTM International.

ASTM D2500 (2017). *Standard Test Method for Cloud Point of Petroleum Products*, Annual Book of Standards. West Conshohocken, Pennsylvania: ASTM International.

ASTM D2502 (2017). *Standard Test Method for Estimation of Mean Relative Molecular Mass of Petroleum Oils from Viscosity Measurements*, Annual Book of Standards. West Conshohocken, Pennsylvania: ASTM International.

ASTM D2503 (2017). *Standard Test Method for Relative Molecular Mass (Molecular Weight) of Hydrocarbons by Thermoelectric Measurement of Vapor Pressure*, Annual Book of Standards. West Conshohocken, Pennsylvania: ASTM International.

ASTM D2569 (2017). *Standard Test Method for Distillation of Pitch (Withdrawn 2006 but Still in Use in some Laboratories)*, Annual Book of Standards. West Conshohocken, Pennsylvania: ASTM International.

ASTM D2622 (2017). *Standard Test Method for Sulfur in Petroleum Products by Wavelength Dispersive X-Ray Fluorescence Spectrometry*, Annual Book of Standards. West Conshohocken, Pennsylvania: ASTM International.

ASTM D2650 (2017). *Standard Test Method for Chemical Composition of Gases by Mass Spectrometry*, Annual Book of Standards. West Conshohocken, Pennsylvania: ASTM International.

ASTM D2710 (2017). *Standard Test Method for Bromine Index of Petroleum Hydrocarbons by Electrometric Titration*, Annual Book of Standards. West Conshohocken, Pennsylvania: ASTM International.

ASTM D2784 (2017). *Standard Test Method for Sulfur in Liquefied Petroleum Gases (Oxy-Hydrogen Burner or Lamp)*, Annual Book of Standards. West Conshohocken, Pennsylvania: ASTM International.

ASTM D2786 (2017). *Standard Test Method for Hydrocarbon Types Analysis of Gas-Oil Saturates Fractions by High Ionizing Voltage Mass Spectrometry*, Annual Book of Standards. West Conshohocken, Pennsylvania: ASTM International.

ASTM D2789 (2017). *Standard Test Method for Hydrocarbon Types in Low Olefinic Gasoline by Mass Spectrometry*, Annual Book of Standards. West Conshohocken, Pennsylvania: ASTM International.

ASTM D287 (2017). *Standard Test Method for API Gravity of Crude Petroleum and Petroleum Products (Hydrometer Method)*, Annual Book of Standards. West Conshohocken, Pennsylvania: ASTM International.

ASTM D2887 (2017). *Standard Test Method for Boiling Range Distribution of Petroleum Fractions by Gas Chromatography*, Annual Book of Standards. West Conshohocken, Pennsylvania: ASTM International.

ASTM D2892 (2017). *Standard Test Method for Distillation of Crude Petroleum (15-Theoretical Plate Column)*, Annual Book of Standards. West Conshohocken, Pennsylvania: ASTM International.

ASTM D3227 (2017). *Standard Test Method for (Thiol Mercaptan) Sulfur in Gasoline, Kerosene, Aviation Turbine, and Distillate Fuels (Potentiometric Method)*, Annual Book of Standards. West Conshohocken, Pennsylvania: ASTM International.

ASTM D323 (2017). *Standard Test Method for Vapor Pressure of Petroleum Products*, Annual Book of Standards. West Conshohocken, Pennsylvania: ASTM International.

ASTM D3239 (2017). *Standard Test Method for Aromatic Types Analysis of Gas-Oil Aromatic Fractions by High Ionizing Voltage Mass Spectrometry*, Annual Book of Standards. West Conshohocken, Pennsylvania: ASTM International.

ASTM D3339 (2017). *Standard Test Method for Acid Number of Petroleum Products by Semi-Micro Color Indicator Titration*, Annual Book of Standards. West Conshohocken, Pennsylvania: ASTM International.

ASTM D3701 (2017). *Standard Test Method for Hydrogen Content of Aviation Turbine Fuels by Low Resolution Nuclear Magnetic Resonance Spectrometry*, Annual Book of Standards. West Conshohocken, Pennsylvania: ASTM International.

ASTM D3710 (2017). *Standard Test Method for Boiling Range Distribution of Gasoline and Gasoline Fractions by Gas Chromatography*, Annual Book of Standards. West Conshohocken, Pennsylvania: ASTM International.

ASTM D3829 (2017). *Standard Test Method for Predicting the Borderline Pumping Temperature of Engine Oil*, Annual Book of Standards. West Conshohocken, Pennsylvania: ASTM International.

ASTM D4006 (2017). *Standard Test Method for Water in Crude Oil by Distillation*, Annual Book of Standards. West Conshohocken, Pennsylvania: ASTM International.

ASTM D4007 (2017). *Standard Test Method for Water and Sediment in Crude Oil by the Centrifuge Method (Laboratory Procedure)*, Annual Book of Standards. West Conshohocken, Pennsylvania: ASTM International.

ASTM D4045 (2017). *Standard Test Method for Sulfur in Petroleum Products by Hydrogenolysis and Rateometric Colorimetry*, Annual Book of Standards. West Conshohocken, Pennsylvania: ASTM International.

ASTM D4052 (2017). *Standard Test Method for Density, Relative Density, and API Gravity of Liquids by Digital Density Meter*, Annual Book of Standards. West Conshohocken, Pennsylvania: ASTM International.

ASTM D4294 (2017). *Standard Test Method for Sulfur in Petroleum Products by Energy-Dispersive X-Ray Fluorescence Spectroscopy*, Annual Book of Standards. West Conshohocken, Pennsylvania: ASTM International.

ASTM D4377 (2017). *Standard Test Method for Water in Crude Oils by Potentiometric Karl Fischer Titration*, Annual Book of Standards. West Conshohocken, Pennsylvania: ASTM International.

ASTM D445 (2017). *Standard Test Method for Kinematic Viscosity of Transparent and Opaque Liquids*, Annual Book of Standards. West Conshohocken, Pennsylvania: ASTM International.
ASTM D4486 (2017). *Standard Test Method for Kinematic Viscosity of Volatile and Reactive Liquids*, Annual Book of Standards. West Conshohocken, Pennsylvania: ASTM International.
ASTM D473 (2017). *Standard Test Method for Sediment in Crude Oils and Fuel Oils by the Extraction Method*, Annual Book of Standards. West Conshohocken, Pennsylvania: ASTM International.
ASTM D4807 (2017). *Standard Test Method for Sediment in Crude Oil by Membrane Filtration*, Annual Book of Standards. West Conshohocken, Pennsylvania: ASTM International.
ASTM D4808 (2017). *Standard Test Methods for Hydrogen Content of Light Distillates, Middle Distillates, Gas Oils, and Residua by Low-Resolution Nuclear Magnetic Resonance Spectroscopy*, Annual Book of Standards. West Conshohocken, Pennsylvania: ASTM International.
ASTM D4928 (2017). *Standard Test Method for Water in Crude Oils by Coulometric Karl Fischer Titration*, Annual Book of Standards. West Conshohocken, Pennsylvania: ASTM International.
ASTM D4929 (2017). *Standard Test Methods for Determination of Organic Chloride Content in Crude Oil*, Annual Book of Standards. West Conshohocken, Pennsylvania: ASTM International.
ASTM D4952 (2017). *Standard Test Method for Qualitative Analysis for Active Sulfur Species in Fuels and Solvents (Doctor Test)*, Annual Book of Standards. West Conshohocken, Pennsylvania: ASTM International.
ASTM D5002 (2017). *Standard Test Method for Density and Relative Density of Crude Oils by Digital Density Analyzer*, Annual Book of Standards. West Conshohocken, Pennsylvania: ASTM International.
ASTM D5236 (2017). *Standard Test Method for Distillation of Heavy Hydrocarbon Mixtures (Vacuum Potstill Method)*, Annual Book of Standards. West Conshohocken, Pennsylvania: ASTM International.
ASTM D5305 (2017). *Standard Test Method for Determination of Ethyl Mercaptan in LP-Gas Vapor*, Annual Book of Standards. West Conshohocken, Pennsylvania: ASTM International.
ASTM D5307 (2017). *Standard Test Method for Determination of Boiling Range Distribution of Crude Petroleum by Gas Chromatography*, Annual Book of Standards. West Conshohocken, Pennsylvania: ASTM International.
ASTM D56 (2017). *Standard Test Method for Flash Point by Tag Closed Tester*, Annual Book of Standards. West Conshohocken, Pennsylvania: ASTM International.
ASTM D5772 (2017). *Standard Test Method for Cloud Point of Petroleum Products (Linear Cooling Rate Method)*, Annual Book of Standards. West Conshohocken, Pennsylvania: ASTM International.

ASTM D5853 (2017). *Standard Test Method for Pour Point of Crude Oils*, Annual Book of Standards. West Conshohocken, Pennsylvania: ASTM International.

ASTM D611 (2017). *Standard Test Methods for Aniline Point and Mixed Aniline Point of Petroleum Products and Hydrocarbon Solvents*, Annual Book of Standards. West Conshohocken, Pennsylvania: ASTM International.

ASTM D6299 (2017). *Standard Practice for Applying Statistical Quality Assurance and Control Charting Techniques to Evaluate Analytical Measurement System Performance*, Annual Book of Standards. West Conshohocken, Pennsylvania: ASTM International.

ASTM D6304 (2017). *Standard Test Method for Determination of Water in Petroleum Products, Lubricating Oils, and Additives by Coulometric Karl Fischer Titration*, Annual Book of Standards. West Conshohocken, Pennsylvania: ASTM International.

ASTM D6352 (2017). *Standard Test Method for Boiling Range Distribution of Petroleum Distillates in Boiling Range from 174 to 700 °C by Gas Chromatography*, Annual Book of Standards. West Conshohocken, Pennsylvania: ASTM International.

ASTM D664 (2017). *Standard Test Method for Acid Number of Petroleum Products by Potentiometric Titration*, Annual Book of Standards. West Conshohocken, Pennsylvania: ASTM International.

ASTM D6792 (2017). *Standard Practice for Quality System in Petroleum Products and Lubricants Testing Laboratories*, Annual Book of Standards. West Conshohocken, Pennsylvania: ASTM International.

ASTM D7397 (2017). *Standard Test Method for Cloud Point of Petroleum Products (Miniaturized Optical Method)*, Annual Book of Standards. West Conshohocken, Pennsylvania: ASTM International.

ASTM D7528 (2017). *Standard Test Method for Bench Oxidation of Engine Oils by ROBO Apparatus*, Annual Book of Standards. West Conshohocken, Pennsylvania: ASTM International.

ASTM D7689 (2017). *Standard Test Method for Cloud Point of Petroleum Products (Mini Method)*, Annual Book of Standards. West Conshohocken, Pennsylvania: ASTM International.

ASTM D86 (2017). *Standard Test Method for Distillation of Petroleum Products at Atmospheric Pressure*, Annual Book of Standards. West Conshohocken, Pennsylvania: ASTM International.

ASTM D92 (2017). *Standard Test Method for Flash and Fire Points by Cleveland Open Cup*, Annual Book of Standards. West Conshohocken, Pennsylvania: ASTM International.

ASTM D93 (2017). *Standard Test Methods for Flash Point by Pensky-Martens Closed Tester*, Annual Book of Standards. West Conshohocken, Pennsylvania: ASTM International.

ASTM D943 (2017). *Standard Test Method for Oxidation Characteristics of Inhibited Mineral Oils*, Annual Book of Standards. West Conshohocken, Pennsylvania: ASTM International.

ASTM D95 (2017). *Standard Test Method for Water in Petroleum Products and Bituminous Materials by Distillation*, Annual Book of Standards. West Conshohocken, Pennsylvania: ASTM International.

ASTM D97 (2017). *Standard Test Method for Pour Point of Petroleum Products*, Annual Book of Standards. West Conshohocken, Pennsylvania: ASTM International.

ASTM D971 (2017). *Standard Test Method for Interfacial Tension of Oil against Water by the Ring Method*, Annual Book of Standards. West Conshohocken, Pennsylvania: ASTM International.

ASTM D974 (2017). *Standard Test Method for Acid and Base Number by Color-Indicator Titration*, Annual Book of Standards. West Conshohocken, Pennsylvania: ASTM International.

ASTM E1252 (2017). *Standard Practice for General Techniques for Obtaining Infrared Spectra for Qualitative Analysis*, Annual Book of Standards. West Conshohocken, Pennsylvania: ASTM International.

ASTM E1316 2017. *Standard Terminology for Nondestructive Examinations.* Annual Book of Standards. ASTM International, West Conshohocken, Pennsylvania.

ASTM E168 (2017). *Standard Practices for General Techniques of Infrared Quantitative Analysis*, Annual Book of Standards. West Conshohocken, Pennsylvania: ASTM International.

ASTM E169 (2017). *Standard Practices for General Techniques of Ultraviolet-Visible Quantitative Analysis*, Annual Book of Standards. West Conshohocken, Pennsylvania: ASTM International.

ASTM E334 (2017). *Standard Practice for General Techniques of Infrared Microanalysis*, Annual Book of Standards. West Conshohocken, Pennsylvania: ASTM International.

ASTM E386 (2017). *Standard Practice for Data Presentation Relating to High-Resolution Nuclear Magnetic Resonance (NMR) Spectroscopy*, Annual Book of Standards. West Conshohocken, Pennsylvania: ASTM International.

ASTM E659 (2017). *Standard Test Method for Autoignition Temperature of Liquid Chemicals*, Annual Book of Standards. West Conshohocken, Pennsylvania: ASTM International.

Drews, A.W. ed. (1998). *Manual on Hydrocarbon Analysis*, 6e. West Conshohocken, Pennsylvania: ASTM International.

Ebert, L.B. (1990). *Fuel Science and Technology International* 8: 563.

Ebert, L.B., Scanlon, J.C., and Mills, D.R. (1984). *Liquid Fuels Technology.* 2: 257.

Ebert, L.B., Mills, D.R., and Scanlon, J.C. (1987). Preprints. *Division of Petroleum Chemistry American Chemical Society* 32 (2): 419.

Fingas, M.F. (1998). Studies on the evaporation of crude oil and petroleum products. II. Boundary layer regulation. *Journal of Hazardous Materials* 57 (1–3): 41–58.
Fowlis, I.A. (1995). *Gas Chromatography*, 2e. New York: Wiley.
Gary, J.G., Handwerk, G.E., and Kaiser, M.J. (2007). *Petroleum Refining: Technology and Economics*, 5e. Boca Raton, Florida: CRC Press, Taylor & Francis Group.
Grob, R.L. (1995). *Modern Practice of Gas Chromatography*, 3e. New York: Wiley.
Hsu, C.S. and Robinson, P.R. ed. (2017). *Handbook of Petroleum Technology*. Cham: Springer International Publishing AG.
Miller, M. ed. (2000). *Encyclopedia of Analytical Chemistry*. Hoboken, New Jersey: Wiley.
Mokhatab, S., William, W.A., and Speight, J.G. (2006). *Handbook of Natural Gas Transmission and Processing*. Amsterdam: Elsevier.
Mushrush, G.W. and Speight, J.G. (1995). *Petroleum Products: Instability and Incompatibility*. New York: Taylor & Francis.
Patnaik, P. ed. (2004). *Dean's Analytical Chemistry Handbook*, 2e. New York: McGraw-Hill.
Rand, S. (2003). *Significance of Tests for Petroleum Products*, 7e. West Conshohocken, Pennsylvania: ASTM International.
Schramm, L.L. ed. (1992). *Emulsions. Fundamentals and Applications in the Petroleum Industry*. Washington, DC: American Chemical Society.
Speight, J.G. (1986). Polynuclear aromatic systems in petroleum. *Preprints – American Chemical Society, Division of Petroleum Chemistry* 31 (4): 818.
Speight, J.G. (1994). Application of spectroscopic techniques to the structural analysis of petroleum. *Applied Spectroscopy Reviews* 29: 269.
Speight, J.G. (2005). *Environmental Analysis and Technology for the Refining Industry*. Hoboken, New Jersey: Wiley.
Speight, J.G. (2007). *Natural Gas: A Basic Handbook*. Houston, Texas: GPC Books, Gulf Publishing Company.
Speight, J.G. (2013). *The Chemistry and Technology of Coal*, 3e. Boca Raton, Florida: CRC Press, Taylor & Francis Group.
Speight, J.G. (2014a). *The Chemistry and Technology of Petroleum*, 5e. Boca Raton, Florida: CRC Press, Taylor & Francis Group.
Speight, J.G. (2014b). *High Acid Crudes*. Oxford: Gulf Professional Publishing, Elsevier.
Speight, J.G. (2014c). *Oil and Gas Corrosion Prevention*. Oxford: Gulf Professional Publishing, Elsevier.
Speight, J.G. (2015). *Handbook of Petroleum Product Analysis*, 2e. Hoboken, New Jersey: Wiley.
Speight, J.G. (2017). *Handbook of Petroleum Refining*. Boca Raton, Florida: CRC Press, Taylor & Francis Group.
Totten, G.E. ed. (2003). *Fuels and Lubricants Handbook: Technology, Properties, Performance, and Testing*. West Conshohocken, Pennsylvania: ASTM International.

7

Constituents of Gas Streams

7.1 Introduction

The defining characteristics of the various gas streams in the contact of this book are that the gases (i) exist in a gaseous state at room temperature and (ii) may contain hydrocarbon constituents with 1–4 carbons, (iii) diluents and inert gases, and (iv) contaminants in the form of nonhydrocarbon constituents. Each constituent of the gas has an effect on the properties. Thus:

Hydrocarbons	Provide the calorific value of natural gas when it is burned
Diluents/inert gases	Typical gases are carbon dioxide, nitrogen, helium, and argon
Contaminants	Present in low concentrations; may affect processing operations

Many hydrocarbon gases do contain C_5 and C_6 hydrocarbon derivatives, and apart from gas streams produced as process by-products in a refinery, the C_{4+} constituents are typically found at lower concentrations (% v/v) in gases than the C_1 to C_4 constituents. There are also a few category members that may contain C_7 and even C_8 hydrocarbons, although such streams would necessarily be at elevated temperature and/or pressure to maintain the C_7 and C_8 constituents in the gaseous state. Hydrocarbon compounds containing C_5, C_6, C_7, and C_8 are typically found predominantly in crude oil-derived naphtha.

Thus, natural gas is a mixture of hydrocarbon and nonhydrocarbon gases found in porous formations beneath the surface of the Earth, often in association with crude oil. Methane can be produced in the laboratory by heating

Handbook of Natural Gas Analysis, First Edition. James G. Speight.

sodium acetate with sodium hydroxide and by the reaction of aluminum carbide (Al_4C_3) with water.

$$Al_4C_3 + 12H_2O \rightarrow 4Al(OH)_3 \downarrow + 3CH_4$$

$$CH_3CO_2Na + NaOH \rightarrow CH_4 + Na_2CO_3$$

Methane is synthesized commercially by heating a mixture of carbon and hydrogen and by distilling bituminous coal. Coal is a combustible rock formed from the remains of decayed vegetation that thrived on Earth about 300 million years ago. It is the only rock containing significant amounts of carbon. The elemental composition of coal varies between 60 and 95% w/w carbon and also contains hydrogen and oxygen, with small concentrations of nitrogen, chlorine, sulfur, and several metals (Speight, 2013). Substances released from bituminous coal when it is distilled, in addition to methane, include water, carbon dioxide, ammonia, benzene, toluene, naphthalene, and anthracene. In addition, the distillation also yields oils, tars, and sulfur-containing products. The nonvolatile component of coal, which remains after distillation, is coke. Coke is almost pure carbon and is an excellent fuel as well as a reducing agent in the smelting of metal ores, especially iron ore. However, coal may contain metals, such as arsenic and lead, that can be serious pollutants if the combustible products are released into the atmosphere (www.scifun.org). In the chemical industry, methane is a raw material for the manufacture of methanol (CH_3OH), formaldehyde (HCHO), nitromethane (CH_3NO_2), chloroform ($CHCl_3$), and carbon tetrachloride (CCl_4).

The principal constituent of most natural gases is methane with minor amounts of heavier hydrocarbons and certain nonhydrocarbon gases such as nitrogen, carbon dioxide, hydrogen sulfide, and helium (Chapter 1). The members of the hydrocarbon gases are primarily alkanes. When inorganic components are present, they consist of asphyxiant gases such as hydrogen. Unlike other categories of crude oil product (such as naphtha, gasoline, diesel fuel, and the higher boiling products), the constituents of hydrocarbon gases can be evaluated, and the results of the constituent evaluation can then be used to estimate the behavior of the gas. The constituents used to evaluate the behavior of the gas are (i) the C_1–C_4 hydrocarbons, (ii) the C_5–C_6 hydrocarbons (although in natural gas the C_1–C_4 constituents predominate), and (iii) the asphyxiant gases, i.e. carbon dioxide, nitrogen, and hydrogen. In general, most gas streams used in this text are composed of predominantly the C_1–C_4 hydrocarbons, which have extremely low melting points and boiling points. These gases also have high vapor pressures and low octanol/water partition coefficients. The aqueous solubility of these components varies, but the solubility of the majority of the hydrocarbons typically falls within a range of $22\,mg\,l^{-1}$ to several hundred parts per million. There are also a few gas streams that may contain C_7 hydrocarbons and even C_8 hydrocarbons, although such streams would necessarily be at elevated temperature and/or pressure to maintain the C_7 constituents and the C_8 constituents in the gaseous state. Hydrocarbon compounds containing C_5, C_6, C_7, and C_8 are found predominantly in low-boiling crude oil naphtha and are also found in gas condensate and natural gasoline (sometimes on occasion erroneously called *casinghead gas*).

By way of recall, in addition to methane, natural gas contains other constituents that are variously referred to as (i) natural gas liquids, (ii) natural gas condensate, and (iii) natural gasoline (Chapter 1). Also, by way of a refresher definition, natural gas liquids are hydrocarbons that occur as gases at atmospheric pressure and as liquids under higher pressures and that can also be liquefied by cooling. The specific pressures and temperatures at which the gases liquefy vary by the type of gas liquids and may be described as *light* or *heavy* according to the number of carbon atoms and hydrogen atoms in the molecule.

Also by way of a further reminder, natural gas condensate (also called *condensate*, or *gas condensate*, or *natural gasoline*) is a low-density mixture of hydrocarbon liquids that are present as gaseous components in the raw natural gas produced from many natural gas fields. Some gas species within the raw natural gas will condense to a liquid state if the temperature is reduced to below the hydrocarbon dew point temperature at a set pressure. There are many condensate sources, and each has its own unique gas condensate composition. In general, gas condensate has a specific gravity ranging from 0.5 to 0.8 and is composed of hydrocarbons such as pentane, hexane, and some higher molecular weight hydrocarbon derivatives.

Blending is the process of mixing gases for a specific purpose where the composition of the resulting mixture is specified and controlled. Natural gas constituents with five or more carbon atoms (such as pentane, or blends of butane, pentane, and other hydrocarbon derivatives with additional carbon atoms) exist as liquids at ambient temperatures. Additionally, condensate may contain additional impurities such as (i) hydrogen sulfide; (ii) mercaptans, also called thiols; and (iii) carbon dioxide, cyclohexane derivatives, and aromatic derivatives, such as benzene (C_6H_6), toluene ($C_6H_5CH_3$, ethylbenzene ($C_6H_5CH_2CH_3$), and the xylene isomers ($H_3CC_6H_4CH_3$).

The primary use for natural gas is as a fuel to produce heat as well as the production of hydrogen (the steam-methane reforming process) and ammonia:

$$CH_4 + H_2O \rightarrow CO + 3H_2 - \text{steam-methane reforming}$$

$$CO + H_2O \rightarrow CO_2 + H_2 - \text{hydrogen production}$$

$$3H_2 + N_2 \rightarrow 2NH_3 - \text{Haber} - \text{Bosch process}$$

The steam–methane reforming process is major source of hydrogen for refineries and other industries (Speight, 2016). In the endothermic process, high-temperature steam (700–1100 °C, 1290–2010 °F) is used to produce hydrogen from a methane source, such as natural gas at pressures on the order of 45–370 psi. Subsequently, in what is called the *water–gas shift reaction*, the carbon monoxide and steam are reacted using a catalyst to produce carbon dioxide and more hydrogen. In a final process step (*pressure-swing adsorption*), carbon dioxide and other impurities are removed from the gas stream, leaving essentially pure hydrogen.

$$CH_4 + H_2O \rightarrow CO + 3H_2 \qquad \text{(steam – methane reforming)}$$

$$CO + H_2O \rightarrow CO_2 + H_2 \qquad \text{(water – gas shift reaction)}$$

The other low molecular weight hydrocarbons in the gas – ethane (C_2), propane (C_3), and the butanes (C_4) isomers, either in the gas phase or liquefied – are also used for heating as well as for motor fuels and as feedstocks for chemical processing. The pentane/pentene derivatives (C_s) are products of natural gas or crude oil fractionation or refinery operations (i.e. reforming and cracking) that are removed for use as chemical feedstocks. The olefins ethylene ($CH_2{=}CH_2$, sometimes called ethene), propylene (propene: $CH_3CH{=}CH_2$), butene isomers (butene-l: $CH_3CH_2CH{=}CH_2$, isobutylene: $(CH_3)_2C{=}CH_2$, butene-2, $CH_3CH{=}CHCH_3$), and butadiene ($CH_2{=}CHCH{=}CH_2$), as well as higher molecular weight olefins are products of the various refining processes involving the use of the saturated (nonolefinic and nonaromatic hydrocarbons as feedstocks. It is only rarely that olefins occur in natural gas, and they are not typically constituents of natural gas. However, olefins do occur in biogas produced by thermal methods (Parkash, 2003; Gary et al., 2007; Speight, 2007, 2017; Hsu and Robinson, 2017).

Mixtures of these hydrocarbons are commonly encountered in gas testing laboratories, and, as might be expected, the composition of the gas varies with dependence on the source of the gas as well as on the intended use of the material. Other nonhydrocarbon constituents of these mixtures are important analytes since they may be useful products or may be undesirable as a source of processing problems. Some of these components are helium, hydrogen, argon, oxygen, nitrogen, carbon monoxide, carbon dioxide, sulfur, and nitrogen containing compounds, as well as heavier hydrocarbons. Desired testing of these hydrocarbon mixtures usually involves the determination of bulk physical or chemical properties and component speciation and quantitation.

The principal quality criterion of natural gas is its heating value (ASTM D1826, 2017). This test method covers the determination with the continuous recording calorimeter of the total calorific (heating) value of fuel gas produced or sold in the natural gas range from 900 to 1200 Btu (standard ft^{-3}). In addition, the gas must be readily transportable through high-pressure pipelines, and, therefore, the water content, as defined by the water dew point (ASTM D1142, 2017), must be considered to prevent the formation of ice or hydrates in the pipeline. Likewise, the amounts of the entrained hydrocarbons having higher molecular weight than ethane, as defined by the hydrocarbon dew point, should be considered to prevent accumulation of condensable liquids that may block the pipeline.

Natural gas and its products of combustion must not be unduly corrosive to the materials with which they come in contact. Thus, the detection and measurement of hydrogen sulfide and total sulfur (ASTM D1072, 2017) are important. The odor of gas must not be objectionable so, in some cases, mercaptan content, even in the ppm range, is significant to give the gas an odor that is detectable by the human nose (ASTM E253, 2017 Standard Terminology Relating to Sensory Evaluation of Materials and Products).

As an example, in the United States, federal regulations (49 CFR Part 192.625) require that a combustible gas in a distribution line must contain a natural odorant or be odorized so that at a concentration in air of one-fifth of the lower explosive limit, the gas is readily detectable by a person with a normal sense of smell. These regulations state further requirements that to ensure the proper concentration of odorant with this section, each operator must conduct periodic sampling of combustible gases using an instrument capable of determining the percentage of gas in air at which the odor becomes readily detectable (ASTM E253, 2017). Additionally, a number of governments have enacted legislation that requires natural gas to be odorized so that it is detectable at concentrations less than one-fifth of the lower explosive limit. The standard test method (ASTM D6273, 2017) describes several procedures for determining the odor intensity of natural gas through the use of instruments that dilute and mix the sampled natural gas with air. The mixed gas stream is then sniffed by the operator for the purpose of determining the threshold detection level or the readily detection level, or both, for odorant in the natural gas stream.

If the gas is to be liquefied by cryogenic processing and stored in liquid form, carbon dioxide will separate out of the cold liquid as a solid and interfere with the refrigeration system. Carbon dioxide can be determined by two test methods (ASTM D1137, 2017; ASTM D1945, 2017), but if the gas stream is to be used as a feedstock to produce hydrogen or petrochemicals, a complete olefin analysis (ASTM D1945, 2017; ASTM D2597, 2017) must be made, because some gases contain various constituents that are deleterious to these processes. One test method (ASTM D1945, 2017) describes the determination of the chemical composition of natural gases and similar gaseous mixtures within the range of composition as defined by the limits of the method. This test method may be abbreviated for the analysis of lean natural gases containing negligible amounts of hexanes and higher hydrocarbons or for the determination of one or more components, as required. This method is of significance for providing data that can be used for the calculation of physical properties of the sample, such as heating value and relative density, or for monitoring the concentrations of one or more of the components in a mixture.

The properties and analysis of liquid constituents such as gas condensate and natural gasoline are presented elsewhere (Chapter 10). The individual components and commercial products that may be fractionated from natural gas for further use (e.g. petrochemical feedstocks) include (i) methane, (ii) ethane, (iii) propane, and (iv) butane. Each of these gases is described in further detail in the following sections.

7.2 Methane

Methane (CH_4) is a colorless, odorless gas with a wide distribution in nature (Table 7.1). It is a naturally occurring substance that is produced over time through biological routes. It is the simplest hydrocarbon in the alkane series

Table 7.1 Properties of methane.

Chemical formula	CH_4
Molar mass	$16.04\,g{\cdot}mol^{-1}$
Appearance	Colorless gas
Odor	Odorless
Density	• $0.656\,gl^{-1}$ (gas, 25 °C, 1 atm) • $0.716\,gl^{-1}$ (gas, 0 °C, 1 atm) • $0.42262\,g/cm^3$ (liquid, −162 °C)
Melting point	−182.5 °C; −296.4 °F; 90.7 °K
Boiling point	−161.49 °C; −258.68 °F; 111.66 °K
Solubility in water	$22.7\,mgl^{-1}$
Solubility	Soluble in ethanol, diethyl ether, benzene, toluene, methanol, acetone
log *P*	1.09
Henry's law constant (k_H)	$14\,nmol{\cdot}Pa^{-1}kg^{-1}$
Magnetic susceptibility (χ)	$-12.2\times10^{-6}\,cm^3mol^{-1}$
Molecular shape	Tetrahedron
Dipole moment	0 D
Specific heat capacity (*C*)	$35.69\,J\,(K\,mol)^{-1}$
Standard molar entropy (S_{298})	$186.25\,J\,(K\,mol)^{-1}$
Standard enthalpy of formation ($\Delta_f H_{298}$)	$-74.87\,kJ\,mol^{-1}$
Standard enthalpy of combustion ($\Delta_c H_{298}$)	−891.1 to $-890.3\,kJ\,mol^{-1}$
Flash point	−188 °C (−306.4 °F; 85.1 °K)
Autoignition temperature	537 °C (999 °F; 810 °K)
Explosive limits	4.4–17%

(Figure 7.1 and 7.2) and is the main constituent of natural gas. The relative abundance of methane on Earth makes it an attractive fuel, though capturing and storing it poses challenges due to its gaseous state under normal conditions for temperature and pressure.

Briefly, in spite of the corruption of the terminology of organic chemistry, a hydrocarbon is an organic chemical compound that is composed only of the elements carbon (C) and hydrogen (H). Hydrocarbons are the principal constituents of crude oil and natural gas and serve as fuels and lubricants as well as raw materials for the production of a range of products including solvents, plastics, fibers, rubber, solvents, and a variety of industrial chemicals (Speight, 2013, 2014, 2017).

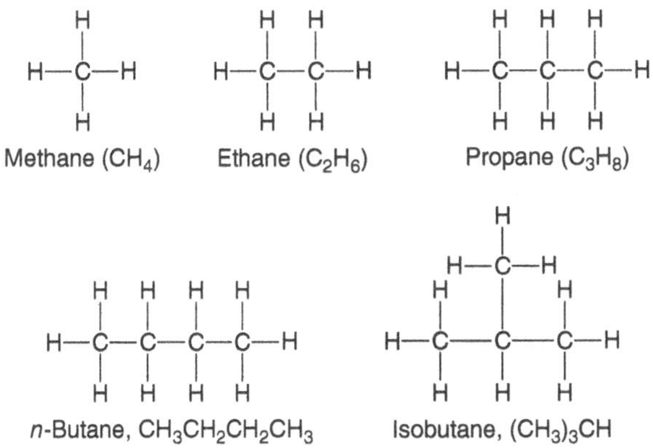

Figure 7.1 Examples of alkane hydrocarbons.

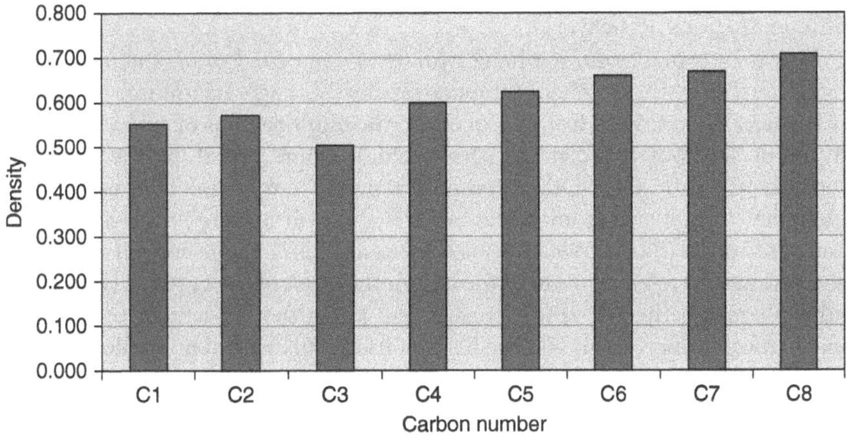

Figure 7.2 Carbon number and density of natural gas hydrocarbons (up to n-octane, C_8H_{18}).

In nature, methane is produced by the anaerobic bacterial decomposition of vegetable matter under water (where it is sometimes called marsh gas or swamp gas) (Swain, 1986). Wetlands are the major natural source of methane produced in this way. In fact, natural wetlands are responsible for approximately 80% v/v of global methane emissions from natural sources. Wetlands provide a habitat conducive to methane-producing bacteria that produce methane during the decomposition of organic material. These bacteria require an anaerobic environment (i.e. an environment with no oxygen) and abundant organic matter, both of which are present in wetland conditions.

Other important natural sources of methane include termites (as a result of the termite digestive processes), volcanoes, vents in the ocean floor, and methane hydrate deposits that occur along continental margins and beneath the continental shelf, Antarctic ice and Arctic permafrost. Methane can be released from the hydrates with changes in temperature, pressure, salt concentrations, and other factors. Overall, the amount of methane stored in these hydrates globally is estimated to be very large with the potential for large releases of methane if there are significant breakdowns in the stability of the deposits. In fact, emissions from gas hydrates (such as methane hydrate) (Chapter 7) on a global scale are estimated to account for approximately 5% v/v of the global methane emissions from natural sources.

Furthermore, the gaseous emissions from animal source (the digestive system of cattle is commonly cited as a source of methane, but other animals such as termites are estimated to be the source of approximately 11% v/v of the global methane emissions from natural sources. Methane is produced in termites as part of their normal digestive process, and the amount generated varies among different species. Ultimately, emissions from termites depend largely on the population of these insects, which can also vary significantly among different regions of the world.

By way of explanation, methane hydrate is composed of crystallized water (ice) that contains methane trapped within a cage (clathrate) structure (Chapter 7). Theses gas hydrates occur in the outer regions of space and deep in the ocean floor and can be represented, to some extent, by the chemical formula $(CH_4)_4(H_2O)_{23}$. Other common names for methane hydrate include methane clathrate, hydromethane, natural gas hydrate, fire ice, methane ice, and gas hydrate (Sloan, 1998a; Belosludov et al., 2007). Large deposits of methane hydrates have been found in the part of the crust of the Earth known as the lithosphere, i.e. the outermost shell of the Earth that is defined by its rigid mechanical properties and on the basis of its chemistry and mineralogy.

Methane also occurs as a component of firedamp (flammable gas) in coal seams (Speight, 2013). The production and combustion of natural gas and coal are the major anthropogenic (human-associated) sources of methane. Activities such as the extraction and processing of natural gas and the destructive distillation of bituminous coal in the manufacture of coal gas and coke-oven gas result in the release of significant amounts of methane into the atmosphere. Other human activities that are associated with methane production include biomass decomposition, livestock farming (the digestive systems of cows produce substantial quantities of methane), and waste management (where bacteria produce methane as they decompose sludge in waste treatment facilities and decaying matter in landfills) (Pohland, 1996).

7.2.1 Chemical Properties

First and foremost in the modern context, methane is a *greenhouse gas* that is 23 times more effective than *carbon dioxide.* It is also less stable and is slowly oxidized by *oxygen* to carbon dioxide and *water.*

Structurally, methane is a *tetrahedral* molecule with four equivalent carbon-hydrogen *(C–H) bonds.* At *room temperature* and *standard pressure,* methane is a colorless, odorless gas. The familiar smell of natural gas as used in homes is achieved by the addition of an *odorant,* usually blends containing a thiol (an organic molecule containing the –SH function), as a safety measure. Methane has a boiling point of −161 °C (−257.8 °F) at a pressure of one *atmosphere.* As a gas, methane is *flammable* over a range of concentrations (5.4–17% v/v) in air at *standard pressure.*

Methane is lighter than air, having a specific gravity of 0.554 (Table 7.1) and is only slightly soluble in water. It burns readily in air, forming carbon dioxide and water vapor; the flame is pale, slightly luminous, and very hot. The boiling point of methane is −162 °C (−259.6 °F), and the melting point is −182.5 °C (−296.5 °F). Methane in general is very stable, but mixtures of methane and air (with the methane content between 5 and 14% v/v) are explosive. Explosions of such mixtures have been a frequent occurrence in coal mines and have been the cause of many mine disasters.

The primary chemical reactions of methane are combustion, steam reforming to synthesis gas (syngas, mixtures of carbon monoxide, and hydrogen), and halogenation. Typically, the chemical reactions of methane are difficult to control. Although there is great interest in converting methane into useful or more easily liquefied compounds, the only practical processes are relatively unselective. In the chemical industry, methane is converted by the steam reforming process to synthesis gas (also called syngas), which is a mixture of carbon monoxide and hydrogen (Speight, 2013, 2014). The process that employs a nickel-based catalyst and requires high temperature is on the order of 700–1100 °C (1290–2010 °F):

$$CH_4 + H_2O \rightarrow CO + 3H_2 - \text{steam reforming process}$$

Similarly, in the Haber–Bosch synthesis of ammonia from air, natural gas (methane) is reduced to a mixture of carbon dioxide, water, and ammonia:

$$CH_4 + H_2O \rightarrow CO + 3H_2 - \text{steam} - \text{methane reforming}$$

$$CO + H_2O \rightarrow CO_2 + H_2 - \text{hydrogen production}$$

$$3H_2 + N_2 \rightarrow 2NH_3 - \text{Haber} - \text{Bosch process}$$

Other commercially viable processes that use methane as a chemical feedstock include the catalytic oxidation to methanol, which is based on the oxidative coupling of methane, and the direct reaction of methane with sulfur trioxide to produce methane sulfonic acid:

$$2CH_4 + O_2 \rightarrow 2CH_3OH$$

$$CH_4 + SO_3 \rightarrow CH_3SO_3H$$

The combustion of methane is an exothermic reaction that can be represented by a simple equation that reflects the thermal oxidation of methane to carbon dioxide and water:

$$CH_4 + 2O_2 \rightarrow CO_2 + 2H_2O$$

The reaction is, however, is a multiple step reaction and can be generally represented by the following equations in which the species M* signifies an energetic third body, from which energy is transferred during a molecular collision. Thus,

$$CH_4 + M^* \rightarrow CH_3 + H + M$$

$$CH_4 + O_2 \rightarrow CH_3 + HO_2$$

$$CH_4 + HO_2 \rightarrow CH_3 + 2OH$$

$$CH_4 + OH \rightarrow CH_3 + H_2O$$

$$O_2 + H \rightarrow O + OH$$

$$CH_4 + O \rightarrow CH_3 + OH$$

$$CH_3 + O_2 \rightarrow CH_2O + OH$$

$$CH_2O + O \rightarrow CHO + OH$$

$$CH_2O + OH \rightarrow CHO + H_2O$$

$$CH_2O + H \rightarrow CHO + H_2$$

$$CHO + O \rightarrow CO + OH$$

$$CHO + OH \rightarrow CO + H_2O$$

$$CHO + H \rightarrow CO + H_2$$

$$H_2 + O \rightarrow H + OH$$

$$H_2 + OH \rightarrow H + H_2O$$

$$CO + OH \rightarrow CO_2 + H$$

$$H + OH + M \rightarrow H_2O + M^*$$

$$H + H + M \rightarrow H_2 + M^*$$

$$H + O_2 + M \rightarrow HO_2 + M^*$$

As illustrated, formaldehyde (HCHO) is an early intermediate product, and oxidation of formaldehyde gives the formyl radical (HCO), which then produce carbon monoxide (CO). Any resulting hydrogen oxidizes to water or other intermediates. Finally, the carbon monoxide is oxidized to carbon dioxide. The overall reaction rate is a function of the concentration of the various entities during the combustion process. The higher the temperature, the greater the concentration of radical species, and the more rapid the combustion process.

In the partial oxidation reaction (Arutyunov, 2007), methane and other hydrocarbons in natural gas react with a limited amount of oxygen (typically from air) that is not enough to completely oxidize the hydrocarbons to carbon dioxide and water. With less than the stoichiometric amount of oxygen available, the reaction products contain primarily hydrogen and carbon monoxide (and nitrogen, if the reaction is carried out with air rather than pure oxygen), and a relatively small amount of carbon dioxide and other compounds. Subsequently, in a water–gas shift reaction, the carbon monoxide reacts with water to form carbon dioxide and more hydrogen:

$$2CH_4 + O_2 \rightarrow 2CO + 2H_2 \quad \text{(partial oxidation of methane)}$$

$$CO + H_2O \rightarrow CO_2 + H_2 \quad \text{(water – gas shift)}$$

The partial oxidation reaction is an exothermic process, and the process is, typically, much faster than steam reforming and requires a smaller reactor vessel. This process initially produces less hydrogen per unit of the input fuel than is obtained by steam reforming of the same fuel.

Reforming low-cost natural gas can provide hydrogen for crude oil refining and for fuel cell electric vehicles (FCEVs) as well as other applications. Over the long term, it is anticipated that hydrogen production from natural gas will be augmented with production from renewable, nuclear, coal (with carbon capture and storage), and other low-carbon, domestic energy resources. The only product from an FCEV tailpipe is water vapor, but even with the upstream process of producing hydrogen from natural gas as well as delivering and storing it for use in FCEV vehicles, the total greenhouse gas emissions are markedly reduced compared with modern gasoline-powered and diesel-powered vehicles.

The most apparent environmental impacts of methane use are in the extraction of natural gas by hydraulic fracturing. The wellhead sites require several acres to house the supplies and equipment necessary for the process and disrupt the land. A prodigious amount of water is required for the fracking fluid and usually has to be hauled into the drill site in convoys of tanker trucks. In locations where water is scarce, fracking can use a substantial fraction of water available for agriculture and other human use. The waste fluid from the well is contaminated and unusable for these purposes and often pumped back into the rock formation. There is some concern that underground freshwater aquifers might be accidentally contaminated. Disruption of large areas of the foundation on which Earth's surface rests can lead to motion of the surface, resulting in tremors (small earthquakes).

A less apparent effect of methane is its role in global warming and climate change. As pointed out above, burning the gas produces carbon dioxide that adds to that produced by burning other fossil fuels. Further, methane is a greenhouse gas in its own right – even more powerful, and www.scifun.org perhaps more important, than carbon dioxide. For the same mass of each gas, methane is at least 100 times more effective as a greenhouse gas than carbon dioxide. Unlike carbon dioxide that persists for centuries in the atmosphere, methane is oxidized in the atmosphere with a half-life of about a decade.

After 20 years an initial mass of methane will have been equivalent in greenhouse effect to about 70 times than mass of carbon dioxide (and the carbon will still be present as carbon dioxide, the methane oxidation product). Natural sources of methane, mostly wetlands, are abundant on the Earth's surface. For almost all the past 10 000 years, the mostly natural atmospheric methane concentration was about 700 ppb (parts per billion). But, since the Industrial Revolution, methane has risen to approximately 1800 ppb as human activities added to natural sources.

At present, this atmospheric methane has about one-fifth the greenhouse warming effect as the atmospheric carbon dioxide. Approximately 33% v/v of the methane emissions from the United States come from wetlands, about one-third of which is from rice fields that are flooded by humans, with the rest from natural marshes and bogs. Another 20% comes from fossil fuel extraction (coal mines and what is not captured from oil and gas wells) as well as leaks in the enormous natural gas distribution system. Another 20% is a result of raising large herds of cattle (for beef and dairy products).

Cellulose makes up a great deal of the food, grass, and grains that cattle (and other ruminant animals like sheep, goats, and camels) eat. Cellulose in feedstocks cannot be digested by many animals, but ruminants (and termites that digest wood, a common source of cellulose) have digestive systems containing microbes that break down the cellulose for them. Unfortunately, one of the by-products of this process is a methane in high yield, which the animals have to get rid of and mostly they do this by a (self-explanatory) process known as *burping*. The solid and liquid wastes all these cattle (and other agricultural animals) produce are often stored to decay in waste ponds, where the decay microbes produce more methane. Similarly, landfills produce methane as the refuse decays. Some municipalities capture some of this methane and use it for heating and power.

7.2.2 Physical Properties

As mentioned earlier, methane is a colorless, odorless gas with a melting point of −182.5 °C and a boiling point of −161.5 °C (−258.7 °F) (Table 7.1). At room temperature, methane is a gas less dense than air. It melts at −182.5 °C (−296.4 °F) and is not very soluble in water. Methane is lighter than air and is only slightly soluble in water but does have a noteworthy solubility in diethyl ether and in ethyl alcohol (ethanol) and other organic solvents (Table 7.1).

In order to meet specification, there is a number of standard test methane that can be applied to methane (Table 7.2). Also, as part of the specification, an odorant has to be added in order to make gas leaks detectable, and a small amount of odoriferous organic sulfur compounds, such as tertiary butyl mercaptan $(CH_3)_3CSH$, is added to sales natural gas to give the gas a detectable odor. An undetected gas leak is dangerous and can result in an explosion; many are reported every year, causing lost lives and occasional spectacular property destruction. In addition, although methane is often specified as a nontoxic gaseous chemical and does not pose an immediate health hazard, high concentrations of the gas will reduce the oxygen percentage in air and lead to asphyxiation.

Table 7.2 ASTM standard test methods used in methane technology.

ASTM D1142: Standard Test Method for Water Vapor Content of Gaseous. Fuels by Measurement of Dew-Point Temperature
ASTM D1945: Standard Test Method for Analysis of Natural Gas by Gas Chromatography
ASTM D1946: Standard Practice for Analysis of Reformed Gas by Gas Chromatography
ASTM D2505: Standard Test Method for Ethylene, Other Hydrocarbons, and Carbon Dioxide in High-Purity Ethylene by Gas Chromatography
ASTM D2650: Standard Test Method for Chemical Composition of Gases by Mass Spectrometry
ASTM D4891: Standard Test Method for Heating Value of Gases in Natural Gas and Flare Gases Range by Stoichiometric Combustion
ASTM D5504: Standard Test Method for Determination of Sulfur Compounds in Natural Gas and Gaseous Fuels by Gas Chromatography and Chemiluminescence
ASTM D5917: Standard Test Method for Trace Impurities in Monocyclic Aromatic Hydrocarbons by Gas Chromatography and External Calibration
ASTM D6159: Standard Test Method for Determination of Hydrocarbon Impurities in Ethylene by Gas Chromatography
ASTM D6345: Standard Guide for Selection of Methods for Active, Integrative Sampling of Volatile Organic Compounds in Air
ASTM D6968: Standard Test Method for Simultaneous Measurement of Sulfur Compounds and Minor Hydrocarbons in Natural Gas and Gaseous Fuels by Gas Chromatography and Atomic Emission Detection
ASTM D7653: Standard Test Method for Determination of Trace Gaseous Contaminants in Hydrogen Fuel by Fourier Transform Infrared (FTIR) Spectroscopy
ASTM D7675: Standard Test Method for Determination of Total Hydrocarbons in Hydrogen by FID-Based Total Hydrocarbon (THC) Analyzer
ASTM D7892: Standard Test Method for Determination of Total Organic Halides, Total Non-Methane Hydrocarbons, and Formaldehyde in Hydrogen Fuel by Gas Chromatography/Mass Spectrometry
ASTM D7941: Standard Test Method for Hydrogen Purity Analysis Using a Continuous Wave Cavity Ring-Down Spectroscopy Analyzer
ASTM D8003: Standard Test Method for Determination of Light Hydrocarbons and Cut Point Intervals in Live Crude Oils and Condensates by Gas Chromatography
ASTM D8098: Standard Test Method for Permanent Gases in C2 and C3 Hydrocarbon Products by Gas Chromatography and Pulse Discharge Helium Ionization Detector

7.3 Ethane

Ethane (C_2H_6, CH_3CH_3) is a colorless, odorless, gaseous hydrocarbon belonging to the paraffin series that is structurally the simplest hydrocarbon that contains a single carbon–carbon bond. Ethane is an important constituent of natural gas that also occurs in dissolved crude oil and as a by-product of crude oil refining, and it is also produced by the carbonization of coal.

Ethane is a normally gaseous paraffinic compound (C_2H_6) having a boiling point of approximately −88.5 °C (−127.4 °F) (Table 7.3) that can be further defined by a series of standard test methods (Table 7.4). Ethane may be handled as a liquid at very high pressures and at temperatures below 32 °C (90 °F). A large proportion of the ethane produced as a separate liquid product are used as a feedstock for the production of ethylene ($CH_2{=}CH_2$) – a -a basic petrochemical (Speight, 2014). Quality criteria for ethane as a petrochemical feedstock are specific to the particular process used for petrochemical manufacture. In general, however, such feedstocks must be very low in moisture content (ASTM D1142, 2017), oxygen (ASTM D1945, 2017), carbon dioxide (ASTM D1945, 2017), sulfur-containing compounds (ASTM D1072, 2017), and other

Table 7.3 Properties of ethane.

Chemical formula	C_2H_6
Molar mass	30.07 g mol^{-1}
Appearance	Colorless gas
Odor	Odorless
Density	• 1.3562 mg cm^{-3} (at 0 °C) • 0.5446 g cm^{-3} (at 184 K)
Melting point	−182.8 °C; −296.9 °F; 90.4 °K
Boiling point	−88.5 °C; −127.4 °F; 184.6 °K
Solubility in water	56.8 mg l^{-1}
Vapor pressure	3.8453 MPa (at 21.1 °C)
Henry's law constant (k_H)	19 nmol Pa^{-1} kg^{-1}
Acidity (pK_a)	50
Basicity (pK_b)	−36
Magnetic susceptibility (χ)	$-37.37 \cdot 10^{-6}$ cm^3 mol^{-1}
Specific heat capacity (C)	52.49 J K^{-1} mol^{-1}
Standard enthalpy of formation ($\Delta_f H^\circ_{298}$)	−84 kJ mol^{-1}
Standard enthalpy of combustion ($\Delta_c H^\circ_{298}$)	−1561.0 to −1560.4 kJ mol^{-1}

Table 7.4 ASTM standard test methods for ethane.

ASTM D1142: Standard Test Method for Water Vapor Content of Gaseous Fuels
ASTM D1838: Standard Test Method for Copper Strip Corrosion by Liquefied Petroleum (LP) Gases
ASTM D1945: Standard Test Method for Analysis of Natural Gas by Gas Chromatography
ASTM D2158: Standard Test Method for Residues in Liquefied Petroleum (LP) Gases
ASTM D2163: Standard Test Method for Determination of Hydrocarbons in Liquefied Petroleum (LP) Gases and Propane/Propene Mixtures by Gas Chromatography
ASTM D2427: Standard Test Method for Determination of C 2 through C 5 Hydrocarbons in Gasolines by Gas Chromatography
ASTM D2504: Standard Test Method for Noncondensable Gases in C 2 and Lighter Hydrocarbon Products by Gas Chromatography
ASTM D2505: Standard Test Method for Ethylene, Other Hydrocarbons, and Carbon Dioxide in High-Purity Ethylene by Gas Chromatography
ASTM D2597: Standard Test Method for Analysis of Demethanized Hydrocarbon Liquid Mixtures Containing Nitrogen and Carbon Dioxide by Gas Chromatography
ASTM D2598: Standard Practice for Calculation of Certain Physical Properties of Liquefied Petroleum (LP) Gases from Compositional Analysis
ASTM D2650: Standard Test Method for Chemical Composition of Gases by Mass Spectrometry
ASTM D2713: Standard Test Method for Dryness of Propane (Valve Freeze Method)
ASTM D4468: Standard Test Method for Total Sulfur in Gaseous Fuels by Hydrogenolysis and Rateometric Colorimetry
ASTM D4891: Standard Test Method for Heating Value of Gases in Natural Gas and Flare Gases Range by Stoichiometric Combustion
ASTM D5228: Standard Test Method for Determination of Butane Working Capacity of Activated Carbon
ASTM D5504: Standard Test Method for Determination of Sulfur Compounds in Natural Gas and Gaseous Fuels by Gas Chromatography and Chemiluminescence
ASTM D5742: Standard Test Method for Determination of Butane Activity of Activated Carbon
ASTM D6159: Standard Test Method for Determination of Hydrocarbon Impurities in Ethylene by Gas Chromatography
ASTM D6968: Standard Test Method for Simultaneous Measurement of Sulfur Compounds and Minor Hydrocarbons in Natural Gas and Gaseous Fuels by Gas Chromatography and Atomic Emission Detection
ASTM D7423: Standard Test Method for Determination of Oxygenates in C2, C3, C4, and C5 Hydrocarbon Matrices by Gas Chromatography and Flame Ionization Detection
ASTM D7653: Standard Test Method for Determination of Trace Gaseous Contaminants in Hydrogen Fuel by Fourier Transform Infrared (FTIR) Spectroscopy
ASTM D7675: Standard Test Method for Determination of Total Hydrocarbons in Hydrogen by FID-Based Total Hydrocarbon (THC) Analyzer
ASTM D7754: Standard Test Method for Determination of Trace Oxygenates in Automotive Spark-Ignition Engine Fuel by Multidimensional Gas Chromatography
ASTM D7892: Standard Test Method for Determination of Total Organic Halides, Total Non-Methane Hydrocarbons, and Formaldehyde in Hydrogen Fuel by Gas Chromatography/Mass Spectrometry

elements that will interfere with the gas-cleaning processes or have an adverse effect on the catalysts when the gas stream is used in manufacture of petrochemical products.

The chief use of ethane is in the chemical industry in the production of ethylene (ethene) by steam cracking. When diluted with steam and briefly heated to very high temperatures (900 °C, 1650 °F or higher), high molecular weight hydrocarbon derivatives decompose down into lower molecular weight hydrocarbon derivatives, and saturated hydrocarbon derivatives (paraffin derivatives) form olefin derivatives. Ethane is favored for ethylene production because the steam cracking of ethane is fairly selective for ethylene production, while the steam cracking of heavier hydrocarbons yields a product mixture poorer in ethylene, and richer in higher molecular weight alkenes (olefins) such as propylene (propene, $CH_3CH{=}CH_2$) and butadiene ($CH_2{=}CHCH{=}CH_2$), and in aromatic hydrocarbons.

7.3.1 Chemical Properties

At standard temperature and pressure, ethane is a colorless, odorless gas (Table 7.3). It has a boiling point of −88.5 °C (−127.3 °F) and melting point of −182.8 °C (−297.0 °F). Solid ethane exists in several modifications. On cooling under normal pressure, the first modification to appear is a plastic crystal, crystallizing in the cubic system. In this form, the positions of the hydrogen atoms are not fixed; the molecules may rotate freely around the long axis. Cooling this ethane below approximately 89.9°K (−183.2 °C; −297.8 °F) changes it to monoclinic metastable ethane. Ethane is only very sparingly soluble in water.

7.3.1.1 Chemical Reactions

Chemically, ethane can be considered as two methyl groups joined, that is, a dimer of methyl groups. The chemistry of ethane involves chiefly free radical reactions. Ethane can react with the halogens, especially chlorine and bromine, by free radical halogenation that proceeds through the propagation of the ethyl radical:

$$C_2H_5\cdot + Cl_2 \rightarrow C_2H_5Cl + Cl\cdot$$

$$Cl\cdot + C_2H_6 \rightarrow C_2H_5\cdot + HCl$$

Because halogenated ethane derivatives can undergo further free radical halogenation, this process results in a mixture of several halogenated products. In the chemical industry, more selective chemical reactions are used for the production of any particular two-carbon haloalkane.

The complete combustion of ethane produces carbon dioxide and water according to the chemical equation:

$$2C_2H_6 + 7O_2 \rightarrow 4CO_2 + 6H_2O$$

Combustion may also occur without an excess of oxygen, forming a mix of amorphous carbon and carbon monoxide:

$$2C_2H_6 + O_2 \rightarrow 4C + 6H_2O$$

$$2C_2H_6 + 5O_2 \rightarrow 4CO + 6H_2O$$

$$2C_2H_6 + 4O_2 \rightarrow 2C + 2CO + 6H_2O$$

Combustion occurs by a complex series of free radical reactions. An important series of reaction in ethane combustion is the combination of an ethyl radical with oxygen and the subsequent breakup of the resulting peroxide into ethoxy and hydroxyl radicals. Thus,

$$C_2H_5\cdot + O_2 \rightarrow C_2H_5OO\cdot$$

$$C_2H_5OO\cdot + HR \rightarrow C_2H_5OOH + R\cdot$$

$$C_2H_5OOH \rightarrow C_2H_5O\cdot + \cdot OH$$

The principal carbon-containing products of incomplete ethane combustion are single-carbon compounds such as carbon monoxide and formaldehyde (HCHO). One important route by which the carbon–carbon bond in ethane is broken, to yield these single-carbon products, is the decomposition of the ethoxy radical into a methyl radical and formaldehyde, which can in turn undergo further oxidation:

$$C_2H_5O\cdot \rightarrow CH_3\cdot + HCHO$$

Minor products in the incomplete combustion of ethane include acetaldehyde (CH_3CHO), methane (CH_4), methanol (CH_3OH), and ethanol (CH_3CH_2OH). At higher temperatures, especially in the range 600–900 °C (1110–1650 °F), ethylene ($CH_2{=}CH_2$) is a significant product:

$$C_2H_5\cdot + O_2 \rightarrow CH_2 = CH_2 + \cdot OOH$$

Similar reactions (with agents other than oxygen as the hydrogen abstractor) are involved in the production of ethylene from ethane in steam cracking (Speight, 2014).

7.3.2 Physical Properties

Like methane and the other gaseous hydrocarbons, ethane is isolated on an industrial scale from natural gas and as a by-product of crude oil refining. In fact, after methane, ethane is the second most abundant constituent of natural gas, but, because of the variance in composition of natural gas (Chapter 1, Chapter 7), gas from different gas fields varies in ethane content from less than 1% v/v to more than 6% v/v.

Prior to the 1960s, ethane and higher molecular weight hydrocarbons were typically not separated from the methane component of natural gas, but simply burnt along with the methane as a fuel. Currently, since the value of ethane as a feedstock for petrochemical production has been established (Speight 2014), ethane is separated from the other components of natural gas in most well-developed gas fields. Ethane can also be separated from process gas that is produced as a by-product of crude oil refining (Parkash, 2003; Gary et al., 2007; Speight, 2007, 2014, 2017; Hsu and Robinson, 2017).

Ethane is most efficiently separated from methane by liquefying it at cryogenic temperatures (Speight 2007, 2014). Various refrigeration strategies exist: the most economical process present in wide use employs a turboexpander and can recover more than 90% of the ethane in natural gas. In this process, chilled gas is expanded through a turbine that reduces the temperature to approximately −100 °C (−148 °F). At this low temperature, gaseous methane can be separated from the liquefied ethane and higher molecular weight hydrocarbon derivatives by distillation. Further distillation then separates ethane from the propane and butane as well as any higher molecular weight hydrocarbon derivatives (pentane derivatives, hexane derivatives, and the like) that are contained in the mixture.

The chief use of ethane is as a feedstock for ethylene production by steam cracking in which the ethane is diluted with steam and briefly heated to very high temperatures (typically 900 °C, 1650 °F, or even higher).

$$CH_3CH_3 \rightarrow CH_2 = CH_2 + H_2$$

7.4 Propane

Propane (C_3H_8, $CH_3CH_2CH_3$) is a three-carbon alkane that is a gas at standard temperature and pressure but compressible to a transportable liquid. It is a gaseous paraffin hydrocarbon (C_3H_8) having a boiling point of −42.3 (−44.0 °F) and a melting point of −187.7 °C (−305.8 °F) (Table 7.5). Propane may be handled as a liquid at ambient temperatures and moderate pressures. Commercial propane as sold on the various markets may include varying amounts of ethane, butanes, and liquefied refinery gases.

Propane is produced as a by-product of two other processes: (i) natural gas processing and (ii) crude oil refining. The processing of natural gas involves removal of butane, propane, and large amounts of ethane from the raw gas to prevent condensation of these volatiles in natural gas pipelines. Additionally, crude oil refineries produce some propane as a by-product of cracking processes (Parkash, 2003; Gary et al., 2007; Speight, 2007, 2014, 2017; Hsu and Robinson, 2017) and as a biofuel by the thermal conversion of various types of biomass (Speight, 2011).

Table 7.5 Properties of propane.

Chemical formula	C_2H_6
Molar mass	30.07 g mol^{-1}
Appearance	Colorless gas
Odor	Odorless
Density	• 1.3562 mg cm^{-3} (at 0 °C) • 0.5446 g cm^{-3} (at 184 K)
Melting point	−182.8 °C; −296.9 °F; 90.4 °K
Boiling point	−88.5 °C; −127.4 °F; 184.6 °K
Solubility in water	56.8 mg l^{-1}
Vapor pressure	3.8453 MPa (at 21.1 °C)
Henry's law constant (k_H)	19 nmol Pa^{-1} kg^{-1}
Acidity (pK_a)	50
Basicity (pK_b)	−36
Magnetic susceptibility (χ)	−37.37·10^{-6} cm^3 mol^{-1}
Specific heat capacity (C)	52.49 J K^{-1} mol^{-1}
Standard enthalpy of formation ($\Delta_f H^o_{298}$)	−84 kJ mol^{-1}
Standard enthalpy of combustion ($\Delta_c H^o_{298}$)	−1561.0 to −1560.4 kJ mol^{-1}

From both crude oil refining and natural gas processing, propane is not produced for its own sake, but is a by-product of these two other processes. Natural gas plant production of propane primarily involves extracting materials such as propane and butane from natural gas to prevent these liquids from condensing and causing operational problems in natural gas pipelines. Similarly, when oil refineries make major products such as motor gasoline, diesel, and heating oil, some propane is produced as a by-product of those processes.

Propane has a wide variety of uses worldwide including small domestic heating applications to large industrial and manufacturing processes. Some of the more common uses of propane are for residential and commercial heating and cooking; motor fuel use in vehicles, irrigation pumps, and power generation; agricultural crop drying and weed control; and as a raw material in the petrochemical industry to make things such as plastics, alcohol, fibers, and cosmetics.

7.4.1 Chemical Properties

Also known as liquefied petroleum gas (LPG), propane is an odorless, nontoxic hydrocarbon (C_3H_8) gas at normal pressures and temperatures (ASTM D2163,

2017). When pressurized, it is a liquid with an energy density 270 times greater than its gaseous form. A gallon of liquid propane has about 25% less energy than a gallon of gasoline.

Propane is a simple asphyxiant, and since, unlike methane, propane is denser than air, it may accumulate in low spaces, such as depressions in the surface of the Earth and near the floor in domestic and industrial building. If a leak in a propane fuel system occurs, the gas will tend to sink into any enclosed area and thus poses a risk of explosion and fire. The typical scenario is a leaking cylinder stored in a basement; the propane leak drifts across the floor to the pilot light on the furnace or water heater and results in an explosion or fire. This property makes propane generally unsuitable as a fuel for boats. One hazard associated with propane storage and transport is known as a boiling liquid expanding vapor explosion (BLEVE).

Propane is stored under pressure at room temperature and, together with its mixtures will flash evaporate at atmospheric pressure and cool well below the freezing point of water. The cold gas, which appears white due to moisture condensing from the air, may cause frostbite.

Propane undergoes dehydrogenation of propylene by a catalytic cracking process:

$$CH_3CH_2CH_3 \rightarrow CH_3CH = CH_2 + H_2$$

It also undergoes combustion reactions in a similar fashion to other alkanes. In the presence of excess oxygen, it burns to form water and carbon dioxide:

$$C_3H_8 + 5O_2 \rightarrow 3CO_2 + 4H_2O$$

When insufficient oxygen is present for complete combustion, incomplete combustion occurs, allowing carbon monoxide and/or soot (carbon) to be formed as well:

$$2C_3H_8 + 9O_2 \rightarrow 4CO_2 + 2CO + 8H_2O$$

$$C_3H_8 + 2O_2 \rightarrow 3C + 4H_2O$$

Propane combustion is much cleaner than gasoline combustion, though not as clean as natural gas (methane) combustion. The presence of carbon–carbon bonds, plus the multiple bonds of propylene ($CH_3CH{=}CH_2$) and butylene ($CH_3CH{=}CHCH_3$, $CH_3CH_2CH{=}CH_2$), creates organic exhausts besides carbon dioxide and water vapor during typical combustion. These bonds also cause propane to burn with a visible flame.

7.4.2 Physical Properties

The properties of propane (Table 7.5) as determined by standard test methods (Table 7.6) are conducive to propane existing in one of two forms: (i) liquid or (ii) gas. Both liquid propane and propane gas are usable but for specific

Table 7.6 ASTM standard test methods for propane.

ASTM D323: Standard Test Method for Vapor Pressure of Petroleum Products (Reid Method)
ASTM D1265: Standard Practice for Sampling Liquefied Petroleum (LP) Gases, Manual Method
ASTM D1267: Standard Test Method for Gage Vapor Pressure of Liquefied Petroleum (LP) Gases (LP-Gas Method)
ASTM D1657: Standard Test Method for Density or Relative Density of Light Hydrocarbons by Pressure Hydrometer
ASTM D1835: Standard Specification for Liquefied Petroleum (LP) Gases
ASTM D1838: Standard Test Method for Copper Strip Corrosion by Liquefied Petroleum (LP) Gases
ASTM D1988: Standard Test Method for Mercaptans in Natural Gas Using Length-of-Stain Detector Tubes
ASTM D2158: Standard Test Method for Residues in Liquefied Petroleum (LP) Gases
ASTM D2163: Standard Test Method for Determination of Hydrocarbons in Liquefied Petroleum (LP) Gases and Propane/Propene Mixtures by Gas Chromatography
ASTM D2420: Standard Test Method for Hydrogen Sulfide in Liquefied Petroleum (LP) Gases (Lead Acetate Method)
ASTM D2593: Standard Test Method for Butadiene Purity and Hydrocarbon Impurities by Gas Chromatography
ASTM D2598: Standard Practice for Calculation of Certain Physical Properties of Liquefied Petroleum (LP) Gases from Compositional Analysis
ASTM D2650: Standard Test Method for Chemical Composition of Gases by Mass Spectrometry
ASTM D2713: Standard Test Method for Dryness of Propane (Valve Freeze Method)
ASTM D4891: Standard Test Method for Heating Value of Gases in Natural Gas and Flare Gases Range by Stoichiometric Combustion
ASTM D5305: Standard Test Method for Determination of Ethyl Mercaptan in LP-Gas Vapor
ASTM D5504: Standard Test Method for Determination of Sulfur Compounds in Natural Gas and Gaseous Fuels by Gas Chromatography and Chemiluminescence
ASTM D6021: Standard Test Method for Measurement of Total Hydrogen Sulfide in Residual Fuels by Multiple Headspace Extraction and Sulfur Specific Detection
ASTM D6159: Standard Test Method for Determination of Hydrocarbon Impurities in Ethylene by Gas Chromatography
ASTM D6667: Standard Test Method for Determination of Total Volatile Sulfur in Gaseous Hydrocarbons and Liquefied Petroleum Gases by Ultraviolet Fluorescence
ASTM D6897: Standard Test Method for Vapor Pressure of Liquefied Petroleum Gases (LPG) (Expansion Method)

Table 7.6 (Continued)

ASTM D6968: Standard Test Method for Simultaneous Measurement of Sulfur Compounds and Minor Hydrocarbons in Natural Gas and Gaseous Fuels by Gas Chromatography and Atomic Emission Detection
ASTM D7423: Standard Test Method for Determination of Oxygenates in C2, C3, C4, and C5 Hydrocarbon Matrices by Gas Chromatography and Flame Ionization Detection
ASTM D7653: Standard Test Method for Determination of Trace Gaseous Contaminants in Hydrogen Fuel by Fourier Transform Infrared (FTIR) Spectroscopy
ASTM D7756: Standard Test Method for Residues in Liquefied Petroleum (LP) Gases by Gas Chromatography with Liquid, On-Column Injection
ASTM D7828: Standard Test Method for Determination of Residue Composition in Liquefied Petroleum Gas (LPG) Using Automated Thermal Desorption/Gas Chromatography (ATD/GC)
ASTM D7901: Standard Specification for Dimethyl Ether for Fuel Purposes
ASTM D7994: Standard Test Method for Total Fluorine, Chlorine, and Sulfur in Liquid Petroleum Gas (LPG) by Oxidative Pyrohydrolytic Combustion Followed by Ion Chromatography Detection (Combustion Ion Chromatography-CIC)

purposes and cannot be used interchangeably. In other words, a propane system designed to use vapor cannot utilize propane in its liquid form and vice versa.

Propane exists in its liquid form at or below its boiling point (−44 °F) as well as when it stored under pressure. To further explain, if the temperature outside is −45 °F, propane will be a liquid and you would be able to pour it out of a bucket. But as soon as the temperature rises to −42 °C (−44 °F), the propane begins to boil and thus give off vapor. If the temperature outside is colder than −42 °C (−44 °F), propane exists as a liquid. It is still propane but it looks a lot like water while at this cold temperature. Thus, it becomes a vapor at temperatures above −42 °C (−44 °F). Propane vapor is heavier than air and will sink to and collect in the lowest point it can find. If propane is vented to the outside air, it will quickly dissipate with the slightest movement of air. Conversely, if it is vented into an airtight structure with no air movement, the propane vapor will collect on the floor and rise vertically if more propane is vented into the structure. This is extremely important to know because if there is a propane leak in a house or building, the propane vapor will seek the lowest possible point where it will collect. Keep in mind that one gallon of propane will produce over 36 ft^3 of vapor, and this vapor will settle in the lowest possible place. If the propane vapor level continues to rise, it may ignite if a source of ignition is present.

Propane is bought and stored in a liquid form (LPG), and thus fuel energy can be stored in a relatively small space. Compressed natural gas (CNG), largely methane, is another gas used as fuel, but it cannot be liquefied by compression at normal temperatures, as these are well above the critical temperature. As a gas, very high pressure is required to store useful quantities. This poses the hazard that, in an accident, just as with any compressed gas cylinder (such as a carbon dioxide cylinder), a CNG cylinder may burst with great force or leak rapidly enough to become a self-propelled missile. Therefore, CNG is much less efficient to store, due to the large cylinder volume required. An alternative means of storing natural gas is as a cryogenic liquid in an insulated container as liquefied natural gas (LNG). This form of storage is at low pressure and is around 3.5 times as efficient as storing it as CNG. Unlike propane, if a spill occurs, CNG (i.e. methane but not all of the gaseous constituents of natural gas) will evaporate and dissipate harmlessly because it is lighter than air.

7.5 Butane

Butane (C_4H_{10}) is a gas at room temperature and atmospheric pressure. The term may refer to either of two structural isomers, *n*-butane ($CH_3CH_2CH_2CH_3$) (Table 7.7) or isobutane ($(CH_3)_3CH$, also called methylpropane) (Table 7.8) or

Table 7.7 Properties of n-butane.

Chemical formula	C_3H_8
Molar mass	44.10 g mol^{-1}
Appearance	Colorless gas
Odor	Odorless
Density	2.0098 kg m^{-3} (at 0 °C, 101.3 kPa)
Melting point	−187.7 °C; −305.8 °F; 85.5 °K
Boiling point	−42.25 to −42.04 °C; −44.05 to −43.67 °F
Solubility in water	47 mg l^{-1} (at 0 °C)
log *P*	2.236
Vapor pressure	853.16 kPa (at 21.1 °C (70.0 °F)
Henry's law constant (k_H)	15 nmol Pa^{-1} kg^{-1}
Magnetic susceptibility (χ)	−40.5·10^{-6} cm^3 mol^{-1}
Specific heat capacity (*C*)	73.60 J K^{-1} mol^{-1}
Standard enthalpy of formation ($\Delta_f H_{298}$)	−105.2 to −104.2 kJ mol^{-1}
Standard enthalpy of combustion ($\Delta_c H_{298}$)	−2.2197 to −2.2187 MJ mol^{-1}

Table 7.8 Properties of isobutane.

Chemical formula	C_4H_{10}
Molar mass	$58.12\,g{\cdot}mol^{-1}$
Appearance	Colorless gas
Odor	Odorless
Density	$2.51\,kg\,m^{-3}$ (at 15 °C, 100 kPa)
Melting point	−159.42 °C (−254.96 °F
Boiling point	−11.7 °C (10.9 °F)
Solubility in water	$48.9\,mg\,l^{-1}$ (at 25 °C, 77 °F)
Vapor pressure	204.8 kPa (at 21 °C, 70 °F)
Henry's law constant (k_H)	$8.6\,nmol\,Pa^{-1}\,kg^{-1}$
Magnetic susceptibility (χ)	$-51.7{\cdot}10^{-6}\,cm^3\,mol^{-1}$
Specific heat capacity (C)	$96.65\,J\,K^{-1}\,mol^{-1}$
Standard enthalpy of formation ($\Delta_f H^{\circ}_{298}$)	−134.8 to $-133.6\,kJ\,mol^{-1}$
Standard enthalpy of combustion ($\Delta_c H_{298}$)	−2.86959 to $-2.86841\,MJ\,mol^{-1}$

to a mixture of these isomers. It is a normally gaseous paraffinic hydrocarbon (C_4H_{10}) that is handled usually as a liquid at ambient temperatures and moderate pressures. It may be fractionated further into normal butane, which has a boiling point of approximately 0 °C (32 °F), and isobutane, which has a boiling point of approximately −12 °C (11 °F).

Two isomers of butane:

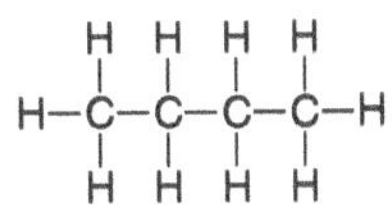

n-Butane

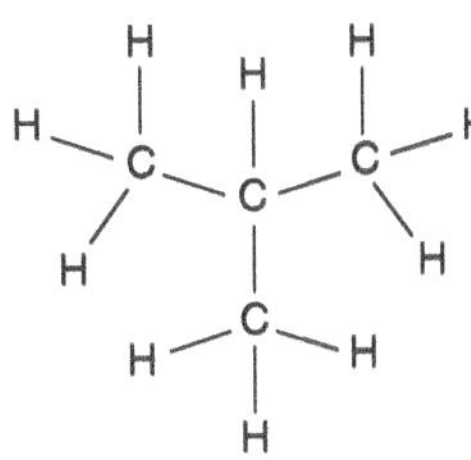

iso-Butane

In the IUPAC system of nomenclature, however, the name "butane" refers only to the *n*-butane isomer. Butanes are highly flammable, colorless, easily liquefied gases that quickly vaporize at room temperature.

The butane isomers present in natural gas can be separated from the large quantities of lower-boiling gaseous constituents, such as methane and ethane, by absorption in a light oil. The butanes thus obtained can be stripped from the absorbent along with propane and marketed as LPG that meets the required specifications (ASTM D1838, 2017; ASTM D2158, 2017; ASTM D2163, 2017; ASTM D2713, 2017), or they can be separated from the propane and then from each other by fractional distillation: *n*-butane boils at −0.5 °C (31.1 °F); isobutane boils at −11.7 °C (10.9 °F) (Table 7.9).

Butanes formed by catalytic cracking and other refinery processes are also recovered by absorption into a light oil. Commercially, *n*-butane can be added to *gasoline* to increase its volatility. Transformed to isobutane in a refinery process known as isomerization, it can be reacted with certain other hydrocarbons such as butylene to form valuable high-octane constituents of gasoline.

7.5.1 Chemical Properties

Propane undergoes combustion reactions in a similar fashion to other alkanes. In the presence of excess oxygen, propane burns to form water and carbon dioxide:

$$C_3H_8 + 5O_2 \rightarrow 3CO_2 + 4H_2O$$

When insufficient oxygen is present for complete combustion, incomplete combustion occurs, allowing carbon monoxide and/or soot (carbon) to be formed as well:

$$2C_3H_8 + 9O_2 \rightarrow 4CO_2 + 2CO + 8H_2O$$

$$C_3H_8 + 2O_2 \rightarrow 3C + 4H_2O$$

Propane combustion is much cleaner than gasoline combustion, though not as clean as natural gas (methane) combustion.

In the presence of excess oxygen, butane burns to form carbon dioxide and water vapor:

$$2C_4H_{10} + 13O_2 \rightarrow 8CO_2 + 10H_2O$$

On the other hand, when the supply of oxygen is limited, carbon (soot) or carbon monoxide may also be formed:

$$2C_4H_{10} + 9O_2 \rightarrow 8CO + 10H_2O$$

Table 7.9 ASTM test methods for butane.

ASTM D1265: Standard Practice for Sampling Liquefied Petroleum (LP) Gases, Manual Method
ASTM D1267: Test Method for Gage Vapor Pressure of Liquefied Petroleum (LP) Gases (LP-Gas Method)
ASTM D1364: Test Method for Water in Volatile Solvents (Karl Fischer Reagent Titration Method)
ASTM D1717: Test Method for Test for Analysis of Commercial Butane-Butene Mixtures and Isobutylene by Gas Chromatography
ASTM D1835: Standard Specification for Liquefied Petroleum Gases
ASTM D1837: Standard Test Method for Volatility of Liquefied Petroleum Gases
ASTM D1838: Test Method for Copper Strip Corrosion by Liquefied Petroleum (LP) Gases
ASTM D1945: Standard Test Method for Analysis of Natural Gas by Gas Chromatography
ASTM D2158: Standard Test Method for Residues in Liquefied Petroleum (LP) Gases
ASTM D2163: Standard Test Method for Determination of Hydrocarbons in Liquefied Petroleum (LP) Gases and Propane/Propene Mixtures by Gas Chromatography
ASTM D2384: Standard Test Methods for Traces of Volatile Chlorides in Butane–Butene Mixtures
ASTM D2427: The determination of C2 to C5 Hydrocarbons in Gasolines by Gas Chromatography
ASTM D2593: Standard Test Method for Butadiene Purity and Hydrocarbon Impurities by Gas Chromatography
ASTM D2598: Standard Practice for Calculation of Certain Physical Properties of Liquefied Petroleum (LP) Gases from Compositional Analysis
ASTM D2650: Standard Test Method for Chemical Composition of Gases by Mass Spectrometry
ASTM D3120: Standard Test Method for Trace Quantities of Sulfur in Light Liquid Petroleum Hydrocarbons by Oxidative Microcoulometry
ASTM D3588: Standard Practice for Calculating Heat Value, Compressibility Factor, and Relative Density of Gaseous Fuels
ASTM D4045: Standard Test Method for Sulfur in Petroleum Products by Hydrogenolysis and Rateometric Colorimetry
ASTM D4307: Standard Practice for Preparation of Liquid Blends for Use as Analytical Standards
ASTM D4424: Standard Test Method for Butylene Analysis by Gas Chromatography
ASTM D4891: Standard Test Method for Heating Value of Gases in Natural Gas and Flare Gases Range by Stoichiometric Combustion
ASTM D5228: Standard Test Method for Determination of the Butane Working Capacity of Activated Carbon

(*Continued*)

Table 7.9 (Continued)

ASTM D5273: Standard Guide for Analysis of Propylene Concentrates
ASTM D5274: Standard Guide for Analysis of 1,3-Butadiene Product
ASTM D5623: Standard Test Method for Sulfur Compounds in Light Petroleum Liquids by Gas Chromatography and Sulfur Selective Detection
ASTM D5742: Standard Test Method for Determination of Butane Activity of Activated Carbon
ASTM D6159: Standard Test Method for Determination of Hydrocarbon Impurities in Ethylene by Gas Chromatography
ASTM D7266: Standard Test Method for Analysis of Cyclohexane by Gas Chromatography (External Standard) Analysis by Gas Chromatography
ASTM D7756: Standard Test Method for Residues in Liquefied Petroleum (LP) Gases by Gas Chromatography with Liquid, On-Column Injection
ASTM E1064: Standard Test Method for Water in Organic Liquids by Coulometric Karl Fischer Titration

n-Butane is the feedstock for the DuPont catalytic process for the preparation of maleic anhydride:

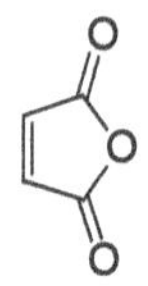

Maleic anhydride

Thus,

$$2CH_3CH_2CH_2CH_3 + 7O_2 \rightarrow 2C_2H_2(CO)_2O + 8H_2O$$

n-Butane, like all hydrocarbons, undergoes free radical chlorination providing both 1-chlorobutane ($CH_3CH_2CH_2CH_2Cl$) and 2-chlorobutane ($CH_3CH_2CH_2CClCH_3$), as well as more highly chlorinated derivatives. The relative rates of the chlorination are partially explained by the differing bond dissociation energy for the two types of C–H bonds.

7.5.2 Physical Properties

n-Butane ($CH_3CH_2CH_2CH_3$) is a colorless gas with a boiling point of −1 °C (30 °F) that, unlike the first three *alkanes*, is very soluble in water (Table 7.8). The principal raw materials for its production are crude oil and LNG. Like *n*-butane, it forms an explosive and flammable mixture with air at low concentrations. Its main uses in industry are as a raw material in the production of butadiene and acetic acid. It is also used as a domestic fuel, as a gasoline blending component, as a solvent and as a refrigerant.

Isobutane ($(CH_3)_2CHCH_3$) is also a colorless gas with a boiling point of −11.7 °C (10.9 °F) that is also soluble in water (Table 7.9). Although its physical properties are similar to that of *n*-butane, it exhibits markedly different chemical behavior. It is obtained by crude oil fractionation of natural gas or by isomerization of butane. It forms an explosive and flammable mixture with air at low concentrations. Its main uses are as a raw material in organic synthesis, for the production of synthetic rubber, and in the production of branched hydrocarbons of high-octane grading.

7.6 Olefins and Diolefins

Olefins derivatives are typical constituents of natural gas but do occur in refinery gases, which can be complex mixtures of hydrocarbon gases and nonhydrocarbon gas (Table 7.10) (Chapter 1). Many low molecular weight olefins (such as ethylene and propylene) and diolefins (such as butadiene) that are produced in the refinery are isolated for petrochemical use (Speight, 2014). The individual products are (i) ethylene, (ii) propylene, and (iii) butadiene.

Table 7.10 Origin of petroleum-related gases.

Gas	Origin
Natural gas	Occurs naturally with or without crude oil
	A gaseous combination of hydrocarbons
	Predominantly C1–C4 hydrocarbons
	May also contain gas condensate or natural gasoline
Refinery gas process gas	A combination of gases produced by distillation
	Products from the cracking of crude oil
	Consists of C2–C4 hydrocarbons including olefin gases
	Boiling range of approximately −51 to −1 °C (−60 to −30 °F)
Tail gas	A combination of hydrocarbons from the distillation of products from catalytically cracked feedstocks
	Predominantly of C1 to C4 hydrocarbons

Ethylene (C_2H_4) is a normally gaseous olefinic compound having a boiling point of approximately −104 °C (−155 °F). It may be handled as a liquid at very high pressures and low temperatures. Ethylene is made normally by cracking an ethane or naphtha feedstock in a high-temperature furnace and subsequent isolation from other components by distillation. The major uses of ethylene are in the production of ethylene oxide, ethylene dichloride, and polyethylene polymers. Other uses include the coloring of fruit, rubber products, ethyl alcohol, and medicine (anesthetic).

Propylene concentrates are mixtures of propylene and other hydrocarbons, principally propane and trace quantities of ethylene, butylenes, and butanes. Propylene concentrates may vary in propylene content from 70% mol up to over 95% mol and may be handled as a liquid at normal temperatures and moderate pressures. Propylene concentrates are isolated from the furnace products mentioned in the preceding paragraph on ethylene. Higher-purity propylene streams are further purified by distillation and extractive techniques. Propylene concentrates are used in the production of propylene oxide, isopropyl alcohol, and polypropylene and in the synthesis of isoprene.

As with any gas stream, propylene-concentrate streams typically require a component analysis, depending upon their final use. The appropriate method for the determination is by gas chromatography (ASTM D2427, 2017). Another gas chromatography method is used to identify major impurities (ASTM D2163, 2017).

Trace quantities of sulfur compounds in hydrocarbon products can be harmful to many catalytic chemical processes in which these products are used. Maximum permissible levels of total sulfur are normally included in specifications for such hydrocarbons. It is recommended that this test method be used to provide a basis for agreement between two laboratories when the determination of sulfur in hydrocarbon gases is important. In the case of LPG, total volatile sulfur is measured on an injected gas sample. One test method (ASTM D3246, 2017) describes a procedure for the determination of sulfur in the range from 1.5 to 100 $mg\,kg^{-1}$ (ppm w/w) in hydrocarbon products that are gaseous at normal room temperature and pressure.

As is the case for ethylene, moisture in propylene is critical. Several field tests and a few laboratory tests are in use by individual firms, but no standard method for moisture has been adopted to date. The problems in sampling for moisture content, especially in the less than 10 ppm range, are hard to overcome. The trace impurities in 90% or better propylene, which is used in polymerization processes, become quite critical. Hydrogen, oxygen, and carbon monoxide are determined by one technique (ASTM D2504, 2017), and acetylene, ethylene, butenes, butadiene, methyl acetylene, and propadiene are determined by using a very sensitive analytical method (ASTM D2712, 2017).

Butylene concentrates are mixtures of butene-1, *cis*- and *trans*-butene-2, and, sometimes, isobutene (2-methyl propylene) (C_4H_8).

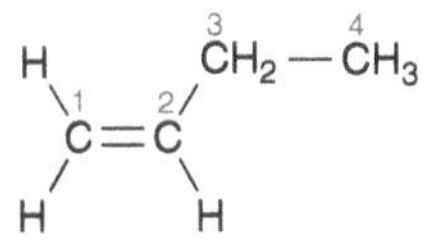

Butene-1

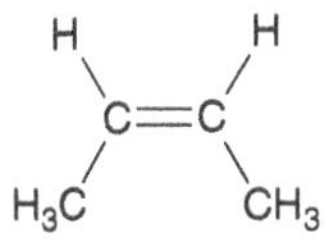

cis-Butene-2

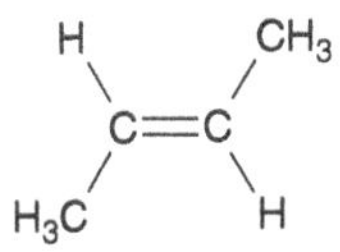

trans-Butene-2

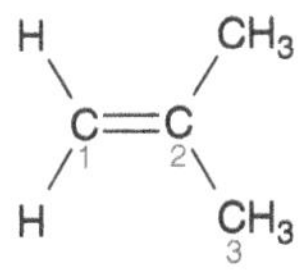

iso-Butene (2-methylpropene, 2-methyl propylene)

These products are stored as liquids at ambient temperatures and moderate pressures. Various impurities such as butane, butadiene, and the C_5 hydrocarbons are generally found in butylene concentrates. Virtually all of the butylene concentrates are used as a feedstock for either (i) an alkylation plant, where isobutane and butylenes are reacted in the presence of either sulfuric acid or hydrofluoric acid to form a mixture of C_7 to C_9 paraffins used in gasoline, or (ii) butylene dehydrogenation reactors for butadiene production.

The major quality criterion for butylene concentrates is the distribution of butylenes that is measured, along with other components, by gas chromatography. Trace impurities generally checked are sulfur-containing derivatives, chlorine-containing derivatives (ASTM D2384, 2017), and acetylene derivatives. These impurities are known catalyst poisons that interfere with (to the point of destroy) catalyst activity or (at best) become unwanted impurities in the final product. When butylene concentrates are used as the feedstock for an alkylation unit, the diolefin content becomes important because of the potential for further reaction to produce unwanted higher molecular weight products (Speight, 2014, 2017).

Butadiene ($CH_2{=}CHCH{=}CH_2$) is a normally gaseous hydrocarbon (C_4H_6) having a boiling point of −4.38 °C (24.1 °F). It may be handled as a liquid at moderate pressure. Ambient temperatures are generally used for long-term storage due to the easy formation of butadiene dimer (4-vinyl cyclohexene-l). Butadiene is produced by two major methods: the catalytic dehydrogenation of butane or butylenes or both and as a by-product from the production of ethylene. In either case, the butadiene must be isolated from other components by extractive distillation techniques and subsequent purification to polymerization-grade specifications by fractional distillation. Butadiene measurement tables (ASTM D1550, 2017) provide the data necessary for proper calculation of commercial quantities of butadiene. The largest end use of butadiene is as a monomer for production of GR-S synthetic rubber. Butadiene is also chlorinated to form 2-chloro butadiene (chloroprene) ($CH_2{=}CHCCl{=}CH_2$) that is a feedstock used to produce neoprene (a polychloroprene rubber).

The major quality criteria for butadiene are the various impurities that may affect the polymerization reactions for which butadiene is used. The gas chromatographic examination of butadiene (ASTM D2593, 2017) can be employed to determine the gross purity as well as C_3, C_4, and C, impurities. Most of these hydrocarbons are innocuous to polymerization reactions, but, some, such as butadiene −1,2 and pentadiene −1,4, are capable of polymer cross-linking.

The above sections have focused on natural gas and refinery gases, which are primarily produced in crude oil refineries as the low-boiling fractions of distillation and cracking processes or in gas plants that separate natural gas and natural gas liquids. Substances in both categories have high vapor pressure and moderate to high water solubility. The gas mixtures are composed primarily of paraffinic and olefinic hydrocarbons, mostly containing one to six carbon atoms (C_1 to C_6 or, in some cases, to C_8). Some of the mixtures may contain varying amounts of other components, including hydrogen, nitrogen, and carbon dioxide. The refinery gas streams also contain olefin constituents that are produced by various cracking processes and some streams also contain varying amounts of other chemicals including ammonia, hydrogen, nitrogen, hydrogen sulfide, mercaptans, carbon monoxide, carbon dioxide, 1,3-butadiene, and/or benzene. There are however, two other gas streams that must be given consideration here because of their increasing importance as fuel gases, and these are (i) biomethane, which falls into the category of biogas, and (ii) landfill gas, which may also be include in the biogas category but for the purpose of this text is included as a separate gas stream.

7.7 Gas Hydrates

Methane hydrates are crystalline solids in which molecules of a guest species occupy and stabilize cages formed by water molecules (Chapters 1 and 7) (Lorenson and Collett, 2000). Methane hydrate forms when methane and water combine at pressure and temperature conditions that are common in the marine sediments of continental margins of the Earth and below about 650 ft depth in permafrost areas. Chemically, gas hydrates are nonstoichiometric compounds formed by a lattice of hydrogen-bonded molecules (host) that engage low molecular weight gases or volatile liquids (guest) with specific properties that differentiate them from ice (Table 7.11) (Bishnoi and Clarke, 2006). No actual chemical bond exists between guest and host molecules. Hydrate formation is favored by low temperature and high pressure (Makogon, 1997; Sloan, 1998a; Carrol, 2003; Seo et al., 2009). The actual process of hydrate formation, which has an effect on the properties of the hydrate, is a heterogeneous process having similarities with crystallization processes (Sloan, 1998b; Collett et al., 2009). Methane clathrates (hydrates) are also commonly formed during natural gas production operations, when liquid water is condensed in the presence of methane at high pressure. Higher molecular weight hydrocarbons such as ethane and propane can also form hydrates, although longer molecules (butane hydrocarbons and pentane hydrocarbons) cannot fit into the water cage structure and, therefore, tend to destabilize the formation of hydrates (Belosludov et al., 2007).

Methane hydrate is a crystalline water-based solid that physically resemble ice in which low molecular weight nonpolar molecules (typically methane, ethane, and the like) or polar molecules with large hydrophobic moieties are trapped inside cages of hydrogen-bonded, frozen water molecules Scientifically, gas hydrates are clathrate compounds in which the host molecule is water and the guest molecule is typically a gas or liquid. Without the support of the trapped molecules, the lattice structure of hydrate clathrates would collapse into conventional ice crystal structure or liquid water. The sequestered molecules are never bonded to the lattice. The formation and decomposition of clathrate hydrates are first-order phase transitions, not

Table 7.11 Miscellaneous properties of ice and methane hydrate.

Property	Ice	Methane hydrate
Dielectric constant at 273 °K	94	58
Isothermal Young's modulus at 268 °K)	9.5	8.4
Poisson's ratio	0.33	0.33
Bulk modulus (272 °K)	8.8	5.6
Shear modulus (272 °K)	3.9	2.4
Bulk density ($g\,cm^{-3}$)	0.916	0.912
Thermal conductivity at 263 °K (W = m-K)	2.23	0.49

chemical reactions. Hydrocarbon clathrates cause problems for the crude oil and natural gas industries because they can form inside pipelines, often resulting in obstructions.

Methane hydrates are similar to ice in appearance and are stable at high pressures and temperatures above freezing (0 °C, 32 °F). Methane is the most common naturally occurring hydrate guest species and methane hydrates are also commonly referred to as gas hydrates. These hydrates are common constituents of the shallow marine geosphere, and they occur in deep sedimentary structures and form outcrops on the ocean floor. These hydrates are believed to form by migration of gas from deep along geological faults, followed by precipitation or crystallization, on contact of the rising gas stream with cold sea water. Thus, when water and methane are mixed at low temperature and elevated pressure, a soft solid is formed in which methane molecules are trapped in cages of water molecules. Methane hydrate is often called clathrate because of the cage effect of the water around the methane (Stern et al., 1996). When allowed to decompose, methane hydrate breaks down and releases up to 160 times its own volume in methane gas (Kim et al., 1987). The methane stored in gas hydrates is of interest and concern to policy makers as a potential alternative energy resource but has a potent greenhouse gas that could be released from sediments to the.

Most methane hydrate deposits also contain small amounts of other hydrocarbon hydrates; these include ethane hydrate and propane hydrate. In fact, gas hydrates of current interest are composed of water and the following molecules: methane, ethane, propane, isobutane, normal butane, nitrogen, carbon dioxide, and hydrogen sulfide. However, other nonpolar components between the sizes of argon (0.35 nm) and ethyl cyclohexane (0.9 nm) can also form hydrates. The hydrates form at temperatures close to 273°K and elevated pressures (Sloan, 1998a). However, for this text, the emphasis is focused on methane hydrates.

The gas content of gas hydrates typically involves the controlled decomposition of the hydrate in a *dissociation chamber* and allowed to decompose at room temperature. The gas pressure of the chamber will increase and reach a stable value over a period of time after which the hydrate gas is ready for analysis. The volume of water that remains in the dissociation chamber is then measured and stored in a glass ampoule for determination of the chemical composition and to ensure that no gases remain dissolved. Gas analysis will show that the gas released from the hydrate is predominantly methane with trace to small (typically <5% v/v) of ethane, propane, isobutane, normal butane, nitrogen, carbon dioxide, and hydrogen sulfide (Gabitto and Tsouris, 2010). Knowledge of gas hydrate morphology and bulk properties will provide the necessary data to determine the mechanism of dissociation (Llamedo et al., 2004).

7.7.1 Structure and Morphology

Methane hydrate ($CH_4 \cdot 5.75H_2O$ or $4CH_4 \cdot 23H_2O$), also called methane clathrate, methane hydrate, hydromethane, methane ice, fire ice, natural gas hydrate, or gas hydrate, is a solid clathrate compound in which a large amount of methane is trapped within a crystal structure of water, forming a solid similar to ice. Deposits of methane clathrate have been found under sediments on the ocean floors of the Earth.

The physical properties of bulk gas hydrates are remarkably close to those of pure ice. As a result, the properties of sediments containing hydrate in the pore space are similar to sediments containing nonhydrate (normal) ice. The morphology of the gas hydrates has large effects on sedimentary physical properties, from seismic velocities on a large scale to borehole electrical resistivity on a smaller scale, and, therefore, gas hydrate morphology impacts the amount of gas hydrate saturation estimated from geophysical data.

The stability of the hydrates is sensitive to temperature changes (Stoll and Bryan, 1979; Stern et al., 2000; Collett, 2001; Belosludov et al., 2007; Collett, 2010). To understand methane release from gas hydrate, the US Geological Survey (USGS) conducted a laboratory investigation of pure methane hydrate thermal properties at conditions relevant to accumulations of naturally occurring methane hydrate. Prior to this work, thermal properties for gas hydrates generally were measured on analog systems such as ice and nonmethane hydrates or at temperatures below freezing; these conditions limit direct comparisons to methane hydrates in marine and permafrost sediment.

More generally, at the molecular level, methane hydrate consists of gas molecules surrounded by cages of water molecules (Bishnoi and Clarke, 2006). Each water cage encloses a space of a particular size, and only a gas molecule small enough to fit within this site can be hosted in that specific hydrate structure. The nominal methane clathrate hydrate composition is $(CH_4)_4(H_2O)_{23}$ or 1 mol of methane for every 5.75 mol of water, corresponding to 13.4% w/w, but the actual composition is dependent on how many methane molecules fit into the various cage structures of the water lattice. Structurally, methane forms a structural hydrate with two dodecahedral (12 water molecules) and six tetradecahedral (14 water molecules) water cages per unit cell. Because of sharing of water molecules between cages, there are only 46 water molecules per unit cell.

The morphology of gas hydrates determines the basic physical properties of the sediment–hydrate matrix. Many remote techniques for gas hydrate detection and quantification are highly dependent on the hydrate morphology. Little attention has been paid to the hydrate morphology until now because previous methods of hydrate collection preserved only the grossest morphologies (e.g. lumps and nodules of hydrate (Holland et al., 2008; Gabitto and Tsouris, 2010; Holland and Schultheiss, 2014; Collett et al., 2015)).

The composition of the hydrate phase seems to be affected by the consumed amount of natural gas, which results in a variation of heating value of retrieved gas from mixed hydrates as a function of formation temperature (Seo et al., 2009). This indicates that the fractionation of the hydrate phase with higher molecular weight hydrocarbon molecules is enhanced in silica gel pores. In addition, when higher molecular weight hydrocarbon molecules are depleted in the vapor phase during the formation of mixed hydrate of different structures can coexist together, The morphology also has a significant effect on sedimentary physical properties, from seismic velocities on a large scale to borehole electrical resistivity on a smaller scale, and will thus impact the amount of gas hydrate saturation estimated from geophysical data (Gabitto and Tsouris, 2010; Du Frane et al., 2011, 2015).

The observed density is on the order of 0.9 g cm^{-3} (Table 7.12), which means that methane hydrate will float to the surface of the sea or of a lake unless it is bound in place by being formed in or anchored to sediment. One volume of fully saturated methane clathrate solid would therefore contain approximately 160 volumes of methane gas at 0 °C (32 °F) at atmosphere pressure or 1 cm^3 of methane clathrate releases approximately 160 cm^3 of gas.

Once formed, hydrates can block pipeline and processing equipment. They are generally then removed by reducing the pressure, heating them, or dissolving them by chemical means (methanol is commonly used). Care must be taken to ensure that the removal of the hydrates is carefully controlled, because of the potential for the hydrate to undergo a phase transition from the solid hydrate to release water and gaseous methane at a high rate when the pressure is reduced. The rapid release of methane gas in a closed system can result in a rapid increase in pressure.

It is preferable to prevent hydrates from forming or blocking equipment. In terms of inhibiting hydrate formation, the most common thermodynamic inhibitors are methanol, monoethylene glycol (MEG), and diethylene glycol (DEG). MEG is preferred over DEG for applications where the temperature is expected to be −10 °C or lower due to high viscosity at low temperatures.

Table 7.12 Density and molar volume of gas hydrates.

Gas	Formula	Molecular weight	Density (g cm^{-3})	Mole volume (cm^3 mol^{-1})
CH_4	$CH_4 \cdot 6H_2O$	124	0.910	136.264
CO_2	$CO_2 \cdot 6H_2O$	152	1.117	136.078
C_2H_6	$C_2H_6 \cdot 7H_2O$	308	0.959	162.669
C_3H_8	$C_3H_8 \cdot 17H_2O$	350	0.866	404.157
iC_4H_{10}	$iC_4H_{10} \cdot 17H_2O$	354	0.91	403.996

The vapor pressure of triethylene glycol (TEG) is too low to be suited as an inhibitor injected into a gas stream. In recent years, development of other forms of hydrate inhibitors have been developed, such as kinetic hydrate inhibitors (which slow the rate of hydrate formation) and anti-agglomerates, which do not prevent hydrates forming, but do prevent them sticking together to block equipment.

When drilling in crude oil-bearing and gas-bearing formations submerged in deep water, the reservoir gas may flow into the wellbore and form gas hydrates owing to the low temperatures and high pressures found during deep water drilling. The gas hydrates may then flow upward with drilling mud or other discharged fluids. When the hydrates rise, the pressure in the annulus decreases and the hydrates dissociate into gas and water. The rapid gas expansion ejects fluid from the well, reducing the pressure further, which leads to more hydrate dissociation and further fluid ejection. The resulting violent expulsion of fluid from the annulus is one potential cause or contributor to a blowout. This behavior is also of value to the analyst who has to handle these hydrates for analytical purposes.

The physical properties of hydrates are dependent upon the morphology of the hydrate. In terms of hydrate in sediment, which is a common means of detecting the presence of the hydrates, all of the physical parameters depend strongly on the hydrate saturation in pore space. This fact significantly reduces the impact of low-hydrate saturations on the measured physical parameters, an effect that is particularly pronounced at the hydrate saturations characteristic of many natural systems (Gabitto and Tsouris, 2010; Du Frane et al., 2011, 2015).

7.7.2 Properties

The properties of methane hydrate (and other gas hydrates) typically focus on the bulk properties of the hydrates, particularly on the dissociation of the hydrates into water and gas. In fact, the physical properties of gas hydrates trapped in sediments are very important in detecting the presence of these compounds, estimating the amount of gas hydrates trapped in the sediments, and developing processes to exploit this resource. Unfortunately, little is known about the physical properties of natural gas hydrate deposits in nature, making their detection by remote geophysical surveys difficult. The presence of gas hydrates in marine sediments dramatically alters some of the normal physical properties of the sediment, which can be detected by field measurements and by downhole logs (Collett, 2001, 2010).

7.7.2.1 Electrical Resistivity

Hydrates normally exclude the salt in the pore fluid from which it forms, and thus they have high electric resistivity just as ice and sediments containing hydrates have a higher resistivity compared with sediments without gas hydrates (Judge, 1982). The unconsolidated sediments in the upper several hundred meters of the marine sediment section (50% porosity) normally have a very low resistivity of about 1 Ω-m. If the saturation in the pore space is on the order of 15–20% v/v saturation (7–10% v/v of the sediment), the resistivity increases by about a factor of 2.

7.7.2.2 Electrical Conductivity

Gas hydrates have low electrical conductivity (σ – this is the inverse of resistivity: $\rho = \sigma^{-1}$) that provides a suitable target for marine controlled source electromagnetic (CSEM) surveys. CSEM sounding measures the amplitude and phase of electromagnetic (EM) energy propagating through the seafloor at one or more frequencies, and these data can be inverted to obtain the spatial distribution of conductivity. The combined use of both seismic and EM methods can help distinguish between gas (low velocity and high resistivity) and gas hydrate (high velocity and high resistivity) to map both the upper and lower boundaries of gas hydrate deposits.

CSEM studies have demonstrated the sensitivity of this method in assessing general gas hydrate concentration, saturation, and distribution patterns (Schwalenberg et al., 2005; Weitemeyer et al., 2006, 2011; Evans, 2007). Quantifying the estimates of hydrate volume, however, requires knowledge of the electrical conductivity of gas hydrates in combination with petrophysical mixing relations established from theory and experiment (Collett and Ladd, 2000; Ellis et al., 2008).

7.7.2.3 Heat of Dissociation

The heat of dissociation (ΔH_d) is defined as the enthalpy change to dissociate the hydrate phase to a vapor and aqueous liquid, with values given at temperatures just above the ice point. Also, the heat of dissociation is a function of the number of crystal hydrogen bonds (generally assumed to be the same as the hydration number) (Liu et al., 2010). However, the value of the heat of dissociation is relatively constant for molecules that occupy the same cavity, within a wide range of component sizes.

The enthalpy of dissociation may be determined via the univariant slopes of phase equilibrium lines (ln P versus $1/T$) using the Clausius–Clapeyron relationship (Sloan, 2006):

$$(\Delta H_d) = -\frac{zR_d(\ln P)}{d(1/T)}$$

7.7.2.4 Specific Heat

Hydrate breakdown is an endothermic process, absorbing heat while the surrounding sediment cools. Because the specific heat, c_p, of methane hydrates is approximately equal to one-half that of water, hydrate-bearing sediment stores less heat that can then be made available to help fuel dissociation (Waite et al., 2006, 2007). When estimating the efficiency of hydrate dissociation, neglecting the reduced contribution of methane hydrates to the host sediment's specific heat results in an overestimate of the dissociation rate and, hence, the methane production rate. There is a temperature and pressure dependence (Table 7.13) (USGS, 2007).

7.7.2.5 Thermal Conductivity

The thermal conductivity, λ, of water is approximately equal to that of methane hydrate, so the thermal conductivity beneath permafrost or in marine settings is essentially independent of methane hydrate content (Ross et al., 1978; Stoll and Bryan, 1979; Cook and Leaist, 1983; Ashworth et al., 1985; Tse and White, 1988). Methane hydrate is generally found in abundance below, rather than in, permafrost, but the presence of methane hydrates in ice-dominated permafrost can measurably increase the geothermal gradient because the thermal conductivity of ice is approximately four times that of methane hydrate.

In contrast with most well-defined crystalline structures, in which the thermal conductivity falls with increasing temperature following a $T-1$ dependence (for $T > 100°K$), the thermal conductivity of clathrate hydrates increases slightly with increasing temperature (Tse and White, 1988). The thermal conductivity of clathrate hydrates is five times lower than that of ice near the melting point and even lower (by a factor > 20) at lower temperatures. The temperature dependence of thermal conductivity in clathrate hydrates is characteristic of an amorphous material (Tse and White, 1988). Also, there is a marked temperature and pressure dependence (Table 7.13) (USGS, 2007).

In sediment-bearing hydrates, the thermal conductivity reflects the competing effects of the thermal conductivity of the phases involved, their volume fraction, and their spatial distribution. The electrical conductivity is controlled

Table 7.13 Temperature and pressure dependence of selected physical properties.

Temperature dependence (pressure held at 31.5 MPa, 4570 psi)	Temperature range
λ (W (m·K)$^{-1}$) = −(2.78 ± 0.05)·10^{-4} (°C) + (0.62 ± 0.02)	−20 to 17 °C
[a]κ(m^2 s^{-1}) = (5.04 ± 0.02)10^{-5}/T (K) + (1.25 ± 0.05) − 10^{-7}	1–17 °C (274–290 K)
c_p (J (kg·K)$^{-1}$) = (6.1 ± 0.3) − T(°C) + (2160 ± 100)	1–17 °C
Pressure dependence (temperature held at 14.4 °C, 58 °F)	
λ (W (m·K)$^{-1}$) = (2.54 ± 0.06)·10^{-4}·P (MPa) + (0.61 ± 0.02)	
κ (m^2 s^{-1}) = (2.87 ± 0.08)·10^{-10}·P (MPa) + (3.1 ± 0.2)·10^{-7}	
c_p (J (kg·K)$^{-1}$) = (3.30 ± 0.06)·P (MPa) + (2140 ± 100)	

[a] The T^{-1} dependence of the κ fit requires input temperatures in Kelvin.

by the availability and mobility of ions. A gradual reduction in conductivity is measured during hydrate formation even though ion exclusion keeps available ions within the unfrozen water. Mechanical properties are strongly influenced by both soil properties and the hydrate loci.

7.7.2.6 Thermal Diffusivity

Because the thermal diffusivity, κ, of water (Waite et al., 2006, 2007) is about half that of methane hydrate, hydrate-bearing sediment can change temperature more rapidly than water-bearing sediment. This characteristic presents a potential geohazard for crude oil production from deepwater production sites overlain by hydrate-bearing layers. If high-temperature hydrocarbons in the wellbore dissociate hydrate in the surrounding sediment, the strength of the sediment decreases, potentially causing well failure or localized submarine landslides. There is a temperature and pressure dependence (Table 7.13) (USGS, 2007).

7.8 Other Gases

7.8.1 Coal Gas

Coal gas is a flammable gaseous fuel produced made from coal and supplied to the user via a piped distribution system. Facilities where the gas was produced were often known as a manufactured gas plant (MGP) or a gasworks. The gas obtained when coal is heated strongly in the absence of air is called coal gas (the carbonization process) (Speight, 2013). Town gas is a more general term referring to manufactured gaseous fuels produced for sale to consumers and municipalities. Coal gas contains a variety of gases including hydrogen, carbon monoxide, methane, and volatile hydrocarbons together with small quantities of noncalorific gases such as carbon dioxide and nitrogen.

Prior to the development of natural gas supply and transmission during the 1940s and 1950s in the United States and during the late 1960s and 1970s in Great Britain and Australia, the majority of the gas used for fuel and lighting was manufactured from coal. Town gas was supplied to households via municipally owned piped distribution systems.

Originally created as a by-product of the coal carbonization process (coking process), the use of coal gas developed during the nineteenth century and early twentieth century. By-products from the production process included coal tar (often used as a precursor of the crude oil-based road asphalt) and ammonia, which were important chemical feedstock for the dye and chemical industry with a wide range of artificial dyes being made from coal gas and coal tar. The expansion of the natural gas industry in the decades after World War II led to the decline of the coal gas industry.

The composition of coal gas varied according to the type of coal and the temperature of the carbonization process. The constituents of coal gas and the approximate composition of the gas are on the order of hydrogen (50% v/v), methane (35% v/v), carbon monoxide (10% v/v), and ethylene (5% v/v).

7.8.2 Biogas

Biomethane falls under the general umbrella term of *biogas* (often predominantly methane, also called biogenic gas) is produced by certain types of bacteria (methanogens) during the process of breaking down organic matter in an oxygen-free environment. However, the name biogas gathers a large variety of gases under the name resulting from specific treatment processes, starting from various organic waste such as livestock manure, food waste, and sewage are all potential sources of biogenic gas, or biogas, which is usually considered a form of renewable energy (Table 7.14).

Small-scale biogas production is a well-established technology in parts of the developing world, particularly Asia, where farmers collect animal manure in vats and capture the methane given off while it decays. The gas is produced in a number of countries for power generation.

The process of biogas production (typically an anaerobic process) is a multistep biological process where the originally complex and big-sized organic

Table 7.14 Composition of various biogas samples.

Constituents	Household waste	Wastewater treatment plant sludge	Agricultural waste
Methane, % v/v	50–60	60–75	60–75
Carbon dioxide, % v/v	38–34	33–19	33–19
Nitrogen, % v/	5–0	1–0	1–0
Oxygen, % v/v	0–1	< 0.5	< 0.5
Water, % v/v	6	6	6
Hydrogen sulfide, $mg\,m^{-3}$	100–900	1000–4000	3000–10 000
Ammonia, $mg\,m^{-3}$	—	—	50–100
Aromatics, $mg\,m^{-3}$	0–200	—	—
Properties:			
Density[a]	0.93		0.85
Wobbe index[b]	6.9		8.1

[a] Natural gas: 0.57.
[b] Natural gas: 14.9.

solid wastes are progressively transformed in simpler and smaller-sized organic compounds by different bacteria strains up to have a final energetically worthwhile gaseous product and a semisolid material (digestate) that is rich in nutrients and thus suitable for its utilization in farming.

Methane produced from biomass or landfills is often referred to as biomethane to save confusion with the methane in natural gas. Biomethane is a naturally occurring gas that is produced by the so-called anaerobic digestion of organic matter such as dead animal and plant material, manure, sewage, and organic waste. Chemically, it is identical to natural gas that is stored deep in the ground and is also produced from dead animal and plant material. However, there are several important differences between biomethane and fossil fuel derived methane despite the fact that both are produced from decayed organic matter.

Natural gas is classified as fossil fuel (Speight, 2007, 2014), whereas biomethane is defined as a nonfossil fuel (Speight, 2011) and is further characterized or described as a green source of energy. Like the name suggests, fossil fuel methane produced ancient sediments that lie buried deep in the ground and that contained the remains of organic matter and are millions of years old. Production of fossil fuel methane, however, depends exclusively on its natural reserves that vary greatly from one country to another and are not available in limitless amounts. Biomethane, on the other hand, is produced from recent organic matter that makes it a renewable source of energy that can be produced worldwide and, in theory, in unlimited amounts.

Noteworthy at this point is that methane, whatever the source (fossil fuel or nonfossil fuel) is approximately 20 times more potent as a greenhouse gas than carbon dioxide when released into the atmosphere. Furthermore, the use of methane for power generation produces heat and emits carbon dioxide and some other gases, but despite that biomethane has a number of environmental benefits that make it a green source of energy. Organic matter from which biomethane is produced would release the carbon dioxide into the atmosphere if simply left to decompose naturally, while other gases that are produced during the decomposition process such as, nitrogen oxide(s) would make an additional contribution to the greenhouse effect.

The production of methane from biological sources does eliminate the release of a great deal of methane and other harmful gases into the atmosphere. This form of methane production eliminates exposure of the decomposing organic matter to the air that prevents methane and other gases from escaping into the atmosphere. In addition, methane production from biological sources reduces the need for fossil fuels by which it further reduces the emissions of greenhouse gases into the air. It is also claimed that the use of organic matter for methane production improves hygienic conditions and quality of life in rural areas and also reduces the risk of water pollution.

Biogas is characterized based on its chemical composition and the physical characteristics that result from it. It is primarily a mixture of methane (CH_4) and inert carbonic gas (CO_2).

Different sources of production lead to different specific compositions. The presence of hydrogen sulfide or carbon dioxide and water make biogas very corrosive and require the use of adapted materials. The composition of a gas issued from a digester depends on the substrate, of its organic matter load, and the feeding rate of the digester.

The anaerobic digestion process by which biogas is produced is utilized to treat organic solid wastes (Esposito et al., 2012). Because of the wide variation on biomass feedstocks and the varying yields of methane from these feedstocks, experimental tests represent the most powerful means to establish the anaerobic digestion as the environmentally and economically most convenient and helpful process to treat organic solids. Among all of the available experimental methods, the biomethane potential (BMP) tests are those that have been most successful, mainly thanks to the relatively easy setup and conduction as well as the useful information obtainable from such test methods (Esposito et al., 2012).

The test methods for the BMP of biomass offer the ability to assess the production of methane by the digestion of organic solids. At present, lack of a standard protocol to carry out these experiments is an important problem for the interpretation of the data, which can result in a serious drawback of the application of such tests. Therefore, efforts are needed to achieve a unique recognized and accredited method to conduct the tests for the BMP. The test methods can be used to study effect that the combination of different organic wastes has on the digestion process (codigestion). Indeed, the concurrent presence in the same anaerobic reactor of different organic wastes can improve the performance of the digestion process (ASTM D5511, 2017; ASTM D5526, 2017; ASTM D6273, 2017; ASTM D7475, 2017) (Esposito et al., 2012).

One test method that is not often discussed in the analysis of natural gas – specifically biogas – is the BMP of biomass. With the move to the use of more biomass feedstocks in various conversion process, this test method is worthy of mention here. The BMP tests are conducted in batch conditions and in bench scale, measuring the maximum amount of biogas or biomethane produced per gram of volatile solids contained in the organics used as substrates in the anaerobic digestion process (Hansen et al., 2004). Furthermore, relevant elements coming from the conduction of such tests are mainly the environmental and operational condition that could lead the process to failure, the time needed to have a complete substrate degradation, the average rate of biomethanation for each substrate investigated, and the evaluation of the digestion kinetics by coupling the test data for the BMP results to a mathematical model simulating the anaerobic digestion.

7.8.3 Landfill Gas

Landfill gas, which is often included under the umbrella definition of biogas, is also produced from the decay of organic wastes (such as municipal solid waste that contains organic materials), but these wastes may not be biomass-type materials (Lohila et al., 2007; Speight, 2011). The gas is a complex mix of different gases created by the action of microorganisms within a landfill. Landfill gas is approximately 40–60% v/v methane, with the remainder being predominantly carbon dioxide. Trace amounts of other volatile organic compounds comprise the remainder (typically, 1–2% v/v or less), and these trace gases include a large array of species, such as low molecular weight hydrocarbon derivatives. Other minor components include hydrogen sulfide, nitrogen oxides, sulfur dioxide, *nonmethane volatile organic compounds*, *polycyclic aromatic hydrocarbon derivatives*, *polychlorinated dibenzodioxin derivatives*, and *polychlorinated dibenzofuran derivatives* (Brosseau, 1994; Rasi et al., 2007). All of the aforementioned agents are harmful to human health at high doses.

The gases produced within a landfill site can be collected for various uses, such as direct utilization on-site in a boiler or any type of combustion system, providing heat. Electricity can also be generated on-site through the use of microturbines, steam turbines, or fuel cells (Sullivan, 2010). The landfill gas can also be sold off-site and sent into natural gas pipelines. This approach requires the gas to be processed into pipeline quality, e.g. by removing various contaminants and components.

As should be expected, the amount of methane that is produced varies significantly based on composition of the waste (Staley and Barlaz, 2009). The efficiency of gas collection at landfills directly impacts the amount of energy that can be recovered – closed landfills (those no longer accepting waste) collect gas more efficiently than open landfills (those that are still accepting waste).

Landfill gas collection is typically accomplished through the installation of wells installed vertically and/or horizontally in the waste mass. Design heuristics for vertical wells call for about one well per acre of landfill surface, whereas horizontal wells are normally spaced about 50–200 ft apart on center. Efficient gas collection can be accomplished at both open and closed landfills, but closed landfills have systems that are more efficient, owing to greater deployment of collection infrastructure since active filling is not occurring. On average, closed landfills have gas collection systems that capture approximately 84% v/v of produced gas, compared with approximately 67% v/v for open landfills.

Landfill gas can also be extracted through horizontal trenches instead of vertical wells. Both systems are effective at collecting. Landfill gas is extracted and piped to a main collection header, where it is sent to be treated or flared. The main collection header can be connected to the leachate collection system to collect condensate forming in the pipes. A blower is needed to pull the gas from the collection wells to the collection header and further downstream.

Landfill gas must be treated to remove impurities, condensate, and particulates. Hence there is a need for analysis to determine the composition of the gas. However, the treatment system depends on the end use: (i) minimal treatment is needed for the direct use of gas in boiler, furnaces, or kilns and (ii) using the gas in electricity generation typically requires more in-depth treatment. Thus, treatment systems are divided into primary and secondary treatment processing. Primary processing systems remove moisture and particulates. Gas cooling and compression are common in primary processing. Secondary treatment systems employ multiple cleanup processes, physical and chemical, depending on the specifications of the end use. Two constituents that may need to be removed are siloxane derivatives and sulfur-containing compounds, which are damaging to equipment and significantly increase maintenance cost. Adsorption and absorption are the most common technologies used in secondary treatment processing.

Landfill gas can be converted to high-Btu gas by reducing the amount of carbon dioxide, nitrogen, and oxygen in the gas. The high-Btu gas can be piped into existing natural gas pipelines or in the form of CNG or liquid natural gas. CNG and liquid natural gas can be used on-site to power hauling trucks or equipment or sold commercially. Three commonly used methods to extract the carbon dioxide from the gas are membrane separation, molecular sieve, and amine scrubbing (Chapter 4). Oxygen and nitrogen are controlled by the design and operation of the landfill since the primary cause for oxygen or nitrogen in the gas is intrusion from outside into the landfill because of a difference in pressure.

7.8.4 Flue Gas

Flue gas is the gas that exits to the atmosphere via some flues, in industry the flue is often referred to as a *smoke stack*, which is a pipe or channel for conveying exhaust gases from a furnace, boiler, or steam generator. In the current context, the flue gas typically refers to the combustion exhaust gas produced at a power plant.

The composition of flue gas depends upon the composition and properties of the fuel being combusted, but, nevertheless, the composition will usually consist predominantly of nitrogen (usually on the order of 60% v/v and higher) derived from the combustion process using air as the oxidant, carbon dioxide, and water vapor, as well as any excess oxygen (also derived from the air). Flue gas can also contain a small percentage of a number of pollutants, such as particulate matter (such as soot), carbon monoxide, oxides of nitrogen (NO_x), and oxides of sulfur (SO_x). The potential for the presence of hydrocarbon derivative is low unless the combustion process used a minimal of air and then also uses a hydrocarbonaceous feedstock.

Flue gases typically, in the context of this book, are produced by the combustion of fossil fuels, such as coal and crude oil that can contain a significant amount of sulfur. When fossil fuels are burned, approximately 90% + w/w of the sulfur in the feedstock is converted to sulfur dioxide, which occurs under normal conditions of temperature in the furnace and the oxygen fed to the combustor. However, when there is an excess of oxygen present, the sulfur dioxide is oxidized to sulfur trioxide (SO_3). In addition, at approximately 800 °C (1470 °F), the formation of sulfur trioxide is favored.

At power plants, flue gas (Chapter 4) is often treated with a series of chemical processes and scrubbers, which remove pollutants such as sulfur dioxide and sulfur trioxide (Speight, 2013). Flue gas desulfurization units capture the sulfur dioxide (and the sulfur trioxide, if present), and electrostatic precipitators or fabric filters remove particulate matter produced by burning fossil fuels, particularly coal. Nitrogen oxides are treated either by modifications to the combustion process to prevent their formation or by high temperature or catalytic reaction with ammonia (NH_3) or urea (H_2NCONH_2). In either case, the aim is to produce nitrogen gas rather than nitrogen oxides. In the United States, there is a rapid deployment of technologies to remove mercury from flue gas. This is typically accomplished by absorption on sorbents or by capture in inert solids as part of the glue gas desulfurization process. Such scrubbing can lead to meaningful recovery of sulfur for further industrial use (Mokhatab et al., 2006; Speight, 2007, 2014).

Technologies based on regenerative capture by amines (usually called *olamine processes*) for the removal of carbon dioxide from flue gas have been deployed to provide high-purity carbon dioxide for further industrial use as well as for enhanced oil recovery (Speight, 2014).

References

Arutyunov, V.S. (2007). Partial oxidation of hydrocarbon gases as a base for new technological processes in gas and power production. *Studies in Surface Science and Catalysis* 167: 269–274.

Ashworth, T., Johnson, L.R., and Lai, L.P. (1985). Thermal conductivity of pure ice and tetrahydrofuran clathrate hydrates. *High Temperatures-High Pressures* 17 (4): 413–419.

ASTM D1072 (2017). *Standard Test Method for Total Sulfur in Fuel Gases by Combustion and Barium Chloride Titration*, Annual Book of Standards. West Conshohocken, Pennsylvania: ASTM International.

ASTM D1142 (2017). *Standard Test Method for Water Vapor Content of Gaseous Fuels by Measurement of Dew-Point Temperature*, Annual Book of Standards. West Conshohocken, Pennsylvania: ASTM International.

ASTM D1550 (2017). *Standard ASTM Butadiene Measurement Tables*, Annual Book of Standards. West Conshohocken, Pennsylvania: ASTM International.
ASTM D1826 (2017). *Standard Test Method for Calorific (Heating) Value of Gases in Natural Gas Range by Continuous Recording Calorimeter*, Annual Book of Standards. West Conshohocken, Pennsylvania: ASTM International.
ASTM D1838 (2017). *Standard Test Method for Copper Strip Corrosion by Liquefied Petroleum (LP) Gases*, Annual Book of Standards. West Conshohocken, Pennsylvania: ASTM International.
ASTM D1945 (2017). *Standard Test Method for Analysis of Natural Gas by Gas Chromatography*, Annual Book of Standards. West Conshohocken, Pennsylvania: ASTM International.
ASTM D2158 (2017). *Standard Test Method for Residues in Liquefied Petroleum (LP) Gases*, Annual Book of Standards. West Conshohocken, Pennsylvania: ASTM International.
ASTM D2163 (2017). *Standard Test Method for Determination of Hydrocarbons in Liquefied Petroleum (LP) Gases and Propane/Propene Mixtures by Gas Chromatography*, Annual Book of Standards. West Conshohocken, Pennsylvania: ASTM International.
ASTM D2427 (2017). *Standard Test Method for Determination of C2 through C5Hydrocarbons in Gasolines by Gas Chromatography*, Annual Book of Standards. West Conshohocken, Pennsylvania: ASTM International.
ASTM D2504 (2017). *Standard Test Method for Noncondensable Gases in C2 and Lighter Hydrocarbon Products by Gas Chromatography*, Annual Book of Standards. West Conshohocken, Pennsylvania: ASTM International.
ASTM D2593 (2017). *Standard Test Method for Butadiene Purity and Hydrocarbon Impurities by Gas Chromatography*, Annual Book of Standards. West Conshohocken, Pennsylvania: ASTM International.
ASTM D2597 (2017). *Test Method for Analysis of Demethanized Hydrocarbon Liquid Mixtures Containing Nitrogen and Carbon Dioxide by Gas Chromatography*, Annual Book of Standards. West Conshohocken, Pennsylvania: ASTM International.
ASTM D2712 (2017). *Standard Test Method for Determination of Hydrocarbon Impurities in High Purity Propylene by Gas Chromatograph*, Annual Book of Standards. West Conshohocken, Pennsylvania: ASTM International.
ASTM D2713 (2017). *Standard Test Method for Dryness of Propane (Valve Freeze Method)*, Annual Book of Standards. West Conshohocken, Pennsylvania: ASTM International.
ASTM D3246 (2017). *Standard Test Method for Sulfur in Petroleum Gas by Oxidative Microcoulometry*, Annual Book of Standards. West Conshohocken, Pennsylvania: ASTM International.
ASTM D5511 (2017). *Standard Test Method for Determining Anaerobic Biodegradation of Plastic Materials Under High-Solids Anaerobic Digestion*

Conditions, Annual Book of Standards. West Conshohocken, Pennsylvania: ASTM International.

ASTM D5526 (2017). *Standard Test Method for Determining Anaerobic Biodegradation of Plastic Materials Under Accelerated Landfill Conditions*, Annual Book of Standards. West Conshohocken, Pennsylvania: ASTM International.

ASTM D6273 (2017). *Standard Test Method for Natural Gas Odor Intensity*, Annual Book of Standards. West Conshohocken, Pennsylvania: ASTM International.

ASTM D7475 (2017). *Standard Test Method for Determining Anaerobic Degradation and Anaerobic Biodegradation of Plastic Materials Under Accelerated Bioreactor Landfill Conditions*, Annual Book of Standards. West Conshohocken, Pennsylvania: ASTM International.

ASTM E253 (2017). *Terminology Relating to Sensory Evaluation of Materials and Products*, Annual Book of Standards. West Conshohocken, Pennsylvania: ASTM International.

Belosludov, V.R., Subbotin, O.S., Krupskii, D.S. et al. (2007). Physical and chemical properties of gas hydrates: theoretical aspects of energy storage application. *Materials Transactions* 48 (4): 704–710.

Bishnoi, P.R. and Clarke, M.A. (2006). Natural gas hydrates. In: *Encyclopedia of Chemical Processing*. Philadelphia, Pennsylvania: Taylor & Francis Publishers.

Brosseau, J. (1994). Trace gas compound emissions from municipal landfill sanitary sites. *Atmospheric-Environment* 28 (2): 285–293.

Carrol, J.J. (2003). *Natural Gas Hydrates*. Burlington, Vermont: Gulf Professional Publishing.

Collett, T.S. (2001). Natural-gas hydrates; resource of the twenty-first century? *Journal of the American Association of Petroleum Geologists* 74: 85–108.

Collett, T.S. (2010). Physical properties of gas hydrates: a review. *Journal of Thermodynamics* 2010: 12 pp., 271291.

Collett, T.S. and Ladd, J.W. (2000). Detection of gas hydrate with downhole logs and assessment of gas hydrate concentrations (saturations) and gas volumes on the Blake Ridge with electrical resistivity log data. *Proceedings Ocean Drilling Program Scientific Results* 164: 179–191.

Collett, T.S., Johnson, A.H., Knapp, C.C., and Boswell, R. (2009). Natural gas hydrates: a review. In: *Natural Gas Hydrates – Energy Resource Potential and Associated Geologic Hazards* (ed. T.S. Collett, A.H. Johnson, C.C. Knapp and R. Boswell) AAPG Memoir No. 89, 146–219. Tulsa, Oklahoma: American Association of Petroleum Geologists.

Collett, T.S., Bahk, J.J., Baker, R. et al. (2015). Methane hydrates in nature – current knowledge and challenges. *Journal of Chemical & Engineering Data* 60 (2): 319–329.

Cook, J.G. and Leaist, D.G. (1983). An exploratory study of the thermal conductivity of methane hydrate. *Geophysical Research Letters* 10 (5): 397–399.

Du Frane, W.L., Stern, L.A., Weitemeyer, K.A. et al. (2011). Electrical properties of polycrystalline methane hydrate. *Geophysics Research Letters* 38: L09313. doi: 10.1029/2011GL047243.

Du Frane, W.L., Stern, L.A., Constable, S. et al. (2015). Electrical properties of methane hydrate + sediment mixtures. *Journal of Geophysical Research: Solid Earth* 120: 4773–4783.

Ellis, M. H., Minshull, T.A., Sinha, M.C., and Best, A.I. (2008). Joint Seismic/electrical effective medium modelling of hydrate-bearing marine sediments and an application to the Vancouver Island Margin. *Proceedings. 6th International Conference on Gas Hydrates, 5586,* Vancouver, Canada.

Esposito, G., Frunzo, L., Liotta, F. et al. (2012). Bio-methane potential tests to measure the biogas production from the digestion and co-digestion of complex organic substrates. *The Open Environmental Engineering Journal* 5: 1–8.

Evans, R.L. (2007). Using CSEM techniques to map the shallow section of the seafloor: from the coastline to the edges of the continental slope. *Geophysics* 72 (2): WA105–WA116.

Gabitto, J. and Tsouris, C. (2010). Physical properties of gas hydrates: a review. *Journal of Thermodynamics* 2010: 271291.

Gary, J.G., Handwerk, G.E., and Kaiser, M.J. (2007). *Petroleum refining: technology and economics,* 5e. Boca Raton, Florida: CRC Press, Taylor & Francis Group.

Hansen, T.L., Schmidt, J.E., Angelidaki, I. et al. (2004). Measurement of methane potentials of solid organic waste. *Waste Management* 24 (4): 393–400.

Holland, M. and Schultheiss, P. (2014). Comparison of methane mass balance and x-ray computed tomographic methods for calculation of gas hydrate content of pressure cores. *Marine and Petroleum Geology* 58 (A): 168–177.

Holland, M., Schultheiss, P., Roberts, J., and Druce, M. (2008). Observed gas hydrate morphologies in marine sediments. *Proceedings. 6th International Conference on Gas Hydrates (ICGH 08).* Vancouver, British Columbia, Canada (July).

Hsu, C.S. and Robinson, P.R. ed. (2017). *Handbook of Petroleum Technology.* Cham: Springer International Publishing AG.

Judge, A. (1982). Natural gas hydrate in Canada. *Proceedings. 4th Canadian Permafrost Conference* (320–328).

Kim, H.C., Bishnoi, P.R., Heidemann, R.A., and Rizvi, S.S.H. (1987). Kinetics of methane hydrate decomposition. *Chemical Engineering Science* 42 (7): 1645–1653.

Liu, C.L., Ye, Y.G., and Meng, Q.C. (2010). Determination of hydration number of methane hydrates using micro-laser raman spectroscopy. *Guang Pu Xue Yu Guang Pu Fen Xi* 30 (4): 963–966. Article in Chinese.

Llamedo, M., Anderson, R., and Tohidi, B. (2004). Thermodynamic prediction of clathrate hydrate dissociation conditions in mesoporous media. *American Mineralogist* 89 (8-9): 1264–1270.

Lohila, A., Laurila, T., Tuovinen, J.-P. et al. (2007). Micrometeorological measurements of methane and carbon dioxide fluxes at a municipal landfill. *Environmental Science & Technology* 41 (8): 2717–2722.

Lorenson, T.D. and Collett, T.S. (2000). Gas content and composition of gas hydrate from sediments of the southeastern north american continental margin. In: *Proceedings of the Ocean Drilling Program, Scientific Results*, vol. 164 (ed. C.K. Paull, R. Matsumoto, P.J. Wallace and W.P. Dillon), 37–46.

Makogon, Y.F. (1997). *Hydrates of Hydrocarbons*. Tulsa, Oklahoma: PennWell Books.

Parkash, S. (2003). *Refining Processes Handbook*. Amsterdam: Gulf Professional Publishing, Elsevier.

Pohland, F.G. (1996). Landfill bioreactors: fundamentals and practice. *Water Quality International* 9 (10): 18–22.

Rasi, S., Veijanen, A., and Rintala, J. (2007). Trace compounds of biogas from different biogas production plants. *Energy* 32: 1375–1380.

Ross, R.G., Anderson, P., and Backstrom, G. (1978). Effects of H and D order on the thermal conductivity of tce phases. *The Journal Chemistry Physics* 68 (9): 3967–3972.

Schwalenberg, K., Willoughby, E., Mir, R., and Edwards, R.N. (2005). Marine gas hydrate electromagnetic signatures in cascadia and their correlation with seismic blank zones. *First Break* 23: 57–63.

Seo, Y., Kang, S.P., and Jang, W. (2009). Structure and composition analysis of natural gas hydrates: 13C NMR spectroscopic and gas uptake measurements of mixed gas hydrates. *The Journal of Physical Chemistry* 113 (35): 9641–9649.

Sloan, E.D. Jr. (1998a). Gas hydrates: review of physical/chemical properties. *Energy & Fuels* 12 (2): 191–196.

Sloan, E.D. Jr. (1998b). *Clathrate Hydrates of Natural Gases*, 2e. New York: Marcel Dekker Inc.

Sloan, E.D. Jr. (2006). *Clathrate Hydrates of Natural Gases*, 3e. New York: Marcel Dekker Inc.

Speight, J.G. (2007). *Natural Gas: A Basic Handbook*. Houston, Texas: GPC Books, Gulf Publishing Company.

Speight, J.G. ed. (2011). *The Biofuels Handbook*. London, United Kingdom: Royal Society of Chemistry.

Speight, J.G. (2013). *The Chemistry and Technology of Coal*, 3e. Boca Raton, Florida: CRC Press, Taylor & Francis Group.

Speight, J.G. (2014). *The Chemistry and Technology of Petroleum*, 5e. Boca Raton, Florida: CRC Press, Taylor & Francis Group.

Speight, J.G. (2016). Hydrogen in refineries. In: *Hydrogen Science and Engineering: Materials, Processes, Systems, and Technology*, vol. 1 (ed. D. Stolten and B. Emonts), 3–18. Weinheim. Chapter: Wiley-VCH Verlag GmbH & Co.

Speight, J.G. (2017). *Handbook of Petroleum Refining*. Boca Raton, Florida: CRC Press, Taylor & Francis Group.

Staley, B. and Barlaz, M.A. (2009). Composition of municipal solid waste in the united states and implications for carbon sequestration and methane yield. *Journal of Environmental Engineering* 135 (10): 901–909.

Stern, L.A., Kirby, S.H., and Durham, W.B. (1996). Peculiarities of methane clathrate hydrate formation and solid-state deformation, including possible superheating of water ice. *Science* 273: 1843–1848.

Stern, L., Kirby, S., Durham, W. et al. (2000). Laboratory synthesis of pure methane hydrate suitable for measurement of physical properties and decomposition behavior. In: *Proceedings. Natural Gas Hydrate in Oceanic and Permafrost Environments* (ed. M.D. Max), 323–348. Dordrecht: Kluwer Academic Publishers.

Stoll, R.G. and Bryan, G.M. (1979). Physical properties of sediments containing gas hydrates. *Journal of Geophysical Research* 84 (B4): 1629–1634.

Sullivan, P. (2010). The Importance of Landfill Gas Capture and Utilization in the U.S. Earth Engineering Center. New York: Columbia University. (6 April). http://www.scsengineers.com/wp-content/uploads/2015/03/Sullivan_Importance_of_LFG_Capture_and_Utilization_in_the_US.pdf (accessed 11 November 2017).

Swain, F.M. (1986). Composition of march gases in the central and eastern United States. *Applied Geochemistry* 1: 301–305.

Tse, J.S. and White, M.A. (1988). Origin of glassy crystalline behavior in the thermal properties of clathrate hydrates: a thermal conductivity study of tetrahydrofuran hydrate. *The Journal of Physical Chemistry* 92 (17): 5006–5011.

USGS (2007). Thermal properties of methane gas hydrates. Fact Sheet 2007–3041. Reston, VA: United States Geological Survey. https://pubs.usgs.gov/fs/2007/3041/ (accessed 17 July 2017).

Waite, W.F., Gilbert, L.Y., Winters, W.J., and Mason, D.H. (2006). Estimating thermal diffusivity and specific heat from needle probe thermal conductivity data. *Review of Scientific Instruments* 77: 044904.

Waite, W.F., Stern, L.A., Kirby, S.H. et al. (2007). Simultaneous determination of thermal conductivity, thermal diffusivity and specific heat in methane hydrate. *Geophysical Journal International* 169: 767–774.

Weitemeyer, K.A., Constable, S.C., Key, K.W., and Behrens, J.P. (2006). First results from a marine controlled-source electromagnetic survey to detect gas hydrates offshore oregon. *Geophysics Research Letters* 33: L03304.

Weitemeyer, K.A., Constable, S.C., and Tréhu, A.M. (2011). A marine electromagnetic survey to detect gas hydrate at hydrate ridge, oregon. *Geophysical Journal International* doi: 10.1111/j.1365-246X.2011.05105.x.

8

Properties of Gas Streams

8.1 Introduction

In consideration of the variability of the composition of gas streams and the properties of the individual constituents, even hydrocarbon constituents only (Chapters 1 and 7), it is not surprising that the bulk properties, and hence the standard test methods to be applied to the streams to determine these properties, are often variable and inconsistent. Without knowledge of the methods by which these gases streams are produced, the sample taker/submitter and the analyst may not be able to determine the most appropriate test methods to be applied to these gas streams.

The gaseous mixtures considered in this volume are mixtures of various constituents that may or may not vary over narrow limits. Typically, these gases fall into the general category of fuel gases, and each gas is any one of several fuels that, under standard conditions of temperature and pressure, are gaseous. Fuel gases consist primarily of hydrocarbons with four or less carbon atoms (as well as lesser amounts of carbon monoxide and hydrogen), which have boiling points that are lower than room temperature, and these products are gases at ambient temperature and pressure. For example, *natural gas* (predominantly *methane*, CH_4) has the lowest boiling and is the least complex of all hydrocarbons (Chapter 1) (Mokhatab et al., 2006; Speight, 2007, 2014). Higher-boiling hydrocarbons up to n-octane (C_8H_{18}) may also be present in which case the gas is referred to as *wet gas*, which is usually processed (refined) to remove the high-boiling hydrocarbons (when isolated, the higher-boiling hydrocarbon mixture is often referred to as *natural gas condensate*). In fact, there are wells from which the predominant product is *condensate*.

Natural gas is the gaseous mixture associated with crude oil reservoirs and is predominantly methane (Chapter 1) (Burrus and Ryder, 2003; Mokhatab et al., 2006; Speight, 2007; Burruss and Ryder, 2014; Speight, 2014). However, although the generic term *natural gas* applies to gases commonly associated with petroliferous (crude oil-producing, crude oil-containing) geologic

Handbook of Natural Gas Analysis, First Edition. James G. Speight.

Table 8.1 General summary of refinery product types and distillation range.

Product	Lower carbon limit	Upper carbon limit	Lower boiling point (°C)	Upper boiling point (°C)	Lower boiling point (°F)	Upper boiling point (°F)
Refinery gas	C_1	C_4	−161	−1	−259	31
Liquefied petroleum gas	C_3	C_4	−42	−1	−44	31
Naphtha	C_5	C_{17}	36	302	97	575
Gasoline	C_4	C_{12}	−1	216	31	421
Kerosene/diesel fuel	C_8	C_{18}	126	258	302	575
Aviation turbine fuel	C_8	C_{16}	126	287	302	548
Fuel oil	C_{12}	$>C_{20}$	216	421	>343	>649
Lubricating oil	$>C_{20}$		>343		>649	
Wax	C_{17}	$>C_{20}$	302	>343	575	>649
Asphalt	$>C_{20}$		>343		>649	
Coke	$>C_{50}$[a]		>1000[a]		>1832[a]	

[a]Carbon number and boiling point difficult to assess; inserted for illustrative purposes only.

formations, it also used to designate the gas derived from coal seams (coal gas) as well as the mixture of gases produced during crude oil refining (refinery gas) (Table 8.1) and gas produced from biomass (biogas) (ASTM D4150, 2017).

Natural gas, like crude oil, is derived from the remains of plants and animals and microorganisms that lived millions and millions of years ago (Speight, 2007, 2014). Natural gas is used primarily as a fuel and as a raw material in manufacturing. It is used in home furnaces, water heaters, and cooking stoves. As an industrial fuel, it is used in brick, cement, and ceramic tile kilns, in glass making, for generating steam in water boilers, and as a clean heat source for sterilizing instruments and processing foods. As a raw material in petrochemical manufacturing, natural gas is used to produce hydrogen, sulfur, carbon black, and ammonia. The ammonia is used in a range of fertilizers and as a secondary feedstock for manufacturing other chemicals, including nitric acid and urea. Ethylene, an important petrochemical, is also produced from natural gas.

Natural gas is found in deep underground rock formations and coal seams (coalbed methane) and generally contains high proportions of methane (CH_4), and some of the higher molecular weight higher paraffins (C_nH_{2n+2}) generally containing up to six or eight carbon atoms may also be present in small

Table 8.2 Constituents of natural gas.

Name	Formula	Vol.%
Methane	CH_4	>85
Ethane	C_2H_6	3–8
Propane	C_3H_8	1–5
Butane	C_4H_{10}	1–2
Pentane$^+$	C_5H_{12}	1–5
Carbon dioxide	CO_2	1–2
Hydrogen sulfide	H_2S	1–2
Nitrogen	N_2	1–5
Helium	He	<0.5

Pentane$^+$: Pentane and higher molecular weight hydrocarbons up to octane as well as benzene and toluene.

quantities (Table 8.2). The hydrocarbon constituents of natural gas are combustible, but nonflammable nonhydrocarbon components such as carbon dioxide, nitrogen, and helium are often present in the minority and are regarded as contaminants that are removed during gas processing (gas cleaning, gas refining) in order to ensure that adverse effects arising from the presence of such contaminants do not occur (Table 8.3) (Mokhatab et al., 2006; Speight, 2007, 2014).

Natural gas is considered as an environmentally friendly clean fuel, offering important environmental benefits when compared with other fossil fuels. Its superior environmental qualities over coal or crude oil are its emissions of sulfur dioxide are negligible and the levels of nitrous oxide and carbon dioxide emissions are lower. This helps reduce problems due to acid rain, ozone layer, or greenhouse gases. Natural gas is also a very safe source of energy when transported, stored, and used.

Natural gas, which might be considered by some observers to be a very uninteresting gas, is colorless and odorless in its pure form. It is a combustible mixture of hydrocarbon gases, and because of major constituents – methane (ASTM D3956, 2017), ethane (ASTM D3984, 2017), propane (ASTM D4362, 2017), butane (ASTM D4650, 2017; ASTM D4651, 2017), and pentane – the composition and properties of natural gas vary widely. It is combustible and when burned produces energy. Natural gas, which typically has 85–95% methane, has been recognized as a plentiful and clean alternative feedstock to crude oil. Currently, the rate of discovery of proven natural gas reserves is increasing faster than the rate of natural gas production. Many of the large natural gas

Table 8.3 Possible constituents of natural gas and their effects.

Constituent	Comment
Hydrocarbons	Provide the calorific value of natural gas when it is combusted
	Hydrocarbon-rich liquid phase can form via condensation from the gas phase
	Liquid phases in the network can cause blockage
	Control of hydrocarbon dew point is the preferred method to control liquid formation
Diluents or inert gases	Typical inert gases are carbon dioxide, nitrogen, and helium
	Very low Wobbe index if a large amount of inert gas is present
	Affects combustion efficiency
Contaminants	May cause corrosion of the pipeline network and restrict the gas flow
Water	Can cause corrosion of the pipeline
	Can lead to formation of hydrates
	Hydrates can block valves
Oxygen	Can promote pipeline corrosion in the presence of water and sulfur
	In underground storage, promotes bacterial activity that can lead to formation of hydrogen sulfide
Hydrogen	Associated with stress corrosion cracking of steel pipelines
Hydrogen sulfide	Hydrogen sulfide reacts with copper piping to form copper sulfide flakes – can cause blockage
Solids or liquids	Can cause corrosion, stress, or abrasion damage
	Can cause blockage

deposits are in areas where abundant crude oil resources lie such as the Middle East. However, huge reserves of natural gas are also found in many other regions of the world, providing oil-deficient countries access to a plentiful energy source. The gas is frequently located in remote areas far from centers of consumption, and pipeline costs can account for as much as one-third of the total natural gas cost. Thus, tremendous strategic and economic incentives exist for gas conversion to liquids especially if this can be accomplished on-site or at a point close to the wellhead where shipping costs become a minor issue.

However, despite reduced prominence and fears about environmental effects, coal technology continues to be a viable option to produce gaseous fuels in the future (Speight, 2013). World crude oil production is expected ultimately to level off and then decline, and despite apparent surpluses of natural gas, production is expected to suffer a similar decline. Coal gasification to synthesis gas is utilized to synthesize liquid fuels in much the same manner as natural gas steam-reforming technology. But the important aspect is to use the natural gas reserves when they are available and to maximize the use of these reserves by conversion of natural gas to liquid fuels. Liquid fuels possess inherent advantageous characteristics in terms of being more readily stored, transported, and metered than natural gas. Liquid fuels also are generally easy to process or clean by chemical and catalytic means. Also, liquid fuels are more compatible with the twentieth-century world fuel infrastructure because most fuel-powered conveyances are designed to function only with relatively clean, low viscosity liquids. Therefore, production of synthetic fuels from alternative feedstocks is based on adjusting the hydrogen-to-carbon ratio to the desired intermediate level.

In addition to the gases obtained by distillation of crude oil, more highly volatile products result from the subsequent processing of naphtha and middle distillate to produce gasoline from desulfurization processes involving hydrogen treatment of naphtha, distillate, and residual fuel and from the coking or similar thermal treatments of vacuum gas oils and residual fuels. The most common processing step in the production of gasoline is the catalytic reforming of hydrocarbon fractions in the heptane (C_7) to decane (C_{10}) range.

Natural gas processing typically begins at the wellhead where the composition of the raw natural gas extracted from producing wells depends on the type, depth, and geology and location of the underground reservoir (Speight, 2014). Thus, gas-processing plants purify raw gas to remove common contaminates such as water, carbon dioxide, and hydrogen. However, some of these have an economic value and are further processed for sale or for use in a refinery petrochemical plant. On the other hand, the gases produced within the refinery are, in a sense, much more complex, and often the composition cannot be predicted. Such gases contain a wider variety of useful hydrocarbons as well as higher amounts of hydrogen sulfide (from the hydrodesulfurization units).

There are many variables in treating natural gas and especially the more complex refinery gas (Table 8.4). The precise area of application of a given process is difficult to define. Several factors must be considered: (i) the types and (ii) concentrations of contaminants in the gas; (iii) the degree of contaminant removal desired, (iv) the selectivity of acid gas removal required; (v) the temperature, (vi) pressure, (vii) volume, and (viii) composition of the gas to be processed; (ix) the carbon dioxide–hydrogen sulfide ratio in the gas; and (x) the desirability of sulfur recovery due to process economics or environmental issues.

Thus, in many cases, the process complexities arise because of the need for recovery of the materials used to remove the contaminants or even recovery of the contaminants in the original, or altered, form. In any case, whatever the source of the gas, careful analysis before processing is necessary (Mokhatab et al., 2006; Speight, 2007, 2014, 2015). Due to the current importance of natural gas and process gas (Mokhatab et al., 2006; Speight, 2007, 2014) and the

Table 8.4 Constituents of refinery gas.

Gas	Molecular weight	Boiling point 1 atm °C (°F)	Density at 60 °F (15.6 °C), 1 atm	
			$g\,l^{-1}$	Relative to air = 1
Methane	16.043	−161.5 (−258.7)	0.6786	0.5547
Ethylene	28.054	−103.7 (−154.7)	1.1949	0.9768
Ethane	30.068	−88.6 (−127.5)	1.2795	1.0460
Propylene	42.081	−47.7 (−53.9)	1.8052	1.4757
Propane	44.097	−42.1 (−43.8)	1.8917	1.5464
1,2-Butadiene	54.088	10.9 (51.6)	2.3451	1.9172
1,3-Butadiene	54.088	−4.4 (24.1)	2.3491	1.9203
1-Butene	56.108	−6.3 (20.7)	2.4442	1.9981
cis-2-Butene	56.108	3.7 (38.7)	2.4543	2.0063
trans-2-Butene	56.108	0.9 (33.6)	2.4543	2.0063
iso-Butene	56.104	−6.9 (19.6)	2.4442	1.9981
n-Butane	58.124	−0.5 (31.1)	2.5320	2.0698
iso-Butane	58.124	−11.7 (10.9)	2.5268	2.0656

increasing importance of biogas and landfill gas, it is necessary to consider the properties and analytical methods that are available for these various gases.

This chapter covers the essential technical points of natural gas as well as the gases from nonpetroliferous sources. With some exceptions, the analytical techniques applied to natural gas can also be applied to alternate fuel gases that are produced from nonpetroliferous gases.

8.2 Composition

A gas can be characterized with consideration of the pure substances that are contained in the gas and there needs to be statements for each substance, such as the proportion of the substance in the gas mixture on a molecule-count basis. To do this, analytical methods are necessary.

Most gases contain hydrocarbons, water, carbon dioxide, hydrogen sulfide, nitrogen, oxygen, and other impurities. The need for the application of analytical techniques to the determination of the properties of natural gas and other gases such as biogas and landfill gas has increased over the past four decades because of the increased use of natural gas. In addition, the need for the analysis of any gas is an essential part of its use and safety.

8.2.1 Petroleum Gases

The term *petroleum gas(es)* in this context is also used to describe the gaseous and liquid phase mixtures composed mainly of methane to butane (C_1–C_4 hydrocarbons) that are dissolved in the crude oil and natural gas, as well gases produced during thermal processes in which the crude oil is converted to other products. It is necessary, however, to acknowledge that in addition to the hydrocarbons, gases such as carbon dioxide, hydrogen sulfide, and ammonia are also produced during crude oil refining and will be constituents of refinery gas that have to be removed. Olefins are also present in the gas streams of various processes and are not included in liquefied petroleum gas, but are removed for use in petrochemical operations (Crawford et al., 1993).

Although covered by a generic term, *natural gas* covers a variety of compositions that are produced from a reservoir. All gas-fired equipment is designed to operate within a particular range, that is, from application to the specifications of the fuel gas. If gases outside this range are combusted, this can lead to a range of problems from poor quality combustion to equipment damage and ultimately dangerous operation. In addition, gas interchangeability relates to more than just a parameter for calorific/heating value. It is governed by gas quality, which is, in turn, a function of gas composition. In fact, gases with the same calorific/heating value are not necessarily interchangeable. This leads to the need for gas analysis.

Liquefied petroleum gas, natural gas, and refinery gas are mixtures of products or naturally occurring materials and fortunately are relative simple mixtures and do not suffer the complexities of the isomeric variations of the higher molecular weight hydrocarbons (Table 8.4) (Drews, 1998; Speight, 2014).

Natural gas is found in crude oil reservoirs as free gas (*associated gas*) or in solution with crude oil in the reservoir (*dissolved gas*) or in reservoirs that contain only gaseous constituents and no (or little) crude oil (*unassociated gas*) (Cranmore and Stanton, 2000; Speight, 2014). The hydrocarbon content varies from mixtures of methane and ethane with very few other constituents (*dry* gas) to mixtures containing all of the hydrocarbons from methane to pentane and even hexane (C_6H_{14}) and heptane (C_7H_{16}) (*wet* gas). In both cases some carbon dioxide (CO_2) and inert gases, including helium (He), are present together with hydrogen sulfide (H_2S) and a small quantity of organic sulfur.

Most gas streams vary greatly in composition (Table 4.1), and the constituents can be two or more of a group of hydrocarbon derivatives and nonhydrocarbon compounds (Table 4.2) (Chapter 1). The treatment required to prepare natural gas for distribution as an industrial or household fuel is specified in terms of the use and environmental regulations.

Briefly, natural gas contains hydrocarbons and nonhydrocarbon gases. Hydrocarbon gases are methane (CH_4), ethane (C_2H_6), propane (C_3H_8), and butanes (C_4H_{10}), with the pentane derivatives (C_5H_{12}), hexane derivatives

(C_6H_{14}), heptane derivatives (C_7H_{16}), and sometimes trace amounts of octane derivative (C_8H_{18}) and higher molecular weight hydrocarbons being more commonly found in gas condensate and natural gasoline. Some aromatics [BTX – benzene (C_6H_6), toluene ($C_6H_5CH_3$), and the xylenes ($CH_3C_6H_4CH_3$)] can also be present in gas condensate and natural gasoline, raising safety issues due to their toxicity. The nonhydrocarbon gas portion of the natural gas contains nitrogen (N_2), carbon dioxide (CO_2), helium (He), hydrogen sulfide (H_2S), water vapor (H_2O), and other sulfur compounds such as carbonyl sulfide (COS) and mercaptans (e.g. methyl mercaptan, CH_3SH) as well as trace amounts of other gases. Carbon dioxide and hydrogen sulfide are commonly referred to as *acid gases* since they form corrosive compounds in the presence of water. Nitrogen, helium, and carbon dioxide are also referred to as *diluents* since none of these burn, and thus they have no heating value. Mercury can also be present either as a metal in vapor phase or as an organometallic compound in liquid fractions. Concentration levels are generally very small, but even at very small concentration levels, mercury can be detrimental due its toxicity and its corrosive properties (reaction with aluminum alloys).

Gas analysis includes the detection of gases and vapors in connection with control of chemical and metallurgical processes and control of environment and in the field of safety control. A whole class of analytical instruments is available using physical or physicochemical detecting methods. Although analyzers are considered to be electrical instruments, some knowledge of chemistry is required to understand their operating principle and to obtain correct and accurate measuring results. The application of standard test methods to gas analysis has particular involvement in (i) combustion optimization or the optimization of combustion processes with the objectives to save energy, to protect the plant (service life extension), and to minimize emissions, (ii) process control or the control of a specified gas composition to monitor a manufacturing process for obtaining certain product specifications and qualities, and (iii) emission control or the control of gas-cleaning plants for correct operation and monitoring of the pollutant concentrations of flue gases for their compliance with the limit value specifications.

Thus, analysis of natural gas is carried out for many reasons: measurement of specific, commonly minor, components (e.g. odorants); identification of source, either by fingerprinting or molar composition, and application areas of increasing importance; calculation of physical properties; and measurement of gas quality against a specification. The amount of detail required depends upon the reason for the analysis, i.e. identification of source by molar composition might be done in some circumstances by an analysis up to pentane with a composite figure for the higher molecular weight hydrocarbons (i.e. C_{6+}). This may also be reasonable for the calculation of calorific value assuming a single bulk contribution from these higher molecular weight hydrocarbons. In the case of phase property calculations, however, e.g. the hydrocarbon dew point, much more detail is needed of the distribution of the higher hydrocarbons.

The objective of a natural gas analysis is to determine the completeness (when used as a fuel gas) of the combustion of the carbon in the fuel and the amount and distribution of the heat losses due to incomplete combustion. The quantities determined by an analysis are the relative proportions by volume of carbon dioxide (CO_2), oxygen (O), and carbon monoxide (CO), sequentially. Thus, before the sale of natural gas, there are several properties that must be determined so that the gas for sale meets the required specification of the buyer. Only those methods and instruments considered are suitable for field operation under the conditions of natural gas transmission and distribution, installed either in-line or online, and do not involve the determination of the gas composition. For example, the density under reference conditions (sometimes referred to as normal, standard, or even base density) is required for conversion of volume data and can be used for other physical properties. Density under operating conditions is measured for mass flow measurement and volume conversion using the observed line density and can be used for other physical properties.

The routine analysis of processed natural gas for calculation of heating value, specific gravity, and compressibility (compression factor, usually given the symbol Z) has now been established as an ordinary and necessary measurement practice by the gas industry worldwide. Automated gas chromatographs, which are now a shelf item from many manufacturers, provide the needed gas composition with acceptable precision (repeatability) and even good accuracy if calibration gas of known uncertainty is used. These routine analyses, however, provide information or only on major and minor components, i.e. nitrogen, carbon dioxide, paraffin derivatives from methane through hexane, and all other hydrocarbons having a higher molecular weight than hexane (C_{6+}), as if they were a single component. The lowest level of concentration afforded by this routine analysis is about 0.01 mol.%.

A natural gas stream traditionally has high proportions of natural gas liquids (NGLs) and is referred to as rich gas. NGLs are constituents such as ethane, propane, butane, and pentanes and higher molecular weight hydrocarbon constituents. The higher molecular weight constituents (i.e. the C_{5+} product) are commonly referred to as natural gasoline. Rich gas will have a high heating value and a high hydrocarbon dew point. When referring to NGLs in the gas stream, the term gallon per thousand cubic feet is used as a measure of high molecular weight hydrocarbon content. On the other hand, the composition of nonassociated gas (sometimes called well gas) is deficient in NGLs. The gas is produced from geological formations that typically do not contain much, if any, hydrocarbon liquids.

By way of explanation, the units that are in use to describe concentration in gases are (i) mass concentration, which is the concentration expressed in terms of mass of substance per unit mass, % w/w; (ii) volume concentration, which is the concentration expressed in terms of gaseous volume of substance per unit volume, % v/v; (iii) mass–volume concentration, which is the concentration expressed in terms of mass of substance per unit mass, % v/w; and (iv) part concentration, which is concentration expressed as number of particles of substance per total number of particles, or particulate matter, in the gas. In gas analysis, both the terms mass concentration and part concentration are common and used in parallel. Items (i) and (ii) are in common use, whereas items (ii) and (iv) are in much less common use. Typically, the mass unit is grams, and the most popular expression for part concentration is ppm (parts per million), which is the number of parts of, say, a contaminant in parts per million parts and is usually used for low concentrations; larger concentrations are expressed in percent by mass (% w/w) or by volume (% v/v).

Thus, because of the lower molecular weight constituents of these gases and their volatility, gas chromatography has been the technique of choice for fixed gas and hydrocarbon speciation, and mass spectrometry is also a method of choice for compositional analysis of low molecular weight hydrocarbons (ASTM D2421, 2017; ASTM D2650, 2017). More recently, piggyback methods (such as gas chromatography/mass spectrometry and other double-technique methods) have been developed for the identification of gaseous and low-boiling liquid constituents of mixtures. The hydrocarbon composition is limited to set to the total amount of ethane, butane, or pentane as well as ethylene and total dienes.

By limiting the amount of hydrocarbons that are lower boiling than the main component, the vapor pressure control is reinforced. Tests are available for vapor pressure 38 °C (100 °F) (ASTM D1267, 2017). The limitation on the amount of higher-boiling hydrocarbons supports the volatility clause. The vapor pressure and volatility specifications will often be met automatically if the hydrocarbon composition is correct. In addition, the amount of ethylene (and similar olefin gases) is limited because it is necessary to restrict the amount of unsaturated components to avoid the formation of deposits caused by the polymerization of the olefin(s). In addition, ethylene (boiling point: −104 °C, −155 °F) is more volatile than ethane (boiling point: −88 °C, −127 °F), and therefore a product with a substantial proportion of ethylene will have a higher vapor pressure and volatility than one that is predominantly ethane. Butadiene is also undesirable because it may also produce polymeric products that form deposits and cause blockage of lines.

Currently the preferred method for the analysis of liquefied petroleum gas, and indeed for most crude oil-related gases, is gas chromatography (ASTM D2163, 2017). This technique can be used for the identification and measurement of both main constituents and trace constituents. Precise compositional data is often needed for chemical feedstock applications and specification analysis. The component distribution data can be used to calculate certain physical properties such as relative density, vapor pressure, and octane number (ASTM D2598, 2017). Thus, this test method describes the approximate determination of the physical characteristics of commercial propane, special duty propane, commercial propane/butane mixtures, and commercial butane (ASTM D1835, 2017), such as (i) vapor pressure, (ii) relative density, and (iii) motor octane number, but this test method (ASTM D2598, 2017) is not applicable to any product exceeding specifications for nonvolatile residues (ASTM D2158, 2017) since there may be inaccuracies in the measurement that are the result of the presence of higher-boiling constituents due to relative volatility of these constituents under the conditions in which the sample is stored.

Capillary column gas chromatography is an even quicker and equally accurate alternative. Mass spectrometry is also suitable for analysis of crude oil gases. Of the other spectroscopic techniques, infrared and ultraviolet absorption may be applied to petroleum gas analysis for some specialized applications. Gas chromatography has also largely supplanted chemical absorption methods of analysis, but again these may have some limited specialized application.

Simple evaporation tests in conjunction with vapor pressure measurement give a further guide to composition. In these tests a liquefied petroleum gas sample can evaporate naturally from an open graduated vessel. Results are recorded based on volume/temperature changes, such as the temperature recorded when 95% has evaporated, or volume left at a specific temperature (ASTM D1071, 2017; ASTM D1837, 2017).

Gases such as ethane that are destined for use as petrochemical feedstocks must adhere to stringent composition controls that are dependent upon the process. For example, moisture content (ASTM D1142, 2017), oxygen content (ASTM D1945, 2017), carbon dioxide content (ASTM D1945, 2017), and sulfur content (ASTM D1072, 2017) must be monitored as they all interfere with catalyst performance in petrochemical processes. The hydrocarbon composition of natural gasoline (although not specifically a gas) for petrochemical use must undergo a compositional analysis (ASTM D2427, 2017) and a test for total sulfur (ASTM D1266, 2017).

Natural gasoline is a mixture of hydrocarbons (extracted from natural gas) that consists mostly of pentane derivatives and higher-boiling hydrocarbon. In the beginning of the natural gasoline industry, the only use for natural gasoline was as motor fuel or as a blending agent in the production of motor fuel (ASTM D4307, 2017). This test method outlines a procedure for the preparation of small volumes of multicomponent liquid blends for use as analytical standards. The method is applicable to components that are typically liquids at ambient temperature and pressure or solids that will form a solution when blended with liquids. Butane derivatives can be included if caution is used when blending such derivatives. Often, the individual components of natural gasoline, namely, isobutane derivatives, *n*-butane derivatives, pentane derivatives, and isopentane derivatives, are separated as base stocks for reforming, alkylation, synthetic rubber, and other petrochemicals.

On the other hand, liquefied refinery gases are mixtures of hydrocarbons that are recovered during the refining of crude oil. These materials are composed principally of propane and butane derivatives and are more like, than natural gas, to contain olefin derivatives such as propylene and butylene derivatives, and they can be stored and handled at ambient temperatures and at moderate pressures.

For use as motor fuel or as a component of motor fuel, the primary criteria for the quality of gas condensate or natural gasoline are its volatility and knock performance. The basic measures of volatility are vapor pressure (ASTM D323, 2017; ASTM D4953, 2017; ASTM D5191, 2017) and distillation (ASTM D6897, 2017). Knock performance is measured by rating in knock test engines by both the motor (ASTM D2700, 2017) and research (ASTM D2699, 2017) methods. Other considerations for natural gasoline used in motor fuels are copper corrosion (ASTM D130, 2017) and gravity (ASTM D1298, 2017). Although gravity has a minor direct relationship with quality, it is necessary to determine gravity for measurement and shipping. When gas condensate and natural gasoline is used as a feedstock for further processing or petrochemicals, the list of quality criteria is almost endless. For nearly all petrochemical uses, composition by hydrocarbon types is needed, and, frequently, a complete analysis of specific components is made (ASTM D2427, 2017). If a catalytic process is involved, total sulfur (ASTM D1266, 2017) is an important (even essential) aspect of the analytical plan because the tendency of sulfur-containing constituents tends to interfere with, even destroy, the activity of the catalyst.

An issue that arises during the characterization of liquefied petroleum gas relates to the accurate determination of higher molecular weight nonvolatile residues (typically, higher molecular weight hydrocarbons and even oils) in the gas. Liquefied petroleum gas (LPG) is a mixture of hydrocarbons, principally propane and butane obtained from either NGLs or refinery gases, which can be stored and handled as liquids at ambient temperatures and moderate pressures. Commercial liquefied petroleum gas also may contain small quantities of ethane and varying quantities of propylene ($CH_3CH{=}CH_2$) or butylene ($CH_3CH_2CH{=}CH_2$ or $CH_3CH{=}CHCH_3$) that are produced during refining and are likely constituents of refinery gases. The original use for liquefied petroleum gas was as a fuel gas in areas where pipeline utility gas was not available. Later, these materials were fractionated into their separate components, principally propane and butane, as preferred feedstock for the petrochemical industry (Speight, 2014).

There are many uses for liquefied petroleum gases, which include (i) as residential and commercial fuels, principally in heating, as engine fuels, and farm uses such as crop drying and tobacco curing, (ii) as chemical feedstock for production into ethylene and propylene, and (iii) as gasoline-blending stocks or manufacture of gasoline components such as alkylate. Special applications may not be covered (ASTM D1835, 2017) because they are usually purchased and manufactured to meet specific needs that may differ from one application to another.

Liquefied petroleum gases are essentially propane or butane or mixtures of the two. As a result, the important characteristics of these products can be defined and controlled by relatively few measurements. The ASTM specifications for liquefied petroleum gases are designed to properly define acceptable products for residential and industrial uses. In many cases it will be found that the products meeting the specifications will be usable in applications other than the ones for which they were designed. The following can be accepted as a general guide to the applications of the various fuel types.

Commercial propane is preferred for residential and commercial uses in geographical areas and in seasons where low ambient temperatures are encountered. It is also suitable for low severity internal combustion engine applications. *Commercial propane* covers a wide range of mixtures and permits the tailoring of fuels to specific needs. These fuels find applications in areas and seasons where low ambient temperatures are not frequently encountered. *Special duty* propane is tailored to meet the needs of internal combustion engines operating under moderate to high engine severity. Special duty propane will be less variable in composition and combustion characteristics than the other fuels (ASTM D1835, 2017).

Test methods using procedures similar to those employed in gas chromatographic simulated distillation are becoming available. In fact, the presence of any component substantially less volatile than the main constituents of the liquefied petroleum gas will give rise to unsatisfactory performance. It is difficult to set limits to the amount and nature of the *residue* that will make a product unsatisfactory. Obviously small amounts of oil-like material can block regulators and valves. In liquid vaporizer feed systems, even gasoline-type material could cause difficulty. The residue as determined by the *end point index* (EPI) (ASTM D2158, 2017) is a measure of the concentration of contaminants boiling above 37.8 °C (100 °F) that may be present in the gas. Other methods that measure residue more directly are available, and for particular applications it may be possible to relate the values obtained to the performance required and to set satisfactory limits.

Analytical methods are available in standard form for determining volatile sulfur content and certain specific corrosive sulfur compounds that are likely to be present. Volatile sulfur determination is made by a combustion procedure (ASTM D126) that uses a modification of the standard wick-fed lamp. Many laboratories use rapid combustion techniques with an oxyhydrogen flame in a Wickbold or Martin–Floret burner (ASTM D2784, 2017).

The *rate of flame propagation* (also referred to as the *burning velocity*) in gas–air mixtures is of importance in utilization problems, including those dealing with burner design and rate of energy release. There are several methods that have been used for measuring such burning velocities in both laminar and turbulent flames. Results by the various methods do not agree, but any one method does give relative values of utility. Maximum burning velocities for turbulent flames are greater than those for laminar flames.

This test method (ASTM D2784, 2017) is valid for sulfur levels in excess of one microgram/g of sulfur in liquefied petroleum gas, but the samples should not contain more than 100 $\mu g\, g^{-1}$ of chlorine. In the test, the sample is bummed in an oxyhydrogen bummer or in a lamp in a closed system in a carbon dioxide–oxygen atmosphere. The latter test method is not recommended for trace quantities of sulfur due to inordinately long combustion times needed. The sulfur oxides produced are absorbed and oxidized to sulfuric acid in a hydrogen peroxide solution, and the sulfate ions are then determined by either titrating with barium perchlorate solution using a thorinmethylene blue mixed indicator or by precipitating as barium sulfate and measuring the turbidity of the precipitate with a photometer.

Trace hydrocarbons that may be regarded as contaminants may be determined by the gas chromatographic methods already discussed. Heavier hydrocarbons in small amounts may not be completely removed from the column. If accurate information is required about the nature and amount of higher molecular weight high-boiling constituents, the temperature programming procedure or a concentration procedure may be used. Analytical methods for determining traces of various other impurities, such as chlorides (ASTM D2384, 2017), are known to be in use. The presence of acetylene derivatives in refinery gases, although unlikely, must still be considered. Acetylene derivatives can be determined using a chemical test method, while carbonyls are determined by the classical hydroxylamine hydrochloride reaction.

The determination of traces of higher-boiling hydrocarbons and oily matter involves the use of a method for residue that involves a preliminary weathering. The residue after weathering is dissolved in a solvent and the solution applied to a filter paper. The presence of residue is indicated by the formation of an oil stain. The procedure is taken further by combining the oil stain observation with other observed values to calculate an *EPI* (ASTM D2158, 2017). The method is not very precise, and work is proceeding in several laboratories to develop a better method for the determination of residue in the form of oily matter.

In liquefied petroleum gas where the composition is such that the hydrocarbon dew point is known to be low, a dew-point method will detect the presence of traces of water (ASTM D1142, 2017). In addition, the odor of liquefied petroleum gas has to be detectable to avoid the risk of explosion. Odor is a very subjective matter and no standard method is available. It is desirable to set up some system in which the concentration of gas can be measured in relation to its explosive limits and in which some variables can be standardized, e.g. flow rate and orifice size. This will ensure that in any one location the liquefied petroleum gas is always being assessed under similar conditions from day to day. Propane (boiling point: −42 °C, −44 °F), isobutane (boiling point: −12 °C, 11 °F), and butane (boiling point: −1 °C, 30 °F) generally constitute this sample type and are used for heating, motor fuels, and as chemical feedstocks (ASTM D2597, 2017; ASTM D2504, 2017; ASTM D2505, 2017).

Hydrocarbon impurities in propylene can be determined by gas chromatographic methods (ASTM D2712, 2017; ASTM D2163, 2017), and another test is available for the determination of traces of methanol in propylene (ASTM Test Method D4864). A gas chromatographic method (ASTM D5303, 2017) is available for the determination of trace amounts of carbonyl sulfide in propylene using a flame photometric detector. Also, sulfur in petroleum gas can be determined by oxidative microcoulometry (ASTM D3246, 2017). In addition, commercial butylene derivatives ($CH_3CH_2CH{=}CH_2$, $CH_3CH{=}CHCH_3$), high-purity butylene derivatives, and butane–butylene mixtures can be analyzed for hydrocarbon constituents (ASTM D4424, 2017), and hydrocarbon impurities in 1,3-butadiene and the presence of butadiene dimer and styrene can also be determined by gas chromatography (ASTM D2593, 2017 and ASTM D2426, 2017, respectively). Carbonyl derivatives in C_4 hydrocarbons are determined by a titrimetric technique (ASTM D4423, 2017) and by the use of a peroxide method (ASTM D5799, 2017).

In general, gas chromatography will undoubtedly continue to be the method of choice for characterization of light hydrocarbon materials. New and improved detection devices and techniques, such as chemiluminescence, atomic emission, and mass spectroscopy, will enhance selectivity, detection limits, and analytical productivity. Laboratory automation through autosampling (ASTM D4307, 2017; ASTM D4177, 2017), computer control, and data handling will provide improved precision and productivity, as well as simplified method operation (ASTM D1265, 2017). Mercury in natural gas is also measured by atomic fluorescence spectroscopy (ASTM D6350, 2017) and by atomic absorption spectroscopy (ASTM D5954, 2017).

Procedures for the determination of hydrogen, helium, oxygen, nitrogen, carbon monoxide, carbon dioxide, methane, ethane, ethylene, propane, butanes, pentanes, and hexanes as well as higher molecular weight hydrocarbons in natural and reformed gases by packed column gas chromatography are available (ASTM D1945, 2017; ASTM D1946, 2017). These compositional analyses are used to calculate many other properties of gases, such as density, heating value, and compressibility. The first five components listed are determined using a molecular sieve column (argon carrier gas), while the remaining components are determined using polydimethylsiloxane partition or porous polymer columns. The hexanes-plus analysis is accomplished by backflushing the column after the elution of pentane or by the use of a backflushed precolumn.

A concentration of constituents that is below 0.01 mol.% typically refers to trace constituents, and there may be a variety of constituents that are collectively classified as *trace constituents*. The major portion of trace constituents are naturally occurring species such as traces of paraffin and aromatic hydrocarbons, hydrogen sulfide, and organic sulfur compounds (such as mercaptans, RSH, the natural odorants) and traces of hydrogen and helium, H_2 and He. In addition, the increasing levels of carbon dioxide, water content, and heavy hydrocarbons present problems in gas-cleaning operations (Mokhatab et al., 2006; Speight, 2007, 2014) that extend to the operators of a liquefied natural gas plant.

Arsenic and mercury have been found in some natural gas deposits. The presence of arsenic was noted when trialkyl arsine sulfides, as a white deposit, was found on pressure regulation devices in natural gas distribution lines. Mercury was detected in gas after having accumulated in NGLs recovery vessels and because of the extensive damage it caused to the aluminum heat exchangers. There are also other trace species in processed natural gas that are inadvertently added by gas-processing operations such as traces of various amines, ammonia, nitrogen oxides (NOx), formaldehyde (HCHO), and mercury (from mercury spills). Formaldehyde and oxides of nitrogen were detected after they were related to odor fading in odorized natural gas. Whatever the sources of these components, it is to the benefit of the gas industry to obtain the required information on the quality of gas to avoid any potential problems.

Important constituents of natural gas not accounted for in these analyses are moisture (water) and hydrogen sulfide, as well as other sulfur compounds (ASTM D1142, 2017; ASTM D1988, 2017; ASTM D4888, 2017; ASTM D5504, 2017; ASTM D5454, 2017; ASTM D6228, 2017). Olefins (ethylene, propylene, butylenes, and pentylene derivatives) that occur in refinery (process) gas have specific characteristics and require specific testing protocols (ASTM D5234, 2017; ASTM D5273, 2017; ASTM D5274, 2017). Thus, hydrocarbon analysis of ethylene is accomplished by two methods (ASTM D2505, 2017; ASTM D6159, 2017), one of which (ASTM D6159, 2017) uses wide-bore (0.53-mm) capillary columns, including an alumina–potassium chloride (Al_2O_3/KCl) PLOT column. Another method (ASTM D2504, 2017) is recommended for the determination of noncondensable gases, and another (ASTM D2505, 2017) for carbon dioxide.

Carbon dioxide (ASTM D1945, 2017; ASTM D4984, 2017) in excess of 3% is normally removed for corrosion prevention (ASTM D1838, 2017). Hydrogen sulfide (ASTM D2420, 2017; ASTM D4084, 2017; ASTM D4810, 2017) is also removed, and the odor of the gas must not be objectionable (ASTM D6273, 2017), so mercaptan content (ASTM D1988, 2017) is important. A simple lead acetate test (ASTM D2420, 2017; ASTM D4084, 2017) is available for detecting the presence of hydrogen sulfide and is an additional safeguard that hydrogen sulfide is not present (ASTM D1835, 2017). The odor of the gases must not be objectionable. Methyl mercaptan, if present, produces a transitory yellow stain on the lead acetate paper that fades completely in less than 5 min. Other sulfur compounds (ASTM D5504, 2017; ASTM D6228, 2017) present in liquefied petroleum gas do not interfere.

In the lead acetate test (ASTM D2420, 2017), the vaporized gas is passed over moist lead acetate paper under controlled conditions. Hydrogen sulfide reacts with lead acetate to form lead sulfide, resulting in a stain on the paper varying in color from yellow to black, depending on the amount of hydrogen sulfide present. Other pollutants can be determined by gas chromatography (ASTM D5504, 2017; ASTM D6228, 2017). The total sulfur content (ASTM D1072, 2017; ASTM D2784, 2017) is normally acceptably low and frequently so low that it needs augmenting by means of alkyl sulfides, mercaptan derivatives, or thiophene derivatives in order to maintain an acceptable safe level of odor.

Natural gas composition analysis is critical for (i) producers and gatherers, (ii) gas distribution companies, (iii) gas utility companies, and (iv) gas engine and appliance manufacturers. Due to variations in natural gas composition, it is important to measure physical properties, especially for custody transfer situations. Compositional analysis can be used to calculate calorific value, specific gravity, and compressibility factor (ASTM D3588, 2017).

At a given pressure and temperature, attractive and repulsive intermolecular forces between gas particles mean that the molar volume is likely to be smaller or larger than for an ideal gas under the same conditions. The compressibility factor is simply the ratio of the molar volume V_m of the gas to the molar volume V°_m of an ideal gas at the same pressure and temperature. Thus,

$$Z = \frac{V_m}{V^{\circ}_m}$$

The value of Z provides information on the dominant types of intermolecular forces acting in a gas. Thus, when $Z = 1$, there are no intermolecular forces and ideal gas behavior; when $Z < 1$, the attractive forces dominate and gas occupies a smaller volume than an ideal gas; when $Z > 1$, the repulsive forces dominate and gas occupies a larger volume than an ideal gas. All gases approach $Z = 1$ at very low pressure, when the spacing between particles is typically large.

Shale gas is natural gas produced from shale formations (Speight, 2013, 2017b) that typically function as both the reservoir and source rocks for the natural gas. In terms of chemical makeup, shale gas is typically a dry gas composed primarily of methane (60–95% v/v), but some formations do produce wet gas Table 8.5. The Antrim and New Albany plays have typically produced water and gas. Shale gas formations are organic-rich shale formations that were previously regarded only as source rocks and seals for gas accumulating in the strata near sandstone and carbonate reservoirs of traditional onshore gas.

Table 8.5 Composition of selected unprocessed shale gas samples.

	(Vol.%)			
Component	**Marcellus**	**Appalachian**	**Haynesville**	**Eagle Ford**
Methane	97.131	79.084	96.323	74.595
Ethane	2.441	17.705	1.084	13.824
Propane	0.095	0.566	0.205	5.425
C_4+	0.014	0.034	0.203	4.462
Hexanes+	0.001	0.000	0.061	0.478
Carbon dioxide	0.040	0.073	1.816	1.536
Nitrogen	0.279	2.537	0.369	0.157
Total inerts (CO_2+N_2)	0.318	2.609	2.184	1.693
Total	**100.0**	**100.0**	**100.1**	**100.5**
HHV (ETU/SCF)	1031.6	1133.2	1009.8	1307.1
Hydrocarbon dew point (°F)	−96.8	−41.3	9.7	119.6
Wobbe number (Btu/SCF)	1367.1	1397.0	1320.1	1490.0

Gas production from unconventional shale gas reservoirs (such as tight shale formations) has become more common in the past decade. Produced shale gas observed to date has shown a broad variation in compositional makeup, with some having wider component ranges, a wider span of minimum and maximum heating values, and higher levels of water vapor and other substances than pipeline tariffs or purchase contracts may typically allow. Indeed, because of these variations in gas composition, each shale gas formation can have unique processing requirements for the produced shale gas to be marketable.

Ethane can be removed by cryogenic extraction, while carbon dioxide can be removed through a scrubbing process. However, it is not always necessary (or practical) to process shale gas to make its composition identical to *conventional* transmission-quality gases. Instead, the gas should be interchangeable with other sources of natural gas now provided to end users. The interchangeability of shale gas with conventional gases is crucial to its acceptability and eventual widespread use in the United States.

Although not highly sour in the usual sense of having high hydrogen sulfide content, and with considerable variation from play to resource to resource and even from well to well within the same resource (due to extremely low permeability of the shale even after fracturing), shale gas often contains varying amounts of hydrogen sulfide with wide variability in the carbon dioxide content. The gas is not ready for pipelining immediately after it has exited the shale formation.

The challenge in treating such gases is the low (or differing) hydrogen sulfide/carbon dioxide ratio and the need to meet pipeline specifications. In a traditional gas-processing plant, the olamine of choice for content for hydrogen sulfide removal is *N*-methyldiethanolamine (MDEA), but whether this olamine will suffice to remove the hydrogen sulfide without removal of excessive amounts of carbon dioxide requires consistent monitoring.

Gas treatment may begin at the wellhead – condensates and free water usually are separated at the wellhead using mechanical separators. Gas, condensate, and water are separated in the field separator and are directed to separate storage tanks and the gas flows to a gathering system. After the free water has been removed, the gas is still saturated with water vapor and, depending on the temperature and pressure of the gas stream, may need to be dehydrated or treated with methanol to prevent hydrates as the temperature drops. But this may not be always the case in actual practice.

8.2.2 Refinery Gas

The terms *refinery gas* or *petroleum gas* are often used to identify liquefied petroleum gas or even gas that emanates from the top of a refinery distillation column. Refinery gas varies in composition and volume, depending on crude origin and on any additions to the crude made at the loading point – in this sense, it is probably the most difficult gas to analyze. Furthermore, it is not uncommon to reinject low molecular weight hydrocarbons such as propane and butane into the crude before dispatch by tanker or pipeline. This results in a higher vapor pressure of the crude, but it allows one to increase the quantity of light products obtained at the refinery. Since light ends in most crude oil markets command a premium, while in the oil field itself, propane and butane may have to be reinjected or flared, the practice of *spiking* crude oil with liquefied petroleum gas is becoming common.

Thus, *refinery gas* (*fuel gas*) is the noncondensable gas that is obtained during distillation of crude oil or treatment (cracking, thermal decomposition) of crude oil. Refinery gas is produced in considerable quantities during the different refining processes and is used as fuel for the refinery itself and as an important feedstock to produce petrochemicals. It consists mainly of hydrogen (H_2), methane (CH_4), ethane (C_2H_6), propane (C_3H_8), butane (C_4H_{10}), and olefins ($RCH{=}CHR^1$, where R and R^1 can be hydrogen or a methyl group) and may also include off-gases from petrochemical processes. Olefins such as ethylene ($CH_2{=}CH_2$, boiling point: −104 °C, −155 °F), propene (propylene, $CH_3CH{=}CH_2$, boiling point: −47 °C, −53 °F), butene (butene-1, $CH_3CH_2CH{=}CH_2$, boiling point: −5 °C, 23 °F), *iso*-butylene ($(CH_3)_2C{=}CH_2$, boiling point: −6 °C, 21 °F), *cis*- and *trans*-butene-2 ($CH_3CH{=}CHCH_3$, boiling point: ca. 1 °C, 30 °F), and butadiene ($CH_2{=}CHCH{=}CH_2$, boiling point: −4 °C, 24 °F) as well as higher-boiling olefins are produced by various refining processes.

Still gas is broad terminology for low-boiling hydrocarbon mixtures and is the lowest-boiling fraction isolated from a distillation (*still*) unit in the refinery. If the distillation unit is separating light hydrocarbon fractions, the still gas will be almost entirely methane with only traces of ethane (CH_3CH_3) and ethylene ($CH_2{=}CH_2$). If the distillation unit is handling higher-boiling fractions, the still gas might also contain propane ($CH_3CH_2CH_3$), butane ($CH_3CH_2CH_2CH_3$), and their respective isomers. *Fuel gas* and still gas are terms that are often used interchangeably, but the term *fuel gas* is intended to denote the product destination to be used as a fuel for boilers, furnaces, or heaters.

In a series of processes commercialized under the names Platforming, Powerforming, Catforming, and Ultraforming, paraffinic and naphthenic (cyclic nonaromatic) hydrocarbons, in the presence of hydrogen and a catalyst, are converted into aromatics or isomerized to more highly branched hydrocarbons. Catalytic reforming processes thus not only result in the formation of a liquid product of higher octane number but also produce substantial quantities of gases. The latter not only are rich in hydrogen but also contain hydrocarbons from methane to butanes, with a preponderance of propane ($CH_3CH_2CH_3$), *n*-butane ($CH_3CH_2CH_2CH_3$), and isobutane [$(CH_3)_3CH$]. The composition of these gases varies in accordance with reforming severity and reformer feedstock. Since all catalytic reforming processes require substantial recycling of a hydrogen stream, it is normal to separate reformer gas into a propane ($CH_3CH_2CH_3$) and/or a butane [$CH_3CH_2CH_2CH_3/(CH_3)_3CH$] stream, which becomes part of the refinery liquefied petroleum gas production, and a lighter gas fraction, part of which, is recycled. In view of the excess of hydrogen in the gas, all products of catalytic reforming are saturated, and there are usually no olefin-type gases present in either gas streams.

A second group of refining operations that contributes to gas production is that of the catalytic cracking processes. Thermofor catalytic cracking, and other process variants, are processes in which heavy gas oils are converted into cracked gas, liquefied petroleum gas, catalytic naphtha, fuel oil, and coke by contacting the heavy hydrocarbon with the hot catalyst. Both catalytic and thermal cracking processes, the latter being now largely used to produce chemical raw materials, result in the formation of unsaturated hydrocarbons, particularly ethylene ($CH_2{=}CH_2$), as well as propylene (propene, $CH_3CH{=}CH_2$), isobutylene [isobutene, $(CH_3)_2C{=}CH_2$], and the *n*-butenes ($CH_3CH_2CH{=}CH_2$ and $CH_3CH{=}CHCH_3$) in addition to hydrogen (H_2), methane (CH_4), and smaller quantities of ethane (CH_3CH_3), propane ($CH_3CH_2CH_3$), and butane isomers [$CH_3CH_2CH_2CH_3$, $(CH_3)_3CH$]. Diolefins such as butadiene ($CH_2{=}CHCH{=}CH_2$) are also present.

Additional gases are produced in refineries with coking or visbreaking facilities for the processing of their heaviest crude fractions. In the visbreaking process, fuel oil is passed through externally fired tubes and undergoes liquid-phase cracking reactions, which result in the formation of lighter fuel oil components.

Oil viscosity is thereby reduced, and some gases, mainly hydrogen, methane, and ethane, are formed. Substantial quantities of both gas and carbon are also formed in coking (both fluid coking and delayed coking) in addition to the middle distillate and naphtha. When coking a residual fuel oil or heavy gas oil, the feedstock is preheated and contacted with hot carbon (coke) that causes extensive cracking of the feedstock constituents of higher molecular weight to produce lower molecular weight products ranging from methane, via liquefied petroleum gas(es) and naphtha, to gas oil and heating oil. Products from coking processes tend to be unsaturated, and olefin-type components predominate in the tail gases from coking processes.

A further source of refinery gas is hydrocracking, a catalytic high-pressure pyrolysis process in the presence of fresh and recycled hydrogen. The feedstock is again heavy gas oil or residual fuel oil, and the process is mainly directed at the production of additional middle distillates and gasoline. Since hydrogen is to be recycled, the gases produced in this process again must be separated into lighter and heavier streams; any surplus recycle gas and the liquefied petroleum gas from the hydrocracking process are both saturated.

Both hydrocracker and catalytic reformer tail gases are commonly used in catalytic desulfurization processes. In the latter, feedstocks ranging from light to vacuum gas oils are passed at pressures of 500–1000 psi with hydrogen over a hydrofining catalyst. This results mainly in the conversion of organic sulfur compounds to hydrogen sulfide:

$$[S]_{feedstock} + H_2 \rightarrow H_2S + \text{hydrocarbons}$$

The process also produces some light hydrocarbons by hydrocracking.

The first and most important aspect of gaseous testing is the measurement of the volume of gas. However, in all cases, it is the composition of the gas in terms of *hydrocarbon type* that is more important in the context of the application. For example, in petrochemical applications, the presence of propylene and butylene above 10% v/v can have an adverse effect on hydrodesulfurization prior to steam reforming. On the other hand, petrochemical processes, such as in the production of *iso*-octane from *iso*-butane and butylene, can require the exclusion of the saturated hydrocarbons. Furthermore, refinery gas specifications will vary depending on the gas quality available and the end use. For fuel uses, gas as specified above presents little difficulty used as supplied. Alternatively, a gas of constant Wobbe index, say for gas turbine use, could readily be produced by the user. Part of the combustion air would be diverted into the gas stream by a Wobbe index controller. This would be set to supply gas at the lowest Wobbe index of the undiluted gas.

Still gas is a term reserved for low-boiling hydrocarbon mixtures that represent the lowest-boiling fraction isolated from an atmospheric distillation unit (*still*) in the refinery. Typically, the still gas will contain methane, ethane

(CH_3CH_3), propane ($CH_3CH_2CH_3$), butane ($CH_3CH_2CH_2CH_3$), and their respective isomers. If cracked products have been recycled to the atmospheric distillation unit, the still gas may also contain volatile olefins such as ethylene ($CH_2{=}CH_2$).

8.2.3 Coalbed Methane

Coal bed methane (coalbed methane, CBM) is predominantly methane extracted from coalbeds (coal seams) and has become an important source of energy in United States, Canada, and other countries. The gas is adsorbed into the solid matrix of the coal and is often referred to as *sweet gas* because of its lack of hydrogen sulfide. Unlike natural gas from conventional reservoirs, CBM contains very little higher molecular weight hydrocarbons such as propane or butane or condensate. It often contains up to a few percent of carbon dioxide. Some coal seams contain little methane, with the predominant coal seam gas being carbon dioxide.

In fact, CBM has been suggested with sufficient justification that the materials comprising a coalbed fall broadly into the following two categories: (i) volatile low molecular weight materials (components) that can be liberated from the coal by pressure reduction, mild heating, or solvent extraction and (ii) materials that will remain in the solid state after the separation of volatile components.

Techniques to detect and deal with CBM in mines have ranged from (i) the classic canary in a cage to detect an oxygen-poor atmosphere, (ii) ventilation fans to force the replacement of a methane-rich environment with outside air, and (iii) drilling CBM wells in front of the coal face to try to degas the coal prior to exposing the mine to the CBM.

With the unique method of storage of CBM, the preponderance of the gas is available only to very low coal-face pressures. The coal-face pressure is set by a combination of flowing wellhead pressure and the hydrostatic head exerted by standing liquid within the wellbore. Effective removal of higher molecular weight hydrocarbons and other constituents (*liquids removal* or *deliquefication*) can reduce or remove the back pressure caused by accumulated liquid.

8.2.4 Coal Gas

Coal gas is any gaseous product that is produced by gasification of coal or by carbonization of coal. Most of these products have been replaced by natural gas (the city and town gas works are an industry of the past) but still find use in some parts of the world. The products of coal gasification are varied insofar as the gas composition varies with the system employed. It is emphasized that the gas product must be first freed from any pollutants such as particulate matter and sulfur compounds before further use, particularly when the intended use is a water–gas shift or methanation.

Low-heat-content gas (low-Btu gas) is produced during the gasification of when the oxygen is not separated from the air, and, thus, the gas product invariably has a low heat content (ca. 150–300 Btu ft^{-3}). Low-heat-content gas is also the usual product of in situ gasification of coal that is used essentially as a technique for obtaining energy from coal without the necessity of mining the coal. The process is a technique for utilization of coal that cannot be mined by other techniques. The nitrogen content of low-heat-content gas ranges from somewhat less than 33% v/v to slightly more than 50% v/v and cannot be removed by any reasonable means; the presence of nitrogen at these levels renders the product gas to be low heat content. The nitrogen also strongly limits the applicability of the gas to chemical synthesis. Two other noncombustible components (water, H_2O, and carbon dioxide, CO) further lower the heating value of the gas; water can be removed by condensation and carbon dioxide by relatively straightforward chemical means. The two major combustible components are hydrogen and carbon monoxide; the H_2/CO ratio varies from approximately 2 : 3 to approximately 3 : 2, but methane may also make an appreciable contribution to the heat content of the gas. Of the minor components hydrogen sulfide is the most significant, and the amount produced is, in fact, proportional to the sulfur content of the feed coal. Any hydrogen sulfide present must be removed by one, or more, of several procedures. Low-heat-content gas is of interest to industry as a fuel gas or even, on occasion, as a raw material from which ammonia, methanol, and other compounds may be synthesized.

Medium-heat-content gas (medium-Btu gas) has a heating value in the range of 300–550 Btu ft^{-3}, and the composition is much like that of low-heat-content gas, except that there is virtually no nitrogen. The primary combustible gases in medium-heat-content gas are hydrogen and carbon monoxide. Medium-heat-content gas is considerably more versatile than low-heat-content gas; like low-heat-content gas, medium-heat-content gas may be used directly as a fuel to raise steam or used through a combined power cycle to drive a gas turbine, with the hot exhaust gases employed to raise steam, but medium-heat-content gas is especially amenable to synthesize methane (by methanation), higher hydrocarbons (by Fischer–Tropsch synthesis), methanol, and a variety of synthetic chemicals. The reactions used to produce medium-heat-content gas are the same as those employed for low-heat-content gas synthesis, the major difference being the application of a nitrogen barrier (such as the use of pure oxygen) to keep diluent nitrogen out of the system.

In medium-heat-content gas, the H_2/CO ratio varies from 2 : 3 to ca. 3 : 1, and the increased heating value correlates with higher methane and hydrogen contents as well as with lower carbon dioxide contents. Furthermore, the very nature of the gasification process used to produce the medium-heat-content gas has a marked effect upon the ease of subsequent processing. For example, the CO_2-acceptor product is quite amenable to use for methane production because it has (i) the desired H_2/CO ratio just exceeding 3 : 1, (ii) an initially high methane content, and (iii) relatively low water and carbon dioxide contents. Other gases may require appreciable shift reaction and removal of large quantities of water and carbon dioxide prior to methanation.

High-heat-content gas (high-Btu gas) is essentially pure methane and often referred to as *synthetic natural gas* (SNG). However, to qualify as synthetic natural gas, a product must contain at least 95% methane; the energy content of synthetic natural gas is 980–1080 Btu ft^{-3}. The commonly accepted approach to the synthesis of high-heat-content gas is the catalytic reaction of hydrogen and carbon monoxide.

$$3H_2 + CO \rightarrow CH_4 + H_2O$$

To avoid catalyst poisoning, the feed gases for this reaction must be quite pure, and, therefore, impurities in the product are rare. The large quantities of water produced are removed by condensation and recirculated as very pure water through the gasification system. The hydrogen is usually present in slight excess to ensure that the toxic carbon monoxide is reacted; this small quantity of hydrogen will lower the heat content to a small degree.

The carbon monoxide/hydrogen reaction is somewhat inefficient as a means of producing methane because the reaction liberates large quantities of heat. In addition, the methanation catalyst is troublesome and prone to poisoning by sulfur compounds, and the decomposition of metals can destroy the catalyst. Thus, hydrogasification may be employed to minimize the need for methanation:

$$C_{coal} + 2H_2 \rightarrow CH_4$$

The product of hydrogasification is far from pure methane, and additional methanation is required after hydrogen sulfide and other impurities are removed.

8.2.5 Biogas

The term biogas (Chapter 1) – a renewable energy source that is of variable composition depending upon the source (Table 8.6) and can be analyzed accordingly (Rasi et al., 2007; Esposito et al., 2012) – typically refers to a flammable mixture of different gases produced by the breakdown of organic matter in the absence of oxygen. Biogas can be produced from raw materials such as agricultural waste, manure, municipal waste, plant material, sewage, green waste, or food waste. Anaerobic digestion is used to produce the biogas – the biomass is converted (digested) inside a closed system or by fermentation of biodegradable materials (Richards et al., 1994).

Table 8.6 Examples of biogas composition.

Constituents	Source-1	Source-2	Source-3
Methane, CH_4 % v/v	50–60	60–75	60–75
Carbon dioxide, CO2 % v/v	38–34	33–19	33–19
Nitrogen, N_2 % v/v	5-0	1-0	1-0
Oxygen, O_2 % v/v	1-0	<0.5	<0.5
Water, H_2O % v/v	6 (40 °C)	6 (40 °C)	6 (40 °C)
Hydrogen sulfide, H_2S, $mg\,m^{-3}$	100–900	1000–4000	3000–10 000
Ammonia, NH_3 $mg\,m^{-3}$	—	—	50–100

Source-1: Household waste.
Source-2: Wastewater treatment plant sludge.
Source-3: Agricultural waste.

Biogas is primarily methane and carbon dioxide and may have small amounts of hydrogen sulfide, moisture, and siloxane derivatives (Table 8.6). The methane, hydrogen sulfide, and carbon monoxide can be combusted or oxidized with oxygen. This energy release allows biogas to be used as a fuel; it can be used for any heating purpose, such as cooking. It can also be used in a gas engine to convert the energy in the gas into electricity and heat.

Raw biogas produced from digestion is approximately roughly 60% v/v methane and 29% v/v carbon dioxide with trace elements of hydrogen sulfide (Rasi et al., 2007). The gas is not of sufficiently high quality to be used as fuel gas for machinery, and the corrosive nature of the hydrogen sulfide is sufficient to corrode and destroy the internals of a plant. In fact, national gas quality specifications define gas quality limits to protect the integrity of the network and ensure that gas supplied to domestic users will combust safely.

The composition of biogas varies depending upon the substrate composition, as well as the conditions within the anaerobic reactor (temperature, pH, and substrate concentration). Advanced waste treatment technologies can produce biogas with 55–75% v/v methane, which for reactors with free liquids can be increased to 80–90% methane using *in situ* gas purification techniques. As produced, biogas contains water vapor – the fractional volume of water vapor is a function of biogas temperature; correction of measured gas volume for water vapor content and thermal expansion is easily done via simple mathematics that yields the standardized volume of dry biogas. In some cases, biogas contains siloxane derivatives that are formed by the anaerobic decomposition of materials commonly found in soaps and detergents (Arnold and Kajolinna, 2010). During combustion of biogas containing siloxanes, silicon is released and can combine with free oxygen or other elements in the combustion gas. Deposits are formed containing predominantly mostly silica (SiO_2) or silicate derivatives (Si_xO_y) and can contain calcium, sulfur, zinc, and phosphorus. These mineral deposits accumulate to a surface thickness of several millimeters and must be removed by chemical or mechanical means. The siloxane derivatives can be removed during gas cleaning (Chapter 4) (Appels et al., 2008; Accettola et al., 2008).

When biogas is sent to processing for impurity removal, the cleaned gas has characteristics of the biogas that are close to the characteristics of as natural gas. In this instance the producer of the biogas can utilize the local gas distribution networks. The gas must be very clean to reach pipeline quality. Water (H_2O), hydrogen sulfide (H_2S), and particulates are removed if present at high levels or if the gas is to be completely cleaned. Carbon dioxide is less frequently removed, but it must also be separated to achieve pipeline quality gas. If the gas is to be used without extensively cleaning, it is sometimes cofired with natural gas to improve combustion. Biogas cleaned up to pipeline quality is called renewable natural gas. In this form the gas can be used in any application in which natural gas is used. The heating value of the raw biogas is on the order of 600 Btu ft^{-3} (5340 cal m^{-3}). The raw gas will burn in an engine, but corrosion can occur unless proper materials of construction are selected.

By analysis, raw biogas produced from digestion is approximately 60% v/v methane and 29% v/v carbon dioxide with trace amounts of hydrogen. The methane component of can be concentrated via a biogas upgrader to the same standards as natural gas, which itself has to go through a cleaning process (Chapter 4). There are four main methods of upgrading: (i) water washing, (ii) pressure swing absorption, (iii) the Selexol absorption process, and (iv) amine gas treating (Chapter 4). In addition to these, the use of membrane separation technology for biogas upgrading is increasing (Mokhatab et al., 2006; Speight, 2007, 2014). However, the most prevalent method of biogas treating is water washing where high-pressure gas flows into a column where the carbon dioxide and other trace elements are scrubbed by cascading water running counterflow to the gas. This arrangement could deliver a gas stream that is 98% v/v methane (Abatzoglou and Boivin, 2009).

8.2.6 Landfill Gas

Landfill gas is produced by wet organic waste decomposing under anaerobic conditions in a biogas. The waste is covered and mechanically compressed by the weight of the material that is deposited above, which prevents oxygen exposure thus allowing anaerobic microbes to thrive. Landfill gas builds up and is slowly released into the atmosphere if the site has not been engineered to capture the gas. The gas released in an uncontrolled way can be hazardous since (like biogas) it can become explosive when the gas escapes from the landfill and mixes with oxygen. Landfill gas typically has methane concentrations on the order of 50% v/v; the LEL is 5% methane, and the upper is 15% methane.

8.2.7 Flue Gas

Exhaust gas generated through combustion processes is called flue gas or stack gas. As with other gases, the composition of flue gas depends on the type of fuel and the combustion conditions, e.g. the air ratio value. Many flue gas components are air pollutants and must therefore, due to governmental regulations, be eliminated or minimized by special cleaning procedures before the gas is released to the atmosphere.

Thus, flue gas contains the exhaust gases of combustion plants containing the reaction products of fuel and combustion air and residual substances such as particulate matter (dust), sulfur oxides, nitrogen oxides, and carbon monoxide. When burning coal, hydrogen chloride and hydrogen fluoride may be present in the flue gas as well as hydrocarbons and heavy metals in case of incineration of waste materials. In many countries, as part of a national environmental protection program, exhaust gases must comply with strict governmental regulations regarding the limit values of pollutants such as dust, sulfur and nitrogen oxides, and carbon monoxide. To meet these limit values, combustion plants are equipped with flue gas cleaning systems such as gas scrubbers and dust filters.

8.3 Properties

Since the composition of natural gas, manufactured gases, and mixed gases can vary so widely, no single set of specifications could cover all situations. The requirements are usually based on performances in burners and equipment, on minimum heat content, and on maximum sulfur content. Gas utilities in most states come under the supervision of state commissions or regulatory bodies, and the utilities must provide a gas that is acceptable to all types of consumers and that will give satisfactory performance in all kinds of consuming equipment. However, particularly relevant are the heating values of the various fuel gases and their constituents. For this reason, measurement of the properties of fuel gases is an important aspect of fuel gas technology.

The different methods for gas analysis include absorption, distillation, combustion, mass spectroscopy, infrared spectroscopy, and gas chromatography. Absorption methods involve absorbing individual constituents one at a time in suitable solvents and recording of contraction in volume measured. Distillation methods depend on the separation of constituents by fractional distillation and measurement of the volumes distilled. In combustion methods, certain combustible elements are caused to burn to carbon dioxide and water, and the volume changes are used to calculate composition. Infrared spectroscopy is useful applications, and for the most accurate analyses, mass spectroscopy and gas chromatography are the preferred methods.

For the analytical chemist, it is necessary to understand that there are four intrinsic, measurable properties of a gas (or, for that matter, any substance): (i) the pressure, *P*, of the gas; (ii) the temperature, *T*, of the gas; (iii) the volume of the gas, *V*, which is, in reality, subject to the container volume; and (iv) the mass, *m*, of the gas or the mole number, *n*. The gas density, *d*, is a derived quantity, which is *m/V*.

Natural gas is colorless and odorless in its pure form. It is a combustible mixture of hydrocarbon gases. While natural gas is formed primarily of methane (CH_4), it can also include ethane (C_2H_6), propane (C_3H_8), butane (C_4H_{10}), and pentane (C_5H_{12}). The composition of natural gas can vary widely (Table 8.2) before it is refined. In its purest form, natural gas is almost pure methane (CH_4) (ASTM D3956, 2017). However, the composition of natural gas varies depending on the field, the formation, or the reservoir from which it is extracted and that are an artifact of its formation. The different hydrocarbons that form natural gas can be separated using their different physical properties as weight, boiling point, or vapor pressure. Depending on its content of higher molecular weight hydrocarbon components, natural gas can be considered as rich (5 or 6 gal or more of recoverable hydrocarbon components per cubic feet) or lean (less than 1 gal of recoverable hydrocarbon components per cubic feet).

Nonassociated gas (sometimes called *well gas*) is produced from geological formations that typically do not contain much, if any, higher boiling hydrocarbons (gas liquids) than methane (Speight, 2007, 2014). It can contain nonhydrocarbon gases such as carbon dioxide and hydrogen sulfide and is directly controllable by the producer; one just turns the valves. The gas flows up the well under the influence of the reservoir own energy, through the wellhead control valves, and along the flow line to the treatment plant. Treatment requires the temperature of the gas to be reduced to a point dependent upon the pressure in the pipeline so that all liquids that would exist at pipeline temperature and pressure condense and are removed (Speight, 2007, 2014).

Associated gas is produced during crude oil production and is the gas that is associated with crude oil and typically has a larger proportion of higher molecular weight hydrocarbon derivatives than that of nonassociated gas (Table 8.7) (Speight, 2007, 2014). Crude oil cannot be produced without producing some of its associated gas, which comes out of solution as the pressure is reduced on the way to and on the surface. Crude oil in the reservoir with minimal or no dissolved associated gas is rare, and as dead crude oil is often difficult to produce, there is little energy to drive it.

Natural gas is considered *dry* when it is almost pure methane, having had most of the other commonly associated hydrocarbons removed. When other hydrocarbons are present, the natural gas is *wet*. The terms *dry gas* and *wet gas* do have quantitative meaning – the term *dry gas* indicates that there is less than 0.1 gal (1 gal, US, = 264.2 m^3) of gasoline vapor (higher molecular weight paraffins) per 1000 ft^3 (1 ft^3 = 0.028 m^3), while the term *wet gas* indicates that there are such paraffins present in the gas, in fact more than 0.1 gal/1000 ft^3.

Table 8.7 Composition of associated natural gas from a petroleum well.

Category	Component	Amount (%)
Paraffinic	Methane (CH_4)	70–98
	Ethane (C_2H_6)	1–10
	Propane (C_3H_8)	Trace–5
	Butane (C_4H_{10})	Trace–2
	Pentane (C_5H_{12})	Trace–1
	Hexane (C_6H_{14})	Trace–0.5
	Heptane and higher (C_7^+)	None-trace
Cyclic	Cyclopropane (C_3H_6)	Traces
	Cyclohexane (C_6H_{12})	Traces
Aromatic	Benzene (C_6H_6), others	Traces
Non-hydrocarbon	Nitrogen (N_2)	Trace–15
	Carbon dioxide (CO_2)	Trace–1
	Hydrogen sulfide (H_2S)	Trace occasionally
	Helium (He)	Trace–5
	Other sulfur and nitrogen compounds	Trace occasionally
	Water (H_2O)	Trace–5

Once brought from underground, the natural gas is refined to remove impurities such as water, other gases, sand, and various other compounds. Some of the constituent hydrocarbons (such as propane and butane) are removed and sold separately. After refining, the clean natural gas is transmitted through a network of pipelines to its point of use.

The amount of energy (energy content, heat content, calorific value) that is obtained from the burning of a volume of natural gas is measured in British thermal units (Btu) that is proportional to the amount of hydrocarbon constituents in the gas (ASTM D3588, 2017). Thus, the value of natural gas is calculated by its Btu content. The energy content of natural gas is variable and depends on its accumulations, which are influenced by the amount and types of energy gases they contain: the more noncombustible gases in a natural gas, the lower the Btu value. In addition, the volume mass of energy gases that are present in a natural gas accumulation also influences the Btu value of natural gas. The more carbon atoms in a hydrocarbon gas, the higher its Btu value. Btu analyses of natural gas are done at each stage of the supply chain. Gas chromatographic process analyzers are used in order to conduct fractional analysis of the natural gas streams, separating natural gas into identifiable components. The components and their concentrations are converted into a gross heating value in Btu-cubic foot (Table 8.8).

Table 8.8 Gross and net heating values of various gases.

	Gross heating value		Net heating value	
Gas[a]	(Btu ft^{-3})[b]	(Btu lb^{-1})	(Btu ft^{-3})	(Btu lb^{-1})
Butane	3225	21 640	2977	19976
Butylene	3077	20780	2876	19420
Carbon monoxide	323	4368	323	4368
Coal gas	149	16500		
Ethane	1783	22198	1630	20295
Ethylene	1631	21884	1530	20525
Hexane	4667	20526	4315	18976
Hydrogen	325	61084	275	51628
Hydrogen sulfide	672	7479		
Methane	1011	23811	910	21433
Natural gas	950	19500	850	17500
Pentane	3981	20908	3679	19322
Propane	2572	21564	2371	19834
Propylene	2336	21042	2185	19683

[a]Listed alphabetically.
[b]1 Btu ft^{-3} = 8.9 kcal m^{-3}.

In addition, gas chromatography can be used to obtain detailed information on the composition of natural gas from which physical properties (such as the Wobbe index and calorific value) of the gas are calculated. The standard test method (ASTM 1945 and 1946, or ISO 6974) defines methods for determining the composition of natural gas and (for that matter) process gases.

Still gas is a term reserved for low-boiling hydrocarbon mixtures that represent the lowest-boiling fraction isolated from an atmospheric distillation unit (*still*) in the refinery (Speight, 2014, 2017a). Typically, the still gas will contain methane, ethane (CH_3CH_3), propane ($CH_3CH_2CH_3$), *n*-butane ($CH_3CH_2CH_2CH_3$), and isobutane [$(CH_3)_3CH$]. If cracked products have been recycled to the atmospheric distillation unit, the still gas may also contain volatile olefins such as ethylene ($CH_2{=}CH_2$). The term *fuel gas* is often used interchangeably with *still gas*, but the term *fuel gas* is intended to denote the destination of the gaseous mixture, such as fuel for boilers, furnaces, or heaters, whereas the term *still gas* may be indicative of the use of the mixture for petrochemical production. Through the data from analysis of different gases, this can lead to tracing gases to national or international standards that cover the composition range of each component that provides information on the linearity of the detector and any bias that will result from using a single-point calibration.

NGLs, often referred to as *natural gasoline*, are often used to designate the hydrocarbons having higher molecular weight than methane that also occur in natural gas (Speight, 2007). Mixtures of liquefied petroleum gas, pentane derivatives, and higher molecular weight hydrocarbons fall into this category. Caution should be taken not to confuse *natural gasoline* with the term *straight-run gasoline* (often also incorrectly referred to as natural gasoline), which is the gasoline distilled unchanged from crude oil. These mixtures of hydrocarbons are extracted in liquid form from raw natural gas in lease separators, field facilities, or gas-processing plants. A significant component of NGLs is condensate. Condensate is similar in characteristics and application to a very light crude oil. These "liquids" may be produced as a "raw mix" or unfractionated stream, fractionated into individual components, or fractionated into various commercial mixtures including liquefied petroleum gas (LP gas) and natural gasoline.

As mentioned earlier, the hydrocarbon constituents of natural gas are combustible, but nonflammable nonhydrocarbon components such as carbon dioxide, nitrogen, and helium are often present in the minority and are regarded as contaminants. In addition to the natural gas found in crude oil reservoirs, there are also those reservoirs in which natural gas may be the sole occupant. In such cases the liquids that occur with the gas are gas condensate, and such reservoirs may also be referred to as *gas condensate reservoirs*. Differences in natural gas composition occur between different reservoirs, and two wells in the same field may also yield gaseous products that are different in composition (Mokhatab et al., 2006; Speight, 2007, 2014).

The principal constituent of this type of natural gas is methane, but other hydrocarbons, such as ethane, propane, and butane, may (but not always) also be present in somewhat lesser amount than natural gas from a crude oil reservoir. Carbon dioxide is also a common constituent of natural gas. Trace amounts of rare gases, such as helium, may also occur, and certain natural gas reservoirs are a source of these rare gases. Just as crude oil can vary in composition, so can natural gas.

Like crude oil resources, natural gas resources are typically divided into two categories: conventional gas and unconventional gas (Mokhatab et al., 2006; Speight, 2007, 2014). However, in a more general sense, unconventional resources also include CBM (which is not a subject of this text but which is included for comparison) and natural gas from shale formations and from tight formations. The latter type of resources (natural gas from shale formations and from tight formations) are the subject of this book. Conventional gas is typically found in reservoirs with a permeability greater than 1 milliDarcy (>1 mD) and can be extracted by means of traditional recovery methods. In contrast, unconventional gas is found in reservoirs with relatively low permeability (<1 mD) and hence cannot be extracted by conventional methods.

In addition, there are several general definitions that have been applied to natural gas. Thus, *lean* gas is gas in which methane is the major constituent. *Wet* gas contains considerable amounts of the higher molecular weight hydrocarbons, and these paraffins are present in the gas, in fact more than 0.1 gal/1000 ft^3. On the other hand, *dry* natural gas indicates that there is less than 0.1 gal (1 gal, US = 264.2 m^3) of gasoline vapor (higher molecular weight paraffins) per 1000 ft^3 (1 ft^3 = 0.028 m^3). Other terms of interest are sour gas, residue gas, and casinghead gas (casing head gas). *Sour* gas contains hydrogen sulfide, whereas *sweet* gas contains very little, if any, hydrogen sulfide. *Residue gas* is natural gas from which the higher molecular weight hydrocarbons have been extracted, and *casinghead gas* is derived from crude oil but is separated at the separation facility at the wellhead.

Thus, *associated natural gas* or *dissolved natural gas* occurs either as free gas or as gas in solution in the crude oil. Gas that occurs as a solution in the crude oil is *dissolved* gas, whereas the gas that exists in contact with the crude oil (*gas cap*) is *associated* gas. Such gas typically has higher proportions of the higher-boiling hydrocarbons that commonly occur as low-boiling crude oil constituents. To further define the terms *dry* and *wet* in quantitative measures, the term *dry* natural gas indicates that there is less than 0.1 gal (1 gal, US = 264.2 m^3) of gasoline vapor (higher molecular weight paraffins) per 1000 ft^3 (1 ft^3 = 0.028 m^3). The term *wet natural gas* indicates that there are such paraffins present in the gas, in fact more than 0.1 gal/1000 ft^3.

The use of term *residue* as it is applied to natural gas is in direct contract to the residue left in the distillation tower during crude oil refining – this residue is the high-boiling fraction of the crude oil from which the lower boiling constituents have been removed as part of the distillation process.

Other components such as carbon dioxide (CO_2), hydrogen sulfide (H_2S), mercaptans (thiols, RSH), and trace amounts of other constituents may also be present (ASTM D5303, 2017; ASTM D5305, 2017). Thus, there is no single composition of components that might be termed *typical* natural gas. Methane and ethane often constitute the bulk of the combustible components; carbon dioxide (CO_2) and nitrogen (N_2) are the major noncombustible (inert) components. Thus, sour gas is natural gas that occurs mixed with higher levels of sulfur compounds (such as hydrogen sulfide, H_2S, and mercaptans or thiols, RSH) and that constitutes a corrosive gas (Speight, 2014). The sour gas requires additional processing for purification (Mokhatab et al., 2006; Speight, 2007).

Natural gas condensate (*gas condensate, natural gasoline*) is a low-density, low-viscosity mixture of hydrocarbon liquids that may be present as gaseous components under reservoir conditions and that occur in the raw natural gas produced from natural wells. The constituents of *condensate* separate from the untreated (raw) gas if the temperature is reduced to below the hydrocarbon dew-point temperature of the raw gas. Briefly, the dew point is the temperature to which a given volume of gas must be cooled, at constant barometric pressure, for vapor to condense into liquid. Thus, the dew point is the saturation point.

On a worldwide scale, there are many gas condensate reservoirs, and each has its own unique gas condensate composition. However, in general, gas condensate has a specific gravity on the order of ranging from 0.5 to 0.8 and is composed of hydrocarbons such as propane, butane, pentane, hexane, heptane, and even octane, nonane, and decane in some cases. In addition, the gas condensate may contain additional impurities such as hydrogen sulfide, thiols (mercaptans, RSH), carbon dioxide, cyclohexane (C_6H_{12}), and low molecular weight aromatics such as benzene (C_6H_6), toluene ($C_6H_5CH_3$), ethylbenzene ($C_6H_5CH_2CH_3$), and xylenes ($H_3CC_6H_4CH_3$) (Mokhatab et al., 2006; Speight, 2007, 2014). Once these impurities are removed through refining, the properties of natural gas can be quite different to the prospects of the unrefined gas (Mokhatab et al., 2006; Speight, 2007, 2014).

When condensation occurs in the reservoir, the phenomenon known as condensate blockage can halt flow of the liquids to the wellbore. Hydraulic fracturing is the most common mitigating technology in siliciclastic reservoirs (reservoirs composed of clastic rocks), and acidizing is used in carbonate reservoirs. Briefly, clastic rocks are composed of fragments, or clasts, of preexisting minerals and rock. A clast is a fragment of geological detritus, chunks, and smaller grains of rock broken off other rocks by physical weathering. Geologists use the term *clastic* with reference to sedimentary rocks as well as to particles in sediment transport whether in suspension or as bed load and in sedimentary deposits.

In addition, production can be improved with less drawdown in the formation. For some gas condensate fields, a lower drawdown means single-phase production above the dew-point pressure can be extended for a longer time. However, hydraulic fracturing does not generate a permanent conduit past a condensate saturation buildup area. Once the pressure drops below the dew point, saturation will increase around the fracture, just as it did around the wellbore. Horizontal or inclined wells are also being used to increase contact area within formations.

Specifications for petroleum gases typically focus on liquefied petroleum gas, propane, and butane, and the specifications generally define physical properties and characteristics of liquefied petroleum gases that make them suitable for private, commercial, or industrial applications (Mokhatab et al., 2006; Speight, 2007, 2014). The composition of liquefied petroleum gas may vary depending on the source and the relative content of propane and butane content. These hydrocarbons exist as gases at atmospheric pressure and ambient temperatures but are readily liquefied under moderate pressures for transportation and utilization. The specifications do not purport to specifically define all possible requirements to meet all possible applications, and the user is cautioned to exercise judgment in formulating final specifications for specific applications.

The specifications will specify the gas allowed in a network and are included in commercial gas trading contracts for energy accounting and to ensure that the gas purchased is suitable for the network. While the calorific value range may be inferred by the interchangeability limits imposed, it is generally not part of a gas quality specification as calorific value is a commercial, energy accounting issue, rather than a quality issue.

Methane in biogas can be concentrated using a biogas upgrader to the same standards as natural gas, which itself must go through a cleaning process (Chapter 4). If the local gas network allows, the producer of the biogas may use their distribution networks. Gas must be very clean to reach pipeline quality and must be of (i) the correct composition for the distribution network to accept and (ii) meet the necessary specifications (Table 8.9). It takes approximately between 3 and 6% of the total energy output in gas to run a biogas upgrading system. On a related note, biomass gasification is a continuous oxygen-deficient process that converts biomass (such as wood chips) and other feedstocks to produce gases of varying composition (Table 8.10) that can provide a challenge for the analyst as well as for application of the various standard test methods.

For process gases, biogas, and landfill gas, the compositional analysis of the gas, including the detection of the various olefin constituents and other constituents (that are not typical constituents of natural gas), is possible (ASTM

Table 8.9 Examples of pipeline specifications for natural gas.

Components (mol%)	Minimum	Maximum
Methane	75	
Ethane		10
Propane		5
Butanes		2
Pentanes plus		0.5
Nitrogen and other inerts		3–4
Carbon dioxide		3–4
Trace components		
Hydrogen sulfide		0.25–1.0 g/100 ft^3
Mercaptan sulfur		0.25–1.0 g/100 ft^3
Total sulfur		5–20 g/100 ft^3
Water vapor		7.0 lbs (mm ft^{-3})
Oxygen		0.2–1.0 ppm v/v
Heating value	950	1150 Btu ft^{-3}

Table 8.10 Composition of different gases.

Gas	Composition, % v/v									
	Carbon dioxide	Carbon monoxide	Methane	Butane	Ethane	Propane	Hydrogen	Hydrogen sulfide	Oxygen	Nitrogen
Coal gas	3.8	28.4	0.2				17			50.6
Coke-oven gas	2	5.5	32				51.9		0.3	4.8
Digester gas	30		64				0.7	0.8		2
Landfill gas	47	0.1	47				0.1	0.01	0.8	3.7
Natural gas	0–0.8	0–0.45	82–93		0–15.8		0–1.8	0–0.18	0–0.35	0.5–8.4
Propane gas				0.5–0.8	2.0–2.2	73–97				

D2163, 2017; ASTM D2426, 2017; ASTM D2504, 2017; ASTM D2505, 2017; ASTM D2593, 2017; ASTM D2712, 2017; ASTM D3246, 2017; ASTM D4423, 2017; ASTM D4424, 2017; ASTM D4864, 2017; ASTM D5303, 2017; ASTM D5799, 2017; ASTM D6159, 2017).

The properties of unrefined natural gas are variable because the composition of natural gas is never constant. Therefore, the properties and behavior of unrefined and refined natural gas (Table 8.11), as determined by a series of standard test methods (Table 8.12) (Mokhatab et al., 2006; Speight, 2014; ASTM, 2017), are best understood by investigating the properties and behavior of the constituents. Determination of the properties of natural gas or, for that matter, any fuel gas gives evidence of efficiency of combustion and is a prime factor in controlling the operation for maximum results and in arriving at improvements in design. It is often desirable (even necessary) to predict the quantity and analysis of the products of combustion to determine flue sizes and furnace pressure and to predict the magnitude of stack or gas losses. There are equations to permit the calculation of the quantities of gases in the

Table 8.11 General properties of unrefined and refined natural gas.

Property	Unrefined	Refined
Relative molar mass	20	
Carbon content (weight %)	73	75
Hydrogen content (weight %)*	27	25
Oxygen content (weight %)	0.4	0
Hydrogen-to-hydrogen atomic ratio	3.5	4
Density relative to air at 15 °C	1.5	0.6
Boiling temperature (°C/1 atm)	−162	
Autoignition temperature (°C)	540	560
Octane number	120	130
Methane number	69	99
Vapor flammability limits (vol.%)	5	15
Flammability limits	0.7	2.1
Lower heating/calorific value (Btu)	900	
Methane concentration (volume %)	100	80
Ethane concentration (volume %)	50	0
Nitrogen concentration (volume %)	15	0
Carbon dioxide concentration (volume %)	5	0
Sulfur concentration (ppm, mass)	5	0

Table 8.12 Examples of standard test methods for natural gas (ASTM, 2017).

ASTM D130	Detection of Copper Corrosion from Petroleum Products by the Copper Strip Tarnish Test
ASTM D216	Distillation of Natural Gasoline
ASTM D323	Vapor Pressure of Petroleum Products (Reid Method)
ASTM D1025	Nonvolatile Residue of Polymerization Grade Butadiene
ASTM D1070	Relative Density (Specific Gravity) of Gaseous Fuels
ASTM D1071	Standard Test Methods for Volumetric Measurement of Gaseous Fuel Samples
ASTM D1072	Total Sulfur in Fuel Gases
ASTM D1089	Carbonyl Content of Butadiene
ASTM D1142	Water Vapor Content of Gaseous Fuels by Measurement of Dew-Point Temperature
ASTM D1265	Sampling Liquefied Petroleum (LP) Gases
ASTM D1266	Sulfur in Petroleum Products (Lamp Method)
ASTM D1267	Vapor Pressure of Liquefied Petroleum (LP) Gases (LP-Gas Method)
ASTM D1298	Density, Relative Density (Specific Gravity), or API Gravity of Crude Petroleum and Liquid Petroleum Products by Hydrometer Method
ASTM D1657	Density or Relative Density of Light Hydrocarbons by Pressure Hydrometer
ASTM D1826	Calorific Value of Gases in Natural Gas Range by Continuous Recording Colorimeter
ASTM D1837	Volatility of Liquefied Petroleum (LP) Gases
ASTM D1838	Copper Strip Corrosion by Liquefied Petroleum (LP) Gases
ASTM D1945	Analysis of Natural Gas by Gas Chromatography
ASTM D1946	Standard Practice for Analysis of Reformed Gas by Gas Chromatography
ASTM D1988	Standard Test Method for Mercaptans in Natural Gas Using Length-of-Stain Detector Tubes
ASTM D2158	Residues in Liquefied Petroleum (LP) Gases
ASTM D2163	Analysis of Liquefied Petroleum (LP) Gases and Propane Concentrates by Gas Chromatography
ASTM D2384	Traces of Volatile Chlorides in Butane-Butene Mixtures
ASTM D2385	Hydrogen Sulfide and Mercaptan Sulfur in Natural Gas (Cadmium Sulfate Iodometric Titration Method)
ASTM D2421	Practice for Interconversion of Analysis of C5 and Lighter Hydrocarbons to Gas Volume, Liquid Volume or Mass Basis
ASTM D2426	Butadiene Dimer and Styrene in Butadiene Concentrates by Gas Chromatography

(Continued)

Table 8.12 (Continued)

ASTM D2427	Determination of C, through C, Hydrocarbons in Gasolines by Gas Chromatography
ASTM D2504	Noncondensable Gases in C_3 and Lighter Hydrocarbon Products by Gas Chromatography
ASTM D2505	Ethylene, Other Hydrocarbons, and Carbon Dioxide in High-Purity Ethylene by Gas Chromatography
ASTM D2593	Butadiene Purity and Hydrocarbon Impurities by Gas Chromatography
ASTM D2597	Test Method for Demethanized Hydrocarbon Liquid Mixtures Containing Nitrogen and Carbon Dioxide by Gas Chromatography
ASTM D2598	Calculation of Certain Physical Properties of Liquefied Petroleum (LP) Gases from Compositional Analysis
ASTM D2623	Knock Characteristics of Liquefied Petroleum (LP) Gases by the Motor (LP) Method
ASTM D2699	Knock Characteristics of Motor Fuels by the Research Method
ASTM D2700	Knock Characteristics of Motor and Aviation Type Fuels by the Motor Method
ASTM D2712	Hydrocarbon Traces in Propylene Concentrates by Gas Chromatography
ASTM D2713	Dryness of Propane (Valve Freeze Method)
ASTM D2725	Hydrogen Sulfide in Natural Gas (Methylene Blue Method)
ASTM D2784	Sulfur in Liquefied Petroleum Gases (Oxy-Hydrogen Burner or Lamp)
ASTM D3031	Total Sulfur in Natural Gas by Hydrogenation
ASTM D3246	Sulfur in Petroleum Gases by Oxidative Microcoulometry
ASTM D3588	Standard Practice for Calculating Heat Value, Compressibility Factor, and Relative Density of Gaseous Fuels
ASTM D3700	Practice for Containing Hydrocarbon Fluid Samples Using a Floating Piston Cylinder
ASTM D4084	Standard Test Method for Analysis of Hydrogen Sulfide in Gaseous Fuels (Lead Acetate Reaction Rate Method
ASTM D4150	Standard Terminology Relating to Gaseous Fuels
ASTM D4891	Standard Test Method for Heating Value of Gases in Natural Gas Range by Stoichiometric Combustion
ASTM D4953	Test Method for Vapor Pressure of Gasoline-Oxygenate Blends (Dry Method)
ASTM D4984	Standard Test Method for Carbon Dioxide in Natural Gas Using Length-of-Stain Detector Tubes
ASTM D5190	Test Method for Vapor Pressure of Petroleum Products (Automatic Method)
ASTM D5191	Test Method for Vapor Pressure of Petroleum Products (Mini Method)
ASTM D5297	Standard Practice for Automatic Sampling of Gaseous Fuels

Table 8.12 (Continued)

ASTM D5454	Standard Test Method for Water Vapor Content of Gaseous Fuels Using Electronic Moisture Analyzers
ASTM D5503	Standard Practice for Natural Gas Sample-Handling and Conditioning Systems for Pipeline Instrumentation
ASTM D5504	Standard Test Method for Determination of Sulfur Compounds in Natural Gas and Gaseous Fuels by Gas Chromatography and Chemiluminescence
ASTM D5954	Standard Test Method for Mercury Sampling and Measurement in Natural Gas by Atomic Absorption Spectroscopy
ASTM D6228	Standard Test Method for Determination of Sulfur Compounds in Natural Gas and Gaseous Fuels by Gas Chromatography and Flame Photometric Detection
ASTM D6273	Standard Test Methods for Natural Gas Odor Intensity
ASTM D6350	Standard Test Method for Mercury Sampling and Analysis in Natural Gas by Atomic Fluorescence Spectroscopy

combustion products of gaseous fuels either by volume or by weight. More recently, mass spectrometry has become a popular method of choice for compositional analysis of low molecular weight and has replaced several older methods (ASTM D2421, 2017; ASTM D2650, 2017). Gas chromatography (ASTM D1945, 2017) is another method of choice for hydrocarbon identification in gases.

Generally, the determination of the bulk properties is easy to perform and, therefore, quick and economical. Several properties may correlate well with certain compositional characteristics and are widely used as a quick and inexpensive means to determine those. The most important properties of a whole crude oil are its boiling point distribution, its density (or API gravity), and its viscosity. The *boiling point distribution, boiling profile,* or *distillation assay* gives the yield of the various distillation cuts, and selected properties of the fractions are usually determined (Table 1.4). In fact, the boiling point distribution of crude oil is a prime property in its own right that indicates how much gasoline and other transportation fuels can be made from crude oil without conversion. Density and viscosity are measured for secondary reasons. The former helps to estimate the paraffinic character of the oil, and the latter permits the assessment of its undesirable residual material that causes resistance to flow. Boiling point distribution, density, and viscosity are easily measured and give a quick first evaluation of crude oil. Sulfur content, another crucial and primary property of a crude oil, is also readily determined. Certain composite characterization values, calculated from density and mid-boiling point, correlate better with molecular composition than density alone (Gary et al., 2007; Speight, 2014, 2017a; Hsu and Robinson, 2017).

Thus, if the natural gas has been cleaned (i.e. any constituents such as carbon dioxide and hydrogen sulfide have been removed) and the only constituents remaining are hydrocarbons, the properties and behavior of natural gas become a study of the properties and behavior of the relevant.

8.3.1 Compression and Expansion

Natural gas, like any gas, can be pressurized using a compressor in which the volume of the gas is decreased. Typically, natural gas is compressed using pressure on the order of 2900–4300 psi, which gives a 200- to 250-fold reduction in the volume of the gas. The compression factor (also known as the compressibility factor or the real gas factor and given the symbol Z) appears in equations governing volumetric metering. Moreover, the conversion of volume at metering conditions to volume under defined reference conditions can properly proceed with an accurate knowledge of Z at both relevant pressure and relevant temperature conditions.

The isothermal gas compressibility, which is given the symbol c_g, is a useful concept that will be used extensively in determining the compressible properties of the reservoir. The isothermal compressibility is also called the bulk modulus of elasticity. Gas usually is the most compressible medium in the reservoir. However, care should be taken so that it is not confused with the gas deviation factor, z, which is sometimes called the supercompressibility factor:

$$c_g = -\frac{1}{V}\left(\frac{\partial V_g}{\partial p}\right)$$

There are many different methods used to compress gasses. When it comes to natural gas, the most common method is the diaphragm compressor, which contains a series of chambers, each one equipped with a specially designed membrane. Gas is pumped into the first chamber, where the volume is constricted by the membrane. Once this chamber is full, it is emptied into the next, smaller, chamber and so on down the line until the natural gas has reached the desired pressure. Once the process is finished, the gas must be kept at pressure in tanks.

Compressor stations are located along a natural gas pipeline and compress the gas to a specified pressure, thereby allowing it to continue traveling along the pipeline to the intended recipient. The total number of compressor station facilities required to move product varies depending on the region and conditions. Generally, compressor stations are located about every 40–70 miles along the pipeline.

There is a wide variation in the pressure within a given section of pipeline compared with other pipelines in other areas. The typical pressure may range anywhere from 200 psi (pounds per square inch) to 1500 psi. This wide variation is also due to the type of area in which the pipeline is operating, its elevation, and the diameter of the pipeline. Because of the change in the environment, compressor stations may compress natural gas at different levels. Supply and demand can also be a factor at times in the level of compression required for the flow of the natural gas.

Compressor stations typically include scrubbers, strainers, or filter separators that remove liquids, dirt, particles, and other impurities from the natural gas. Though natural gas is considered "dry" as it passes through the pipeline, water and other hydrocarbons may condense out of the gas as it travels. Thus, compressor stations will also remove these impurities from the gas so that they can be disposed of or sold as desired.

As the compressed natural gas is unloaded from the pipeline or from the transportation vehicle, it is depressurized to ensure a consistent flow without interruption. Due to the Joule–Thomson effect, where gas temperature will drop when the pressure drops, this letdown of pressure means heat must be put back into the gas stream to prevent freezing of sensitive instruments and pipelines. For convenience and safety, the trailers also serve to house the compressed natural gas, thereby saving customers the cost of on-site storage.

Biogas (including and landfill gas) can be compressed, in the same way as natural gas is compressed to produce compressed natural gas, and used to power motor vehicles. Biogas can be cleaned and upgraded to natural gas standards to produce biomethane. It is a renewable resource because its production-and-use cycle is continuous, and it generates no net carbon dioxide. Organic material grows, is converted and used, and then regrows in a continually repeating cycle. From a carbon perspective, as much carbon dioxide is absorbed from the atmosphere in the growth of the primary bioresource as is released when the material is ultimately converted to energy.

8.3.2 Copper Corrosion

Copper strip corrosion (ASTM D1838, 2017) limits provide assurance that difficulties will not be experienced in the deterioration of copper and copper alloy fittings and connections commonly used in liquefied petroleum gas systems. As little as 1 ppm of hydrogen sulfide can cause a copper strip test failure. Thus, the copper strip corrosion test is an extremely sensitive test that will detect virtually all species of corrosive sulfur, including minute traces of hydrogen sulfide. It is critically important that the product being tested does not contain any additives that may diminish the reaction with the copper strip.

Reactive sulfur compounds such as elemental sulfur, hydrogen sulfide, and some other sulfur species are corrosive to copper. Another test method (ASTM D1835, 2017) is an additional safeguard that hydrogen sulfide is not present. A simple pass–fail test for hydrogen sulfide (ASTM D2420, 2017) is based on the discoloration of a lead acetate solution on a filter paper exposed to propane vapors.

8.3.3 Density and Formation Volume Factor

Density is defined as mass per unit volume of a fluid. It is a state function and for a pure compound depends on both temperature and pressure. Liquid densities decrease as temperature increases, but the effect of pressure on liquid densities at moderate pressures is usually negligible. At low and moderate pressures (less than a few bars), saturated liquid density is nearly the same as actual density at the same temperature.

The density of light hydrocarbons can be determined by several methods (ASTM D1070, 2017) including a hydrometer method (ASTM D1298, 2017) or by a pressure hydrometer method (ASTM D1657, 2017). The specific gravity (relative density) (ASTM D1070, 2017; ASTM D1657, 2017) by itself has little significance compared with its use for higher molecular weight liquid petroleum products and can only give an indication of quality characteristics when combined with values for volatility and vapor pressure. It is important for stock quantity calculations and is used in connection with transport and storage.

Density is a physical property of matter and is a measure of the relative molecular weight (often referred to as the *heaviness*) of hydrocarbons and other chemicals at a constant volume, and each constituents of natural gas has a unique density associated with it (Table 8.13). Another term, *specific gravity*, is commonly used in relation to the properties of hydrocarbons. The specific gravity of a substance is a comparison of its density to that of water. For most chemical compounds (i.e. those that are solid or liquid), the density is

Table 8.13 Density of gases at 0°C (32°F) and atmospheric pressure.

Gas	Symbol	Vapor density air = 1.0000	weight in pounds	One pound, Volume in cubic feet	Vapor density Hydrogen = 1
Oxygen	O	1.053	0.08922	11.208	15.87
Nitrogen	N	0.9673	0.07829	12.773	13.92
Hydrogen	H	0.0696	0.005621	177.9	1
Carbon dioxide	CO_2	1.5291	0.12269	8.151	21.83
Carbon monoxide	CO	0.9672	0.07807	12.809	13.89
Methane	CH_4	0.5576	0.04470	22.371	7.95
Ethane	C_2H_6	1.075	0.08379	11.935	14.91
Sulfur dioxide	SO_2	2.2639	0.17862	5.598	31.96
Air	—	1	0.08071	12.39	—

measured relative to water (1.00). For gases, the density is more likely to be compared with the density of air (also given the number 1.00, but this is arbitrary and bears no relationship to the density of water). As a comparison, the density of liquefied natural gas (LNG) is approximately 0.41–0.5 kg l^{-1}, depending on temperature, pressure, and composition; in comparison the density of water is 1.0 kg l^{-1}. The liquefaction process requires the removal of some of the nonmethane components such as water and carbon dioxide from the produced natural gas to prevent them from forming solids when the gas is cooled to about LNG temperature (−161 °C, −256 °F). As a result, liquefied natural gas is typically made up mostly of methane.

Density values including those of natural gas hydrocarbons (Figure 8.1) are given at room temperature unless otherwise indicated by a superscript figure. For example, 2.487^{15} indicates a density of 2.487 g cm^{-1} at 15 °C. The density of any gas compared with the density of air is the *vapor density* and is a very important characteristic of the constituents of natural gas and natural gas constituents (Figure 8.2). Furthermore, the statement *natural gas is lighter than air* is often made because of the continued insistence by engineers and scientists that the properties of a mixture are determined by the mathematical average of the properties of the individual constituents of the mixture. Such mathematical bravado and inconsistency of thought is *detrimental to safety* and needs to be qualified.

The formation volume factor of gas is the volume of gas at the reservoir temperature and pressure to the volume at the standard or surface temperature and pressure (p_s and T_s). It is given the symbol B_g and is a dimensionless quantity but is often expressed in either cubic feet or barrels of reservoir volume per

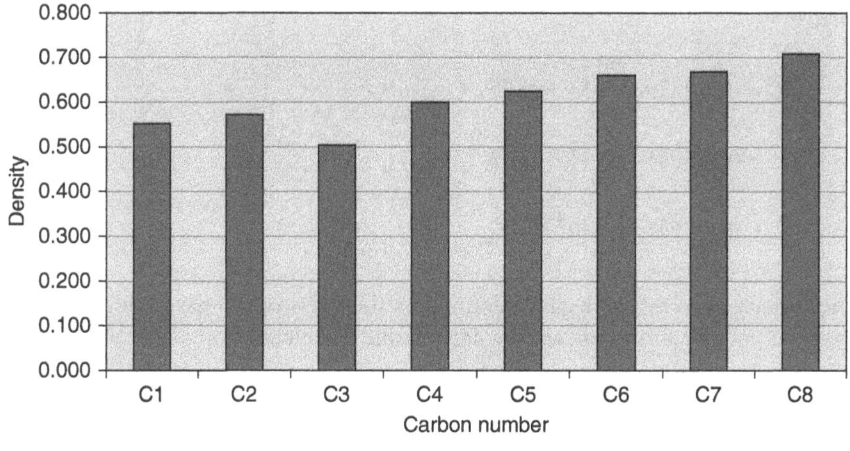

Figure 8.1 Carbon number and density of natural gas hydrocarbons (up to *n*-octane, C_8H_{18}).

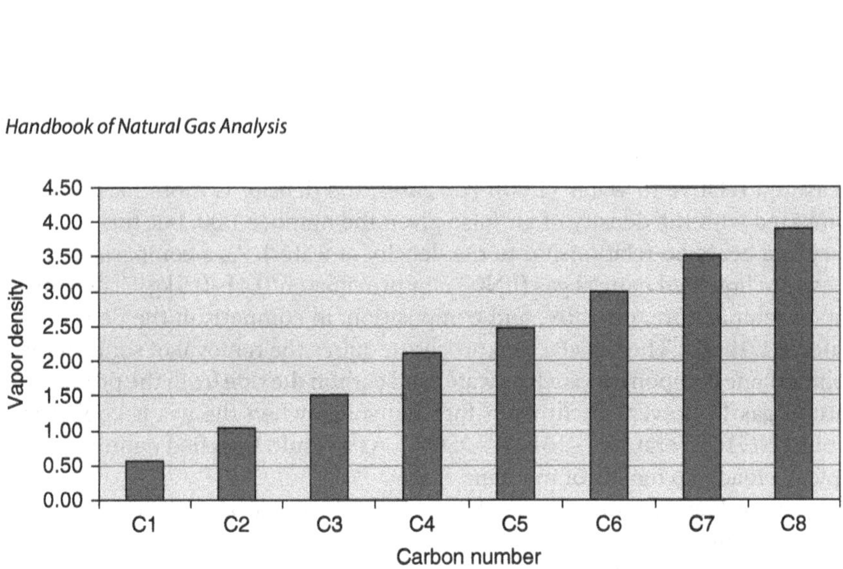

Figure 8.2 Carbon number and vapor density (relative to air = 1.0) of natural gas hydrocarbons (up to *n*-octane, C_8H_{18}).

standard cubic foot of gas. The gas deviation factor is unity under standard conditions; hence the equation for the gas formation volume factor can be calculated using the real gas equation:

$$B_g = \frac{V_R}{V_s} = \frac{znRT}{p}\frac{p_s}{z_s nRT_s} = \frac{p_s zT}{T_s p}$$

When p_s is 1 atm (14.6959 psia or 101.325 kPa) and T_s is 60 °F (519.67 °R or 288.71 °K), this equation can be written in three well-known standard forms:

$$\begin{aligned} B_g &= 0.0282793\frac{zT}{p}\text{RCF / SCF} \\ &= 0.00503676\frac{zT}{p}\text{RB / SCF} \\ &= 0.350958\frac{zT}{p}\text{Rm}^3\text{ / Sm}^3 \end{aligned}$$

The density of a reservoir gas is defined as the mass of the gas divided by its reservoir volume and so can also be derived and calculated from the real gas law:

$$\begin{aligned} \rho_g &= \frac{m_g}{V_R} = \frac{nM_g}{znRT / p} = \frac{nM_{air}\gamma_g}{znRT / p} \\ &= \frac{28.97\gamma_g p}{zRT} \end{aligned}$$

8.3.4 Dew Point

The hydrocarbon dew point is reduced to such a level that retrograde condensation, i.e. condensation resulting from pressure drop, cannot occur under the worst conditions likely to be experienced in the gas transmission system. Similarly, the water dew point is reduced to a level sufficient to preclude formation of C_1–C_4 hydrates in the system. Once the composition of a mixture has been determined, it is possible to calculate various properties such as specific gravity, vapor pressure, calorific value, and dew point. Since dew point can be calculated from composition, direct determination of dew point for a liquefied petroleum gas sample is a measure of composition. It is, of course, of more direct practical value, and if there are small quantities of higher molecular weight material present, it is preferable to use a direct measurement.

The natural gas after appropriate treatment for acid gas reduction, odorization, and hydrocarbon and moisture dew-point adjustment (ASTM D1142, 2017) would then be sold within prescribed limits of pressure, calorific value, and possibly Wobbe index (cv/(sp. gr.).

The most important of these are the Wobbe index (or Wobbe number = calorific value/specific gravity) and the flame speed, usually expressed as a factor or an arbitrary scale on which that of hydrogen is 100. This factor can be calculated from the gas analysis. In fact, calorific value and specific gravity can be calculated from compositional analysis (ASTM D3588, 2017).

The Wobbe number gives a measure of the heat input to an appliance through a given aperture at a given gas pressure. Using this as a vertical coordinate and the flame speed factor as the horizontal coordinate, a combustion diagram can be constructed for an appliance, or a whole range of appliances, with the aid of appropriate test gases. This diagram shows the area within which variations in the Wobbe index of gases may occur for the given range of appliances without resulting in either incomplete combustion, flame lift, or the lighting back of pre-aerated flames. This method of prediction of combustion characteristics is not sufficiently accurate to eliminate entirely the need for the practical testing of new gases.

Another important combustion criterion is the Gas Modulus, $M = P/W$, where P is the gas pressure and W is the Wobbe number of the gas. This must remain constant if a given degree of aeration is to be maintained in a pre-aerated burner using air at atmospheric pressure.

8.3.5 Flammability and Explosive Properties

In order for combustion of a hydrocarbon to occur, three elements are required: (i) hydrocarbon vapor; (ii) oxygen, i.e. air; and (iii) a source to initiate the combustion. One important parameter to have a good combustion is the ratio of air to hydrocarbon fuel. The combustion does not occur if there is too much air (little fuel) or too little air (too much fuel). This suggests that combustion occurs when hydrocarbon concentration in the air is within a certain range. This range is called *flammability range* and is usually expressed in terms of lower and upper volume percent in the mixture of hydrocarbon vapor and air. If the volume of hydrocarbon in the air is within the flammability range, then the mixture is flammable by a spark or flame.

The boiling point (boiling temperature) of a substance is the temperature at which the vapor pressure of the substance is equal to atmospheric pressure. At the boiling point, a substance changes its state from liquid to gas. A stricter definition of boiling point is the temperature at which the liquid and vapor (gas) phases of a substance can exist in equilibrium. When heat is applied to a liquid, the temperature of the liquid rises until the *vapor pressure* of the liquid equals the pressure of the surrounding atmosphere (gases). At this point there is no further rise in temperature, and the additional heat energy supplied is absorbed as *latent heat* of vaporization to transform the liquid into gas. This transformation occurs not only at the surface of the liquid (as in the case of *evaporation*) but also throughout the volume of the liquid, where bubbles of gas are formed. The boiling point of a liquid is lowered if the pressure of the surrounding atmosphere (gases) is decreased. On the other hand, if the pressure of the surrounding atmosphere (gases) is increased, the boiling point is raised. For this reason, it is customary when the boiling point of a substance is given to include the pressure at which it is observed, if that pressure is other than standard, i.e. 760 mm of mercury or 1 atm (STP, standard temperature and pressure).

The boiling points of crude oil fractions are rarely, if ever, distinct temperatures. It is, in fact, more correct to refer to the boiling ranges of the various fractions; the same is true of natural gas. To determine these ranges, the material in question is tested in various methods of distillation, either at atmospheric pressure or at reduced pressure. Thus, the boiling points of the hydrocarbon constituents of natural gas increase with molecular weight, and the initial boiling point of natural gas corresponds to the boiling point of the most volatile constituents (i.e. methane) (Figure 8.3).

Purified natural gas is neither corrosive nor toxic, its ignition temperature is high, and it has a narrow flammability range, making it an apparently safe fossil fuel compared with other fuel sources. In addition, purified natural gas (i.e. methane) having a specific gravity (0.60) lower than that of air (1.00) rises if escaping and dissipates from the site of any leak. However, methane is highly flammable, burns easily and almost completely. Therefore, natural gas can also be hazardous to life and property through an explosion. When natural gas is confined, such as within a house or in a coal mine, concentration of the gas can reach explosive mixtures that, if ignited, results in blasts that could destroy buildings.

The *flash point* of crude oil or a crude oil product, including natural gas, is the temperature to which the product must be heated under specified

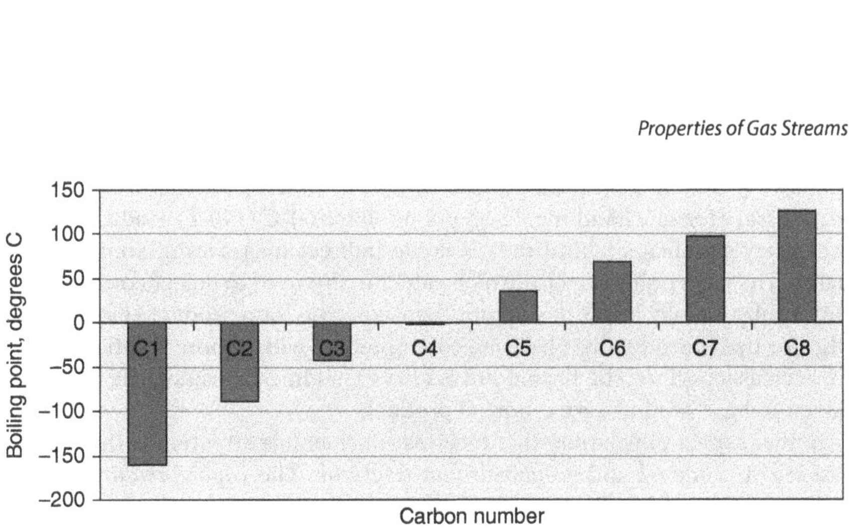

Figure 8.3 Carbon number and boiling points of natural gas hydrocarbons (up to *n*-octane, C_8H_{18}).

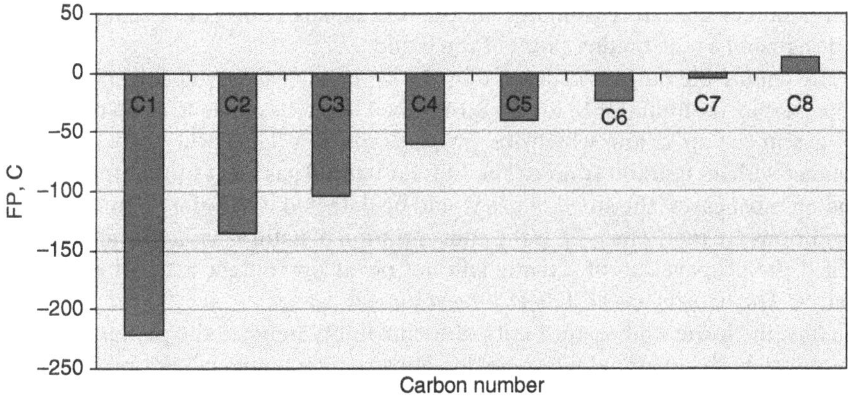

Figure 8.4 Carbon number and flash point of natural gas hydrocarbons (up to *n*-octane, C_8H_{18}).

conditions to give of sufficient vapor to form a mixture with air that can be ignited momentarily by a specified flame (ASTM D56, D92, and D93). As with other properties, the flash point is dependent on the composition of the gas and the presence of other hydrocarbon constituents (Figure 8.4). The *fire point* is the temperature to which the gas must be heated under the prescribed conditions of the method to burn continuously when the mixture of vapor and air is ignited by a specified flame (ASTM D92).

From the viewpoint of safety, information about the flash point is of most significance at or slightly above the maximum temperatures (30–60 °C, 86–140 °F) that may be encountered in storage, transportation, and use of liquid crude oil products, in either closed or open containers. In this temperature range the relative fire and explosion hazard can be estimated from the flash point. For products with flash point below 40 °C (104 °F), special precautions are necessary for safe handling. Flash points above 60 °C (140 °F) gradually lose their safety significance until they become indirect measures of some other quality. The flash point of a crude oil product is also used to detect contamination. A substantially lower flash point than expected for a product is a reliable indicator that a product has become contaminated with a more volatile product, such as gasoline. The flash point is also an aid in establishing the identity of a natural gas product or a crude oil product.

Another aspect of volatility that receives considerable attention is the vapor pressure of crude oil and its constituent fractions. The *vapor pressure* is the force exerted on the walls of a closed container by the vaporized portion of a liquid. Conversely, it is the force that must be exerted on the liquid to prevent it from vaporizing further (ASTM D323, 2017). The vapor pressure increases with temperature for any given gasoline, liquefied petroleum gas, or other product. The temperature at which the vapor pressure of a liquid, either a pure compound or a mixture of many compounds, equals 1 atm (14.7 psi, absolute) is designated as the boiling point of the liquid.

The flammable range is expressed by the lower explosive limit (LEL) and the upper explosive limit (UEL) (Table 8.14). The LEL is the concentration of natural gas in the air below which the propagation of a flame will not occur on contact with an ignition source. The LEL for natural gas is 5% by volume in air, and, in most cases, the smell of gas would be detected well before combustion conditions are met. The UEL is the concentration of natural gas in the air above which the propagation of a flame will not occur on contact with an ignition source. The natural gas UEL is 15% by volume in air.

Thus, the lower and upper limits of flammability indicate the percentage of combustible gas in air below which and above which flame will not propagate. When flame is initiated in mixtures having compositions within these limits, it will propagate, and therefore the mixtures are flammable. A knowledge of flammable limits and their use in establishing safe practices in handling gaseous fuels is important, for example, when purging equipment used in gas service, in controlling factory or mine atmospheres, or in handling liquefied gases.

The calculation of flammable limits is accomplished by Le Chatelier's modification of the mixture law, which (expressed in the simplest form) is

$$L = \frac{100}{p_1 / N_1 + p_2 / N_2 \cdots + p_n / N_n}$$

L is the volume percentage of fuel gas in a limited mixture of air and gas; p_1, $p_2, \ldots, p_n$ are the volume percentages of each combustible gas present in the fuel gas, calculated on an air- and inert-free basis so that $p_1 + p_2 + \cdots + p_n = 100$; and N_1, N_2, ..., N_n are the volume percentages of each combustible gas in a limit mixture of the individual gas and air. The foregoing relation may be applied to

Table 8.14 Flammability limits of the constituents of fuel gases.

Gas[a]	LFL % v/v in air	UFL % v/v in air
n-Butane	1.6	8.4
Butylene (1-butene)	1.7	9.7
Carbon monoxide	12.5	74.2
Carbonyl sulfide	12.5	74
2,2-Dimethyl propane	1.4	7.5
Ethane	3	12.4
Ethylene	2.7	36
n-Heptane	1	7
n-Hexane	1	7.5
Hydrogen	4	75
Hydrogen sulfide	4	46
Isobutane	1.82	9.6
Methane	5	15
Methyl mercaptan	3.9	21.8
n-Octane	1	7
Isooctane	0.6	6
n-Pentane	1.4	7.8
Isopentane	1.3	9.2
Propane	2.1	9.5
Propylene	2	11.1

[a]Listed alphabetically.

gases with inert content of 10% or less without introducing an absolute error of more than 1 or 2% in the calculated limits.

Explosions caused by natural gas leaks occur a few times each year. Individual homes, small businesses, and boats are most frequently affected when an internal leak builds up gas inside the structure. Frequently, the blast will be enough to significantly damage a building but leave it standing. Occasionally, the gas can collect in high enough quantities to cause a deadly explosion, disintegrating one or more buildings causing serious injury or death to the building occupants.

In any form, a minute amount of odorant that has an obvious smell is added to the otherwise colorless and odorless gas, so that leaks can be detected before a fire or explosion occurs. Odorants are considered nontoxic in the extremely low concentrations occurring in natural gas delivered to the end user.

8.3.6 Gas Laws

There are two types of gases: an ideal gas and a nonideal gas. An ideal gas has the following properties: (i) There are no intermolecular forces between the gas particles, (ii) the volume occupied by the particles is negligible compared with the volume of the container they occupy, and (iii) the only interactions between the particles and the container walls are perfectly elastic collisions.

In terms of properties and behavior, natural gas is a nonideal gas and obeys the gas law:

$$PV = nZRT$$

P is the pressure, V is the volume, T is the absolute temperature (degree Kelvin), Z is the compressibility, n is the number of kilo-moles of the gas, and R is the gas constant. For example, if all other factors remained constant, when the volume of a certain mass of gas is reduced by 50%, the pressure would double and so on. As a gas, it would expand to fill any volume it is in. However, the compressibility, Z, is the factor that differentiates natural gas from an ideal gas. For methane, Z is 1 at 1 atmosphere pressure (14.7 psi).

8.3.7 Heat of Combustion

The *heat of combustion* (*energy content, calorific value*) of natural gas is the amount of energy that is obtained from the burning of a specified volume of natural gas that is measured in Btu. The value of natural gas is calculated by its Btu content. One Btu is the quantity of heat required to raise the temperature of 1 lb of water of 1 °F at atmospheric pressure. All of the hydrocarbon constituents of natural gas are combustible, but nonflammable nonhydrocarbon components (carbon dioxide, hydrogen sulfide, nitrogen, and helium) detract slightly from the heating value of natural gas. The calorific value (heat of combustion) gives an indication of the satisfactory combustion of hydrocarbon gases, which depends upon the matching of burner and appliance design with certain gas characteristics. Various types of test methods are available for the direct determination of calorific value (ASTM D1826, 2017; ASTM D3588, 2017). Typically, the heat value of gases is generally determined at constant pressure in a flow calorimeter in which the heat released by the combustion of a definite quantity of gas is absorbed by a measured quantity of water or air. A continuous recording calorimeter is available for measuring heat values of natural gas.

The complete oxidation of gaseous hydrocarbon fuels forms carbon dioxide (CO_2) from all of the carbon and water (H_2O) from the hydrogen, that is, for a gaseous hydrocarbon fuel with the general composition C_nH_m:

$$C_nH_m + (n + m/4)O_2 \rightarrow nCO_2 + (m/2)H_2O$$

Moreover, even in the case of complete combustion, the accounting of all species present in combustion exhaust involves more than simply measuring the carbon dioxide and water emissions. Since a gaseous fuel is burned in air rather than in pure oxygen, the nitrogen in the air may participate in the combustion process to produce nitrogen oxides. Also, if the gas should, by chance, contain elements other than carbon and hydrogen, these foreign elements will also be transformed during the combustion process. In addition, the combustion of the gas and the noncarbon and nonhydrogen components is not always complete, and the effluent gases will, in such cases, contain unburned and partially burned products in addition to carbon dioxide and water. The test method for determining the heat of combustion of natural gas and the potential for interference in the result must also be taken into account. In fact, knowing the heat of combustion of methane can give an indication of the composition of the natural gas from the heat of combustion of the gas. Thus, a cubic foot of natural gas has an energy content of approximately 1031 Btu, but the range of values is between 500 and 1500 Btu, depending upon the composition of the gas.

In fact, gas chromatography in one form or another will continue to be the method of choice for characterization of low molecular weight hydrocarbons. Developments in high-speed test methods using gas chromatographic instrumentation and data processing are already leading to new and revised test methods.

In the United States, at retail, natural gas is often sold in units of therms (th, 1 th = 100 000 Btu). Wholesale transactions are generally done in decatherms (Dth), in thousand decatherms (MDth), or in million decatherms (MMDth). A million decatherms is roughly a billion cubic feet of natural gas.

The gross heats of combustion of crude oil and its products are given with fair accuracy by the equation

$$Q = 12400 - 2100d^2$$

where d is the 60/60 °F specific gravity. Deviation from the formula is generally less than 1%.

8.3.8 Liquefied Natural Gas

When natural gas is cooled to a temperature of approximately −160°C (approximately −260 °F) at atmospheric pressure, it condenses to a liquid (*liquefied natural gas*; LNG). One volume of this liquid takes up about 1/600th of the volume of natural gas. Liquefied natural gas weighs less than one-half that of water, approximately 45% as much. Liquefied natural gas is odorless, colorless, noncorrosive, and nontoxic. When vaporized it burns only in concentrations of 5–15% v/v when mixed with air (Section 8.2.3). Neither liquefied natural gas nor its vapor can explode in an unconfined environment. Since liquefied natural gas takes less volume and weight, it presents more convenient options for storage and transportation.

8.3.9 Methane Number

The methane number (Table 8.11) is a measure of the knock resistance of the natural gas and hence an essential design parameter for engines that use compressed natural gas as the fuel. A high methane number allows engines to be optimized in terms of higher performance and efficiency. The optimization range is limited by the minimum methane number to be considered by the engine developers. This issue is also crucial for the aspect of bi-fuel operation with natural gas and gasoline. Since gasoline has a lower knock resistance than natural gas, the need to ensure smooth and reliable engine operation with petrol is another restriction when it comes to optimizing the engine for natural gas. The knock-resistant methane has, by definition, a methane number of 100, while the knock-prone hydrogen has the methane number of 0. Gases with higher knock resistance than methane are compared with methane and carbon dioxide mixtures.

Typically, a gas with a low calorific value is accompanied by a high inert gas content, and a gas with a high calorific value is accompanied by a high content of higher molecular weight hydrocarbons (C_{2+}) content. With respect to the methane number, inert gases are favorable for the methane number level, whereas the higher molecular weight hydrocarbons (C_{2+}) reduce the methane number.

8.3.10 Odorization

Odorization is a technique that is applied for safety reasons to the largest consumer product – natural gas. Since natural gas as delivered to pipelines has practically no odor, the addition of an odorant is required by most regulations in order that the presence of the gas can be detected readily in case of accidents and leaks. This odorization is provided by the addition of trace amounts of some organic sulfur compounds to the gas before it reaches the consumer. The standard requirement is that a user will be able to detect the presence of the gas by odor when the concentration reaches 1% of gas in air. Since the lower limit of flammability of natural gas is approximately 5%, this 1 percent requirement is essentially equivalent to one-fifth the lower limit of flammability. The combustion of these trace amounts of odorant does not create any serious problems of sulfur content or toxicity.

8.3.11 Relative Density

Specific gravity of a substance is the ratio of the density of a substance to the density of a reference substance. Also, it is the ratio of the mass of a substance to the mass of a reference substance for the same given volume. The apparent specific gravity is the ratio of the weight of a volume of the substance to the weight of an equal volume of the reference substance. For liquid hydrocarbons, the reference substance is typically water at its densest (4 °C), but for gases, the relative substance is air at room temperature (25 °C). Nonetheless, the temperature and pressure must be specified for both the sample and the reference. Pressure is nearly always 1 atm (14.7 psi, 101.325 kPa).

To determine the density or specific gravity of liquefied petroleum gas in its liquid state, there are two methods, using a metal pressure pycnometer. A pressure hydrometer may be used (ASTM D1267, 2017) for the relative density that may also be calculated from compositional analysis (ASTM D2598, 2017). Various procedures, manual and recording, for specific gravity or density in the gaseous state are given in two methods (ASTM D1070, 2017). Calculation of the density is also possible using any one of four models, depending upon the composition of the gas (ASTM D4784, 2017).

The *relative density* (*specific gravity*) is the ratio of the density (mass of a unit volume) of a substance to the density of a given reference material. Specific gravity usually means relative density (of a liquid) with respect to water.

$$\text{Relative density} = \left[\rho(\text{substance})\rho(\text{reference})\right]$$

As it pertains to gases, particularly in relation to safety considerations at commercial and industrial facilities in the United States, the relative density of a gas is usually defined with respect to air, in which air is assigned a *vapor density* of one (unity) (Table 8.15). With this definition, the vapor density indicates whether a gas is denser (greater than 1) or less dense (less than 1) than air. It has implications for container storage and personnel safety – if a container can release a dense gas, its vapor could sink and, if flammable, collect until it is at a concentration sufficient for ignition. Even if not flammable, it could collect

Table 8.15 Relative density (specific gravity) of natural gas hydrocarbons relative to air.

Gas	Specific gravity
Air	**1.000**
Methane, CH_4	0.5537
Ethane, C_2H_6	1.0378
Propane, C_3H_8	1.5219
Butane, C_4H_{10}	2.0061
Pentane, C_5H_{12}	2.487
Hexane, C_6H_{14}	2.973

in the lower floor or level of a confined space and displace air, possibly presenting a smothering hazard to individuals entering the lower part of that space.

Gases can be divided into two groups based upon their vapor density: (i) gases that are heavier than air and (ii) gases that are as light as air or lighter than air. Gases that have a vapor density greater than one will be found in the bottom of storage containers and will tend to migrate downhill and accumulate in low-lying areas. Gases that have a vapor density that is the same or less than the vapor density of air will disperse readily into the surrounding environment. Additionally, chemicals that have the same vapor density as air (1.0) tend to disperse uniformly into the surrounding air when contained, and, when released into the open air, chemicals that are lighter than air will travel up and away from the ground.

Methane is the only hydrocarbon constituent of natural gas that is lighter than air. The higher molecular weight hydrocarbons have a higher vapor density than air and are likely, after a release, to accumulate in low-lying areas and represent a danger to investigator (of the release). However, the other hydrocarbon constituents of unrefined natural gas (i.e. ethane, propane, butane, etc.) are denser than air (Table 8.16; Figure 8.2). Therefore, should a natural gas leak occur in field operations, especially where the natural gas contains constituents other than methane, only methane dissipates readily into the air, whereas the other hydrocarbon constituents that are heavier than air do not readily dissipate into the atmosphere. This poses considerable risk if these constituents of natural gas accumulate or pool at ground level when it has been erroneously assumed that natural gas is lighter than air.

Relative density is of importance in measurement and custody transfer calculations. It is not a specification property but is usually determined for other reasons. Relative density is determined by the use of a pressure hydrometer (ASTM D1267, 2017) and may also be calculated from a compositional analysis (ASTM D2598, 2017). The vaporization and combustion characteristics of liquefied petroleum gases are completely defined for normal applications by vapor pressure, volatility, and relative density. There are other limits imposed on the liquefied petroleum gas that provide assurance that the product will perform dependably under more extreme conditions of use.

Table 8.16 Boiling point and density of methane relative to air and water.

- Boiling point (760 mmHg): −161.5 °C (−258.7 °F)
- Gas specific gravity: 0.55–0.64 (air = 1.00)
- Specific gravity of liquefied natural gas: 0.42–0.46 (water = 1.00)
- Gas density (varies slightly): 0.0438 lb scf^{-1}

8.3.12 Residue

The standard test method for residue in a gas stream (ASTM D2158, 2017) is a measure of the concentration of contaminants boiling above 37.8 °C (100 °F) that may be present in the liquefied petroleum gas. The contaminants are usually compressor oils, lubricants from valves, plasticizers from hoses, corrosion inhibitors, and other crude oil products from pumps, pipelines, and storage vessels that are used in multiple service applications. Contaminants, whatever the source, can be particularly troublesome in liquid withdrawal systems such as those used in internal combustion engine fuel systems where the materials accumulate in the vaporizer and will ultimately plug the fuel system.

8.3.13 Sulfur Content

The manufacturing processes for liquefied petroleum gas are designed so that the majority, if not all, of the sulfur compounds are removed. The total sulfur level is therefore considerably lower than for other crude oil-based fuels, and a maximum limit for sulfur content helps to define the product more completely. The sulfur compounds that are mainly responsible for corrosion are hydrogen sulfide, carbonyl sulfide, and, sometimes, elemental sulfur. Hydrogen sulfide and mercaptans have distinctive unpleasant odors. A control of the total sulfur content, hydrogen sulfide, and mercaptans ensures that the product is not corrosive or nauseating. Stipulating a satisfactory copper strip test further ensures the control of the corrosion.

Total sulfur in gas can be determined by combustion (ASTM D1072, 2017), by the lamp method (ASTM D1266, 2017), or by hydrogenation (ASTM D4468, 2017). Trace total organic and bound nitrogen is determined (ASTM D4629, 2017). The current test method for higher molecular weight and relatively nonvolatile residues in liquefied petroleum gas (ASTM D2158, 2017) involves evaporation of a liquefied petroleum gas sample, measuring the volume of residue and observing the residue for oil stain on a piece of filter paper.

Corrosive sulfur compounds can be detected by their effect on copper and the form in which the general copper strip corrosion test (ASTM D1838, 2017) for crude oil products is applied to liquefied petroleum gas. Hydrogen sulfide can be detected by its action on moist lead acetate paper, and a procedure is also used as a measure of sulfur compounds. The method follows the principle of the standard doctor test.

8.3.14 Vapor Pressure

The vapor pressure of hydrocarbon derivatives (Table 8.17) is an indirect measure of the low-temperature conditions under which initial vaporization can be expected to take place. It can also be considered a semiquantitative

Table 8.17 Vapor pressures of the common hydrocarbon derivatives found in natural gas, process gas, gas condensate, and natural gasoline.

Compound	Formula	Molecular Weight	Vapor pressure (100 °F, psia)
Methane	CH_4	16.043	–5000
Ethane	C_2H_6	30.07	–800
Propane	C_3H_8	44.097	188
n-Butane	C_4H_{10}	58.124	51.54
Isobutane	C_4H_{10}	58.124	72.39
n-Pentane	C_5H_{12}	72.151	15.575
Isopentane	C_5H_{12}	72.151	20.4444
Neopentane	C_5H_{12}	72.151	36.66
n-Hexane	C_6H_{14}	86.178	4.96
2-Methylpentane	C_6H_{14}	86.178	6.767
3-Methylpentane	C_6H_{14}	86.178	6.103
Neohexane	C_6H_{14}	86.178	9.859
2,3-Dimethylbutane	C_6H_{14}	86.178	7.406
n-Heptane	C_7H_{16}	100.205	1.62
2-Methylhexane	C_7H_{16}	100.205	2.2719
3-Methylhexane	C_7H_{16}	100.205	2.131
3-Ethylpentane	C_7H_{16}	100.205	2.013
2,2-Dimethylpentane	C_7H_{16}	100.205	3.494
2.4-Dimethylpentane	C_7H_{16}	100.205	3.293
3,3-Dimethylpentane	C_7H_{16}	100.205	2.774
n-Octane	C_8H_{18}	114.232	0.537
Isooctane	C_8H_{18}	114.232	1.709
n-Nonane	C_9H_{20}	128.259	0.1796
n-Decane	$C_{10}H_{22}$	142.286	0.0609
Cyclopentane	C_5H_{10}	70.135	9.914
Methylcyclopentane	C_6H_{12}	84.162	4.503
Cyclohexane	C_6H_{12}	84.162	3.266
Methylcyclohexane	C_7H_{14}	98.189	1.6093
Propylene	C_3H_6	42.081	227.6
1-Butene	C_4H_8	56.108	62.1
cis-2-Butene	C_4H_8	56.108	45.95
trans-2-Butene	C_4H_8	56.108	49.94

Table 8.17 (Continued)

Compound	Formula	Molecular Weight	Vapor pressure (100 °F, psia)
Isobutene	C_4H_8	56.108	63.64
1-Pentene	C_5H_{10}	70.135	19.117
1,3-Butadiene	C_4H_6	54.092	59.4
Benzene	C_6H_6	78.114	3.225
Toluene	$C_6H_5CH_3$	92.141	1.033
Ethylbenzene	C_8H_{10}	106.168	0.376
o-Xylene	C_8H_{10}	106.168	0.263
m-Xylene	C_8H_{10}	106.168	0.325
p-Xylene	C_8H_{10}	106.168	0.3424

measure of the amount of the most volatile material present in the product. The predominant volatile component in propane is ethane, and that in butane is propane. Vapor pressure is also used as a means for predicting the maximum pressures that may be experienced in storage and assures trouble-free performance in commercial equipment.

For most purposes, liquefied petroleum gases will be stored, handled, and transported as a liquid and used as a gas. In order to handle, store, and transport it safely, the vapor pressure must be known (ASTM D1267, 2017). Vapor pressure can also be calculated from a compositional analysis (ASTM D2598, 2017). The accuracy of the vapor pressure calculation is entirely dependent on the accuracy of the compositional analysis. It should also be noted that the vapor pressure is also an effective limit on the amount of ethane in propane.

8.3.15 Volatility

Volatility (ASTM D1837, 2017) is expressed in terms of the 95% evaporated temperature and is a measure of the amount of least volatile fuel components present in the product. This specification controls the heavy ends in the fuel and is, in effect, a restriction on higher-boiling fractions that will not vaporize for use at system temperatures. When coupled with the vapor pressure, single-component products for propane and butane and two-component products for butane–propane mixtures are ensured.

Volatility is expressed in terms of the temperature at which 95% of the sample is evaporated and presents a measure of the least volatile component present (ASTM D1837, 2017). Vapor pressure is, therefore, a measure of the most extreme low-temperature conditions under which initial vaporization can take place. By setting limits to vapor pressure and volatility jointly, the specification serves to ensure essentially single-component products for the butane and propane grades (ASTM D1267, 2017; ASTM D2598, 2017). By combining vapor pressure/volatility limits with specific gravity for propane–butane mixtures, essentially two-component systems are ensured. As mentioned earlier, the residue (ASTM D1025, 2017; ASTM D2158, 2017), i.e. nonvolatile matter, is a measure of the concentration of contaminants boiling above 37.8 °C (100 °F) that may be present in the gas.

As stated above, for natural gasoline, the primary criteria are volatility (vapor pressure) and knock performance. The determination of the vapor pressure (ASTM D323, 2017; ASTM D4953, 2017; ASTM D5191, 2017) and distillation profile (ASTM D6897, 2017) is essential. Knock performance is determined by rating in knock test engines by both the motor method (ASTM D2700, 2017) and the research method (ASTM D2699, 2017). The knock characteristics of liquefied petroleum gases can also be determined. Other considerations for natural gasoline are copper corrosion (ASTM D130, 2017) and specific gravity (ASTM D1298, 2017), the determination of the latter being necessary for measurement and transportation of the gas.

8.3.16 Viscosity

Just as the compressibility of natural gas is much higher than that of oil, water, or rock, the viscosity of natural gas is usually several orders of magnitude lower than oil or water. This makes gas much more mobile in the reservoir than either oil or water. Reliable correlation charts are available to estimate gas viscosity, and the viscosity of gas mixtures at one atmosphere and reservoir temperature can be determined from the gas mixture composition:

$$\mu_{ga} = \frac{\sum_{i=1}^{N} y_i \mu_i \sqrt{M_{gi}}}{\sum_{i=1}^{N} \sqrt{M_{gi}}}$$

μ_{ga} is the viscosity of the gas mixture at the desired temperature and atmospheric pressure, y_i is the mole fraction of the ith component, μ_{ga} is the viscosity of the ith component of the gas mixture at the desired temperature and atmospheric pressure, M_{gi} is the molecular weight of the ith component of the gas mixture, and N is the number of components in the gas mixture.

This viscosity is then multiplied by the viscosity ratio to obtain the viscosity at reservoir temperature and pressure.

8.3.17 Water Content

It is a fundamental requirement that liquefied petroleum gas should not contain free water (ASTM D2713, 2017). Dissolved water may give trouble by forming hydrates and giving moisture vapor in the gas phase. Both of these will lead to blockages. Therefore, test methods are available to determine the presence of water using electronic moisture analyzers (ASTM D5454, 2017), dew-point temperature (ASTM D1142, 2017), and length-of-stain detector tubes (ASTM D4888, 2017).

Water content (moisture content, ASTM D2713, 2017) is a measure of the approximate saturation of the product with water. This measurement is a requirement only in commercial and special duty propane types of LP gas. Moisture content is controlled to provide assurance that pressure reducing regulators and similar equipment will operate consistently without freeze-ups caused by the separation of dissolved water from the product.

Moisture content must also be controlled to low concentrations to avoid hydrate formation in pipelines. Gas hydrates are formed as a granular solid or slushy substance when liquid water is intimately mixed with some of the light hydrocarbons, principally methane, ethane, and propane under certain conditions of temperature and pressure. Hydrates can, and do, form in pipelines well above the freezing point of water.

8.4 Environmental Effects

In order to fully evaluate the environmental effects of natural gas, the general properties of the constituents (Table 8.18) must also be considered in addition to the properties related to combustion and the effects of combustion products on the environment.

Currently, natural gas represents approximately one quarter of the energy consumed in the United States with increases in use projected for the next decade. These increases are expected because emissions of greenhouse gases are much lower with the consumption of natural gas relative to other fossil fuel consumption (Speight and Islam, 2016). For example, natural gas, when burned, emits lower quantities of greenhouse gases and criteria pollutants per unit of energy produced than other fossil fuels. This occurs in part because natural gas is fully combusted more easily and in part because natural gas contains fewer impurities than any other fossil fuel. However, the major constituent of natural gas, methane, also contributes directly to the greenhouse effect through venting or leaking of natural gas into the atmosphere (Speight, 2005).

Purified natural gas (methane) is the cleanest of all the fossil fuels. The main products of the combustion of natural gas are carbon dioxide and water vapor. Coal and crude oil release higher levels of harmful emissions, including a higher ratio of carbon emissions, nitrogen oxides (NO_x), and sulfur dioxide (SO_2). Coal and fuel oil also release ash particles into the environment, substances that do not burn, but instead are carried into the atmosphere and contribute to pollution. The combustion of purified natural gas, on the other hand, releases very small amounts of sulfur dioxide and nitrogen oxides, virtually no

Table 8.18 General properties of the constituents of natural gas up to and including *n*-octane (C_8H_{18}) as well as (for comparison) toluene, xylene, and xylene.

Constituent	Molecular weight	Specific gravity	Vapor density	Boiling point, °C	Ignition temperature, °C	Flash point, °C
Methane	16	0.553	0.56	−160	537	−221
Ethane	30	0.572	1.04	−89	515	−135
Propane	44	0.504	1.50	−42	468	−104
Butane	58	0.601	2.11	−1	405	−60
Pentane	72	0.626	2.48	36	260	−40
Hexane	86	0.659	3.00	69	225	−23
Benzene	78	0.879	2.80	80	560	−11
Heptane	100	0.668	3.50	98	215	−4
Octane	114	0.707	3.90	126	220	13
Toluene	92	0.867	3.20	161	533	4
o-Xylene	106	0.88	3.70	144	463	32
m-Xylene	106	0.86	3.70	139	527	27
p-Xylene	106	0.861	3.70	138	5284	27
Ethylbenzene	106	0.867	3.70	136	432	15

ash or particulate matter, and lower levels of carbon dioxide, carbon monoxide, and other reactive hydrocarbons.

Natural gas has no known toxic or chronic physiological effects (that is, it is not poisonous), but it is dangerous insofar as an atmosphere rich in natural gas will result in asphyxiation and death to humans and animals. But concentration of the gas is key. Exposure to a moderate concentration of natural gas may result in a headache or similar symptoms due to oxygen deprivation, but it is likely that the smell (through the presence of the odorant) would be detected well in advance of concentrations being high enough for this to occur.

In fact, in the natural gas and refining industries (Speight, 2005), as in other industries, air emissions include point and nonpoint sources. Point sources are emissions that exit stacks and flares and, thus, can be monitored and treated. Nonpoint sources are *fugitive emissions* that are difficult to locate and capture. Fugitive emissions occur throughout refineries and arise from, for example, the thousands of valves, pipe connections, seals in pumps and compressors, storage tanks, pressure relief valves, and flanged joints.

While individual leaks are typically small, the sum of all fugitive leaks at a gas processing plant can be one of its largest emission sources. These leaks can release methane and volatile constituents of natural gas into the air. Companies can minimize *fugitive emissions* by designing facilities with the fewest possible components and connections and avoiding components known to cause significant fugitive emissions. When companies quantify fugitive emissions, this provides them with important information they can then use to design the most effective leak repair program for their company. Directed inspection and maintenance programs are designed to identify the source of these leaks and to prioritize and plan their repair in a timely fashion. A reliable and effective directed inspection and maintenance plan for an individual facility will be composed of a number of components, including methods of leak detection, a definition of what constitutes a leak, set schedules and targeted devices for leak surveys, and allowable repair time.

A directed inspection and maintenance program begins with a baseline survey to identify and quantify leaks. Quantification of the leaks is critical because this information is used to determine which leaks are serious enough to justify their repair costs. Repairs are then made only to the leaking components that are cost effective to fix. Subsequent surveys are then scheduled and designed based on information collected from previous surveys, permitting operators to concentrate on the components that are more likely to leak. Some natural gas companies have demonstrated that directed inspection and maintenance programs can profitably eliminate as much as 95% of gas losses from equipment leaks.

The environmental effects produced by biogas (and landfill gas) are similar to the effects produced by natural gas, but with an additional risk from the toxicity of the hydrogen sulfide component and any unburned methane that escapes is a potent and long-lived greenhouse gas. Also, biogas can be explosive when mixed in the ratio of one part of biogas to 8–20 parts of air. Special safety precautions have to be taken for entering an empty biogas digester for maintenance work, and it is important that a biogas system never has negative pressure as this could cause an explosion. Negative gas pressure can occur if too much gas is removed or leaked. Because of this, biogas should not be used at pressures below one column inch of water, measured by a pressure gauge.

References

Abatzoglou, N. and Boivin, S. (2009). A review of biogas purification processes. *Biofuels, Bioproducts and Biorefining* 3 (1): 42–71.

Accettola, F., Guebitz, G.M., and Schoeftner, R. (2008). Siloxane removal from biogas by biofiltration. Biodegradation studies. *Clean Technologies and Environmental Policy* 10 (2): 211–218.

Appels, L., Baeyens, J., and Dewil, R. (2008). Siloxane removal from biosolids by peroxidation. *Energy Conversion and Management* 49 (10): 2859–2864.

Arnold, M. and Kajolinna, T. (2010). Development of on-line measurement techniques for siloxanes and other trace compounds in biogas. *Waste Management* 30 (6): 1011–1017.

ASTM (2017). *Annual Book of Standards*. West Conshohocken, Pennsylvania: ASTM international.

ASTM D1025 (2017). *Standard Test Method for Nonvolatile Residue of Polymerization-Grade Butadiene*, Annual Book of Standards. West Conshohocken, Pennsylvania: ASTM International.

ASTM D1070 (2017). *Standard Test Methods for Relative Density of Gaseous Fuels*, Annual Book of Standards. West Conshohocken, Pennsylvania: ASTM International.

ASTM D1071 (2017). *Standard Test Methods for Volumetric Measurement of Gaseous Fuel Samples*, Annual Book of Standards. West Conshohocken, Pennsylvania: ASTM International.

ASTM D1072 (2017). *Standard Test Method for Total Sulfur in Fuel Gases by Combustion and Barium Chloride Titration*, Annual Book of Standards. West Conshohocken, Pennsylvania: ASTM International.

ASTM D1142 (2017). *Standard Test Method for Water Vapor Content of Gaseous Fuels by Measurement of Dew-Point Temperature*, Annual Book of Standards. West Conshohocken, Pennsylvania: ASTM International.

ASTM D1265 (2017). *Standard Practice for Sampling Liquefied Petroleum (LP) Gases. Manual Method*, Annual Book of Standards. West Conshohocken, Pennsylvania: ASTM International.

ASTM D1266 (2017). *Standard Test Method for Sulfur in Petroleum Products (Lamp Method)*, Annual Book of Standards. West Conshohocken, Pennsylvania: ASTM International.

ASTM D1267 (2017). *Standard Test Method for Gage Vapor Pressure of Liquefied Petroleum (LP) Gases (LP-Gas Method)*, Annual Book of Standards. West Conshohocken, Pennsylvania: ASTM International.

ASTM D1298 (2017). *Standard Test Method for Density, Relative Density, or API Gravity of Crude Petroleum and Liquid Petroleum Products by Hydrometer Method*, Annual Book of Standards. West Conshohocken, Pennsylvania: ASTM International.

ASTM D130 (2017). *Standard Test Method for Corrosiveness to Copper from Petroleum Products by Copper Strip Test*, Annual Book of Standards. West Conshohocken, Pennsylvania: ASTM International.

ASTM D1657 (2017). *Standard Test Method for Density or Relative Density of Light Hydrocarbons by Pressure Hydrometer*, Annual Book of Standards. West Conshohocken, Pennsylvania: ASTM International.

ASTM D1826 (2017). *Standard Test Method for Calorific (Heating) Value of Gases in Natural Gas Range by Continuous Recording Calorimeter*, Annual Book of Standards. West Conshohocken, Pennsylvania: ASTM International.

ASTM D1835 (2017). *Standard Specification for Liquefied Petroleum (LP) Gases*, Annual Book of Standards. West Conshohocken, Pennsylvania: ASTM International.

ASTM D1837 (2017). *Standard Test Method for Volatility of Liquefied Petroleum (LP) Gases*, Annual Book of Standards. West Conshohocken, Pennsylvania: ASTM International.
ASTM D1838 (2017). *Standard Test Method for Copper Strip Corrosion by Liquefied Petroleum (LP) Gases*, Annual Book of Standards. West Conshohocken, Pennsylvania: ASTM International.
ASTM D1945 (2017). *Standard Test Method for Analysis of Natural Gas by Gas Chromatography*, Annual Book of Standards. West Conshohocken, Pennsylvania: ASTM International.
ASTM D1946 (2017). *Standard Practice for Analysis of Reformed Gas by Gas Chromatography*, Annual Book of Standards. West Conshohocken, Pennsylvania: ASTM International.
ASTM D1988 (2017). *Standard Test Method for Mercaptans in Natural Gas Using Length-of-Stain Detector Tubes*, Annual Book of Standards. West Conshohocken, Pennsylvania: ASTM International.
ASTM D2158 (2017). *Standard Test Method for Residues in Liquefied Petroleum (LP) Gases*, Annual Book of Standards. West Conshohocken, Pennsylvania: ASTM International.
ASTM D2163 (2017). *Standard Test Method for Determination of Hydrocarbons in Liquefied Petroleum (LP) Gases and Propane/Propene Mixtures by Gas Chromatography*, Annual Book of Standards. West Conshohocken, Pennsylvania: ASTM International.
ASTM D2384 (2017). *Standard Test Methods for Traces of Volatile Chlorides in Butane-Butene Mixtures*, Annual Book of Standards. West Conshohocken, Pennsylvania: ASTM International.
ASTM D2420 (2017). *Standard Test Method for Hydrogen Sulfide in Liquefied Petroleum (LP) Gases (Lead Acetate Method)*, Annual Book of Standards. West Conshohocken, Pennsylvania: ASTM International.
ASTM D2421 (2017). *Standard Practice for Interconversion of Analysis of C5 and Lighter Hydrocarbons to Gas-Volume, Liquid-Volume, or Mass Basis*, Annual Book of Standards. West Conshohocken, Pennsylvania: ASTM International.
ASTM D2426 (2017). *Standard Test Method for Butadiene Dimer and Styrene in Butadiene Concentrates by Gas Chromatography*, Annual Book of Standards. West Conshohocken, Pennsylvania: ASTM International.
ASTM D2427 (2017). *Standard Test Method for Determination of C2 through C5 Hydrocarbons in Gasolines by Gas Chromatography*, Annual Book of Standards. West Conshohocken, Pennsylvania: ASTM International.
ASTM D2504 (2017). *Standard Test Method for Noncondensable Gases in C2 and Lighter Hydrocarbon Products by Gas Chromatography*, Annual Book of Standards. West Conshohocken, Pennsylvania: ASTM International.
ASTM D2505 (2017). *Standard Test Method for Ethylene, Other Hydrocarbons, and Carbon Dioxide in High-Purity Ethylene by Gas Chromatography*, Annual Book of Standards. West Conshohocken, Pennsylvania: ASTM International.

ASTM D2593 (2017). *Standard Test Method for Butadiene Purity and Hydrocarbon Impurities by Gas Chromatography*, Annual Book of Standards. West Conshohocken, Pennsylvania: ASTM International.

ASTM D2597 (2017). *Standard Test Method for Analysis of Demethanized Hydrocarbon Liquid Mixtures Containing Nitrogen and Carbon Dioxide by Gas Chromatography*, Annual Book of Standards. West Conshohocken, Pennsylvania: ASTM International.

ASTM D2598 (2017). *Standard Practice for Calculation of Certain Physical Properties of Liquefied Petroleum (LP) Gases from Compositional Analysis*, Annual Book of Standards. West Conshohocken, Pennsylvania: ASTM International.

ASTM D2650 (2017). *Standard Test Method for Chemical Composition of Gases by Mass Spectrometry*, Annual Book of Standards. West Conshohocken, Pennsylvania: ASTM International.

ASTM D2699 (2017). *Standard Test Method for Research Octane Number of Spark-Ignition Engine Fuel*, Annual Book of Standards. West Conshohocken, Pennsylvania: ASTM International.

ASTM D2700 (2017). *Standard Test Method for Motor Octane Number of Spark-Ignition Engine Fuel*, Annual Book of Standards. West Conshohocken, Pennsylvania: ASTM International.

ASTM D2712 (2017). *Standard Test Method for Hydrocarbon Traces in Propylene Concentrates by Gas Chromatography*, Annual Book of Standards. West Conshohocken, Pennsylvania: ASTM International.

ASTM D2713 (2017). *Standard Test Method for Dryness of Propane (Valve Freeze Method)*, Annual Book of Standards. West Conshohocken, Pennsylvania: ASTM International.

ASTM D2784 (2017). *Standard Test Method for Sulfur in Liquefied Petroleum Gases (Oxy-Hydrogen Burner or Lamp)*, Annual Book of Standards. West Conshohocken, Pennsylvania: ASTM International.

ASTM D323 (2017). *Standard Test Method for Vapor Pressure of Petroleum Products (Reid Method)*, Annual Book of Standards. West Conshohocken, Pennsylvania: ASTM International.

ASTM D3246 (2017). *Standard Test Method for Sulfur in Petroleum Gas by Oxidative Microcoulometry*, Annual Book of Standards. West Conshohocken, Pennsylvania: ASTM International.

ASTM D3588 (2017). *Standard Practice for Calculating Heat Value, Compressibility Factor, and Relative Density of Gaseous Fuels*, Annual Book of Standards. West Conshohocken, Pennsylvania: ASTM international.

ASTM D3956 (2017). *Standard Specification for Methane Thermophysical Property Tables*, Annual Book of Standards. West Conshohocken, Pennsylvania: ASTM International.

ASTM D3984 (2017). *Standard Specification for Ethane Thermophysical Property Tables*, Annual Book of Standards. West Conshohocken, Pennsylvania: ASTM International.

ASTM D4084 (2017). *Standard Test Method for Analysis of Hydrogen Sulfide in Gaseous Fuels (Lead Acetate Reaction Rate Method)*, Annual Book of Standards. West Conshohocken, Pennsylvania: ASTM International.

ASTM D4150 (2017). *Standard Terminology Relating to Gaseous Fuels*, Annual Book of Standards. West Conshohocken, Pennsylvania: ASTM International.

ASTM D4177 (2017). *Standard Practice for Automatic Sampling of Petroleum and Petroleum Products*, Annual Book of Standards. West Conshohocken, Pennsylvania: ASTM International.

ASTM D4307 (2017). *Standard Practice for Preparation of Liquid Blends for Use as Analytical Standards*, Annual Book of Standards. West Conshohocken, Pennsylvania: ASTM International.

ASTM D4362 (2017). *Standard Specification for Propane Thermophysical Property Tables*, Annual Book of Standards. West Conshohocken, Pennsylvania: ASTM International.

ASTM D4423 (2017). *Standard Test Method for Determination of Carbonyls in C4 Hydrocarbons*, Annual Book of Standards. West Conshohocken, Pennsylvania: ASTM International.

ASTM D4424 (2017). *Standard Test Method for Butylene Analysis by Gas Chromatography*, Annual Book of Standards. West Conshohocken, Pennsylvania: ASTM International.

ASTM D4468 (2017). *Standard Test Method for Total Sulfur in Gaseous Fuels by Hydrogenolysis and Rateometric Colorimetry*, Annual Book of Standards. West Conshohocken, Pennsylvania: ASTM International.

ASTM D4629 (2017). *Standard Test Method for Trace Nitrogen in Liquid Petroleum Hydrocarbons by Syringe/Inlet Oxidative Combustion and Chemiluminescence Detection*, Annual Book of Standards. West Conshohocken, Pennsylvania: ASTM International.

ASTM D4650 (2017). *Standard Specification for Normal Butane Thermophysical Property Tables*, Annual Book of Standards. West Conshohocken, Pennsylvania: ASTM International.

ASTM D4651 (2017). *Standard Specification for Isobutane Thermophysical Property Tables*, Annual Book of Standards. West Conshohocken, Pennsylvania: ASTM International.

ASTM D4784 (2017). *Standard for LNG Density Calculation Models*, Annual Book of Standards. West Conshohocken, Pennsylvania: ASTM International.

ASTM D4810 (2017). *Standard Test Method for Hydrogen Sulfide in Natural Gas Using Length-of-Stain Detector Tubes*. West Conshohocken, Pennsylvania: Annual Book of Standards. ASTM International.

ASTM D4864 (2017). *Standard Test Method for Determination of Traces of Methanol in Propylene Concentrates by Gas Chromatography*, Annual Book of Standards. West Conshohocken, Pennsylvania: ASTM International.

ASTM D4888 (2017). *Standard Test Method for Water Vapor in Natural Gas Using Length-of-Stain Detector Tubes*, Annual Book of Standards. West Conshohocken, Pennsylvania: ASTM International.

ASTM D4953 (2017). *Standard Test Method for Vapor Pressure of Gasoline and Gasoline-Oxygenate Blends (Dry Method)*, Annual Book of Standards. West Conshohocken, Pennsylvania: ASTM International.

ASTM D4984 (2017). *Standard Test Method for Carbon Dioxide in Natural Gas Using Length-of-Stain Detector Tubes*, Annual Book of Standards. West Conshohocken, Pennsylvania: ASTM International.

ASTM D5191 (2017). *Standard Test Method for Vapor Pressure of Petroleum Products (Mini Method)*, Annual Book of Standards. West Conshohocken, Pennsylvania: ASTM International.

ASTM D5234 (2017). *Standard Guide for Analysis of Ethylene Product*, Annual Book of Standards. West Conshohocken, Pennsylvania: ASTM International.

ASTM D5273 (2017). *Standard Guide for Analysis of Propylene Concentrates*, Annual Book of Standards. West Conshohocken, Pennsylvania: ASTM International.

ASTM D5274 (2017). *Standard Guide for Analysis of 1,3-Butadiene Product*, Annual Book of Standards. West Conshohocken, Pennsylvania: ASTM International.

ASTM D5303 (2017). *Standard Test Method for Trace Carbonyl Sulfide in Propylene by Gas Chromatography*, Annual Book of Standards. West Conshohocken, Pennsylvania: ASTM International.

ASTM D5305 (2017). *Standard Test Method for Determination of Ethyl Mercaptan in LP-Gas Vapor*, Annual Book of Standards. West Conshohocken, Pennsylvania: ASTM International.

ASTM D5454 (2017). *Standard Test Method for Water Vapor Content of Gaseous Fuels Using Electronic Moisture Analyzers*, Annual Book of Standards. West Conshohocken, Pennsylvania: ASTM International.

ASTM D5504 (2017). *Standard Test Method for Determination of Sulfur Compounds in Natural Gas and Gaseous Fuels by Gas Chromatography and Chemiluminescence*, Annual Book of Standards. West Conshohocken, Pennsylvania: ASTM International.

ASTM D5799 (2017). *Standard Test Method for Determination of Peroxides in Butadiene*, Annual Book of Standards. West Conshohocken, Pennsylvania: ASTM International.

ASTM D5954 (2017). *Standard Test Method for Mercury Sampling and Measurement in Natural Gas by Atomic Absorption Spectroscopy*, Annual Book of Standards. West Conshohocken, Pennsylvania: ASTM International.

ASTM D6159 (2017). *Standard Test Method for Determination of Hydrocarbon Impurities in Ethylene by Gas Chromatography*, Annual Book of Standards. West Conshohocken, Pennsylvania: ASTM International.

ASTM D6228 (2017). *Standard Test Method for Determination of Sulfur Compounds in Natural Gas and Gaseous Fuels by Gas Chromatography and Flame Photometric Detection*, Annual Book of Standards. West Conshohocken, Pennsylvania: ASTM International.

ASTM D6273 (2017). *Standard Test Methods for Natural Gas Odor Intensity*, Annual Book of Standards. West Conshohocken, Pennsylvania: ASTM International.

ASTM D6350 (2017). *Standard Test Method for Mercury Sampling and Analysis in Natural Gas by Atomic Fluorescence Spectroscopy*, Annual Book of Standards. West Conshohocken, Pennsylvania: ASTM International.

ASTM D6897 (2017). *Standard Test Method for Vapor Pressure of Liquefied Petroleum Gases (LPG) (Expansion Method)*, Annual Book of Standards. West Conshohocken, Pennsylvania: ASTM International.

Burruss, R.C., and Ryder, R.T. 2003. Composition of crude oil and natural gas produced from 14 wells in the Lower Silurian "Clinton" Sandstone and Medina Group, Northeastern Ohio and Northwestern Pennsylvania. *Open-File Report 03-409*. Reston, VA: United States Geological Survey.

Burruss, R.C. and Ryder, R.T. (2014). Composition of natural gas and crude oil produced from 10 wells in the Lower Silurian "Clinton" Sandstone, Trumbull County, Ohio. In: *Coal and Petroleum Resources in The Appalachian Basin; Distribution, Geologic Framework, And Geochemical Character*. L.F. Ruppert and R.T. Ryder (Editors). Professional Paper 1708, Reston, Virginia: United States Geological Survey.

Cranmore, R.E. and Stanton, E. (2000). Chapter 9. In: *Modern Petroleum Technology. Volume 1: Upstream* (ed. R.A. Dawe). New York: Wiley.

Crawford, D.B., Durr, C.A., Finneran, J.A., and Turner, W. (1993). Chemicals from natural gas. In: *Chemical Processing Handbook* (ed. J.J. McKetta), 2. New York: Marcel Dekker Inc.

Drews, A.W. (1998). Introduction. In: *Manual on Hydrocarbon Analysis*, 6e (ed. A.W. Drews). West Conshohocken, Pennsylvania: American Society for Testing and Materials.

Esposito, G., Frunzo, L., Liotta, F. et al. (2012). Bio-methane potential tests to measure the biogas production from the digestion and co-digestion of complex organic substrates. *The Open Environmental Engineering Journal* 5: 1–8.

Gary, J.G., Handwerk, G.E., and Kaiser, M.J. 2007. *Petroleum Refining: Technology and Economics*, 5 e. CRC Press, Taylor & Francis Group, Boca Raton, Florida.

Hsu, C.S. and Robinson, P.R. ed. (2017). *Handbook of Petroleum Technology*. Cham, Switzerland: Springer International Publishing AG.

Mokhatab, S., Poe, W.A., and Speight, J.G. (2006). *Handbook of Natural Gas Transmission and Processing*. Amsterdam, The Netherlands: Elsevier.

Rasi, S., Veijanen, A., and Rintala, J. (2007). Trace compounds of biogas from different biogas production plants. *Energy* 32: 1375–1380.

Richards, B., Herndon, F.G., Jewell, W.J. et al. (1994). In situ methane enrichment in methanogenic energy crop digesters. *Biomass and Bioenergy* 6 (4): 275–282.

Speight, J.G. (2005). *Environmental Analysis and Technology for the Refining Industry*. Hoboken, New Jersey: Wiley.

Speight, J.G. (2007). *Natural Gas: A Basic Handbook. GPC Books.* Houston, Texas: Gulf Publishing Company.

Speight, J.G. (2013). *Shale Gas Production Processes.* Oxford, United Kingdom: Gulf Professional Publishing, Elsevier.

Speight, J.G. 2014. *The Chemistry and Technology of Petroleum.* 5e CRC Press, Taylor & Francis Group, Boca Raton, Florida.

Speight, J.G. 2015. *Handbook of Petroleum Product Analysis* 2e. Wiley, Hoboken, New Jersey.

Speight, J.G. (2017a). *Handbook of Petroleum Refining.* Boca Raton, Florida: CRC Press, Taylor & Francis Group.

Speight, J.G. (2017b). *Deep Shale Oil and Gas.* Oxford, United Kingdom: Gulf Professional Publishing, Elsevier.

Speight, J.G. and Islam, M.R. (2016). *Peak Energy – Myth or Reality.* Beverly, Massachusetts: Scrivener Publishing.

9

Analysis of Gas Streams

9.1 Introduction

The analysis of gas streams, compared with liquid streams, is relatively simple because bulk characterization of a single phase is implicit. However, when condense matter (such as gas condensate) is present, the analysis is more complicated. In the case of gas condensate, besides bulk analysis, there may be interest on surface composition (often quite distinct to that of the bulk phase). Compositional analysis in which the components of the mixture are identified may be achieved by (i) physical means, which is measurement of physical properties; (ii) by pure chemical means, which is measurement of chemical properties; or more commonly (iii) by physicochemical means. Gas analysis is even more dangerous and difficult if the composition is a complete unknown. However, when some main constituents are known to occur, the analysis gains accuracy (and may be easier) if the known component is removed; this is particularly important in the case of water vapor, which may condense on the instruments, or constituents when the molecular behavior may complicate spectral analyses.

Crude oil-related gases (including natural gas) and refinery gases (process gases) as well as product gases produced from petroleum refining, upgrading, or natural gas processing facilities are a category of saturated and unsaturated gaseous hydrocarbons, predominantly C_1 to C_6. Some gases may also contain inorganic compounds, such as hydrogen, nitrogen, hydrogen sulfide, carbon monoxide, and carbon dioxide. As such, petroleum and refinery gases (unless produced as a saleable product that must meet specifications prior to sale) are often unknown or variable composition and toxic (API, 2009). The site-restricted petroleum and refinery gases (i.e. those not produced for sale) often serve as fuels consumed on-site, as intermediates for purification and recovery of various gaseous products, or as feedstocks for isomerization and alkylation processes within a facility.

Handbook of Natural Gas Analysis, First Edition. James G. Speight.

As with petroleum, natural gas from different wells varies widely in composition and analyses (Mokhatab et al., 2006; Speight, 2007, 2014), and the proportion of nonhydrocarbon constituents can vary over a very wide range. The nonhydrocarbon constituents of natural gas can be classified as two types of materials: (i) diluents, such as nitrogen, carbon dioxide, and water vapors, and (ii) contaminants, such as hydrogen sulfide and/or other sulfur compounds. Thus, a particular natural gas field could require production, processing, and handling protocols different from those used for gas from another field.

The diluents are noncombustible gases that reduce the heating value of the gas and are on occasion used as *fillers* when it is necessary to reduce the heat content of the gas. On the other hand, the contaminants are detrimental to production and transportation equipment in addition to being obnoxious pollutants. Thus, the primary reason for gas refining is to remove the unwanted constituents of natural gas and to separate the gas into its various constituents. The processes are analogous to the distillation unit in a refinery where the feedstock is separated into its various constituent fractions before further processing to products. The major diluents or contaminants of natural gas are (i) acid gas, which is predominantly hydrogen sulfide although carbon dioxide does occur to a lesser extent; (ii) water, which includes all entrained free water or water in condensed forms; (iii) liquids in the gas, such as higher-boiling hydrocarbons as well as pump lubricating oil, scrubber oil, and, on occasion, methanol; and (iv) any solid matter that may be present, such as fine silica (sand) and scaling from the pipe.

In addition to the gases obtained by distillation of crude petroleum, more highly volatile products result from the subsequent processing of naphtha and middle distillate to produce gasoline; from desulfurization processes involving hydrogen treatment of naphtha, distillate, and residual fuel; and from the coking or similar thermal treatments of vacuum gas oils and residual fuels. The most common processing step in the production of gasoline is the catalytic reforming of hydrocarbon fractions in the heptane (C_7) to decane (C_{10}) range.

Like any other refinery product, the gas must be processed to prepare it for final use and to ascertain the extent of contaminants that could cause environmental damage. Furthermore, gas processing is a complex industrial process designed to clean raw (dirty, contaminated) gas by separating impurities and various nonmethane and fluids to produce what is known as *pipeline quality* dry natural gas.

Natural gas processing typically begins at the wellhead where the composition of the raw natural gas extracted from producing wells depends on the type, depth, and the geology and location of the underground reservoir (Speight, 2014, 2017). Thus, gas processing plants purify raw gas to remove common contaminates such as water, carbon dioxide, and hydrogen. However, some of the substances that contaminate gas have economic value and are further processed for sale or for use in a refinery petrochemical plant. On the other hand, the gases produced within the refinery from the refinery processes are, in a sense, much more complex, and often the composition cannot be predicted. Such gases contain a wider variety of useful hydrocarbons as well as higher amounts of hydrogen sulfide (from the hydrodesulfurization units). Gases such as ethane that are destined for use as petrochemical feedstocks must adhere to stringent composition controls that are dependent upon the process. For example, moisture content (ASTM D1142, 2017), oxygen content (ASTM D1945, 2017), carbon dioxide content (ASTM D1945, 2017), and sulfur content (ASTM D1072, 2017) must be monitored as they all interfere with catalyst performance in petrochemical processes.

There are many variables in treating natural gas and refinery gas. The precise area of application of a given process is difficult to define. Several factors must be considered: (i) the types of contaminants in the gas, (ii) the concentrations of contaminants in the gas, (iii) the degree of contaminant removal desired, (iv) the selectivity of acid gas removal required, (v) the temperature of the gas to be processed, (vi) the pressure of the gas to be processed, (vii) the volume of the gas to be processed, (viii) the composition of the gas to be processed, (ix) the carbon dioxide–hydrogen sulfide ratio in the gas, and (x) the desirability of sulfur recovery due to process economics or environmental issues.

Thus, in many cases, the process complexities arise because of the need for recovery of the materials used to remove the contaminants or even recovery of the contaminants in the original, or altered, form. In any case, whatever the source of the gas, careful analysis before processing is necessary (Mokhatab et al., 2006; Speight, 2007, 2014).

9.2 Types of Gases

The gaseous products that occur in a refinery comprise mixtures that vary from natural gas to gases produced during refining (*refinery gas, process gas*). The constituents of each type of gas may be similar (except for the olefin-type gases produced during thermal processes), but the variations of the amounts of these constituents cover wide ranges. Thus, the gas products of a refinery are (i) *liquefied petroleum gas* (LPG), (ii) *natural gas*, (iii) *shale gas*, and (iv) *refinery gas*, which also includes *still gas* and *process gas*. Each type of gas may be analyzed by similar methods although the presence of high-boiling hydrocarbons and nonhydrocarbon species such as carbon dioxide and hydrogen sulfide may require slight modifications to the various analytical test methods (Table 9.1).

However, at the onset, it is necessary that the analyst determine the type of gas to be analyzed. Mixtures of the various constituents named above are commonly encountered in material testing, and the composition varies depending upon the source and intended use of the material. Other nonhydrocarbon constituents of these mixtures are important analytes since they may be useful

Table 9.1 Standard test methods commonly applied to the determination of gas quality.

ASTM D792. Standard Test Methods for Density and Specific Gravity (Relative Density) of Plastics by Displacement.
ASTM D1434. Standard Test Method for Determining Gas Permeability Characteristics of Plastic Film and Sheeting.
ASTM D1505. Standard Test Method for Density of Plastics by the Density-Gradient Technique, D1505. West Conshohocken, PA.
ASTM D1945. American Society for Testing and Materials. Standard Test Method for Analysis of Natural Gas by Gas Chromatography.
ASTM D3588. Standard Practice for Calculating Heat Value, Compressibility Factor, and Relative Density of Gaseous Fuels.
ASTM D1826. Standard Test Method for Calorific (Heating) Value of Gases in Natural Gas Range by Continuous Recording Calorimeter.
ASTM D1070. Standard Test Methods for Relative Density of Gaseous Fuels.
ASTM D4084. Standard Test Method for Analysis of Hydrogen Sulfide in Gaseous Fuels – Lead Acetate Reaction Rate Method.
ASTM 5199. Standard Test Methods for Measuring the Nominal Thickness of Geosynthetics.
ASTM D5454. Standard Test Method for Water Vapor Content of Gaseous Fuels Using Electronic Moisture Analyzer.

products or may be undesirable as a source of processing problems. Some of these components are helium, hydrogen, argon, oxygen, nitrogen, carbon monoxide, carbon dioxide, sulfur, and nitrogen-containing compounds, as well as higher molecular weight hydrocarbons. Thus, the desired testing of these hydrocarbon mixtures involves (i) identification of the type of gas for component speciation and quantitation and (ii) the influence of the composition not only on the bulk physical or chemical properties but, more important, the influence on the composition on the performance of the test method.

9.2.1 Natural Gas

Natural gas is the gaseous mixture associated with petroleum reservoirs and is predominantly methane but does contain other combustible hydrocarbon compounds as well as nonhydrocarbon compounds (Mokhatab et al., 2006; Speight, 2007, 2014). In fact, associated natural gas is believed to be the most economical form of ethane. The gas occurs in the porous rock of the crust of the Earth either alone or with accumulations of petroleum (Chapter 1). In the latter case, the gas forms the gas cap, which is the mass of gas trapped between the liquid petroleum and the impervious cap rock of the petroleum reservoir.

When the pressure in the reservoir is sufficiently high, the natural gas may be dissolved in the petroleum and is released upon penetration of the reservoir as a result of drilling operations.

Natural gas (predominantly *methane*) denoted by the chemical structure CH_4 is the lowest boiling and least complex of all hydrocarbons. Natural gas from an underground reservoir, when brought to the surface, can contain other higher-boiling hydrocarbons and is often referred to as *wet gas*. Wet gas is usually processed to remove the entrained hydrocarbons that are higher boiling than methane, and, when isolated, the higher-boiling hydrocarbons sometimes liquefy and are called *natural gas condensate.*

Natural gas is found in petroleum reservoirs as free gas (*associated gas*) or in solution with petroleum in the reservoir (*dissolved gas*) or in reservoirs that contain only gaseous constituents and no (or very little) petroleum (*unassociated gas*) (Cranmore and Stanton, 2000; Speight, 2014). The hydrocarbon content varies from mixtures of methane and ethane with very few other constituents (*dry* gas) to mixtures containing all of the hydrocarbons from methane to pentane and even hexane (C_6H_{14}) and heptane (C_7H_{16}) (*wet* gas). In both cases some carbon dioxide (CO_2) and inert gases, including helium (He), are present together with hydrogen sulfide (H_2S) and a small quantity of organic sulfur.

Other terminologies that are of interest (even necessity) to understand are (i) dry gas and (ii) wet gas. There are *dry gas* or *lean gas*, which is mostly methane, and *wet gas*, which contains considerable amounts of higher molecular weight and higher-boiling hydrocarbons (Mokhatab et al., 2006; Speight, 2007, 2014). *Sour gas* contains high proportions of much hydrogen sulfide, whereas *sweet gas* contains little or no hydrogen sulfide. *Residue gas* is the gas remaining (mostly methane) after the higher molecular weight paraffins have been extracted. *Casinghead gas* is the gas derived from an oil well by extraction at the surface. Natural gas has no distinct odor and the main use is for fuel, but it can also be used to make chemicals and LPG.

The term *petroleum gas(es)* in this context is also used to describe the gaseous phase and liquid-phase mixtures comprised mainly of methane to butane (C_1 to C_4 hydrocarbons) that are dissolved in the crude oil and natural gas, as well as gases produced during thermal processes in which the crude oil is converted to other products. It is necessary, however, to acknowledge that in addition to the hydrocarbons, gases such as carbon dioxide, hydrogen sulfide, and ammonia are also produced during petroleum refining and will be constituents of refinery gas that have to be removed. Olefins are also present in the gas streams of various processes and are not included in LPG but are removed for use in petrochemical operations (Crawford et al., 1993).

Raw natural gas varies greatly in composition (Table 9.2), and the constituents can be several of a group of started hydrocarbons from methane to butane (Table 9.3) and nonhydrocarbons. The treatment required to prepare natural

Table 9.2 Composition of associated natural gas from a petroleum well.

Category	Component	Amount (%)
Paraffinic	Methane (CH_4)	70–98
	Ethane (C_2H_6)	1–10
	Propane (C_3H_8)	Trace–5
	Butane (C_4H_{10})	Trace–2
	Pentane (C_5H_{12})	Trace–1
	Hexane (C_6H_{14})	Trace–0.5
	Heptane and higher (C_{7+})	None–trace
Cyclic	Cyclopropane (C_3H_6)	Traces
	Cyclohexane (C_6H_{12})	Traces
Aromatic	Benzene (C_6H_6), others	Traces
Nonhydrocarbon	Nitrogen (N_2)	Trace–15
	Carbon dioxide (CO_2)	Trace–1
	Hydrogen sulfide (H_2S)	Trace occasionally
	Helium (He)	Trace–5
	Other sulfur and nitrogen compounds	Trace occasionally
	Water (H_2O)	Trace–5

Table 9.3 Possible constituents of natural gas and refinery gas.

			Density at 60 °F (15.6 °C), 1 atm	
Gas	Molecular weight	Boiling point 1 atm °C (°F)	$g\,l^{-1}$	Relative to air = 1
Methane	16.043	−161.5 (−258.7)	0.6786	0.5547
Ethylene	28.054	−103.7 (−354.7)	1.1949	0.9768
Ethane	30.068	−88.6 (−127.5)	1.2795	1.0460
Propylene	42.081	−47.7 (−53.9)	1.8052	1.4757
Propane	44.097	−42.1 (−43.8)	1.8917	1.5464
1,2-Butadiene	54.088	10.9 (51.6)	2.3451	1.9172
1,3-Butadiene	54.088	−4.4 (24.1)	2.3491	1.9203
1-Butene	56.108	−6.3 (20.7)	2.4442	1.9981
cis-2-Butene	56.108	3.7 (38.7)	2.4543	2.0063
trans-2-Butene	56.108	0.9 (33.6)	2.4543	2.0063
iso-Butene	56.104	−6.9 (19.6)	2.4442	1.9981
n-Butane	58.124	−0.5 (31.1)	2.5320	2.0698
iso-Butane	58.124	−11.7 (10.9)	2.5268	2.0656

gas for distribution as an industrial or household fuel is specified in terms of the use and environmental regulations. Briefly, natural gas contains hydrocarbons and nonhydrocarbon gases. Hydrocarbon gases are methane (CH_4), ethane (C_2H_6), propane (C_3H_8), butanes (C_4H_{10}), pentanes (C_5H_{12}), hexane (C_6H_{14}), heptane (C_7H_{16}), and sometimes trace amounts of octane (C_8H_{18}) and higher molecular weight hydrocarbons. Some aromatics (BTX – benzene [C_6H_6], toluene [$C_6H_5CH_3$], and the xylenes [$CH_3C_6H_4CH_3$]) can also be present, raising safety issues due to their toxicity. The nonhydrocarbon gas portion of the natural gas contains nitrogen (N_2), carbon dioxide (CO_2), helium (He), hydrogen sulfide (H_2S), water vapor (H_2O), and other sulfur compounds, such as carbonyl sulfide (COS) and mercaptans (e.g. methyl mercaptan, CH_3SH) and trace amounts of other gases.

Propane, *iso*-butane (boiling point: −12 °C, 11 °F), and butane generally constitute this sample type and are used for heating, for motor fuels, and as chemical feedstocks (ASTM D2504, 2017; ASTM D2505, 2017; ASTM D2597, 2017). Other important constituents of natural gas that are often not accounted for in these analyses are moisture (water) and hydrogen sulfide, as well as other sulfur compounds (ASTM D1142, 2017; ASTM D1988, 2017; ASTM D4888, 2017; ASTM D5454, 2017; ASTM D5504, 2017; ASTM D6228, 2017). In addition, the trace gases are typically one or more of the following gases: argon, hydrogen, and helium. Generally, the hydrocarbons having a higher molecular weight than methane, carbon dioxide, and hydrogen sulfide are removed from natural gas prior to its use as a fuel. Gases produced in a refinery contain methane, ethane, ethylene, propylene, hydrogen, carbon monoxide, carbon dioxide, and nitrogen, with low concentrations of water vapor, oxygen, and other gases.

Carbon dioxide and hydrogen sulfide are commonly referred to as *acid gases* since they form corrosive compounds in the presence of water. Nitrogen, helium, and carbon dioxide are also referred to as *diluents* since none of these burn, and thus they have no heating value. Mercury can also be present either as a metal in vapor phase or as an organometallic compound in liquid fractions. Concentration levels are generally very small, but even at very small concentration levels, mercury can be detrimental due to its toxicity and its corrosive properties (reaction with aluminum alloys). Mercury in natural gas is also measured by atomic fluorescence spectroscopy (ASTM D6350, 2017) and by atomic absorption spectroscopy (ASTM D5954, 2017).

A natural gas stream traditionally has high proportions of natural gas liquids (NGLs) and is referred to as rich gas. NGLs are constituents such as ethane, propane, butane, and pentanes and higher molecular weight hydrocarbon constituents. The higher molecular weight constituents (i.e. the C5+ product) are commonly referred to as natural gasoline. Rich gas will have a high heating value and a high hydrocarbon dew point (ASTM D1142, 2017). In LPG where the composition is such that the hydrocarbon dew point is known to be low, a dew-point method will detect the presence of traces of water (ASTM D1142, 2017).

When referring to NGLs in the gas stream, the term gallon per thousand cubic feet is used as a measure of high molecular weight hydrocarbon content. On the other hand, the composition of nonassociated gas (sometimes called well gas) is deficient in NGLs. The gas is produced from geological formations that typically do not contain much, if any, hydrocarbon liquids.

Carbon dioxide (ASTM D1945, 2017; ASTM D4984, 2017) in excess of 3% is normally removed for reasons of corrosion prevention (ASTM D1838, 2017). The first test method (ASTM D1945, 2017) is of significance for providing data for calculating physical properties of the sample, such as heating value and relative density, or for monitoring the concentrations of one or more of the components in a mixture. The method covers the determination of the chemical composition of natural gases and similar gaseous mixtures and may be abbreviated for the analysis of lean natural gases containing negligible amounts of hexane derivatives and higher hydrocarbon derivatives, or for the determination of one or more components, as required. The second test method (ASTM D4984, 2017) describes a rapid and simple field determination of carbon dioxide in natural gas pipelines. Available detector tubes provide a total measuring range of 100 ppm (parts per million) up to 60% v/v, although the majority of applications will be on the lower end of this range (that is, under 5% v/v). The third text method of this group (ASTM D1838, 2017) takes a different approach by detecting the presence of components in LPG that can be corrosive to copper. The copper corrosion limits provide assurance that difficulties will not be experienced in deterioration of the copper and copper alloy fittings and connections that are commonly used in many types of utilization, storage, and transportation equipment.

Hydrogen sulfide (ASTM D2420, 2017; ASTM D4084, 2017; ASTM D4810, 2017) is also removed, and the odor of the gas must not be objectionable (ASTM D6273, 2017), so mercaptan content (ASTM D1988, 2017) is important. A simple lead acetate test (ASTM D2420, 2017; ASTM D4084, 2017) is available for detecting the presence of hydrogen sulfide and is an additional safeguard that hydrogen sulfide not be present (ASTM D1835, 2017). This test method covers the specification for LPG consisting of propane, propylene (propene), butane, and mixtures of these gases. The method is especially important when the products are intended for use as domestic, commercial and industrial heating, and engine fuels. However, care must be taken to in sampling of the liquefied gases for test results to be reliable and significant (Chapter 5). All four types of LPG covered by this specification should conform to the specified requirements for vapor pressure, volatile residue, residue matter, relative density, and corrosion.

Many sources of natural and petroleum gases contain sulfur compounds that are odorous, corrosive, and poisonous to catalysts used in gaseous fuel processing. In fact, sulfur odorants are added (in the ppm range, i.e. 1–4 ppm v/v) to natural gas and LPG for safety purposes. Some odorants are unstable and react to form compounds having lower odor thresholds. Quantitative analysis of these odorized gases ensures that odorant injection equipment is performing to specification. However, mercaptan derivatives, if present in a gas stream, may only produce a transitory yellow stain on the lead acetate paper that fades completely in less than 5 min (ASTM D2420, 2017; ASTM D4084, 2017). In the lead acetate test (ASTM D2420, 2017), the vaporized gas is passed over moist lead acetate paper under controlled conditions. Hydrogen sulfide reacts with lead acetate to form lead sulfide, resulting in a stain on the paper varying in color from yellow to black, depending on the amount of hydrogen sulfide present. Other pollutants can be determined by gas chromatography (GC) (ASTM D5504, 2017; ASTM D6228, 2017).

The first test method (ASTM D5504, 2017), although not intended for application to gases other than natural gas and related fuels, has been successfully applied to fuel-type gases including refinery, landfill, cogeneration, and sewage digester gas. Refinery, landfill, sewage digester, and other related fuel-type gases inherently contain volatile sulfur compounds that are subject to federal, state, or local control. The methane fraction of these fuel-type gases is occasionally sold to distributors of natural gas. For these reasons, both regulatory agencies and production and distribution facilities may require the accurate determination of sulfur to satisfy regulatory, production, or distribution requirements. Fuel gases are also used in energy production or are converted to new products using catalysts that are poisoned by excessive sulfur in the feed gas. As a refresher note, GC is commonly used in the determination of fixed gas and organic composition of natural gas (ASTM D1945, 2017). Other standard test methods for the analysis of sulfur in fuel gases include test methods for total sulfur content (ASTM D4468, 2017) and standard test methods for hydrogen sulfide content.

The second test method (ASTM D6228, 2017) describes a technique to determine individual sulfur species in gaseous fuel and the total sulfur content by calculation. GC is used commonly and extensively to determine other components in gaseous fuels including fixed gas and organic components (ASTM D1945, 2017). This test method dictates the use of a specific GC technique with one of the more common detectors for measurement. The test method covers the determination of individual volatile sulfur-containing compounds in gaseous fuels by GC with a flame photometric detector (FPD) or a pulsed flame photometric detector (PFPD). The detection range for sulfur compounds is from 20 to 20 000 picograms (pg) of sulfur (1 pg = 1×10^{-12} g = 0.001 ng).

This test method (ASTM D1142, 2017) covers the determination of the water vapor content of gaseous fuels by measurement of the dew-point temperature and the calculation therefrom of the water vapor content. The method is of considerable importance when it is necessary (on a contract basis) for application to the pipeline transmission of natural gas to contain specifications limiting the maximum concentration of water vapor allowed. Excess water vapor can cause corrosive conditions, degrading pipelines, and equipment, and, in addition, the water can also condense and freeze or form methane hydrates causing blockages. Water vapor content also affects the heating value of natural gas, thus influencing the quality of the gas. Since dew point can be calculated from composition, direct determination of dew point for an LPG sample is a measure of composition. It is, of course, of more direct practical value, and if there are small quantities of higher molecular weight material present, it is preferable to use a direct measurement.

The hydrocarbon dew point is reduced to such a level that retrograde condensation, i.e. condensation resulting from pressure drop, cannot occur under the worst conditions likely to be experienced in the gas transmission system. Similarly, the water dew point is reduced to a level sufficient to preclude formation of C_1 to C_4 hydrates in the system. The natural gas after appropriate treatment for acid gas reduction, odorization, and hydrocarbon and moisture dew-point adjustment (ASTM D1142, 2017) would then be sold within prescribed limits of pressure, calorific value, and possibly Wobbe index (cv/[sp. gr.]) (Chapter 8).

9.2.2 Shale Gas

Natural gas is also associated with shale formations, and such gas is commonly referred to as *shale gas* – to define the origin of the gas rather than the character and properties (Speight, 2013b). Thus, shale gas is natural gas produced from shale formations that typically function as both the reservoir and source rocks for the natural gas (Speight, 2013b). In terms of chemical makeup, shale gas is typically a dry gas composed primarily of methane (60–95% v/v), but some formations do produce wet gas. The Antrim and New Albany plays have typically produced water and gas. Gas shale formations are organic-rich shale formations that were previously regarded only as source rocks and seals for gas accumulating in the strata near sandstone and carbonate reservoirs of traditional onshore gas.

The test methods applied to the analysis of shale gas are those same methods that are applied to the analysis on gas from typical petroleum-type reservoirs. Similarly, but with the appropriate modifications due to the relative amounts of the hydrocarbon and nonhydrocarbon gases present, the same test methods can also be applied to coal gas, biogas, and landfill gas (Speight, 2011, 2013a).

Gas production from unconventional shale gas reservoirs (such as tight shale formations) has become more common in the past decade. Produced shale gas observed to date has shown a broad variation in compositional makeup, with some having wider component ranges, a wider span of minimum and maximum heating values, and higher levels of water vapor and other substances than pipeline tariffs or purchase contracts may typically allow. Indeed, because of these variations in gas composition, each shale gas formation can have unique processing requirements for the produced shale gas to be marketable.

Ethane can be removed by cryogenic extraction, while carbon dioxide can be removed through a scrubbing process. However, it is not always necessary (or practical) to process shale gas to make its composition identical to *conventional* transmission-quality gases. Instead, the gas should be interchangeable with other sources of natural gas now provided to end users. The interchangeability of shale gas with conventional gases is crucial to its acceptability and eventual widespread use in the United States.

Although not highly sour in the usual sense of having high hydrogen sulfide content, and with considerable variation from play to resource to resource and even from well to well within the same resource (due to extremely low permeability of the shale even after fracturing) (Speight, 2013b), shale gas often contains varying amounts of hydrogen sulfide with wide variability in the carbon dioxide content. The gas is not ready for pipelining immediately after it has exited the shale formation.

The challenge in treating such gases is the low (or differing) hydrogen sulfide/carbon dioxide ratio and the need to meet pipeline specifications. In a traditional gas processing plant, the olamine of choice for content for hydrogen sulfide removal is *N*-methyldiethanolamine (MDEA) (Mokhatab et al., 2006; Speight, 2007, 2014), but whether or not this olamine will suffice to remove the hydrogen sulfide without removal of excessive amounts of carbon dioxide is another issue.

Gas treatment may begin at the wellhead – condensates and free water usually are separated at the wellhead using mechanical separators. In this form of processing, the wellhead gas is usually separated at as high a pressure as possible, reducing compression costs when the gas is to be used for gas lift or delivered to a pipeline. After gas removal, lighter hydrocarbons and hydrogen sulfide are removed as necessary to obtain a product having a suitable vapor pressure for transport yet retaining most of the natural gasoline constituents. In addition to composition and thermal content (Btu/scf, Btu/ft^3), natural gas can also be characterized on the basis of the mode of the natural gas found in reservoirs where there is no, or at best only minimal amounts of, petroleum.

The gas, condensate, and water are separated in the field separator and are directed to separate storage tanks and the gas flows to a gathering system. After the free water has been removed, the gas is still saturated with water vapor and, depending on the temperature and pressure of the gas stream, may need to be dehydrated or treated with methanol to prevent hydrates as the temperature drops. But this may not be always the case in actual practice.

9.2.3 Liquefied Petroleum Gas

Liquefied petroleum gas is the term applied to certain specific hydrocarbons and their mixtures, which exist in the gaseous state under atmospheric ambient conditions but can be converted to the liquid state under conditions of moderate pressure at ambient temperature. Typically, fuel gas with four or less carbon atoms in the hydrogen–carbon combination have boiling points that are lower than room temperature, and these products are gases at ambient temperature and pressure.

Liquefied petroleum gas or liquid petroleum gas (LP gas), also referred to as simply propane or butane, are flammable mixtures of hydrocarbon gases used as fuel in heating appliance and vehicles. Propylene and butylene derivatives as well as various other hydrocarbons are usually also present in small concentrations.

LPG is prepared by a crude oil refinery or at a gas processing plant and is almost entirely derived from fossil fuel sources, being manufactured during the refining of petroleum (crude oil) or extracted from petroleum or natural gas streams as they emerge from the ground. As its boiling point is below room temperature, LPG will evaporate quickly at normal temperature and pressure (STP) and is usually supplied in pressurized steel vessels. The pressure at which LPG becomes liquid is the vapor pressure that varies depending on composition and temperature. LPG is heavier than air and, unlike methane, will flow along floors and tend to settle in low spots, such as basement or depressions in the earth. This can lead to explosion if the mixture of the LPG and air is within the explosive limits and there is an ignition source. Also, LPG can cause suffocation when it displaces air, causing a decrease in oxygen concentration.

The constituents of LPG are produced during natural gas refining, petroleum stabilization, and petroleum refining (Gary et al., 2007; Speight, 2014, 2017; Hsu and Robinson, 2017). Thus, LPG is a hydrocarbon mixture containing propane ($CH_3CH_2CH_3$, boiling point: −42 °C, −44 °F), butane ($CH_3CH_2CH_2CH_3$, boiling point: 0 °C, 32 °F), *iso*-butane ($CH_3CH(CH_3)CH_3$, boiling point: −11.7 °C (10.9 °F)), and to a lesser extent propylene ($CH_3CH{=}CH_2$, boiling point: −47.6 °C (−53.7 °F)) or butylene ($CH_3CH_2CH{=}CH_2$, boiling point: −6.47 °C (20.35 °F)). The most common commercial products are propane, butane, or some mixture of the two and are generally extracted from natural gas or crude petroleum. The propane and butane can be derived from natural gas or from refinery operations, but, in this latter case, substantial proportions of the corresponding olefins will be present and need to be separated. The hydrocarbons are normally liquefied under pressure for transportation and storage.

Propylene and butylene isomers result from cracking other hydrocarbons in a petroleum refinery and are two important chemical feedstocks. The presence of propylene and butylenes in LPG used as fuel gas is not critical. The vapor pressures of these olefins are slightly higher than those of propane and butane, and the flame speed is substantially higher, but this may be an advantage since the flame speeds of propane and butane are slow. However, one issue that often limits the amount of the olefins in LPG is the propensity of the olefins to form soot.

In addition, LPG is usually available in different grades (usually specified as commercial propane, commercial butane, commercial propane–butane [P—B] mixtures, and special-duty propane). During the use of LPG, the gas must vaporize completely and burn satisfactorily in the appliance without causing any corrosion or producing any deposits in the system.

Commercial propane consists predominantly of propane and/or propylene, while commercial butane is mainly composed of butanes and/or butylenes. Both must be free from harmful amounts of toxic constituents (API, 2009) and free from mechanically entrained water (that may be further limited by specifications) (ASTM D1835, 2017).

Commercial propane–butane mixtures are produced to meet sales specifications such as volatility, vapor pressure, specific gravity, hydrocarbon composition, sulfur and its compounds, corrosion of copper, residues, and water content. These mixtures are used as fuels in areas and at times where low ambient temperatures are less frequently encountered. Analysis by GC is possible (ASTM D5504, 2017; ASTM D6228, 2017).

Special-duty propane is intended for use in spark-ignition engines, and the specification includes a minimum *motor octane number* (MON) to ensure satisfactory antiknock performance. Propylene ($CH_3CH{=}CH_2$) has a significantly lower octane number than propane, so there is a limit to the amount of this component that can be tolerated in the mixture. Analysis by GC is possible (ASTM D5504, 2017; ASTM D6228, 2017).

LPG and liquefied natural gas can share the facility of being stored and transported as a liquid and then vaporized and used as a gas. In order to achieve this, LPG must be maintained at a moderate pressure but at ambient temperature. The liquefied natural gas can be at ambient pressure but must be maintained at a temperature of roughly −1 to 60°C (30–140°F). In fact, in some applications it is actually economical and convenient to use LPG in the liquid phase. In such cases, certain aspects of gas composition (or quality such as the ratio of propane to butane and the presence of traces of heavier hydrocarbons, water, and other extraneous materials) may be of lesser importance compared with the use of the gas in the vapor phase.

For normal (gaseous) use, the contaminants of LPG are controlled at a level at which they do not corrode fittings and appliances or impede the flow of the gas. For example, hydrogen sulfide (H_2S) and carbonyl sulfide (COS) should be absent. Organic sulfur to the level required for adequate odorization (ASTM D5305, 2017), or *stenching*, is a normal requirement in LPG; dimethyl sulfide (CH_3SCH_3) and ethyl mercaptan (C_2H_5SH) are commonly used at a concentration of up to 50 ppm.. Natural gas is similarly treated possibly with a wider range of volatile sulfur compounds.

The presence of water in LPG (or in natural gas) is undesirable since it can produce hydrates that will cause, for example, line blockage due to the formation of hydrates under conditions where the water *dew point* is attained (ASTM D1142, 2017). If the amount of water is above acceptable levels, the addition of a small methanol will counteract any such effect.

In addition to other gases, LPG may also be contaminated by higher-boiling constituents (residua) such as the constituents of middle distillates to lubricating oil. These contaminants become included in the gas during handling and must be prevented from reaching unacceptable levels. Olefins and especially diolefins are prone to polymerization and should be removed.

Control over the amount of residue in LPG is essential in end-use applications of LPG (ASTM D1835, 2017). In fact, an oily residue in LPG is a contamination that can lead to problems during production, transportation, storage, or when in use. For example, when the liquefied gas is used for automotive fuel application, a residue can lead to troublesome deposits that will accumulate and corrode or plug the LPG fuel filter, the low-pressure regulators, the fuel mixer, or the control solenoids. LPG can be contaminated with oily residues during its production or transport. Transport contamination can be a result of shared pipelines, valves, and trucks used for the distribution of other products. Production sources such as the desulfurization process may contribute sulfur absorbent oil to the LPG stream. Commercial LPG, especially for automotive applications, should comply with current fuel specifications.

Fuel specifications for oily residue in LPG use a method known as the oil stain method (ASTM D2158, 2017). In the method 100 ml of the LPG is evaporated and the remaining volume of residue is read from the glass evaporation tube. This volume is one quantification used for fuel specifications. In addition, the residue is dissolved in a solvent and the resulting solution is slowly dripped on the adsorption paper. The size and persistence of the stain that remains on the paper after the solvent evaporates is the other empirical quantification of the oily residue in the LPG sample. The accuracy of both quantifications is subject to question.

On the other hand, the liquefied has injector method (ASTM D7756, 2017) in which a dedicated sampler (the liquefied gas injector) is used to inject the liquefied gas at room temperature directly on to the gas chromatographic column. This test method is quicker and much more sensitive than manual methods (such as ASTM D2158, 2017), which is based on evaporation of large sample volumes followed by visual or gravimetric estimation of residue content. In addition, this method provides enhanced sensitivity in measurements of heavier (oily) residues, with a quantification limit of 10 mg kg^{-1} total residue. This test method gives both quantitative results and information about contaminant composition such as boiling point range and fingerprint, which can be very useful in tracing the source of a particular contaminant. The method covers the determination, by GC, of soluble hydrocarbon materials, sometimes called "oily residue," which can be present in LPG and which are substantially less volatile than the LPG itself. Also, the method offers quantitative data over the range of 10–600 mg kg^{-1} (ppm w/w), the residue with a boiling point between 174 °C (345 °F) and 522 °C (970 °F) (decane to tetracontane, i.e. C_{10} to C_{40}) in LPG. Higher-boiling materials, or materials that adhere permanently to the chromatographic column, will not flow though the column and, therefore, will not be detected.

Finally, most methods for the compositional analysis of LPG recommend the use of a liquid sampling valve (LSV) for sample introduction into the split inlet of the gas chromatographic instrument.

9.2.4 Refinery Gas

The terms *petroleum gas* or *refinery gas* are often used to identify LPG or even gas that emanates from the top of a refinery distillation column. For the purpose of this text, petroleum gas not only describes LPG but also natural gas and refinery gas (Mokhatab et al., 2006; Gary et al., 2007; Speight, 2014, 2017; Hsu and Robinson, 2017). In this chapter, each gas is, in turn, referenced by its name rather than the generic term *petroleum gas* (ASTM D4150, 2017). However, the composition of each gas varies (Table 9.4), and recognition of this is essential before the relevant testing protocols are selected and applied.

Table 9.4 General summary of product types and distillation range.

Product	Lower carbon limit	Upper carbon limit	Lower boiling point (°C)	Upper boiling point (°C)	Lower boiling point (°F)	Upper boiling point (°F)
Refinery gas	C_1	C_4	−161	−1	−259	31
Liquefied petroleum gas	C_3	C_4	−42	−1	−44	31
Naphtha	C_5	C_{17}	36	302	97	575
Gasoline	C_4	C_{12}	−1	216	31	421
Kerosene/diesel fuel	C_8	C_{18}	126	258	302	575
Aviation turbine fuel	C_8	C_{16}	126	287	302	548
Fuel oil	C_{12}	$>C_{20}$	216	421	>343	>649
Lubricating oil	$>C_{20}$		>343		>649	
Wax	C_{17}	$>C_{20}$	302	>343	575	>649
Asphalt	$>C_{20}$		>343		>649	
Coke	$>C_{50}$[a]		>1000[a]		>1832[a]	

[a]Carbon number and boiling point difficult to assess; inserted for illustrative purposes only.

Refinery gas varies in composition and volume, depending on crude origin and on any additions to the crude made at the loading point – in this sense, it is probably the most difficult gas to analyze. Furthermore, it is not uncommon to reinject low molecular weight hydrocarbons such as propane and butane into the crude before dispatch by tanker or pipeline. This results in a higher vapor pressure of the crude, but it allows one to increase the quantity of light products obtained at the refinery. Since light ends in most petroleum markets command a premium, while in the oil field itself propane and butane may have to be reinjected or flared, the practice of *spiking* crude oil with LPG is becoming fairly common.

Thus, *refinery gas* (*fuel gas*) is the noncondensable gas that is obtained during distillation of crude oil or treatment (cracking, thermal decomposition) of petroleum (Robinson and Faulkner, 2000; Speight, 2014). Refinery gas is produced in considerable quantities during the different refining processes and is used as fuel for the refinery itself and as an important feedstock for the production of petrochemicals. It consists mainly of hydrogen (H_2), methane (CH_4), ethane (C_2H_6), propane (C_3H_8), butane (C_4H_{10}), and olefins ($RCH{=}CHR^1$, where R and R^1 can be hydrogen or a methyl group) and may also include off-gases from petrochemical processes (Table 9.3). Olefins such as ethylene ($CH_2{=}CH_2$, boiling point: −104 °C, −155 °F), propene (propylene, $CH_3CH{=}CH_2$, boiling point: −47 °C, −53 °F), butene (butene-1, $CH_3CH_2CH{=}CH_2$, boiling point: −5 °C, 23 °F), *iso*-butylene ($(CH_3)_2C{=}CH_2$, −6 °C, 21 °F), *cis*- and *trans*-butene-2 ($CH_3CH{=}CHCH_3$, boiling point: ca. 1 °C, 30 °F), and butadiene ($CH_2{=}CHCH{=}CH_2$, boiling point: −4 °C, 24 °F) as well as higher-boiling olefins are produced by various refining processes.

Still gas is broad terminology for low-boiling hydrocarbon mixtures and is the lowest boiling fraction isolated from a distillation (*still*) unit in the refinery. If the distillation unit is separating light hydrocarbon fractions, the still gas will be almost entirely methane with only traces of ethane (CH_3CH_3) and ethylene ($CH_2{=}CH_2$). If the distillation unit is handling higher-boiling fractions, the still gas might also contain propane ($CH_3CH_2CH_3$) and butane ($CH_3CH_2CH_2CH_3$) and their respective isomers. *Fuel gas* and still gas are terms that are often used interchangeably, but the term *fuel gas* is intended to denote the product's destination – to be used as a fuel for boilers, furnaces, or heaters.

In a series of processes commercialized under the names Platforming, Powerforming, Catforming, and Ultraforming, paraffinic and naphthenic (cyclic nonaromatic) hydrocarbons, in the presence of hydrogen and a catalyst, are converted into aromatics or isomerized to more highly branched hydrocarbons. Catalytic reforming processes thus not only result in the formation of a liquid product of higher octane number but also produce substantial quantities of gases. The latter are rich in hydrogen but also contain hydrocarbons from methane to butanes, with a preponderance of propane ($CH_3CH_2CH_3$), *n*-butane ($CH_3CH_2CH_2CH_3$), and isobutane [$(CH_3)_3CH$]. The composition of these gases varies in accordance with reforming severity and reformer feedstock. Since all catalytic reforming processes require substantial recycling of a hydrogen stream, it is normal to separate reformer gas into a propane ($CH_3CH_2CH_3$) and/or a butane [$CH_3CH_2CH_2CH_3/(CH_3)_3CH$] stream, which becomes part of the refinery LPG production, and a lighter gas fraction, part of which is recycled. In view of the excess of hydrogen in the gas, all products of catalytic reforming are saturated, and there are usually no olefin-type gases present in either gas stream.

A second group of refining operations that contributes to gas production is that of the catalytic cracking processes, that is,. thermofor catalytic cracking and other variants in which heavy gas oils are converted into cracked gas, LPG, catalytic naphtha, fuel oil, and coke by contacting the heavy hydrocarbon with the hot catalyst (Gary et al., 2007; Speight, 2014, 2017; Hsu and Robinson, 2017). Both catalytic and thermal cracking processes, the latter being now largely used for the production of chemical raw materials, result in the formation of unsaturated hydrocarbons, particularly ethylene ($CH_2{=}CH_2$) but also propylene (propene, $CH_3CH{=}CH_2$), isobutylene (isobutene, $(CH_3)_2C{=}CH_2$), and the *n*-butenes ($CH_3CH_2CH{=}CH_2$ and $CH_3CH{=}CHCH_3$) in addition to hydrogen (H_2), methane (CH_4), and smaller quantities of ethane (CH_3CH_3), propane ($CH_3CH_2CH_3$), and butane isomers ($CH_3CH_2CH_2CH_3$, $(CH_3)_3CH$). Diolefins such as butadiene ($CH_2{=}CHCH{=}CH_2$) are also present.

Additional gases are produced in refineries with coking or visbreaking facilities for the processing of their heaviest crude fractions. In the visbreaking process, fuel oil is passed through externally fired tubes and undergoes liquid-phase cracking reactions, which result in the formation of lighter fuel oil components. Oil viscosity is therefore reduced, and some gases, mainly hydrogen, methane, and ethane, are formed. Substantial quantities of both gas and carbon are also formed in coking (both fluid coking and delayed coking) in addition to the middle distillate and naphtha. When coking a residual fuel oil or heavy gas oil, the feedstock is preheated and contacted with hot carbon (coke), which causes extensive cracking of the feedstock constituents of higher molecular weight to produce lower molecular weight products ranging from methane, via LPG(s) and naphtha, to gas oil and heating oil. Products from coking processes tend to be unsaturated, and olefin-type components predominate in the tail gases from coking processes.

A further source of refinery gas is hydrocracking, a catalytic high-pressure pyrolysis process in the presence of fresh and recycled hydrogen. The feedstock is again heavy gas oil or residual fuel oil, and the process is mainly directed at the production of additional middle distillates and gasoline. Since hydrogen is to be recycled, the gases produced in this process again have to be separated into lighter and heavier streams; any surplus recycle gas and the LPG from the hydrocracking process are both saturated.

Both hydrocracker and catalytic reformer tail gases are commonly used in catalytic desulfurization processes. In the latter, feedstocks ranging from light to vacuum gas oils are passed at pressures of 500–1000 psi (3.5–7.0 × 10^3 kPa) with hydrogen over a hydrofining catalyst. This results mainly in the conversion of organic sulfur compounds to hydrogen sulfide:

$$[S]_{feedstock} + H_2 \rightarrow H_2S + \text{hydrocarbons}$$

The process also produces some light hydrocarbons by hydrocracking.

The first and most important aspect of gaseous testing is the measurement of the volume of gas (ASTM D1071, 2017). In this test method, several techniques are described and may be employed for any purpose where it is necessary to know the quantity of gaseous fuel. In addition, the thermophysical properties of methane (ASTM D3956, 2017), ethane (ASTM D3984, 2017), propane (ASTM D4362, 2017), *n*-butane (ASTM D4650, 2017), and *iso*-butane (ASTM D4651, 2017) should be available for use and consultation (Stephenson and Malanowski, 1987).

In all cases, it is the composition of the gas in terms of ***hydrocarbon type*** that is more important in the context of the application. For example, in petrochemical applications, the presence of propylene and butylene above 10% v/v can have an adverse effect on hydrodesulfurization prior to steam reforming. On the other hand, petrochemical processes, such as in the production of *iso*-octane from *iso*-butane and butylene, can require the exclusion of the saturated hydrocarbons.

Refinery gas specifications will vary according to the gas quality available and the end use (Johansen, 1998). For fuel uses, gas as specified above presents little difficulty used as supplied. Alternatively, a gas of constant Wobbe index, say, for gas turbine use, could readily be produced by the user. Part of the combustion air would be diverted into the gas stream by a Wobbe index controller. This would be set to supply gas at the lowest Wobbe index of the undiluted gas.

Residual refinery gases, usually in more than one stream, which allows a degree of quality control, are treated for hydrogen sulfide removal, and gas sales are usually on a thermal content (calorific value, heating value) basis with some adjustment for variation in the calorific value and hydrocarbon type (Cranmore and Stanton, 2000; Speight, 2007, 2014).

9.2.5 Coal Gas

Gases of high calorific value are obtained by low-temperature or medium-temperature carbonization of coal. The gases obtained by the carbonization of any given coal change in a progressive manner with increasing temperature (Table 9.5). The composition of coal gas also changes during the course of carbonization at a given temperature, and secondary reactions of the volatile products are important in determining gas composition (Speight, 2013a).

The standard test methods for coal gas are instrumental in the testing and chemical analysis of natural gas and other gaseous fuels (ASTM, 2017). These

Table 9.5 Effect of carbonization temperature on gas composition.

Gas composition and yields (% v/v)	Temperature of carbonization					
	500 °C	600 °C	700 °C	800 °C	900 °C	1000 °C
CO_2	5.7	5.0	4.4	4.0	3.2	2.5
Unsaturates	3.2	4.0	5.2	5.1	4.8	4.5
CO	5.8	6.4	7.5	8.5	9.5	11.0
H_2	20.0	29.0	40.0	47.0	50.0	51.0
CH_4	49.5	47.0	36.0	31.0	29.5	29.0
C_2H_6	14.0	5.3	4.5	3.0	1.0	0.5
Yield ($m^3\,tonne^{-1}$)	62.3	102	176	238	278	312
CV ($MJ\,m^{-3}$)	39.0	29.0	26.5	24.4	22.3	22.3
Yield ($MJ\,tonne^{-1}$)	2118	2960	4660	5810	6200	6960

standards also include the thermophysical property tables for methane, ethane, propane, normal butane, and isobutane, and the test methods also allow laboratories and other chemical facilities to examine and assess these fuels to ensure safe handling and use.

With the onset of renewable energy programs, it would be remiss not to mention prominent gases produced from biomass and waste materials, viz. biogas and landfill gas. Both types of gas contain methane and carbon dioxide as well as various other constituents and are often amenable to application of the testing method that are applied to natural gas.

9.2.6 Biogas

The term *biogas* typically refers to the mixture of gases produced by the breakdown of organic matter in the absence of oxygen. Biogas can be produced from raw materials such as agricultural waste, manure, municipal waste, plant material, sewage, green waste, or food waste. The gas can be produced by (i) anaerobic digestion with anaerobic organisms, which digest material inside a closed system, or (ii) by fermentation of biodegradable materials. Biogas can be cleaned and upgraded to natural gas standards, when it is referred to as biomethane.

The composition of biogas varies depending upon the substrate composition, as well as on the conditions within the anaerobic reactor (temperature, pH, and substrate concentration). Advanced waste treatment technologies can produce biogas with 55–75% v/v methane, which for reactors with free liquids can be increased to 80–90% v/v methane using *in situ* gas purification techniques. As produced, biogas contains water vapor. The fractional volume of water vapor is a function of biogas temperature; correction of measured gas volume for water vapor content and thermal expansion is easily done via simple mathematics that yields the standardized volume of dry biogas.

9.2.7 Landfill Gas

Landfill gas is produced by wet organic waste decomposing under anaerobic conditions in a biogas. In fact, landfill gas is a product of three processes: (i) evaporation of volatile organic compounds such as low-boiling solvents, (ii) chemical reactions between waste components, and (iii) microbial action, especially methanogenesis. The first two processes depend strongly on the nature of the waste – the most dominant process in most landfills is the third process whereby anaerobic bacteria decompose organic waste to produce biogas, which consists of methane and carbon dioxide together with traces of other compounds. Despite the heterogeneity of waste, the evolution of gases follows well-defined kinetic pattern in which the formation of methane and carbon dioxide commences approximately 6 months after depositing the landfill material. The evolution of landfill gases reaches a maximum at approximately 20 years and then declines over the course of several decades.

In the process, the waste is covered and mechanically compressed by the weight of the material that is deposited above. This material prevents oxygen exposure, thus allowing anaerobic microbes to thrive. Biogas builds up and is slowly released into the atmosphere if the site has not been engineered to capture the gas. Landfill gas released in an uncontrolled way can be hazardous since it can become explosive when it escapes from the landfill and mixes with oxygen.

9.2.8 Flue Gas

Another gas that requires analysis is flue gas, which is the exhaust gases of combustion plants containing the reaction products of fuel and combustion air and residual substances such as particulate matter (dust), sulfur oxides, nitrogen oxides, and carbon monoxide (Table 9.6). When burning coal, hydrogen chloride and hydrogen fluoride may be present in the flue gas as well as hydrocarbons and heavy metals in case of incineration of waste materials. In many countries, as part of a national environmental protection program, exhaust gases must comply with strict governmental regulations regarding the limit values of pollutants such as dust, sulfur and nitrogen oxides, and carbon monoxide. To meet these limit values, combustion plants are equipped with flue gas cleaning systems such as gas scrubbers and dust filters.

An important aspect of gas analysis is knowledge of fuel composition for an optimum and economical combustion process. Increasing percentage of

Table 9.6 Constituents of flue gas.

Constituents	Comment
Nitrogen	The main constituent (79 Vol.%) of air; fed to the combustion as part of the combustion air but is not involved directly in the combustion process. Minor quantities of this combustion air-related nitrogen are, together with the nitrogen released from the fuel, responsible for the formation of the nitrogen oxides.
Carbon dioxide	A colorless and odorless gas with a slightly sour taste; produced during all combustion processes including respiration
Water vapor	Hydrogen contained in the fuel will react with oxygen and form water (H_2O); together with the water content of the fuel and the combustion air exists either as flue gas humidity (at higher temperatures) or as condensate (at lower temperatures).
Oxygen	The oxygen that has not been consumed by the combustion process remains as part of the flue gas and is a measure for the efficiency of the combustion.
Carbon Monoxide	A colorless, odorless, toxic gas; formed predominantly during incomplete combustion of carbonaceous fuels.
Oxides of nitrogen	Formed in combustion processes nitrogen of the fuel and, at high temperatures, also of the combustion air reacts to a certain amount with oxygen of the combustion air and forms first nitric oxide (fuel NO and thermal NO)
Sulfur dioxide	A colorless, toxic gas with a pungent smell. It is formed through oxidation of sulfur that is present in the fuel; with water or condensate, forms sulfurous acid and sulfuric acid.
Hydrogen sulfide	A toxic odorous gas; a component of crude oil and natural gas and is therefore present in refineries and natural gas plants; also generated during some other industrial processes.
Hydrocarbons	An extensive group of chemical compounds that are composed of hydrogen and carbon; occur in crude oil, natural gas, and coal; formed through incomplete combustion processes
Hydrocyanic acid	A very toxic liquid with a boiling point of only 25.6 °C; it may exist in flue gases of incineration plants
Ammonia	Relevant in flue gases in connection with denitrification plants.
Hydrogen halides	Occur in flue gas from the combustion of coal and/or waste material; the formation of the hydrogen halides hydrogen chloride and hydrogen fluoride may result, which form aggressive acids in humid atmospheres.
Solids (dust, soot)	Solid pollutants in flue gases originate from the incombustible components of solid or liquid fuels; include oxides of silica, aluminum, and calcium in case of coal.

incombustible (inert) fuel components reduces the gross and net calorific value of the fuel and increases contamination of the furnace walls. Increasing water content raises the water dew point and consumes energy to evaporate water in the flue gas. The sulfur contained in the fuel is burned (oxidized) to sulfur dioxide and sulfur trioxide that, at temperatures below the dew point, may lead to the formation of the corrosive and environmentally hazardous sulfurous acid and sulfuric acid.

Still gas is a term reserved for low-boiling hydrocarbon mixtures that represent the lowest boiling fraction isolated from an atmospheric distillation unit (*still*) in the refinery (Speight, 2014, 2017; Hsu and Robinson, 2017). By way of recall, still gas (which is often grouped with other gases under the general term *process gas*) typically will contain methane, ethane (CH_3CH_3), propane ($CH_3CH_2CH_3$), and butane ($CH_3CH_2CH_2CH_3$) and their respective isomers. If cracked products have been recycled to the atmospheric distillation unit, the still gas may also contain volatile olefins such as ethylene ($CH_2{=}CH_2$). The term *fuel gas* is often used interchangeably with *still gas*, but the term *fuel gas* is intended to denote the destination of the gaseous mixture, such as fuel for boilers, furnaces, or heaters, whereas the term *still gas* may be indicative of the use of the mixture for petrochemical production.

Petroleum-related gases such as natural gas and refinery gases (process gases) as well as product gases produced from petroleum refining, upgrading, or natural gas processing facilities are a category of saturated and unsaturated light hydrocarbons, predominantly C_1 to C_6 as well as, in some cases, as high as C_{10}. Some gases may also contain inorganic compounds, such as hydrogen, nitrogen, hydrogen sulfide, carbon monoxide, and carbon dioxide. As such, petroleum and refinery gases (unless produced as a saleable product that must meet specifications prior to sale) are of unknown or variable composition and toxic (API, 2009). The site-restricted petroleum and refinery gases (i.e. those not produced for sale) often serve as fuels consumed on-site, as intermediates for purification and recovery of various gaseous products, or as feedstocks for isomerization and alkylation processes within a facility. Thus, the principal types of gaseous fuels are oil (distillation) gas, reformed natural gas, and reformed propane or LPG. *Mixed gas* is a gas prepared by adding natural gas or LPG to a manufactured gas, giving a product of better utility and higher heat content or British thermal unit (Btu) value.

The importance of natural gas as a commodity and the cost to consumers has made the accuracy of analytical determination for the components of natural gas very important. Pricing of natural gas is based on the heating value of the gas determined from either calorimetry measurements or calculations based on individual component concentrations determined by GC. Due to the expense of accurate calibration standards, many analysts and laboratories will use a single calibration standard to perform natural gas analysis (Rhoderick, 2003). In addition, there is the modern need to analyze natural gas and its component products so that they can be judged as suitable for final use and to ascertain the extent of contaminants that could cause environmental damage. Furthermore, gas processing is a complex industrial process designed to clean raw (dirty, contaminated) gas by separating impurities and various nonmethane and fluids to produce what is known as *pipeline quality* dry natural gas.

More specifically, gas analysis finds wide application in the petrochemical, chemical, and energy industries, especially since gases such as carbon monoxide, carbon dioxide, and nitrogen are common in refinery gases. Understanding the concentration of these components is an important aspect for controlling manufacturing processes and production quality, and there are several common standard test methods that can be used to determine gas quality (Table 9.1). For example, impurities such as carbon monoxide and carbon dioxide *in* petrochemical-grade propylene and ethylene are deleterious to certain catalysts.

Several methods for gas analysis based on packed columns are standardized, for example, the determination of hydrogen, nitrogen, oxygen, and carbon monoxide at the parts per million (ppm v/v) level in ethane and lighter hydrocarbon products (ASTM D2504, 2017), while a separate test method (ASTM D1946, 2017) covers the analysis of gases such as methane, ethane, and ethylene.

In summary, there is no guarantee of the following composition at any location or as an overall system average. Since the different gas supplies enter a natural gas system at different locations, the exact composition at any site will vary among the different regions and is to be determined by analysis at that site. The system average heating value will depend on the mix of gas supplies and, therefore, can vary from any so-called typical (or average) value.

9.3 Analytical Methods

Because of the wide range of chemical and physical properties, a wide range of tests have been (and continue to be) developed to provide an indication of the means by which a particular gas should be processed (Table 9.7) although certain of these test methods are in more common use than others (Table 9.1). Initial inspection of the nature of the petroleum will provide deductions about the most logical means of refining or correlation of various properties to structural types present and hence attempted classification of the petroleum. Proper interpretation of the data resulting from the inspection of crude oil requires an understanding of their significance. Having decided what characteristics are necessary, it then remains to describe the product in terms of a specification. This entails selecting suitable test methods and setting appropriate limits. Many specifications in widespread use have evolved usually by the addition of extra clauses (rarely is a clause deleted). This has resulted in unnecessary restrictions that, in turn, result in increased cost of the products specified.

Table 9.7 Test methods for gaseous fuels and low-boiling hydrocarbons.

ASTM D130	Detection of Copper Corrosion from Petroleum Products by the Copper Strip Tarnish Test
ASTM D216	Distillation of Natural Gasoline
ASTM D323	Vapor Pressure of Petroleum Products (Reid Method)
ASTM D1025	Nonvolatile Residue of Polymerization-Grade Butadiene
ASTM D1070	Relative Density (Specific Gravity) of Gaseous Fuels
ASTM D1072	Total Sulfur in Fuel Gases
ASTM D1089	Carbonyl Content of Butadiene
ASTM D1142	Water Vapor Content of Gaseous Fuels by Measurement of Dew-Point Temperature
ASTM D1265	Sampling Liquefied Petroleum (LP) Gases
ASTM D1266	Sulfur in Petroleum Products (Lamp Method)
ASTM D1298	Density, Relative Density (Specific Gravity), or API Gravity of Crude Petroleum and Liquid Petroleum Products by Hydrometer Method
ASTM D1826	Calorific Value of Gases in Natural Gas Range by Continuous Recording Colorimeter
ASTM D1945	Analysis of Natural Gas by Gas Chromatography
ASTM D2385	Hydrogen Sulfide and Mercaptan Sulfur in Natural Gas (Cadmium Sulfate Iodometric Titration Method)
ASTM D2725	Hydrogen Sulfide in Natural Gas (Methylene Blue Method)
ASTM D3031	Total Sulfur in Natural Gas by Hydrogenation
ASTM D1265	Sampling Liquefied Petroleum (LP) Gases
ASTM D1267	Vapor Pressure of Liquefied Petroleum (LP) Gases (LP-Gas Method)
ASTM D1657	Density or Relative Density of Light Hydrocarbons by Pressure Hydrometer
ASTM D1837	Volatility of Liquefied Petroleum (LP) Gases
ASTM D1838	Copper Strip Corrosion by Liquefied Petroleum (LP) Gases
ASTM D2158	Residues in Liquefied Petroleum (LP) Gases
ASTM D2163	Analysis of Liquefied Petroleum (LP) Gases and Propane Concentrates by Gas Chromatography
ASTM D2384	Traces of Volatile Chlorides in Butane-Butene Mixtures
ASTM D2421	Practice for Interconversion of Analysis of C5 and Lighter Hydrocarbons to Gas-Volume, Liquid Volume, or Mass Basis
ASTM D2426	Butadiene Dimer and Styrene in Butadiene Concentrates by Gas Chromatography
ASTM D2427	Determination of C, through C, Hydrocarbons in Gasoline by Gas Chromatography

Table 9.7 (Continued)

ASTM D2504	Noncondensable Gases in C_3 and Lighter Hydrocarbon Products by Gas Chromatography
ASTM D2505	Ethylene, Other Hydrocarbons, and Carbon Dioxide in High-Purity Ethylene by Gas Chromatography
ASTM D2593	Butadiene Purity and Hydrocarbon Impurities by Gas Chromatography
ASTM D2597	Analysis of Demethanized Hydrocarbon Liquid Mixtures Containing Nitrogen and Carbon Dioxide by Gas Chromatography
ASTM D2598	Calculation of Certain Physical Properties of Liquefied Petroleum (LP) Gases from Compositional Analysis
ASTM D2623	Knock Characteristics of Liquefied Petroleum (LP) Gases by the Motor (LP) Method
ASTM D2699	Knock Characteristics of Motor Fuels by the Research Method
ASTM D2700	Knock Characteristics of Motor and Aviation Type Fuels by the Motor Method
ASTM D2712	Hydrocarbon Traces in Propylene Concentrates by Gas Chromatography
ASTM D2713	Dryness of Propane (Valve Freeze Method)
ASTM D2784	Sulfur in Liquefied Petroleum Gases (Oxy-Hydrogen Burner or Lamp)
ASTM D3246	Sulfur in Petroleum Gas by Oxidative Microcoulometry
ASTM D3700	Practice for Containing Hydrocarbon Fluid Samples Using a Floating Piston Cylinder
ASTM D4953	Test Method for Vapor Pressure of Gasoline and Gasoline-Oxygenate Blends (Dry Method)
ASTM D5190	Test Method for Vapor Pressure of Petroleum Products (Automatic Method)
ASTM D5191	Test Method for Vapor Pressure of Petroleum Products (Mini Method)

Test methods for gaseous fuels have been developed over many years, extending back into the 1930s. Bulk physical property tests, such as density and heating value, as well as some compositional tests, such as the Orsat analysis and the mercuric nitrate method for the determination of unsaturation, were widely used. More recently, mass spectrometry has become a popular method of choice for compositional analysis of low molecular weight and has replaced several older methods (ASTM D2421, 2017; ASTM D2650, 2017). GC (ASTM D1945, 2017) is another method of choice for hydrocarbon identification in gases. In additional, compositional analysis can be used to calculate calorific value, specific gravity, and compressibility factor (ASTM D3588, 2017).

In fact, GC in one form or another will continue to be the method of choice for characterization of low molecular weight hydrocarbons. Developments in high-speed test methods using gas chromatographic instrumentation and data processing are already leading to new and revised test methods. Other detection techniques such as chemiluminescence, atomic emission, and mass spectrometry will undoubtedly enhance selectivity, detection limits, and analytical productivity. Furthermore, laboratory automation using autosampling and data handling will provide improved precision and productivity accompanied by simplified method operation.

Specifications for petroleum gases typically focus on LPG, propane, and butane, and the specifications generally define physical properties and characteristics of LPG that make them suitable for private, commercial, or industrial applications. The specifications do not purport to specifically define all possible requirements to meet all possible applications, and the user is cautioned to exercise judgment in formulating final specifications for specific applications.

The Gas Processors Association, its management, and supporting companies claim no specific knowledge of how manufacturers and users will produce, handle, store, transfer, or consume the products defined herein and, therefore, are not responsible for any claims, causes of action, liabilities, losses, or expenses resulting from injury or death of persons and/or damage to property arising directly or indirectly from the use of LPG or these specifications relating to LPG (Table 9.8) (GPA, 1997).

Hydrocarbon gases are amenable to analytical techniques, and there has been the tendency, and it remains, for the determination of both major constituents and trace constituents than is the case with the heavier hydrocarbons. The complexity of the mixtures that are evident as the boiling point of petroleum fractions and petroleum products increases makes identification of many of the individual constituents difficult, if not impossible. In addition, methods have been developed for the determination of physical characteristics such as calorific value, specific gravity, and enthalpy from the analyses of mixed hydrocarbon gases, but the accuracy does suffer when compared with the data produced by methods for the direct determination of these properties.

Bulk physical property tests, such as density and heating value, as well as some compositional tests, such as the Orsat analysis (the measurement of the quantitative amounts by selective absorption of carbon dioxide, oxygen, and carbon monoxide) and the mercuric nitrate method for the determination of unsaturation, are still used.

However, the choice of a particular test rests upon the decision of the analyst that, then, depends upon the nature of the gas under study. For example, judgment by the analyst is necessary whether or not a test that is applied to LPG is suitable for natural gas insofar as inference from the nonhydrocarbon constituents will be minimal.

Specifications are normally prescriptive on concentrations for impurities, such as maximum values for oxygen, total sulfur, and hydrogen sulfide and maximum values for dew points of water and hydrocarbons. A limited number

Table 9.8 Examples of GPA test methods for natural gas.

GPA 1167: GPA Glossary – Definition of Words and Terms Used in the Gas Processing Industry
GPA 2100: Qualitative Determination of Carbonyl Sulfide in Propane Using Length of Stain Tubes
GPA 2103: Tentative Method for the Analysis of Natural Gas Condensate Mixtures Containing Nitrogen and Carbon Dioxide by Gas Chromatography
GPA 2108: GPA Fractionation Grade Product Specifications
GPA 2140: Liquefied Petroleum Gas Specifications and Test Methods
GPA 2145: Table of Physical Constants of Paraffin Hydrocarbons and Other Components of Natural Gas
GPA 2165: Method for Analysis of Natural Gas Liquid Mixtures by Gas Chromatography
GPA 2166: Obtaining Natural Gas Samples for Analysis by Gas Chromatography
GPA 2172: Calculation of Gross Heating Value, Relative Density, and Compressibility Factor for Natural Gas Mixtures from Compositional Analysis
GPA 2174: Obtaining Liquid Hydrocarbons Samples for Analysis by Gas Chromatography
GPA 2177: Analysis of Demethanized Hydrocarbon Liquid Mixtures Containing Nitrogen and Carbon Dioxide by Gas Chromatography
GPA 2186: Method for the Extended Analysis of Hydrocarbon Liquid Mixtures Containing Nitrogen and Carbon Dioxide by Temperature Programmed Gas Chromatography
GPA 2187: Method for the Determination of Ammonia in Liquid Propane
GPA 2188: Method for Determination of Ethyl Mercaptan in LP-Gas Using Length of Stain Tubes
GPA 2194: Low-Pressure Field Method for Determining Ethyl Mercaptan Odorant in LP-Gas Using Length of Stain Tubes
GPA 2198: Selection, Preparation, Validation, Care and Storage of Natural Gas and Natural Gas Liquids Reference Standard Blends
GPA 2199: Determination of Specific Sulfur Compounds by Capillary Gas Chromatography and Sulfur Chemiluminescence Detection
GPA 2261: Analysis for Natural Gas and Similar Gaseous Mixtures by Gas Chromatography
GPA 2286: Tentative Method of Extended Analysis for Natural Gas and Similar Gaseous Mixtures by Temperature Programmed Gas Chromatography
GPA 2290: Tentative Method for the Determination of Carbonyl Sulfide (COS) in Unoderized Liquid Propane
GPA 2377: Test for Hydrogen Sulfide and Carbon Dioxide in Natural Gas Using Length of Stain Tubes
GPA 8173: Standard for Converting Mass of Natural Gas Liquids and Vapors to Equivalent Liquid Volumes
GPA 8182: Standard for Mass Measurement of Natural Gas Liquids
GPA 8186: Measurement of Liquid Hydrocarbons by Truck Scales
GA 8195: Tentative Standard for Converting Net Vapor Space Volumes to Equivalent Liquid Volumes.

of countries have introduced specifications for maximum hydrogen content. These impurity specifications are necessary for protection of pipeline systems and/or customer facilities from corrosion and mechanical or other damage. Often, gas quality specifications also include a general "impurity clause" aimed at proscribing (trace) components that are not monitored but could, if present in the gas, cause operational and/or other problems to the end user.

Substances may be added to the gas to protect the pipework or instrumentation on the network. Cast iron pipes are jointed with lead and yarn that is kept swollen to maintain the seal by adding ethylene glycol. Mechanical or rubber joints are kept swollen by injecting distillate into the gas flow. With the emergence of polyethylene pipe, the need for such additives is steadily reducing. Other additives include valve-flush agents that are used during maintenance procedures.

9.3.1 Gas Chromatography

GC is used to give detailed information on the composition of natural gas from which physical properties (such as the gas interchangeability parameters Wobbe index and calorific value) of the gas are calculated. The ASTM defines methods that should be used to evaluate the performance of a gas chromatograph. Presenting the instrument with a number of gases of known composition that are traceable to national or international standards and that cover the composition range of each component provides information on the linearity of the detector and any bias that will result from using a single-point calibration.

9.3.2 Infrared Absorption

Infrared (IR) radiation (wavelength range is some _m) is absorbed by gases such as CO, CO_2, SO_2, or NO at a wavelength that is specific for the gas. As IR radiation passes through a measuring cell filled with the gas to be analyzed, an increase in the gas concentration causes a corresponding increase in IR absorption and thus a decrease in the radiation intensity received by the IR detector. To ensure selectivity for just one gas component, two different principles are used: The **dispersive** principle uses radiation, which has been dispersed by a prism or grating before entering the measuring cell. Only two wavelengths (or narrow wavelength ranges) of the entire spectrum are used, one that the component in question absorbs and one that is not, as a reference. The ratio of absorption at these two wavelengths is a measure for the gas concentration.

The nondispersive (NDIR) principle uses a broad band of radiation from a lamp or glower without dispersion before entering the measuring cell. When a component that does absorb IR radiation is present in the sample gas, the radiation intensity reaching the detector will be reduced at the specific part of the spectrum. Two methods are in use for detecting the level of absorption:

- A *gas detector* filled specifically with the gas of interest will, from the broadband radiation leaving the measuring cell, detect only the component of interest resp. its variations caused by the varying concentration of the component in the sample gas passing through the measuring cell. Radiation absorption affects the gas pressure in the detector cell; these pressure variations are used to generate a detector signal as measure of the concentration of the gas of interest.
- A *solid-state IR detector* is used together with a narrowband interference filter that allows from the broadband spectrum of IR radiation only the radiation range of the component of interest (e.g. CO_2) to pass through. As the IR radiation passes through the measuring cell, a variation in CO_2 concentration will cause a proportional variation in the IR absorption and thus in the detector.

9.3.3 Ultraviolet Absorption

Certain gas components, e.g. SO_2 or NO, absorb also ultraviolet (UV) radiation. For these components the UV method competes with the IR method offering the major advantage that water does not cause any interference effect in the UV absorption range. In contrast to IR, however, UV light sources are more expensive, and therefore the UV method is used preferably for particular applications.

9.3.4 Chemiluminescence Detector

This method is mainly used for the determination of low concentrations of NO and NO_2. It is based on the feature of NO to react with ozone (O_3), resulting in the emission of characteristic light radiation (chemiluminescence). The intensity of the emitted radiation is proportional to the NO concentration. Such analyzers (chemiluminescence detectors [CLD]) are very common for analysis of NO and NO_2 in car exhaust gases. They comprise essentially an ozone generator, an enrichment and reaction chamber, and a photomultiplier detector.

9.3.5 Paramagnetic Method

This method is applied particularly for the analysis of oxygen. Oxygen is the only gas with a considerably high value of paramagnetism with the result that oxygen molecules are attracted by a magnetic field. This effect is used to the determine oxygen concentration in gases. Different analyzer designs are available.

9.3.6 The Dumbbell Analyzer

Two balls made of glass and filled with nitrogen are mounted at the ends of a thin bar, forming a dumbbell, which is suspended in the sample gas stream in a strong inhomogeneous magnetic field. With oxygen present in the sample gas, the oxygen molecules will be attracted by the magnetic field and get the dumbbell swinging. This effect is compensated with a prestressed spring, resulting in a measuring signal.

9.3.7 The Differential Pressure Analyzer

The pressure produced at the contact surface of two gases with more and less magnetic characteristics is used here. Between the sample gas containing more or less paramagnetic oxygen molecules and an oxygen-free reference gas, e.g. nitrogen, a pressure difference is generated, resulting in a gas flow that is a measure of the oxygen concentration.

9.3.8 Thermal Conductivity Method

This principle (mostly applied for the analysis of H_2) uses the different values of thermal conductivity of two gases (sample and reference gas). In the analyzer a Wheatstone bridge circuit is built in the filaments that act at the same time as source of heat and as detector of temperature changes. The bridge is positioned with one part in the sample gas and with the other part in the reference gas. The different coolings of the filaments due to the different gas compositions result in an electrical signal that is used for further processing.

9.3.9 Catalytic Filament Method

Oxidation of combustible gases at a heated catalytic filament is used to detect the *total* concentration of combustibles in a sample gas. The combustibles react with a preheated catalytic material (wire or a bead, supported on a platinum coil) and increase, through their combustion heat, the temperature of the wire, resulting in a change of its electrical conductivity. This process will only work with sufficient oxygen available in the gas for the combustion.

The change in conductivity is a measure for the content of *all* combustibles present in the sample gas. These are mainly hydrocarbons (acronym: HC or C_xH_y) but also CO or H_2. Cross sensitivities therefore exist between these components and must be considered in data evaluation.

9.3.10 Flame Ionization Detection

The sample gas together with the hydrocarbons is introduced into a flame (usually hydrogen/air) where the hydrocarbons are ionized. The positively charged ions are collected by a negatively charged collecting electrode, generating a current proportional to the hydrocarbon concentration. The different sensitivity of this principle to different hydrocarbon compounds is taken into account by applying specific response factors during data evaluation.

9.3.11 Solid-state Electrolyte Method

A very common oxygen measuring principle uses a solid-state electrolyte sensor (made of zircon oxide ceramics) coated with layers (electrodes) of porous platinum as catalyst on two surface sides. At high temperatures >500 °C, oxygen molecules coming in contact with the platinum electrodes become oxygen *ions* with great mobility that intend to move through the molecular "holes" of the zirconium cell. As the two electrodes are in contact with gases of different oxygen concentrations (sample gas and reference gas, e.g. ambient air), oxygen ions will move through the cell from one electrode to the other and generate a differential voltage across the electrodes. From that, with a known oxygen concentration of the reference gas, the sample gas oxygen content can be determined.

Such analyzers are known as oxygen or *zirconium probes*. They are capable to measure correctly even under very harsh conditions and are, therefore, very often installed directly in the process flow (*in situ*). They measure the gas *in its original state including the content of humidity!* Because of the dilution effect caused by the humidity content, the oxygen values measured by zirconium probes are generally lower than of analyzers that measure dry sample gas.

9.4 Properties of Gases

The characterization of natural gas depends on the characterization and properties of pure hydrocarbons. Calculation of the properties of a mixture depends on the properties of its constituents. However, calculation of the property of a mixture based on an average calculation neglects any interactions between the constituents.

In this chapter, first basic parameters and properties of pure compounds are defined. These properties are either temperature independent or values of some basic properties at a fixed temperature. However, the composition of natural gas is not universally constant, as it is normally drawn from several production fields. Variations in the composition of gas delivered by pipelines can be caused by (i) variations in the proportion of the contribution from various sources at a given supply location and (ii) time variations within a given supply source. These variations involve mainly the composition of the hydrocarbon mix and are not regulated by the general-purpose pipeline gas standards beyond keeping within the limits of the heating value and the Wobbe index. For example, variations in composition across the United

States could produce variations on the order of 14% in heating value, 14% in density, 20% in the Wobbe index, and 25% in the stoichiometric air–fuel ratio. Thus, the composition of natural gas from a particular source or at a particular location can also vary over time that can cause difficulties in resolving the data from the application of standard test methods (Klimstra, 1978; Liss and Thrasher, 1992).

The actual composition of natural gas depends primarily on the production field from which it is extracted, and limited variations in composition must therefore be accepted. Worldwide there are major differences in natural gas quality specifications. These specifications are mainly intended to meet pipeline requirements and the needs of industrial and domestic consumers. The gas quality specifications are based on gas composition and other properties that are determined by application of a series of standard test methods to the gas. Changes in gas quality can be used to ensure better gas industry operational efficiencies, but quality standards must take into account the impact on end-use gas-fired equipment performance. Some common gas quality parameters include (i) water content; (ii) hydrocarbon dew point, below which the gas will begin to condense; (iii) sulfur content; and (iv) the Wobbe index, also called the Wobbe number.

Natural gas is mainly used or burned in combustion processes but is also an important feedstock in chemical production. Combustion is a complex chemical and physical phenomenon. The chemical energy contained within the fuel is released, producing heat and light, and the fuel is converted to more stable products. To ensure that different natural gases behave similarly in combustion systems, they must have similar combustion parameters (for example, calorific value, Wobbe index, and relative density). Gas quality is known to have an impact on combustion operation and efficiency, and systems are optimized to use one particular composition of the gas. If the gas quality changes, the system performance may deteriorate. Interchangeability of gases is achieved if combustion is not materially changed when one natural gas is substituted for another, particularly with regard to safety, efficiency, and emissions.

There are two major technical aspects to which gas quality relates: (i) the pipeline specification in which stringent specifications for water content and hydrocarbon dew point are stated along with limits for contaminants such as sulfur – the objective is to ensure pipeline material integrity for reliable gas transportation purpose, and (ii) the interchangeability specification that may include analytical data such as calorific value and relative density that are specified to ensure satisfactory performance of end-use equipment.

Gas interchangeability is a subset of the gas quality specification ensuring that gas supplied to domestic users will combust safely and efficiently. The Wobbe number is a common, but not universal, measure of interchangeability and is used to compare the rate of combustion energy output of different composition fuel gases in combustion equipment. For two fuels with identical Wobbe indices, the energy output will be the same for given pressure and valve settings.

Finally, in terms of properties (and any test methods that are applied to natural gas), it is necessary to recognize the other constituents of a natural gas stream that is produced from a reservoir. Thus, NGLs are products other than methane from natural gas: ethane, butane, *iso*-butane, and propane. Natural gasoline may also be included in this group.

NGLs are, in fact, separate and distinct hydrocarbons contained within some streams of natural gas. Streams that contain commercial quantities of NGLs are called *wet gas*, and those with little or no liquids present are known as *dry gas* (see above). Chemical manufacturers use ethane in making *ethylene*, an important petrochemical. Butane and propane, and mixtures of the two, are classified as LPG that is used chiefly as a heating fuel in industry and homes. Pentane, hexane, and heptane are collectively referred to as *gas condensate* (*natural gasoline, casinghead gasoline, natural gas gasoline*). However, at high pressures, such as those existing in the deeper fields, the density of the gas increases, and the density of the oil decreases until they form a single phase in the reservoir.

Wet natural gas contains natural gasoline in vapor form. The wet gas, also known as *casinghead gas*, is chiefly a mixture of methane, ethane, and the volatile hydrocarbons propane, butane, pentane (C_5H_{12}), hexane (C_6H_{14}), and heptane (C_7H_{16}). The latter three hydrocarbons form the main constituents of natural gasoline, which is recovered in refineries in liquid form mainly by absorption or compression processes. Pentane, hexane, and heptane are liquids under normal atmospheric conditions and are the chief components of ordinary refinery gasoline. Natural gasoline is used as blending stock for refinery gasoline and may be *cracked* to produce lower-boiling products, such as ethylene, propylene, and butylene. Caution should be taken not to confuse *natural gasoline* with *straight-run gasoline* (often also incorrectly referred to as natural gasoline), which is the gasoline distilled unchanged from petroleum. The various tests that are applied to specifications for this group of low-boiling liquids will be referenced in the chapters dealing with LPG and gas condensate (Chapters 8, 9, and 10).

It is therefore essential that the type of gas be identified, and it is even more essential that that gas has received some form (at least a preliminary from) of separation (from gas liquids and gas condensate) so that the appropriate test can be applied to determination of the properties and behavior of the gas.

9.4.1 Chemical Properties

Natural gas is found in petroleum reservoirs as free gas (*associated gas*) or in solution with petroleum in the reservoir (*dissolved gas*) or in reservoirs that contain only gaseous constituents and no (or little) petroleum (*unassociated gas*) (Cranmore and Stanton, 2000; Speight, 2014). The hydrocarbon content varies from mixtures of methane and ethane with very few other constituents (*dry* gas) to mixtures containing all of the hydrocarbons from methane to pentane and even hexane (C_6H_{14}) and heptane (C_7H_{16}) (*wet* gas). In both cases some carbon dioxide (CO_2) and inert gases, including helium (He), are present together with hydrogen sulfide (H_2S) and a small quantity of organic sulfur.

The term *petroleum gas(es)* in this context is also used to describe the gaseous phase and liquid-phase mixtures comprised mainly of methane to butane (C_1 to C_4 hydrocarbons) that are dissolved in the crude oil and natural gas, as well as gases produced during thermal processes in which the crude oil is converted to other products. It is necessary, however, to acknowledge that in addition to the hydrocarbons, gases such as carbon dioxide, hydrogen sulfide, and ammonia are also produced during petroleum refining and will be constituents of refinery gas that must be removed. Olefins are also present in the gas streams of various processes and are not included in LPG but are removed for use in petrochemical operations (Crawford et al., 1993).

Raw natural gas varies greatly in composition (Table 9.2), and the constituents can be several of a group of saturated hydrocarbons from methane to butane (Table 9.3) and nonhydrocarbons. The treatment required to prepare natural gas for distribution as an industrial or household fuel is specified in terms of the use and environmental regulations.

Carbon dioxide and hydrogen sulfide are commonly referred to as *acid gases* since they form corrosive compounds in the presence of water. Nitrogen, helium, and carbon dioxide are also referred to as *diluents* since none of these burn, and thus they have no heating value. Mercury can also be present either as a metal in vapor phase or as an organometallic compound in liquid fractions. Concentration levels are generally very small, but even at very small concentration levels, mercury can be detrimental due to its toxicity and its corrosive properties (reaction with aluminum alloys).

Nonassociated natural gas, which is found in reservoirs in which there is no, or at best only minimal amounts of, petroleum (Chapter 1). Nonassociated gas is usually richer in methane but is markedly leaner in terms of the higher molecular weight hydrocarbons and condensate. Conversely there is also *associated* natural gas (*dissolved* natural gas) that occurs either as free gas or as gas in solution in the petroleum. Gas that occurs as a solution with the crude petroleum is *dissolved gas*, whereas the gas that exists in contact with the crude petroleum (*gas cap*) is *associated gas* (Chapter 1). Associated gas is usually leaner in methane than the nonassociated gas but is richer in the higher molecular weight constituents.

The most preferred type of natural gas is the nonassociated gas. Such gas can be produced at high pressure, whereas associated, or dissolved, gas must be separated from petroleum at lower separator pressures, which usually involves increased expenditure for compression. Thus, it is not surprising that such gas (under conditions that are not economically favorable) is often flared or vented.

Typically, natural gas samples are analyzed for molecular composition by GC and for stable isotopic composition by isotope ratio mass spectrometry. Carbon isotopic composition was determined for methane (CH_4), ethane (C_2H_6), propane (C_3H_8), and butane, particularly *iso*-butane (C_4H_{10}) (ASTM D1945, 2017). Carbon dioxide (ASTM D1945, 2017; ASTM D4984, 2017) more than 3% is normally removed for reasons of corrosion prevention (ASTM D1838, 2017). Hydrogen sulfide (ASTM D2420, 2017; ASTM D4084, 2017; ASTM D4810, 2017) is also removed, and the odor of the gas must not be objectionable (ASTM D6273, 2017), so mercaptan content (ASTM D1988, 2017) is important. A simple lead acetate test (ASTM D2420, 2017; ASTM D4084, 2017) is available for detecting the presence of hydrogen sulfide and is an additional safeguard that hydrogen sulfide not be present (ASTM D1835, 2017). The odor of the gases must not be objectionable. Methyl mercaptan, if present, produces a transitory yellow stain on the lead acetate paper that fades completely in less than 5 min. Other sulfur compounds (ASTM D5504, 2017; ASTM D6228, 2017) present in LPG do not interfere.

In the lead acetate test (ASTM D2420, 2017), the vaporized gas is passed over moist lead acetate paper under controlled conditions. The total sulfur content (ASTM D1072, 2017; ASTM D2784, 2017) is normally acceptably low and frequently so low that it needs augmenting by means of alkyl sulfide derivatives, mercaptan derivatives, or thiophene derivatives to maintain an acceptable safe level of odor. The relevant test method (ASTM D1072, 2017) is used for the determination of total sulfur in combustible fuel gases and is applicable to natural gases, manufactured gases, mixed gases, and other miscellaneous gaseous fuels. The method is particularly applicable for the determination of total sulfur in combustible fuel gases, when present in sulfur concentrations between approximately 25 and 700 $mg\,m^{-3}$.

Another important property of the gas streams discussed in this text is the hydrocarbon dew point. The hydrocarbon dew point is reduced to such a level that retrograde condensation, i.e. condensation resulting from pressure drop, cannot occur under the worst conditions likely to be experienced in the gas transmission system. Similarly, the water dew point is reduced to a level sufficient to preclude formation of C_1 to C_4 hydrates in the system. Generally, pipeline owners prefer that the specifications for the transmission of natural gas limit the maximum concentration of water vapor allowed. Excess water vapor can cause corrosive conditions, degrading pipelines and equipment. The water can also condense and freeze or form methane hydrates (Chapter 7) causing blockages. Water vapor content also affects the heating value of natural gas, thus influencing the quality of the gas.

One relevant text method is ASTM D1142 (2017). This test method permits the determination of water content of natural gas. The method covers the determination of the water vapor content of gaseous fuels by measurement of the dew-point temperature and the calculation therefrom of the

Table 9.9 Number of isomers for the selected hydrocarbons.

Carbon atoms	Number of isomers
1	1
2	1
3	1
4	2
5	3
6	5
7	9
8	18
9	35
10	75
15	4 347
20	366 319
25	36 797 588
30	4 111 846 763
40	62 491 178 805 831

water vapor content. However, if the dew point of the condensable hydrocarbons is higher than the water vapor dew point, when such hydrocarbon derivatives are present in large amounts, they may interfere with the performance of the test method.

LPG, natural gas, and refinery gas are mixtures of products or naturally occurring materials and, fortunately, are relative simple mixtures and do not suffer the complexities of the isomeric variations of the higher molecular weight hydrocarbons (Table 9.9) (Drews, 1998; Speight, 2014). Thus, because of the lower molecular weight constituents of these gases and their volatility, GC has been the technique of choice for fixed gas and hydrocarbon speciation, and mass spectrometry is also a method of choice for compositional analysis of low molecular weight hydrocarbons (ASTM D2421, 2017; ASTM D2650, 2017).

Test method ASTM D2421 (2017) describes the procedure for the interconversion of the analysis of C_5 and lighter hydrocarbon mixtures to gas volume (mole), liquid volume, or mass basis. For custody transfer and other purposes, it is frequently necessary to convert a component analysis of light hydrocarbon mixture from one basis (either gas volume, liquid volume, or mass) to another. The component distribution data of low-boiling hydrocarbon mixtures can be used to calculate physical properties such as relative density, vapor pressure, and calorific value. More recently, piggyback methods (such as GC/mass spectrometry and other double technique methods) have been developed for the identification of gaseous and low-boiling liquid constituents of mixtures. The hydrocarbon composition is limited to set to the total amount of ethane, butane, or pentane as well as ethylene and total dienes.

By limiting the number of hydrocarbons that are lower boiling than the main component, the vapor pressure control is reinforced. The limitation on the amount of higher-boiling hydrocarbons supports the volatility clause. The vapor pressure and volatility specifications will often be met automatically if the hydrocarbon composition is correct. Information on the vapor pressures of LPG products is pertinent to selection of properly designed storage vessels, shipping containers, and customer utilization equipment to ensure safe handling of these products.

The determination of the vapor pressure of LPG is important for safety reasons to ensure that the maximum operating design pressures of storage, handling, and fuel systems will not be exceeded under normal operating temperature conditions. For LPG, vapor pressure is an indirect measure of the most extreme low temperature conditions under which initial vaporization can be expected to occur. It can be considered a semiquantitative measure of the amount of the most volatile material present in the product. On test method (ASTM D1267, 2017) 113 °F covers the determination of the gage vapor pressures of LPG products at temperatures of 37.8 °C (100 °F) up to and including a test temperature of 70 °C (158 °F).

The amount of ethylene in a gas stream is limited because it is necessary to restrict the number of unsaturated components to avoid the formation of deposits caused by the polymerization of the olefin constituents. In addition, ethylene (boiling point: −104 °C, −155 °F) is more volatile than ethane (boiling point: −88 °C, −127 °F), and therefore a product with a substantial proportion of ethylene will have a higher vapor pressure and volatility than one that is predominantly ethane. Butadiene is also undesirable because it may also produce polymeric products that form deposits and cause blockage of lines.

Currently the preferred method for the analysis of LPG, and indeed for most petroleum-related gases, is GC (ASTM D2163, 2017). This test method covers the quantitative determination of individual hydrocarbons in LPG and mixtures of propane and propene, excluding high-purity propene in the range of C_1–C_5. Component concentrations are determined in the range of 0.01–100 vol.%. This method can be used for the identification and measurement of both main constituents and trace constituents. However, the method does not fully determine hydrocarbons heavier than C_5 and nonhydrocarbon constituents, and additional tests may be necessary to fully characterize an LPG sample. Thus, there may be some accuracy issues that arise in the measurement of the higher-boiling constituents due to relative volatility under the conditions in which the sample is held.

Capillary column GC is an even quicker and equally accurate alternative. Mass spectrometry is also suitable for analysis of petroleum gases. Of the other spectroscopic techniques, IR and UV absorption may be applied to petroleum gas analysis for some specialized applications. GC has also largely supplanted chemical absorption methods of analysis, but again these may have some limited specialized application. Once the composition of a mixture has been determined, it is possible to calculate various properties such as specific gravity, vapor pressure, calorific value, and dew point.

Simple evaporation tests in conjunction with vapor pressure measurement give a further guide to composition. In these tests, an LPG sample is allowed to evaporate naturally from an open graduated vessel. Results are recorded based on volume–temperature changes, such as the temperature recorded when 95% v/v has evaporated, or volume left at a particular temperature (ASTM D1837, 2017). This test method is a measure of the relative purity of the various types of LPG and is a benefit in the assurance of volatility performance. The test results, when properly related to vapor pressure and density of the product, can be used to indicate the presence of butane and higher molecular weight constituents in propane-type LPG and pentane and higher molecular weight constituents in propane–butane-type fuel gases as well as in butane-type fuel gases. The presence of hydrocarbon compounds less volatile than those of which the LPG is primarily composed is indicated by an increase in the 95% v/v evaporated temperature. When the type and concentration of higher-boiling components is required, chromatographic analysis should be used.

Specific gravity again can be calculated, but if it is necessary to measure it, several pieces of apparatus are available. For determining the density or specific gravity of LPG in its liquid state, there are two methods, using a metal pressure pycnometer. A pressure hydrometer may be used (ASTM D1267, 2017) for the relative density that may also be calculated from compositional analysis (ASTM D2598, 2017). Various procedures, manual and recording, for specific gravity or density in the gaseous state are given in two methods (ASTM D1070, 2017). Calculation of the density is also possible using any one of four models, depending upon the composition of the gas (ASTM D4784, 2017).

The hydrocarbon composition of natural gasoline (although not specifically a gas) for petrochemical use must undergo a compositional analysis (ASTM D2427, 2017) and a test for total sulfur (ASTM D1266, 2017). An issue that arises during the characterization of LPG relates to the accurate determination of heavy residues (i.e. higher molecular weight hydrocarbons and even oils) in the gas. Test methods using procedures similar to those employed in gas chromatographic simulated distillation are becoming available. In fact, the presence of any component substantially less volatile than the main constituents of the LPG will give rise to unsatisfactory performance. It is difficult to set limits to the amount and nature of the *residue* that will make a product unsatisfactory. Obviously small amounts of oil-like material can block regulators and valves. In liquid vaporizer feed systems, even gasoline-type material can cause difficulty. The residue as determined by the *end point index* (EPI) (ASTM D2158, 2017) is a measure of the concentration of contaminants boiling above 37.8 °C (100 °F) that may be present in the gas. Other methods are available that measure residue more directly, and for particular applications, it may be possible to relate the values obtained to the performance required and so set satisfactory limits.

Analytical methods are available in standard form for determining volatile sulfur content and certain specific corrosive sulfur compounds that are likely to be present. Volatile sulfur determination is made by a combustion procedure (ASTM D1266, 2017) that uses a modification of the standard wick-fed lamp. Many laboratories use rapid combustion techniques with an oxyhydrogen flame in a Wickbold or Martin–Floret burner (ASTM D2784, 2017).

The first test method (ASTM D1266, 2017) provides a method for monitoring the sulfur level of various petroleum products and additives. The test method covers the determination of total sulfur in liquid petroleum products in concentrations from 0.01 to 0.4% w/w. The direct burning procedure described in the test method is applicable to the analysis of such materials as naphtha, gasoline, kerosene, and other liquids that can be burned completely in a wick lamp. The blending procedure is applicable to the analysis of gas oils and distillate fuel oils, high sulfur content petroleum products, and many other materials that cannot be burned satisfactorily by the direct burning procedure. Phosphorus compounds normally present in commercial gasoline do not interfere. A correction is given for the small amount of acid resulting from the combustion of the lead antiknock fluids in gasolines. Appreciable concentrations of acid-forming or base-forming elements from other sources interfere when the titration procedure is employed since no correction is provided in these cases.

A special sulfate analysis procedure is also described in the test that permits the determination of sulfur in concentrations as low as 5 $mg\,kg^{-1}$. The data can be used to predict performance, handling, or processing properties. In some cases, the presence of sulfur components is beneficial to the product, and monitoring the depletion of sulfur compounds provides useful information. In other cases, the presence of sulfur compounds is detrimental to the processing or use of the product.

The comparable lamp method for the determination of sulfur in LPG (ASTM D2784, 2017) is also suitable for the determination of sulfur. For the determination of sulfur in higher-boiling petroleum products that cannot be burned in a lamp, alternate standard test methods are available (ASTM D129, 2017; ASTM D1552, 2017).

The second test method (ASTM D2784, 2017) is valid for sulfur levels on the order of >1 $\mu g\,g^{-1}$ of sulfur in LPG, but the samples should not contain more than 100 $\mu g\,g^{-1}$ of chlorine. In the test, the sample is bummed in an oxyhydrogen bummer or in a lamp in a closed system in a carbon dioxide–oxygen atmosphere. The latter is not recommended for trace quantities of sulfur due to inordinately long combustion times needed. The sulfur oxides produced are absorbed and oxidized to sulfuric acid in a hydrogen peroxide solution. The sulfate ions are then determined by either titrating with barium perchlorate solution using a thorin–methylene blue mixed indicator or precipitating as barium sulfate and measuring the turbidity of the precipitate with a photometer.

Trace hydrocarbons that may be regarded as contaminants may be determined by the gas chromatographic methods already discussed. Heavier hydrocarbons in small amounts may not be completely removed from the column. If accurate information is required about the nature and amount of higher-boiling constituents, the temperature programming or a concentration procedure may be used.

Analytical methods for determining traces of various other impurities, such as chlorides (ASTM D2384, 2017), are known to be in use. Such information is valuable in cases where chloride is deleterious in the use of this product; also, chloride contributes to corrosion problems in processing units in instances where further processing of this material is involved. The presence of acetylene derivatives in refinery gases, although unlikely, must still be considered. Acetylene derivatives can be determined using a chemical test method, while carbonyls are determined by the classical hydroxylamine hydrochloride reaction.

The determination of traces of higher-boiling hydrocarbons and oily matter involves the use of a method for residue that involves a preliminary weathering. The residue after weathering is dissolved in a solvent and the solution applied to a filter paper. The presence of residue is indicated by the formation of an oil stain. The procedure is taken further by combining the oil stain observation with other observed values to calculate an EPI (ASTM D2158, 2017). The method is not very precise, and work is proceeding in several laboratories to develop a better method for the determination of residue in the form of oily matter.

The odor of LPG has to be detectable to avoid the risk of explosion. Odor is a very subjective matter and no standard method is available. It is desirable to set up some system in which the concentration of gas can be measured in relation to its explosive limits and in which some variables can be standardized, e.g. flow rate and orifice size. This will ensure that in any one location the LPG is always being assessed under similar conditions from day to day.

Propane, *iso*-butane (boiling point: −12 °C, 11 °F), and butane generally constitute this sample type and are used for heating, for motor fuels, and as chemical feedstocks (ASTM D2504, 2017; ASTM D2505, 2017; ASTM D2597, 2017). Test method ASTM D2597 (2017) covers the analysis of demethanized liquid hydrocarbon streams containing nitrogen/air and carbon dioxide and purity products such as an ethane–propane mix that fall within the compositional ranges. This test method is limited to mixtures containing less than 5 mol% of heptanes and heavier fractions. The heptane derivatives and higher-boiling fraction, when present in the sample, are analyzed by either (i) reverse flow of carrier gas after hexane and peak grouping or (ii) precut column to elute heptanes and heavier first as a single peak. For purity mixes without heptanes and heavier, no reverse of carrier flow is required. In the case of unknown samples with a relatively large fraction of the higher-boiling constituents and where precise results are important, it is desirable to determine the molecular weight (or other pertinent physical properties) of these fractions. Since this test method makes no provision for determining physical properties, the physical properties needed can be determined by an extended analysis or agreed to by the contracting parties.

Procedures for the determination of hydrogen, helium, oxygen, nitrogen, carbon monoxide, carbon dioxide, methane, ethane, ethylene, propane, butanes, pentanes, and hexanes as well as higher molecular weight hydrocarbons in natural and reformed gases by packed column GC are available (ASTM D1945, 2017; ASTM D1946, 2017). These compositional analyses from these two test methods can be used to calculate many other properties of gases, such as density, heating value, and compressibility. The first five components listed (above) are determined using a molecular sieve column (argon carrier gas), while the remaining components are determined using polydimethylsiloxane partition or porous polymer columns. The hexanes-plus analysis is accomplished by backflushing the column after the elution of pentane or by the use of a backflushed precolumn.

Important constituents of natural gas not accounted for in these analyses are moisture (water) and hydrogen sulfide, as well as other sulfur compounds (ASTM D1142, 2017; ASTM D1988, 2017; ASTM D4888, 2017; ASTM D5454, 2017; ASTM D5504, 2017; ASTM D6228, 2017). Olefins (ethylene, propylene, butylene derivatives, and pentylene derivatives) that occur in refinery (process) gas have specific characteristics and require specific testing protocols (ASTM D5234, 2017; ASTM D5273, 2017; ASTM D5274, 2017).

Test method ASTM D5234 (2017) provides an overview of the typical concentrations of the possible components found in ethylene product, the methods used in analysis, and the units of measure. This overview is intended to be used to improve the consistency of methods and the units reported so that errors are minimized. Each producer and user of ethylene product should immediately review this guide to improve their awareness of the various analytical methods in use, the units of measure, and concentration levels of the possible components. Test method ASTM D5273 (2017) provides information on the likely composition of propylene concentrates and on probable ways to test them. Since there are currently no ASTM test methods for determining all components of interest, this method also provides information on other potentially available test methods. Although this method is not to be used for specifications, it can provide a starting point for parties to develop mutually agreed-upon specifications that meet their respective requirements. It can also be used as a starting point in finding suitable test methods for determining various components of propylene. Finally, test method ASTM D5274 (2017) covers the analysis of 1,3-butadiene products produced in North America. It includes possible components and test methods, both ASTM and other, either actually used or believed to be in use, to test for these components. In addition, there are several standard test methods that cover the analysis of ethylene (ASTM D2505, 2017; ASTM D6159, 2017), propylene (ASTM D2163, 2017; ASTM D2712, 2017), and butylene (ASTM D4424, 2017), and the hydrocarbon impurities in 1,3-butadiene can also be determined by GC (ASTM D2593, 2017).

The hydrocarbon analysis of ethylene is accomplished by two methods (ASTM D2505, 2017; ASTM D6159, 2017), one of which (ASTM D6159, 2017) uses wide-bore (0.53 mm) capillary columns, including an alumina–potassium chloride (Al_2O_3/KCl) PLOT column. Another method (ASTM D2504, 2017) is recommended for determination of noncondensable gases, and yet another (ASTM D2505, 2017) is used for the determination of carbon dioxide.

Hydrocarbon impurities in propylene can be determined by gas chromatographic methods (ASTM D2712, 2017; ASTM D2163, 2017), and another test is available for determination of traces of methanol in propylene (ASTM D4864, 2017). A gas chromatographic method (ASTM D5303, 2017) is available for the determination of trace amounts of carbonyl sulfide in propylene using an FPD. Also, sulfur in petroleum gas can be determined by oxidative microcoulometry (ASTM D3246, 2017).

Commercial butylenes, high-purity butylenes, and butane–butylene mixtures are analyzed for hydrocarbon constituents (ASTM D4424, 2017), and hydrocarbon impurities in 1,3-butadiene can also be determined by GC (ASTM D2593, 2017). The presence of butadiene dimer and styrene is determined in butadiene by GC (ASTM D2426, 2017). Also, carbonyl derivatives in the C4 hydrocarbons can be determined by a titrimetric technique (ASTM D4423, 2017) and by the use of a peroxide-based method (ASTM D5799, 2017).

In general, GC will undoubtedly continue to be the method of choice for characterization of light hydrocarbon materials. New and improved detection devices and techniques, such as chemiluminescence, atomic emission, and mass spectroscopy, will enhance selectivity, detection limits, and analytical productivity. Laboratory automation through autosampling, computer control, and data handling will provide improved precision and productivity, as well as simplified method operation.

9.4.2 Physical Properties

Specifications for petroleum gases typically focus on LPG, propane, and butane, and the specifications generally define physical properties and characteristics of LPG that make them suitable for private, commercial, or industrial applications. The specifications do not purport to specifically define all possible requirements to meet all possible applications, and the user is cautioned to exercise judgment in formulating final specifications for specific applications. The Gas Processors Association, its management, and supporting companies claim no specific knowledge of how manufacturers and users will produce, handle, store, transfer, or consume the products defined herein and, therefore, are not responsible for any claims, causes of action, liabilities, losses, or expenses resulting from injury or death of persons and/or damage to property arising directly or indirectly from the use of LPG or these specifications relating to LPG (GPA, 1997).

LPG is composed of hydrocarbon compounds, predominately propane and butane, produced during the processing of natural gas and also in the conventional processing of crude oil (Mokhatab et al., 2006; Speight, 2007, 2014). The composition of LPG may vary depending on the source and the relative content of propane and butane content. These hydrocarbons exist as gases at atmospheric pressure and ambient temperatures but are readily liquefied under moderate pressures for transportation and utilization. There are many uses for LPG, the major ones being as (i) petrochemical, synthetic rubber, and motor gasoline feedstocks and as (ii) commercial, domestic, and industrial fuel.

The following may be accepted as a general guide for the common uses for the four fuel types covered by these specifications: (i) Commercial propane is the preferred fuel type for domestic, commercial, and industrial fuels; it is also a suitable fuel for low severity internal combustion engines. (ii) Commercial butane is used principally as feedstock for petrochemicals, as synthetic rubber, and as blending stocks or feedstocks in the manufacture of motor gasoline; its use as a fuel is generally limited to industrial applications where vaporization problems are not encountered; however, small quantifies are used as domestic fuel. (iii) Commercial butane–propane mixtures cover a broad range of mixtures, which permits the tailoring of fuels or feedstocks to specific needs. (iv) Propane is less variable in composition and combustion characteristics than other products covered by these specifications; it is also suitable as a fuel for internal combustion engines operating at moderate to high engine severity.

9.4.2.1 Calorific Value

The calorific value (heat of combustion) gives an indication of the satisfactory combustion of hydrocarbon gases, which depends upon the matching of burner and appliance design with certain gas characteristics. For use as heating agents, the relative merits of gases from different sources and having different compositions can be compared readily on the basis of their heating values. Therefore, the heating value is used as a parameter for determining the price of gas in custody transfer as well as an essential factor in calculating the efficiencies of energy conversion devices such as gas-fired turbines. The heating values of a gas depend not only upon the temperature and pressure but also upon the degree of saturation with water vapor. However, some calorimetric methods for measuring heating values are based upon the gas being saturated with water at the specified conditions.

The heating value of natural gas (or any fuel gas) may be determined experimentally using a calorimeter in which fuel is burned in the presence of air at constant pressure. The products are allowed to cool to the initial temperature, and a measurement is made of the energy released during complete combustion. All fuels that contain hydrogen release water vapor as a product of combustion, which is subsequently condensed in the calorimeter. The resulting measurement of the heat released is the higher heating value (HHV), also known as the gross heating value, and includes the heat of vaporization of water. The lower heating value (LHV), also known as the net heating value, is calculated by subtracting the heat of vaporization of water from the measured HHV and assumes that all products of combustion including water remain in the gaseous phase. The US system of measurement uses Btu per pound or Btu per standard cubic foot when expressed on a volume basis. This property is an indicator of the performance and torque potential of the gas for a defined engine configuration. Various types of test methods are available for the direct determination of calorific value (ASTM D1826, 2017; ASTM D3588, 2017). The calorific value is an important property insofar as it is used to calculate the Wobbe index (Wobbe index) (Section 9.4.2.11). Test method ASTM D1826 (2017) 4.1 provides an accurate and reliable method to measure the total calorific value of a fuel gas, on a continuous basis, which is used for regulatory compliance, custody transfer, and process control. The method covers the determination with the continuous recording calorimeter of the total calorific (heating) value of fuel gas produced or sold in the natural gas range from 900 to 1200 Btu/standard ft^3. Test method ASTM D3588 (2017) covers procedures for calculating heating value, relative density, and compressibility factor at base conditions (14.696 psia and 60 °F [15.6 °C]) for natural gas mixtures from compositional analysis. The method applies to all common types of utility gaseous fuels, for example, dry natural gas, reformed gas, oil gas (both high and low Btu), propane–air, carbureted water gas, coke oven gas, and retort coal gas, for which suitable methods of analysis as described are available. Both the HHV and LHV may also be calculated from the gas compositional analysis (ASTM D3588, 2017).

There are two types of calorific value: the superior calorific value, which includes condensation heat of the water content in the exhaust gases and is used for gas billing, and the inferior calorific value, which does not include this condensation heat and tends to be the preferred variable for engine applications. The lower calorific value is also often referred as the lower heating value, and the higher calorific value named the higher heating value.

9.4.2.2 Composition

LPG, natural gas, and refinery gas are mixtures of products or naturally occurring materials and, fortunately, are relative simple mixtures and do not suffer the complexities of the isomeric variations of the higher molecular weight hydrocarbons (Table 9.9) (Drews, 1998; Speight, 2014).

Thus, because of the lower molecular weight constituents of these gases and their volatility, GC has been the technique of choice for fixed gas and hydrocarbon speciation, and mass spectrometry is also a method of choice for compositional analysis of low molecular weight hydrocarbons (ASTM D2421, 2017; ASTM D2650, 2017).

One method (ASTM D2421, 2017) describes the procedure for the interconversion of the analysis of C_5 and lighter hydrocarbon mixtures to gas volume (mole), liquid volume, or mass basis. The computation procedures described assume that gas volume percentages have already been corrected for nonideality of the components as a part of the analytical process by which they have been obtained. The procedure assumes the absence of nonadditivity corrections for mixtures of the pure liquid compounds. This is approximately true only for mixtures of hydrocarbons of the same number of carbon atoms and in the absence of diolefins and acetylenic compounds. On the other hand, test method ASTM D2650 (2017) covers the quantitative analysis of gases containing specific combinations of the following components: hydrogen, hydrocarbons with up to six carbon atoms per molecule, carbon monoxide, carbon dioxide, mercaptans with one or two carbon atoms per molecule, hydrogen sulfide, and air (nitrogen, oxygen, and argon). This test method cannot be used for the determination of constituents present in amounts less than 0.1 mol%.

More recently, piggyback methods (such as GC/mass spectrometry and other double technique methods) have been developed for the identification of gaseous and low-boiling liquid constituents of mixtures. The hydrocarbon composition is limited to set to the total amount of ethane, butane, or pentane as well as ethylene and total dienes.

By limiting the amount of the different hydrocarbons that are lower boiling than the main component, the vapor pressure control is reinforced. Tests are available for vapor pressure at 38 °C (100 °F) and at 45 °C (113 °F) (ASTM D1267, 2017). Information on the vapor pressures of LPG products under temperature conditions from 37.8 to 70 °C (100–158 °F) is pertinent to selection of properly designed storage vessels, shipping containers, and customer utilization equipment to ensure safe handling of these products. Thus, determination of the vapor pressure of LPG is important for safety reasons to ensure that the maximum operating design pressures of storage, handling, and fuel systems will not be exceeded under normal operating temperature conditions. For LPG, vapor pressure is an indirect measure of the most extreme low temperature conditions under which initial vaporization can be expected to occur. The vapor pressure, as determined by this test method, can be considered to be a semiquantitative measure (an estimate or indication) of the amount of the most volatile material present in the product. Boiling hydrocarbons supports the volatility clause. The vapor pressure and volatility specifications will often be met automatically if the hydrocarbon composition is in order with the specification.

As stated above, the amount of ethylene in the mixture is limited because it is necessary not only to restrict the amount of the unsaturated components (olefins) so as to avoid the formation of deposits caused by the polymerization of the olefin(s) but also to control the volatility of the sample. Ethylene (boiling point: −104 °C, −155 °F) is more volatile than ethane (boiling point: −88 °C, −127 °F), and therefore a product with a substantial proportion of ethylene will have a higher vapor pressure and volatility than one that is predominantly ethane. Butadiene is also undesirable because it may also produce polymeric products that form deposits and cause blockage of lines.

The hydrocarbon component distribution of LPG and propene mixtures is often required for end-use sale of this material. Applications such as chemical feedstocks or fuel require precise compositional data to ensure uniform quality. Trace amounts of some hydrocarbon impurities in these materials can have adverse effects on their use and processing. The component distribution data of LPG and propene mixtures can be used to calculate physical properties such as relative density, vapor pressure, and motor octane (ASTM D2598, 2017). Precision and accuracy of compositional data are extremely important when these data are used to calculate various properties of these petroleum products.

Currently the preferred method for the analysis of LPG, and indeed for most petroleum-related gases, is GC (ASTM D2163, 2017). This test method covers the quantitative determination of individual hydrocarbons in LPG and mixtures of propane and propene, excluding high-purity propene in the range of C_1–C_5. Component concentrations are determined in the range of 0.01–100 vol.%. This technique can be used for the identification and measurement of both main constituents and trace constituents. However, there may be some accuracy issues that arise in the measurement of the higher-boiling constituents due to relative volatility under the conditions in which the sample is held.

However, assuming the reliability of the data from this test method, the hydrocarbon component distribution of LPG and propylene mixtures is often required for end-use sale of this material. Applications such as chemical feedstocks or fuel require precise compositional data to ensure uniform quality. Trace amounts of some hydrocarbon impurities in these materials can have adverse effects on their use and processing. In addition, the distribution of the constituents of LPG and propene mixtures can be used to calculate physical properties such as relative density, vapor pressure, and motor octane (ASTM D2598, 2017). This test method covers, through compositional analysis, the approximate determination of the following physical characteristics of commercial propane, special-duty propane, commercial propane–butane mixtures, and commercial butane (ASTM D1835, 2017): vapor pressure, relative density, and MON.

Vapor pressure is an important specification property of commercial propane, special-duty propane, propane–butane mixtures, and commercial butane that assures adequate vaporization, safety, and compatibility with commercial appliances. Relative density, while not a typical criterion contained in specifications, is necessary for determination of filling densities and custody transfer. The MON is useful in determining the products' suitability as a fuel for internal combustion engines. Precision and accuracy of compositional data are extremely important when these data are used to calculate various properties of these petroleum products.

Capillary column GC is an even quicker and equally accurate alternative. Mass spectrometry is also suitable for analysis of petroleum gases. Of the other spectroscopic techniques, IR and UV absorption may be applied to petroleum gas analysis for some specialized applications. GC has also largely supplanted chemical absorption methods of analysis, but again these may have some limited specialized application. Once the composition of a mixture has been determined, it is possible to calculate various properties such as specific gravity, vapor pressure, calorific value, and dew point. This test method covers the determination of the water vapor content of gaseous fuels by measurement of the dew-point temperature and the calculation therefrom of the water vapor content.

The hydrocarbon dew point is often considered to be the most important factor when performing any type of gas sampling. In the simplest terms, hydrocarbon dew point is the point at which the gas components begin to change phase from gaseous to liquid. When phase change occurs, certain components of the gas stream drop out and form liquids, thereby making an accurate gas sample impossible to obtain. The hydrocarbon dew point is a function of gas composition and pressure. A hydrocarbon dew-point curve is a reference chart that determines the specific pressure and temperature at which condensation occurs. No two hydrocarbon dew-point curves are alike due to differing gas compositions. Since dew point can be calculated from composition, direct determination of dew point for a particular LPG sample is a measure of composition. It is, of course, of more direct practical value, and if there are small quantities of higher molecular weight material present, it is preferable to use a direct measurement.

Simple evaporation tests in conjunction with vapor pressure measurement give a further guide to composition. In these tests an LPG sample is allowed to evaporate naturally from an open graduated vessel. Results are recorded on the basis of volume/temperature changes, such as the temperature recorded when 95% v/v has evaporated, or volume left at a particular temperature (ASTM D1837, 2017). This test method is a measure of the relative purity of the various types of LPG and is a benefit in the assurance of volatility performance. The test results, when properly related to vapor pressure and density of the product, can be used to indicate the presence of butane and higher molecular weight constituents in propane-type LPG and pentane and higher molecular weight constituents in propane–butane-type fuel gases as well as in butane-type fuel gases. The presence of hydrocarbon compounds less volatile than those of which the LPG is primarily composed is indicated by an increase in the 95% v/v evaporated temperature. When the type and concentration of higher-boiling components are required, chromatographic analysis should be used.

Specific gravity again can be calculated, but if it is necessary to measure, it several pieces of apparatus are available. For determining the density or specific gravity of LPG in its liquid state, there are two methods, using a metal pressure pycnometer. A pressure hydrometer may be used (ASTM D1267, 2017) for the relative density that may also be calculated from compositional analysis (ASTM D2598, 2017). Various procedures, manual and recording, for specific gravity or density in the gaseous state are given in two methods (ASTM D1070, 2017). In this test method (ASTM D1070, 2017), accurate and reliable test protocols are provided to measure the relative density of gaseous fuels on an intermittent or continuous basis. The test protocols cover the determination of relative density of gaseous fuels, including LPG, in the gaseous state at normal temperatures and pressures. The methods specified are sufficiently varied in nature so that one or more may be used for laboratory, control, reference, or gas measurement, or in fact, for any purpose in which it is desired to know the relative density of gas or gases as compared with the density of dry air at the same temperature and pressure. These measurements are frequently used for regulatory or contract compliance custody transfer and process control.

A knowledge of the composition of lower-boiling hydrocarbon derivatives in the overhead from the depentanization process is useful in converting analyses of the depentanized fraction to a total sample basis. Thus, the presence of pentane and lighter (lower-boiling) hydrocarbon derivatives in gasoline, naphtha, and similar petroleum distillates can have an adverse effect of the result of a test method (ASTM D2789, 2017). Pentane and lower-boiling hydrocarbon derivatives are separated by this test method (ASTM D2001, 2017) so that the depentanized residue can be analyzed and so the pentane and lighter hydrocarbons can be analyzed by other methods, if desired.

The hydrocarbon composition of natural gasoline (although not specifically a gas) for petrochemical use must undergo a compositional analysis (ASTM D2427, 2017) and a test for total sulfur (ASTM D1266, 2017). The method for compositional analysis of the fraction (ASTM D2427, 2017) covers the determination of the two (C_2) through five(C_5^-) carbon paraffins and monoolefins in gasolines. The concentrations by volume or mass (weight) of the following components are generally reported. The hydrocarbon-type analyses of naphtha and gasoline fractions (low-boiling and highly volatile fuels) are accomplished after stabilization by depentanization (ASTM D2001, 2017) prior to analysis.

Test method (ASTM D2001, 2017) covers the removal of pentane derivatives and lower-boiling hydrocarbon derivatives from gasoline, naphtha, and similar petroleum distillates to prepare samples suitable for the determination of hydrocarbon types (ASTM D2789, 2017). In addition, this test method determines the volume percent of bottoms remaining after depentanization. Under the conditions specified in the test method (ASTM D2001, 2017), some C_5 and lower-boiling hydrocarbon derivatives remain in the bottom fraction, while some C_6 and higher-boiling hydrocarbon derivatives carry over to the overhead. Expressed as volume percent of charge, the amounts are typically 2% v/v or less, which is considered adequate for the purpose designated in the test method. However, it should be recognized that when expressed as volume percent of overhead or of bottoms, the percentages can be higher, making this test method unsuitable for any purposes not designated in the scope of the test method.

An issue that arises during the characterization of LPG relates to the accurate determination of heavy residues (i.e. higher molecular weight hydrocarbons and even oils) in the gas. Test methods using procedures similar to those employed in gas chromatographic simulated distillation are becoming available. In fact, the presence of any component substantially less volatile than the main constituents of the LPG will give rise to unsatisfactory performance. It is difficult to set limits to the amount and nature of the *residue* that will make a product unsatisfactory. For example, LPG that contain certain anti-icing additives can give erroneous results by this test method.

Control over the residue content (ASTM D1835, 2017) is of considerable importance in end-use applications of LPG. In liquid feed systems, residues can lead to troublesome deposits, and in vapor withdrawal systems, residues that are carried over can foul regulating equipment. Any residue that remains in the vapor withdrawal systems will accumulate, can be corrosive, and will contaminate subsequent product. Water, particularly if alkaline, can cause failure of regulating equipment and corrosion of metals. Obviously small amounts of oil-like material can block regulators and valves. In liquid vaporizer feed systems, the gasoline-type material could cause difficulty.

The residue as determined by the EPI (ASTM D2158, 2017) is a measure of the concentration of contaminants boiling above 37.8 °C (100 °F) that may be present in the gas. The extraneous materials will generally be dissolved in the LPG but may have phase separated in some instances. Other methods are available that measure residue more directly, and for particular applications, it may be possible to relate the values obtained to the performance required and so set satisfactory limits.

Analytical methods are available in standard form for determining volatile sulfur content and certain specific corrosive sulfur compounds that are likely to be present. Volatile sulfur determination is made by a combustion procedure (ASTM D126) that uses a modification of the standard wick-fed lamp. Many laboratories use rapid combustion techniques with an oxyhydrogen flame in a Wickbold or Martin–Floret burner (ASTM D2784, 2017).

This test method (ASTM D2784, 2017) is valid for sulfur levels of >1 $\mu g\,g^{-1}$ of sulfur in LPG, but the samples should not contain more than 100 $\mu g\,g^{-1}$ of chlorine. In the test, the sample is bummed in an oxyhydrogen bummer or in a lamp in a closed system in a carbon dioxide–oxygen atmosphere. The latter is not recommended for trace quantities of sulfur due to inordinately long combustion times needed. The sulfur oxides produced are absorbed and oxidized to sulfuric acid in a hydrogen peroxide solution. The sulfate ions are then determined either by titrating with barium perchlorate solution using a thorin–methylene blue mixed indicator or by precipitating as barium sulfate and measuring the turbidity of the precipitate with a photometer.

Trace hydrocarbons that may be regarded as contaminants may be determined by the gas chromatographic methods already discussed. Heavier hydrocarbons in small amounts may not be completely removed from the column. If accurate information is required about the nature and amount of the higher-boiling constituents, then a temperature programming procedure or a procedure involving the concentration of the constituents may be used.

Analytical methods for determining traces of various other impurities, such as chlorides (ASTM D2384, 2017), are known to be in use. As an example, this test method (ASTM D2384, 2017) covers the determination of the total volatile organic chlorides in concentrations from 10 to 100 ppm in butane–butene mixtures. The finish is not directly applicable in the presence of other substances that combine with silver ion or oxidize chloride ion in dilute acid solution. Bromides, sulfides, ammonia, tobacco smoke, and more than 25 μg of hydrogen peroxide in the test solution interfere in the spectrophotometric procedure. By way of definition, amperometric means relating to or being a chemical titration in which the measurement of the electric current flowing under an applied potential difference between two electrodes in a solution is used for detecting the end point.

Procedures for the determination of hydrogen, helium, oxygen, nitrogen, carbon monoxide, carbon dioxide, methane, ethane, ethylene, propane, butanes, pentanes, and hexanes as well as higher molecular weight hydrocarbons in natural and reformed gases by packed column GC are available (ASTM D1945, 2017; ASTM D1946, 2017). These compositional analyses are used to calculate many other properties of gases, such as density, heating value, and compressibility. The first five components listed are determined using a molecular sieve column (argon carrier gas), while the remaining components are determined using polydimethylsiloxane partition or porous polymer columns. The hexanes-plus analysis is accomplished by backflushing the column after the elution of pentane or by the use of a backflushed precolumn.

Olefins (ethylene, propylene, butylene derivatives, and pentylene derivatives) that occur in refinery (process) gas have specific characteristics and require specific testing protocols (ASTM D5234, 2017; ASTM D5273, 2017; ASTM D5274, 2017). Ethylene is one of the highest volume chemicals produced in the world, with global production exceeding 100 million metric tons annually. Ethylene is primarily used in the manufacture of polyethylene, ethylene oxide, and ethylene dichloride, as well as many other lower volume products. Most of these production processes use various catalysts to improve product quality and process yield. Impurities in ethylene can damage the catalysts, resulting in significant replacement costs, reduced product quality and process downtime and decreased yield.

Ethylene is typically manufactured through the use of steam cracking. In this process, gaseous or light liquid hydrocarbons are combined with steam and heated to 750–950°C (1380–1740°F) in a pyrolysis furnace. Numerous free radical reactions are initiated, and larger hydrocarbons are converted (cracked) into smaller hydrocarbons. In addition, the high temperatures used in steam cracking promote the formation of unsaturated or olefin compounds, such as ethylene. Ethylene feedstocks must be tested to ensure that only high-purity ethylene is delivered for subsequent chemical processing.

Samples of high-purity ethylene typically contain only two minor impurities, methane and ethane, which can be detected in low ppm v/v concentrations. However, steam cracking can also produce higher molecular weight hydrocarbons, especially when propane, butane, or light liquid hydrocarbons are used as starting materials. Although fractionation is used in the final production stages to produce a high-purity ethylene product, it is still important to be able to identify and quantify any other hydrocarbons present in an ethylene sample. Achieving sufficient resolution of all of these compounds can be challenging due to their similarities in boiling point and chemical structure.

The analysis of ethylene is accomplished by two methods (ASTM D2505, 2017; ASTM D6159, 2017); one of which (ASTM D6159, 2017) uses wide-bore (0.53 mm) capillary columns, including an alumina–potassium chloride (Al_2O_3/KCl) PLOT column. Another method (ASTM D2504, 2017) is recommended for determination of noncondensable gases, and yet another (ASTM D2505, 2017) is used for the determination of carbon dioxide.

Hydrocarbon impurities in propylene can be determined by gas chromatographic methods (ASTM D2163, 2017; ASTM D2712, 2017), and another test is available for determination of traces of methanol in propylene (ASTM D4864, 2017). A gas chromatographic method (ASTM D5303, 2017) is available for the determination of trace amounts of carbonyl sulfide in propylene using an FPD. Also, sulfur in petroleum gas can be determined by oxidative microcoulometry (ASTM D3246, 2017).

Commercial butylenes, high-purity butylenes, and butane–butylene mixtures are analyzed for hydrocarbon constituents (ASTM D4424, 2017), and hydrocarbon impurities in 1,3-butadiene can also be determined by GC (ASTM D2593, 2017). The presence of butadiene dimer and styrene are determined in butadiene by GC (ASTM D2426, 2017).

As for the determination of the chemical properties of gases and low-boiling liquids, GC will undoubtedly continue to be the method of choice for characterization of light hydrocarbon materials. New and improved detection devices and techniques, such as chemiluminescence, atomic emission, and mass spectroscopy, will enhance selectivity, detection limits, and analytical productivity. Laboratory automation through autosampling, computer control, and data handling will provide improved precision and productivity, as well as simplified method operation.

9.4.2.3 Density and Relative Density

The density of low-boiling hydrocarbon derivatives can be determined by several methods (ASTM D1070, 2017) including a hydrometer method (ASTM D1298, 2017) or by a pressure hydrometer method (ASTM D1657, 2017). The specific gravity (relative density) (ASTM D1070, 2017; ASTM D1657, 2017) by itself has little significance compared with its use for higher molecular weight liquid petroleum products) and can only give an indication of quality characteristics when combined with values for volatility and vapor pressure. It is important for stock quantity calculations and is used in connection with transport and storage.

The relative density typically relates to the density of natural as relative to the density of air (Table 9.10) although, in some cases, the density of hydrogen may be used for comparison with the density of natural gas. The relative density, as a measure of gas density relative to air at reference conditions, is used for interchangeability specifications to limit the higher hydrocarbon content of the gas. An increased higher hydrocarbon content could lead to combustion problems

Table 9.10 Relative density (specific gravity) of natural gas hydrocarbons relative to air.

Gas	Specific gravity
Air	1.000
Methane, CH_4	0.5537
Ethane, C_2H_6	1.0378
Propane, C_3H_8	1.5219
Butane, C_4H_{10}	2.0061
Pentane, C_5H_{12}	2.487
Hexane, C_6H_{14}	2.973

such as increased carbon monoxide emissions, soot formation, engine knock, or spontaneous ignition on gas turbines even at the same Wobbe index value.

The statement is often made that natural gas is lighter than air. This statement often arises because of the continued insistence by engineers and scientists that the properties of a mixture are determined by the mathematical average of the properties of the individual constituents of the mixture. Such mathematical bravado and inconsistency of thought is detrimental to safety and needs to be qualified. The *relative density* (*specific gravity*) is the ratio of the density (mass of a unit volume) of a substance to the density of a given reference material. Specific gravity usually means relative density (of a liquid) with respect to water.

$$\text{Relative density} = \left[\rho(\text{substance})\rho(\text{reference})\right]$$

As it pertains to gases, particularly in relation to safety considerations at commercial and industrial facilities in the United States, the relative density of a gas is usually defined with respect to air, in which air is assigned a *vapor density* of one (unity). With this definition, the vapor density indicates whether a gas is denser (greater than one) or less dense (less than one) than air. The vapor density has implications for container storage and personnel safety – if a container can release a dense gas, its vapor could sink and, if flammable, collect until it is at a concentration sufficient for ignition. Even if not flammable, it could collect in the lower floor or level of a confined space and displace air, possibly presenting a smothering hazard to individuals entering the lower part of that space.

Gases can be divided into two groups based upon their vapor density: (i) gases that are heavier than air and (ii) gases that are as light as air or lighter than air. Gases that have a vapor density greater than one will be found in the bottom of storage containers and will tend to migrate downhill and accumulate in low-lying areas. Gases that have a vapor density that is the same or less than the vapor density of air will disperse readily into the surrounding environment. Additionally, chemicals that have the same vapor density as air (1.0) tends to disperse uniformly into the surrounding air when contained, and, when released into the open air, chemicals that are lighter than air will travel up and away from the ground.

Methane is the only hydrocarbon constituent of natural gas that is lighter than air (Table 9.10). The higher molecular weight hydrocarbons have a higher vapor density than air and are likely, after a release, to accumulate in low-lying areas and represent a danger to investigator (of the release). However, the other hydrocarbon constituents of unrefined natural gas (i.e. ethane, propane, butane, etc.) are denser than air. Therefore, should a natural gas leak occur in field operations, especially where the natural gas contains constituents other than methane, only methane dissipates readily into the air, whereas the other hydrocarbon constituents that are heavier than air do not readily dissipate into the atmosphere. This poses considerable risk if these constituents of natural gas accumulate or pool at ground level when it has been erroneously assumed that natural gas is lighter than air.

9.4.2.4 Dew Point

The dew point or dew-point temperature of a gas is the temperature at which the water vapor (ASTM D2029, 2017) or low-boiling hydrocarbon derivative contained in the gas is transformed into the liquid state. The formed liquid (condensate) exists as a liquid below the dew-point temperature, but above the dew point the liquid is a gaseous component of the gas.

9.4.2.5 Flammability

The flammability range is the range of temperature over which natural gas is flammable. The flammability limits of natural gas in air as an expression of the percent concentration in air (by volume) are given for the lower and upper limits. These values give an indication of relative flammability. The limits are sometimes referred to as *lower explosive limit* (LEL) and *upper explosive limit* (UEL).

By way of explanation, before a fire or explosion can occur, three conditions must be met simultaneously. A fuel (i.e. combustible gas) and oxygen (air) must exist in certain proportions, along with an ignition source, such as a spark or flame. The ratio of fuel and oxygen that is required varies with each combustible gas or vapor. The minimum concentration of a particular combustible gas or vapor necessary to support its combustion in air is defined as the lower explosive limit for that gas. Below this level, the mixture is too lean to burn. The maximum concentration of a gas or vapor that will burn in air is defined as the upper explosive limit. Above this level, the mixture is too rich to burn. The range between the LEL and UEL is known as the flammable range for that gas or vapor. Typically, the values given for the UEL and the LEL (Table 9.11) are valid only for the conditions under which they were determined (usually room temperature and atmospheric pressure using a 2 in. tube with spark ignition). The flammability range of most materials expands as temperature, pressure, and container diameter increase.

The vapor density has implications for flammability during storage. If a container can release a dense gas, its vapor could sink and, if flammable, collect until it is at a concentration sufficient for ignition. Even if not flammable, it could collect in the lower floor or level of a confined space and displace air, possibly presenting a smothering hazard to individuals entering the lower part of that space.

9.4.2.6 Higher Molecular Weight Hydrocarbons

Some gas utilities may add propane–air mixtures to natural gas during peak demand periods. Propane has a low vapor pressure, and if present in significant quantities, it will form a liquid phase at elevated pressures and low temperatures. Fuel variability due to vaporization of this liquid condensate at reduced storage

Table 9.11 Lower explosive limits (LEL) and upper explosive limits (UEL) for various constituents of gases, gas condensate, and natural gasoline.

Constituent	LEL	UEL
Benzene	1.3	7.9
1,3-Butadiene	2	12
Butane	1.8	8.4
n-Butanol	1.7	12
1-Butene	1.6	10
cis-2-Butene	1.7	9.7
trans-2-Butene	1.7	9.7
Carbon monoxide	12.5	74
Carbonyl sulfide	12	29
Cyclohexane	1.3	7.8
Cyclopropane	2.4	10.4
Diethylbenzene	0.8	
2,2-Dimethylpropane	1.4	7.5
Ethane	3	12.4
Ethyl benzene	1	6.7
Ethylene	2.7	36
Gasoline	1.2	7.1
Heptane	1.1	6.7
Hexane	1.2	7.4
Hydrogen	4	75
Hydrogen sulfide	4	44
Isobutane	1.8	8.4
Isobutylene	1.8	9.6
Methane	5	15
3-Methyl-1-butene	1.5	9.1
Methyl mercaptan	3.9	21.8
Pentane	1.4	7.8
Propane	2.1	9.5
Propylene	2.4	11
Toluene	1.2	7.1
Xylene	1.1	6.6

cylinder pressure can lead to difficulty in controlling the air–fuel ratio. In addition, the significant presence of the heavier hydrocarbons in the gas mixture lowers its knock rating and can lead to potential engine damage.

There may also be substantial amounts of oil that can be added to the gas during compression, which can subsequently condense and interfere with the operation of compressed natural gas (CNG) engine components such as gas pressure regulators. On the other hand, a minimum level of carryover oil is required for durable gas injector operation. Various injector manufacturers recommend different minimum oil levels.

Oil in the gas at the compressor outlet is commonly removed by coalescing filters; however, they are insufficient in many cases, as up to 50% of the carryover oil exists in vapor form in the warm (or hot) compressor outlet gas. Additional measures will need to be considered, for example, by additional cooling of the discharge gas or by using synthetic oil or mineral oil or a combination of mineral oil and a suitable adsorption filter downstream of the coalescing filter (Czachorski et al., 1995).

9.4.2.7 Methane Number

The main parameter for rating the knock resistance of gaseous fuels is the methane number (MN), which is analogous to the octane number for gasoline, and suitable test methods for determining the MN of a gaseous fuel are being developed (Malenshek and Olsen, 2009).

Different scales have been used to rate the knock resistance of CNG including the MON and the MN. The differences in these ratings are the reference fuel blends used for comparison to the natural gas. MN uses a reference fuel blend of methane, with an MN of 100, and hydrogen, with an MN of 0. Correlations have been generated between the reactive hydrogen–carbon ratio (H/C) and the MON and between MON and MN:

$$\text{MON} = -406.14 + 508.04 \times (\text{H}/\text{C}) - 173.55 * (\text{H}/\text{C})2 + 20.17 * (\text{H}/\text{C})3\,\text{MN}$$
$$= 1.624 * \text{MON} - 119.1$$

Thus, if a gas mixture has an MN of 70, its knock resistance is equivalent to that of a gas mixture of 70% methane and 30% hydrogen.

To ensure safe engine operation, the MN must always be at least equal to the methane number requirement (MNR) of the gas engine. The MN required by the engine is affected by design and operating parameters, with the adjustment of the MNR being achieved by changing engine operation. Changes in ignition timing, air–fuel ratio, and output are effective measures to reduce the MNR.

9.4.2.8 Sulfur Content

Sulfur compounds in natural gas are in the form of mercaptans, hydrogen sulfide, and odorants. The first two are naturally occurring at source (gas fields) and have already been reduced by treatment at the gas processing plant.

The manufacturing processes for LPG are designed so that the majority, if not all, of the sulfur compounds are removed. The total sulfur level is therefore considerably lower than for other petroleum fuels, and a maximum limit for sulfur content helps to define the product more completely. The sulfur compounds that are mainly responsible for corrosion are hydrogen sulfide, carbonyl sulfide, and, sometimes, elemental sulfur. Hydrogen sulfide and mercaptans have distinctive unpleasant odors. Control of the total sulfur content, hydrogen sulfide, and mercaptans ensures that the product is not corrosive or nauseating. Stipulating a satisfactory copper strip test further ensures the control of the corrosion.

Total sulfur in gas can be determined by combustion (ASTM D1072, 2017), by the lamp method (ASTM D1266, 2017), or by hydrogenation (ASTM D4468, 2017). Trace total organic and bound nitrogen is determined (ASTM D4629, 2017). The current test method for heavy residues in LPG (ASTM D2158, 2017) involves evaporation of an LPG sample, measuring the volume of residue, and observing the residue for oil stain on a piece of filter paper.

Corrosive sulfur compounds can be detected by their effect on copper and the form in which the general copper strip corrosion test (ASTM D1838, 2017) for petroleum products is applied to LPG. Hydrogen sulfide can be detected by its action on moist lead acetate paper, and a procedure is also used as a measure of sulfur compounds. The method follows the principle of the standard doctor test (ASTM D4952, 2017).

9.4.2.9 Volatility and Vapor Pressure

The vaporization and combustion characteristics of LPG are defined for normal applications by volatility, vapor pressure, and, to a lesser extent, specific gravity.

Volatility is expressed in terms of the temperature at which 95% of the sample is evaporated and presents a measure of the least volatile component present (ASTM D1837, 2017). Vapor pressure is, therefore, a measure of the most extreme low temperature conditions under which initial vaporization can take place. By setting limits to vapor pressure and volatility jointly, the specification serves to ensure essentially single component products for the butane and propane grades (ASTM D1267, 2017; ASTM D2598, 2017). By combining vapor pressure/volatility limits with specific gravity for propane–butane mixtures, essentially two-component systems are ensured. The residue (ASTM D1025, 2017; ASTM D2158, 2017), i.e. nonvolatile matter, is a measure of the concentration of contaminants boiling above 37.8 °C (100 °F) that may be present in the gas.

In a closed container, the vapor pressure of a pure compound is the force exerted per unit area of walls by the vaporized portion of the liquid. Vapor pressure can also be defined as a pressure at which the vapor phase and the liquid phase of a pure substance are in equilibrium with each other. The vapor pressure is also called saturation pressure and the corresponding temperature is called saturation temperature. In an open air under atmospheric pressure, a liquid at any temperature below its boiling point has its own vapor pressure that is less than 1 atm. When vapor pressure reaches 1 atm (14.7 psi), the saturation temperature becomes the normal boiling point. Vapor pressure increases with temperature, and the highest value of vapor pressure for a substance is its critical pressure in which the corresponding temperature is the critical temperature.

Vapor pressure is a very important thermodynamic property of any substance and it is a measure of the volatility of a fluid. Compounds with a higher tendency to vaporize have higher vapor pressures. More volatile compounds are those that have lower boiling points and are called light compounds. For example, propane (C_3H_8) has boiling point less than that of *n*-butane (nC_4H_{10}), and as a result it is more volatile. At a fixed temperature, vapor pressure of propane is higher than that of butane. In this case, propane is called the light compound (more volatile) and butane the heavy compound. Generally, more volatile compounds have higher critical pressure and lower critical temperature, and lower density and lower boiling point than those of less volatile (heavier) compounds, although this is not true for the case of some isomeric compounds.

Vapor pressure is a useful parameter in calculations related to hydrocarbon losses and flammability of hydrocarbon vapor in the air. More volatile compounds are more ignitable than heavier compounds. For example, *n*-butane is added to gasoline to improve its ignition characteristics. Low vapor pressure compounds reduce evaporation losses and chance of vapor lock. Therefore, for a fuel there should be a compromise between low and high vapor pressure. However, one of the major applications of vapor pressure is in calculation of equilibrium ratios for phase equilibrium calculations. For pure hydrocarbons, values of vapor pressure at the reference temperature of 100 °F (38 °C) are provided by the API. For natural gas, the method of Reid is used to measure vapor pressure at 100 °F. *Reid vapor pressure* (RVP) is measured by the ASTM test method D323, and it is approximately equivalent to vapor pressure at 100 °F (38 °C).

For natural gasoline, the primary criterion is volatility (vapor pressure and knock performance). Determination of the vapor pressure (ASTM D323, 2017; ASTM D4953, 2017; ASTM D5191, 2017) and distillation profile (ASTM D216, 2017) is essential. Knock performance is determined is determined by rating in knock test engines by both the motor method (ASTM D2700, 2017) and the research method (ASTM D2699, 2017). The knock characteristics of LPG can also be determined. Other considerations for natural gasoline are copper corrosion (ASTM D130, 2017) and specific gravity (ASTM D1298, 2017), the latter determination being necessary for measurement and transportation.

9.4.2.10 Water Content

The use of natural gas with high water content can result in the formation of liquid water, ice particles, or hydrates at low operating temperatures and high pressure that will interfere with consistently smooth flow of fuel into the engine and cause problems such as poor drivability or even engine stoppage. Thus, it is a fundamental requirement that LPG should not contain free water (ASTM D2713, 2017). Dissolved water may give trouble by forming hydrates and giving moisture vapor in the gas phase. Both of these will lead to blockages. Therefore, test methods are available to determine the presence of water using electronic moisture analyzers (ASTM D5454, 2017), dew-point temperature (ASTM D1142, 2017), and length-of-stain detector tubes (ASTM D4888, 2017).

9.4.2.11 Wobbe Index

The Wobbe index gives a measure of the heat input to an appliance through a given aperture at a given gas pressure. Using this as a vertical coordinate and the flame speed factor as the horizontal coordinate, a combustion diagram can be constructed for an appliance, or a whole range of appliances, with the aid of appropriate test gases. The concept behind defining the Wobbe index is to have a measure for the interchangeability of gases, i.e. gases with the same Wobbe index (under defined pressure conditions) generate the same output during combustion. However, a distinction is made between what is referred to as the higher Wobbe index and the lower Wobbe index, depending on whether the superior or the inferior calorific value is used in the formula.

The natural gas after appropriate treatment for acid gas reduction, odorization, and hydrocarbon and moisture dew-point adjustment (ASTM D1142, 2017) would then be sold within prescribed limits of pressure, calorific value, and possibly Wobbe index (also called the Wobbe index, which equals the calorific value divided by the specific gravity) and the flame speed. The Wobbe index is typically expressed as a factor or an arbitrary scale on which that of hydrogen is 100. This factor can be calculated from the gas analysis. In fact, calorific value and specific gravity can be calculated from compositional analysis (ASTM D3588, 2017). Thus

$$\mathrm{WI} = \frac{\mathrm{CV}}{\sqrt{\mathrm{SG}}}$$

where WI is the Wobbe index or Wobbe index that is typically expressed in Btu/scf, CV is the calorific value (HHV), and SG is the specific gravity.

Gas turbines can operate with a large range of fuels, but the fuel variation that a particular installation can cope with is limited. The modified Wobbe index (MWI) is used particularly by gas turbine manufacturers because it takes into account the temperature of the fuel. The MWI is the ratio of the LHV to the square root of the product of the specific gravity and the absolute gas temperature:

$$\mathrm{MWI} = \frac{\mathrm{LHV}}{\sqrt{\mathrm{SG}_{\mathrm{gas}}}} \times \mathrm{T}_{\mathrm{gas}}$$

This is equivalent to:

$$MWI = \frac{LHV}{\sqrt{MW_{gas} / 28.96}} \times T_{gas}$$

where LHV is the lower heating value of the fuel gas (Btu/scf), SG_{gas} is the specific gravity of the fuel gas relative to air, MW_{gas} is the molecular weight of the fuel gas, T_{gas} is the absolute temperature of the fuel gas (in degrees Rankine), and 28.96 is the molecular weight of dry air.

Any change in the heating value of the gas will require a corresponding change in the fuel's flow rate to the machine, and incorporation of temperature effects is important in calculating energy flows in turbines where a large input temperature variation is possible. The allowable MWI range is established to ensure that required fuel nozzle pressure ratios are maintained during all combustion/turbine modes of operation. For older, diffusion-type combustors, the gas turbine control system can typically accommodate variations in the MWI as large as ±15%. But for newer, dry low NOx (DLN) combustors, variations in the MWI of only ±3% could cause problems. Fuel instability can be caused by velocity changes through a precisely sized fuel nozzle orifice that can cause flame instability, resulting in pressure pulsations and/or combustion dynamics that can, in the worst case, destroy the combustion system.

Also, since the Wobbe index is an indicator of the interchangeability of fuel gases, it (alone or with other analyses) can be used to control blending of fuel gases. Since the Wobbe index and the Btu value of fuel gases make similar curves, either could be used to control blending of fuel gases, thereby controlling the amount of nitrogen in the blended fuel (Segers et al., 2011). Another important combustion criterion is the gas modulus, $M = P/W$, where P is the gas pressure and W the Wobbe index of the gas. This must remain constant if a given degree of aeration is to be maintained in a preaerated burner using air at atmospheric pressure.

Finally, despite the common acceptance of Wobbe index as the main interchangeability parameter, a variety of units and reference temperatures are used across the world.

References

API (2009). Refinery Gases Category Analysis and Hazard Characterization. Submitted to the EPA by the American Petroleum Institute, Petroleum HPV Testing Group. HPV Consortium Registration # 1100997 United States Environmental Protection Agency, Washington, DC. 10 June.

ASTM (2017). Annual Book of Standards. West Conshohocken, Pennsylvania: ASTM International.

ASTM D1025 (2017). *Standard Test Method for Nonvolatile Residue of Polymerization-Grade Butadiene*, Annual Book of Standards. West Conshohocken, Pennsylvania: ASTM International.
ASTM D1070 (2017). *Standard Test Methods for Relative Density of Gaseous Fuels*, Annual Book of Standards. West Conshohocken, Pennsylvania: ASTM International.
ASTM D1071 (2017). *Standard Test Methods for Volumetric Measurement of Gaseous Fuel Samples*, Annual Book of Standards. West Conshohocken, Pennsylvania: ASTM International.
ASTM D1072 (2017). *Standard Test Method for Total Sulfur in Fuel Gases by Combustion and Barium Chloride Titration*, Annual Book of Standards. West Conshohocken, Pennsylvania: ASTM International.
ASTM D1142 (2017). *Standard Test Method for Water Vapor Content of Gaseous Fuels by Measurement of Dew-Point Temperature*, Annual Book of Standards. West Conshohocken, Pennsylvania: ASTM International.
ASTM D1265 (2017). *Standard Practice for Sampling Liquefied Petroleum (LP) Gases, Manual Method*, Annual Book of Standards. West Conshohocken, Pennsylvania: ASTM International.
ASTM D1266 (2017). *Standard Test Method for Sulfur in Petroleum Products (Lamp Method)*, Annual Book of Standards. West Conshohocken, Pennsylvania: ASTM International.
ASTM D1267 (2017). *Standard Test Method for Gage Vapor Pressure of Liquefied Petroleum (LP) Gases (LP-Gas Method)*, Annual Book of Standards. West Conshohocken, Pennsylvania: ASTM International.
ASTM D129 (2017). *Standard Test Method for Sulfur in Petroleum Products (General High Pressure Decomposition Device Method)*, Annual Book of Standards. West Conshohocken, Pennsylvania: ASTM International.
ASTM D1298 (2017). *Standard Test Method for Density, Relative Density, or API Gravity of Crude Petroleum and Liquid Petroleum Products by Hydrometer Method*, Annual Book of Standards. West Conshohocken, Pennsylvania: ASTM International.
ASTM D130 (2017). *Standard Test Method for Corrosiveness to Copper from Petroleum Products by Copper Strip Test*, Annual Book of Standards. West Conshohocken, Pennsylvania: ASTM International.
ASTM D1552 (2017). *Standard Test Method for Sulfur in Petroleum Products by High Temperature Combustion and Infrared (IR) Detection or Thermal Conductivity Detection (TCD)*, Annual Book of Standards. West Conshohocken, Pennsylvania: ASTM International.
ASTM D1657 (2017). *Standard Test Method for Density or Relative Density of Light Hydrocarbons by Pressure Hydrometer*, Annual Book of Standards. West Conshohocken, Pennsylvania: ASTM International.
ASTM D1826 (2017). *Standard Test Method for Calorific (Heating) Value of Gases in Natural Gas Range by Continuous Recording Calorimeter*, Annual Book of Standards. West Conshohocken, Pennsylvania: ASTM International.

ASTM D1835 (2017). *Standard Specification for Liquefied Petroleum (LP) Gases*, Annual Book of Standards. West Conshohocken, Pennsylvania: ASTM International.

ASTM D1837 (2017). *Standard Test Method for Volatility of Liquefied Petroleum (LP) Gases*, Annual Book of Standards. West Conshohocken, Pennsylvania: ASTM International.

ASTM D1838 (2017). *Standard Test Method for Copper Strip Corrosion by Liquefied Petroleum (LP) Gases*, Annual Book of Standards. West Conshohocken, Pennsylvania: ASTM International.

ASTM D1945 (2017). *Standard Test Method for Analysis of Natural Gas by Gas Chromatography*, Annual Book of Standards. West Conshohocken, Pennsylvania: ASTM International.

ASTM D1946 (2017). *Standard Practice for Analysis of Reformed Gas by Gas Chromatography*, Annual Book of Standards. West Conshohocken, Pennsylvania: ASTM International.

ASTM D1988 (2017). *Standard Test Method for Mercaptans in Natural Gas Using Length-of-Stain Detector Tubes*, Annual Book of Standards. West Conshohocken, Pennsylvania: ASTM International.

ASTM D2001 (2017). *Standard Test Method for Depentanization of Gasoline and Naphthas*, Annual Book of Standards. West Conshohocken, Pennsylvania: ASTM International.

ASTM D2029 (2017). *Standard Test Methods for Water Vapor Content of Electrical Insulating Gases by Measurement of Dew Point*, Annual Book of Standards. West Conshohocken, Pennsylvania: ASTM International.

ASTM D2158 (2017). *Standard Test Method for Residues in Liquefied Petroleum (LP) Gases*, Annual Book of Standards. West Conshohocken, Pennsylvania: ASTM International.

ASTM D216 (2017). *Standard Practice for Conversion of Kinematic Viscosity to Saybolt Universal Viscosity or to Saybolt Furol Viscosity*, Annual Book of Standards. West Conshohocken, Pennsylvania: ASTM International.

ASTM D2163 (2017). *Standard Test Method for Determination of Hydrocarbons in Liquefied Petroleum (LP) Gases and Propane/Propene Mixtures by Gas Chromatography*, Annual Book of Standards. West Conshohocken, Pennsylvania: ASTM International.

ASTM D2384 (2017). *Standard Test Methods for Traces of Volatile Chlorides in Butane-Butene Mixtures*, Annual Book of Standards. West Conshohocken, Pennsylvania: ASTM International.

ASTM D2420 (2017). *Standard Test Method for Hydrogen Sulfide in Liquefied Petroleum (LP) Gases (Lead Acetate Method)*, Annual Book of Standards. West Conshohocken, Pennsylvania: ASTM International.

ASTM D2421 (2017). *Standard Practice for Interconversion of Analysis of C5 and Lighter Hydrocarbons to Gas-Volume, Liquid-Volume, or Mass Basis*, Annual Book of Standards. West Conshohocken, Pennsylvania: ASTM International.

ASTM D2426 (2017). *Standard Test Method for Butadiene Dimer and Styrene in Butadiene Concentrates by Gas Chromatography*, Annual Book of Standards. West Conshohocken, Pennsylvania: ASTM International.

ASTM D2427 (2017). *Standard Test Method for Determination of C2 through C5 Hydrocarbons in Gasolines by Gas Chromatography*, Annual Book of Standards. West Conshohocken, Pennsylvania: ASTM International.

ASTM D2504 (2017). *Standard Test Method for Non-Condensable Gases in C2 and Lighter Hydrocarbon Products by Gas Chromatography*, Annual Book of Standards. West Conshohocken, Pennsylvania: ASTM International.

ASTM D2505 (2017). *Standard Test Method for Ethylene, Other Hydrocarbons, and Carbon Dioxide in High-Purity Ethylene by Gas Chromatography*, Annual Book of Standards. West Conshohocken, Pennsylvania: ASTM International.

ASTM D2593 (2017). *Standard Test Method for Butadiene Purity and Hydrocarbon Impurities by Gas Chromatography*, Annual Book of Standards. West Conshohocken, Pennsylvania: ASTM International.

ASTM D2597 (2017). *Standard Test Method for Analysis of Demethanized Hydrocarbon Liquid Mixtures Containing Nitrogen and Carbon Dioxide by Gas Chromatography*, Annual Book of Standards. West Conshohocken, Pennsylvania: ASTM International.

ASTM D2598 (2017). *Standard Practice for Calculation of Certain Physical Properties of Liquefied Petroleum (LP) Gases from Compositional Analysis*, Annual Book of Standards. West Conshohocken, Pennsylvania: ASTM International.

ASTM D2650 (2017). *Standard Test Method for Chemical Composition of Gases by Mass Spectrometry*, Annual Book of Standards. West Conshohocken, Pennsylvania: ASTM International.

ASTM D2699 (2017). *Standard Test Method for Research Octane Number of Spark-Ignition Engine Fuel*, Annual Book of Standards. West Conshohocken, Pennsylvania: ASTM International.

ASTM D2700 (2017). *Standard Test Method for Motor Octane Number of Spark-Ignition Engine Fuel*, Annual Book of Standards. West Conshohocken, Pennsylvania: ASTM International.

ASTM D2712 (2017). *Standard Test Method for Hydrocarbon Traces in Propylene Concentrates by Gas Chromatography*, Annual Book of Standards. West Conshohocken, Pennsylvania: ASTM International.

ASTM D2713 (2017). *Standard Test Method for Dryness of Propane (Valve Freeze Method)*, Annual Book of Standards. West Conshohocken, Pennsylvania: ASTM International.

ASTM D2784 (2017). *Standard Test Method for Sulfur in Liquefied Petroleum Gases (Oxy-Hydrogen Burner or Lamp)*, Annual Book of Standards. West Conshohocken, Pennsylvania: ASTM International.

ASTM D2789 (2017). *Standard Test Method for Hydrocarbon Types in Low Olefinic Gasoline by Mass Spectrometry*, Annual Book of Standards. West Conshohocken, Pennsylvania: ASTM International.

ASTM D323 (2017). *Standard Test Method for Vapor Pressure of Petroleum Products (Reid Method)*, Annual Book of Standards. West Conshohocken, Pennsylvania: ASTM International.
ASTM D3246 (2017). *Standard Test Method for Sulfur in Petroleum Gas by Oxidative Microcoulometry*, Annual Book of Standards. West Conshohocken, Pennsylvania: ASTM International.
ASTM D3588 (2017). *Standard Practice for Calculating Heat Value, Compressibility Factor, and Relative Density of Gaseous Fuels*, Annual Book of Standards. West Conshohocken, Pennsylvania: ASTM International.
ASTM D3700 (2017). *Standard Practice for Obtaining LPG Samples Using a Floating Piston Cylinder*, Annual Book of Standards. West Conshohocken, Pennsylvania: ASTM International.
ASTM D3956 (2017). *Standard Specification for Methane Thermophysical Property Tables*, Annual Book of Standards. West Conshohocken, Pennsylvania: ASTM International.
ASTM D3984 (2017). *Standard Specification for Ethane Thermophysical Property Tables*, Annual Book of Standards. West Conshohocken, Pennsylvania: ASTM International.
ASTM D4084 (2017). *Standard Test Method for Analysis of Hydrogen Sulfide in Gaseous Fuels (Lead Acetate Reaction Rate Method)*, Annual Book of Standards. West Conshohocken, Pennsylvania: ASTM International.
ASTM D4150 (2017). *Standard Terminology Relating to Gaseous Fuels*, Annual Book of Standards. West Conshohocken, Pennsylvania: ASTM International.
ASTM D4362 (2017). *Standard Specification for Propane Thermophysical Property Tables*, Annual Book of Standards. West Conshohocken, Pennsylvania: ASTM International.
ASTM D4424 (2017). *Standard Test Method for Butylene Analysis by Gas Chromatography*, Annual Book of Standards. West Conshohocken, Pennsylvania: ASTM International.
ASTM D4468 (2017). *Standard Test Method for Total Sulfur in Gaseous Fuels by Hydrogenolysis and Rateometric Colorimetry*, Annual Book of Standards. West Conshohocken, Pennsylvania: ASTM International.
ASTM D4629 (2017). *Standard Test Method for Trace Nitrogen in Liquid Petroleum Hydrocarbons by Syringe/Inlet Oxidative Combustion and Chemiluminescence Detection*, Annual Book of Standards. West Conshohocken, Pennsylvania: ASTM International.
ASTM D4650 (2017). *Standard Specification for Normal Butane Thermophysical Property Tables*, Annual Book of Standards. West Conshohocken, Pennsylvania: ASTM International.
ASTM D4651 (2017). *Standard Specification for Isobutane Thermophysical Property Tables*, Annual Book of Standards. West Conshohocken, Pennsylvania: ASTM International.
ASTM D4784 (2017). *Standard for LNG Density Calculation Models*, Annual Book of Standards. West Conshohocken, Pennsylvania: ASTM International.

ASTM D4810 (2017). *Standard Test Method for Hydrogen Sulfide in Natural Gas Using Length-of-Stain Detector Tubes*, Annual Book of Standards. West Conshohocken, Pennsylvania: ASTM International.

ASTM D4864 (2017). *Standard Test Method for Determination of Traces of Methanol in Propylene Concentrates by Gas Chromatography*, Annual Book of Standards. West Conshohocken, Pennsylvania: ASTM International.

ASTM D4888 (2017). *Standard Test Method for Water Vapor in Natural Gas Using Length-of-Stain Detector Tubes*, Annual Book of Standards. West Conshohocken, Pennsylvania: ASTM International.

ASTM D4952 (2017). *Standard Test Method for Qualitative Analysis for Active Sulfur Species in Fuels and Solvents (Doctor Test)*, Annual Book of Standards. West Conshohocken, Pennsylvania: ASTM International.

ASTM D4953 (2017). *Standard Test Method for Vapor Pressure of Gasoline and Gasoline-Oxygenate Blends (Dry Method)*, Annual Book of Standards. West Conshohocken, Pennsylvania: ASTM International.

ASTM D4984 (2017). *Standard Test Method for Carbon Dioxide in Natural Gas Using Length-of-Stain Detector Tubes*, Annual Book of Standards. West Conshohocken, Pennsylvania: ASTM International.

ASTM D5191 (2017). *Standard Test Method for Vapor Pressure of Petroleum Products (Mini Method)*, Annual Book of Standards. West Conshohocken, Pennsylvania: ASTM International.

ASTM D5234 (2017). *Standard Guide for Analysis of Ethylene Product*, Annual Book of Standards. West Conshohocken, Pennsylvania: ASTM International.

ASTM D5273 (2017). *Standard Guide for Analysis of Propylene Concentrates*, Annual Book of Standards. West Conshohocken, Pennsylvania: ASTM International.

ASTM D5274 (2017). *Standard Guide for Analysis of 1,3-Butadiene Product*, Annual Book of Standards. West Conshohocken, Pennsylvania: ASTM International.

ASTM D5303 (2017). *Standard Test Method for Trace Carbonyl Sulfide in Propylene by Gas Chromatography*, Annual Book of Standards. West Conshohocken, Pennsylvania: ASTM International.

ASTM D5305 (2017). *Standard Test Method for Determination of Ethyl Mercaptan in LP-Gas Vapor*, Annual Book of Standards. West Conshohocken, Pennsylvania: ASTM International.

ASTM D5454 (2017). *Standard Test Method for Water Vapor Content of Gaseous Fuels Using Electronic Moisture Analyzers*, Annual Book of Standards. West Conshohocken, Pennsylvania: ASTM International.

ASTM D5504 (2017). *Standard Test Method for Determination of Sulfur Compounds in Natural Gas and Gaseous Fuels by Gas Chromatography and Chemiluminescence*, Annual Book of Standards. West Conshohocken, Pennsylvania: ASTM International.

ASTM D5799 (2017). *Standard Test Method for Determination of Peroxides in Butadiene*, Annual Book of Standards. West Conshohocken, Pennsylvania: ASTM International.

ASTM D5954 (2017). *Standard Test Method for Mercury Sampling and Measurement in Natural Gas by Atomic Absorption Spectroscopy*, Annual Book of Standards. West Conshohocken, Pennsylvania: ASTM International.

ASTM D6159 (2017). *Standard Test Method for Determination of Hydrocarbon Impurities in Ethylene by Gas Chromatography*, Annual Book of Standards. West Conshohocken, Pennsylvania: ASTM International.

ASTM D6228 (2017). *Standard Test Method for Determination of Sulfur Compounds in Natural Gas and Gaseous Fuels by Gas Chromatography and Flame Photometric Detection*, Annual Book of Standards. West Conshohocken, Pennsylvania: ASTM International.

ASTM D6273 (2017). *Standard Test Methods for Natural Gas Odor Intensity*, Annual Book of Standards. West Conshohocken, Pennsylvania: ASTM International.

ASTM D6350 (2017). *Standard Test Method for Mercury Sampling and Analysis in Natural Gas by Atomic Fluorescence Spectroscopy*, Annual Book of Standards. West Conshohocken, Pennsylvania: ASTM International.

ASTM D7756 (2017). *Standard Test Method for Residues in Liquefied Petroleum (LP) Gases by Gas Chromatography with Liquid, on-Column Injection*, Annual Book of Standards. West Conshohocken, Pennsylvania: ASTM International.

Cranmore, R.E. and Stanton, E. (2000). *Modern Petroleum Technology. Volume 1: Upstream* (ed. R.A. Dawe). New York: Wiley Chapter 9.

Crawford, D.B., Durr, C.A., Finneran, J.A., and Turner, W. (1993). Chemicals from natural gas. In: *Chemical Processing Handbook* (ed. J.J. McKetta), 2. New York: Marcel Dekker Inc.

Czachorski, M., Blazek, C., Chao, S., et al. (1995). NGV Fueling Station Compressor Oil Carryover Measurement and Control, Report No. GRI-95/0483, 1995. Gas Research Institute, Chicago, Illinois.

Drews, A.W. (1998). Introduction. In: *Manual on Hydrocarbon Analysis*, 6e (ed. A.W. Drews). West Conshohocken, Pennsylvania: American Society for Testing and Materials.

Gary, J.G., Handwerk, G.E., and Kaiser, M.J. (2007). *Petroleum Refining: Technology and Economics*, 5e. Boca Raton, Florida: CRC Press, Taylor & Francis Group.

GPA (1997). *Liquefied Petroleum Gas Specifications and Test Methods. GPA Standard 2140*. Tulsa, Oklahoma: Gas Processors Association.

Hsu, C.S. and Robinson, P.R. ed. (2017). *Handbook of Petroleum Technology*. Cham, Switzerland: Springer International Publishing AG.

Johansen, N.G. (1998). Chapter 1. In: *Manual on Hydrocarbon Analysis*, 6e (ed. A.W. Drews). West Conshohocken, Pennsylvania: American Society for Testing and Materials.

Klimstra, J. (1978). Interchangeability of Gaseous Fuels – The Importance of the Wobbe-Index, Report No. SAE 861578. Society of Automotive Engineers, SAE International, Warrendale, Pennsylvania.

Liss, W.E., and Thrasher, W.R. (1992. Variability of Natural Gas Composition in Select Major Metropolitan Areas of the U.S, Report No. GRI-92/0123. Gas Research Institute, Chicago, Illinois.

Malenshek, M. and Olsen, D.B. (2009). Methane number testing of alternative gaseous fuels. *Fuel* 88: 650–656.

Mokhatab, S., Poe, W.A., and Speight, J.G. (2006). *Handbook of Natural Gas Transmission and Processing*. Amsterdam, the Netherlands: Elsevier.

Rhoderick, G.C. (2003). Analysis of natural gas: the necessity of multiple standards for calibration. *Journal of Chromatography A* 1017 (1–2): 131–139.

Robinson, J.D. and Faulkner, R.P. (2000). Chapter 1. In: *Modern Petroleum Technology. Volume 2: Downstream* (ed. A.G. Lucas). New York: Wiley.

Segers, M., Sanchez, R., Cannon, P. et al. (2011). Blending fuel gas to optimize use of off-spec natural gas. Proceedings Presented at ISA Power Industry Division 54th Annual I&C Symposium, Concord, North Carolina.

Speight, J.G. (2007). *Natural Gas: A Basic Handbook*. Houston, Texas: GPC Books, Gulf Publishing Company.

Speight, J.G. ed. (2011). *The Biofuels Handbook*. London, UK: The Royal Society of Chemistry.

Speight, J.G. (2013a). *The Chemistry and Technology of Coal*, 5e. Boca Raton, Florida: CRC Press, Taylor & Francis Group.

Speight, J.G. (2013b). *Shale Gas Production Processes*. Oxford, UK: Gulf Professional Publishing, Elsevier.

Speight, J.G. (2014). *The Chemistry and Technology of Petroleum*, 5e. Boca Raton, Florida: CRC Press, Taylor & Francis Group.

Speight, J.G. (2017). *Handbook of Petroleum Refining. Petroleum Refining Processes*. Boca Raton, Florida: CRC Press, Taylor & Francis Group.

Stephenson, R.M. and Malanowski, S. (1987). *Handbook of the Thermodynamics of Organic Compounds*. New York: Elsevier.

10

Properties and Analysis of Gas Condensate

10.1 Introduction

Briefly, and by way of review in the context of this chapter, raw natural gas may come from any one of three types of gas wells:

i) Crude oil wells. The natural gas that is produced from crude oil wells is called *associated gas*, a gas that can exist separate from the crude oil in the underground formation or dissolved in the crude oil. Also, condensate produced from oil wells is often referred to as *lease condensate*.
ii) Dry gas wells. Raw natural gas that does not contain any hydrocarbon liquids is called *nonassociated* gas; also condensate from dry gas is extracted at gas processing plants and is often called *plant condensate*.
iii) Condensate wells. These are wells that produce raw natural gas along with natural gas liquid (NGL); also, such gas is also called *associated* gas and may also be referred to as *wet gas*.

Previously, natural gas reservoirs have been divided into three groups: dry gas, wet gas, and retrograde-condensate gas (Chapter 1). A dry gas reservoir is a reservoir that produces a single composition of gas that is constant in the reservoir, wellbore, and lease-separation equipment throughout the life of a field. Some liquids may be recovered by processing in a gas plant. A wet gas reservoir is a reservoir that produces a single gas composition to the producing well perforations throughout its life. Gas condensate will form either while flowing to the surface or in lease-separation equipment (Thornton, 1946). A retrograde-condensate gas reservoir initially contains a single-phase fluid, which changes to two phases (condensate and gas) in the reservoir when the reservoir pressure decreases. Additional condensate forms with changes in pressure and temperature in the tubing and during lease separation. From a reservoir standpoint, dry and wet gas can be treated similarly in terms of producing characteristics, pressure behavior, and recovery potential. Studies of

Handbook of Natural Gas Analysis, First Edition. James G. Speight.

retrograde-condensate gas reservoirs must consider changes in condensate yield as reservoir pressure declines, the potential for decreased well deliverability as liquid saturations increase near the wellbore, and the effects of two-phase flow on wellbore hydraulics.

The term *gas condensate* (or *condensate*) is often applied to any liquid composed of low-boiling hydrocarbons produced from a gas well (Tables 10.1, 10.2, and 10.3). However, the term *condensate reservoir* should be applied only to those reservoirs in which condensate is formed in the reservoir because of retrograde behavior. Wet-gas reservoirs can always be treated as containing single-phase gas in the reservoir, while retrograde-condensate reservoirs may not. Wet-gas reservoirs generally produce low-boiling liquids with gravities similar to those for retrograde condensates, but with yields less than approximately 20 standard stock barrels per million cubic feet (STB/MMscf).

However, although the difference between retrograde-condensate and wet gases is notable, there is much less distinction between wet and dry gases. For both wet and dry gases, reservoir engineering calculations are based on a single-phase reservoir gas. The only issue is whether there is a sufficient volume of produced liquid to be considered in such calculations as material balance or wellbore hydraulics. Retrograde systems require more complex calculations

Table 10.1 Typical chemical and physical properties of condensate.

Property	Comment/value
Appearance	Amber to dark brown
Physical form	Liquid
Odor	Sulfurous
Vapor pressure	5–15 psia (Reid VP) at 37.8 °C (100 °F)
Vapor density (air = 1)	1
Initial boiling point/range	−29 to 427 °C (−20 to 800 °F)
Solubility in water	Negligible
Specific gravity (water = 1)	0.6–0.8 at 15.6 °C (60 °F)
Bulk density	6.25 lbs gal^{-1}
VOC content (% v/v)	50
Evaporation rate (*n*-butyl acetate = 1)	1
Flash point	−46 °C (−51 °F)
Lower explosive limits (% v/v in air)	1.1
Upper explosive limits (% v/v in air)	6.0
Autoignition temperature	310 C (590 °F

Table 10.2 Properties of gas condensate from Bibiyana gas field, Bangladesh (Sujan et al., 2015).

Property	Method	Data
Range of boiling point, °C	ASTM D1160	21–345
Naphtha, % v/v	50	
Kerosene, % v/v	23	
Diesel, % v/v	24	
Density at 15 °C, kg/l	ASTMD1298	0.8184
API gravity	ASTM D287	43
Kinematic viscosity, cSt, 70 °F	ASTM D445	1.43
Kinematic viscosity, 122 °F	ASTM D445	0.999
Pour point, °C	ASTM D97	< −20
Aniline point, °C	ASTM D611	43
Flash point, °C	ASTM D93	22
Sulfur content, mass %	ASTM D129	0.245
Carbon residue, % w/w	ASTM D189	1.50
Ash content, % w/w	ASTM D482	0.0075

Table 10.3 Distillation products from gas condensate (Bibiyana Gas Field, Bangladesh) (Sujan et al., 2015).

Yield, % v/v	Temperature (°C)
Initial boiling point	21
10%	85
20%	96
30%	104
40%	118
50%	142
60%	162
70%	190
80%	222
90%	286
95%	342
Residue	2
Loss	3

using equation of state that depend upon data produced in an analytical or research laboratory.

Gas condensate (sometimes referred to as *naphtha, low-boiling naphtha,* or *ligroin*) contains significant amounts of C_{5+} components (usually to C_8 or C_{10} components depending upon the source), and they exhibit the phenomenon of *retrograde condensation* at reservoir conditions; in other words, as pressure decreases, increasing amounts of liquid condenses in the reservoir (down to about 2000 psia). This results in a significant loss of *in situ* condensate reserves that may only be partially recovered by revalorization at lower pressures. Gas condensate reservoirs exhibit producing gas–oil ratios from 2500 to 50 000 SCF/STB (400 to 10 STB/MMSCF). Gas cycling projects designed to avoid liquid loss from retrograde condensation can usually be justified for fluids with liquid content higher than about 50–100 STB/MMSCF, whereas offshore, the minimum liquid content to justify cycling is about 100 STB/MMSCF.

However, when production of a hydrocarbon reservoir commences, analysis plays a major role in determining the types of fluids that are present, along with their main physicochemical characteristics (Table 10.4). Generally, that information is obtained by performing a pressure–volume–temperature (PVT) analysis on a fluid sample of the reservoir as well as other physical analyses that relate to phase behavior and other phenomena. Conventional production measurements, such as a drill stem test (DST), are the only parameters that can be measured almost immediately after a well is completed (Breiman and Friedman, 1985; Kleyweg, 1989; Dandekar and Stenby, 1997).

From an engineering point of view, the issues that must be addressed (by analysis) in a gas condensate reservoir are (Whitson et al., 1999) (i) the variation of the condensate yield that will vary during the life of a reservoir and (ii) the manner in which the two-phase gas–oil flow near to the wellbore affects *gas* productivity. Both of these issues are strongly related to the PVT properties of the fluid system (though productivity is more affected by relative permeability effects).

It is in the development of gas condensate reservoirs that analytical data reach a high level of importance. Compositional grading in gas condensate reservoirs is important to the design of well placement, estimation of in-place surface volumes, reserves, and prediction of fluid communication vertically (between geologic layers) and horizontally (between fault blocks) (Organick and Golding, 1952; Niemstschik et al., 1993). Prediction of a potential underlying oil is often required in discovery wells that are drilled up-structure and encounter only gas that is near saturated. In such cases, accurate sampling, accurate analysis (though the use of standard test methods) (Table 10.4), and PVT modeling are of paramount importance (Pedersen and Fredenslund, 1987; McCain, 1990; Marruffo et al., 2001). A PVT model should describe accurately the key phase, volumetric, and viscosity behavior dictating the key processes affecting rate–time performance and ultimate recovery of gas and oil at the wellhead.

Table 10.4 Suite of methods for the analysis of natural gas and gas condensate (in alphabetical order).

- API gravity – ASTM D287
- API/density/specific gravity (densitometer) – ASTM D5002, D4052
- Carbon residue (microcarbon) – ASTM 4530
- Chloride, organic – ASTM D7536
- Composition – ASTM D1945, ASTM D2427, ASTM D2650, ASTM D5134
- Density and relative density (specific gravity) of liquids – ASTM D1217, ASTM D1298, ASTM D1657, ASTM D4052
- Depentanization –ASTM D2001
- Distillation, atmospheric – ASTM D86
- Electrical conductivity – ASTM D4308
- Flash point – ASTM D56, ASTM D92, ASTM D93, ASTM D3828, ASTM D6450
- Freezing points of high-purity hydrocarbons – ASTM D1015, ASTM D1016
- Heat of combustion – ASTM D240
- High-temperature simulated distillation – ASTM D7169
- Hydrogen content – ASTTM D4808
- Hydrogen sulfide in vapor phase – ASTM D5705 MOD, ITM 3468
- Light ends analysis by GC for crude oils – ASTM D7900, ASTM D6730
- Mass conversion – ASTM D2421
- Metal analysis by ICP, Ni, V, Fe – ASTM D5708 Method B
- Pour point – ASTM D97
- Purity – ASTM D2268
- Residue – ASTM D1015, ASTM D2158, ASTM D7756
- Sampling – ASTM D4057
- Sediment and water – ASTM D1796, ASTM D4007, ASTM D4177
- Sediment by extraction – ASTM D473
- Solubility of gases – ASTM D3429, ASTM D3827
- Sulfur content – ASTM D4294, ASTM D5623
- Thermal conductivity – ASTM 2717
- Vapor pressure – ASTM D323, ASTM D5191, ASTM D6377
- Vapor pressure – crude oil, ASTM D6377
- Viscosity – ASTM D4486, ASTM 7042
- Viscosity, Saybolt – ASTM D2161
- Volume conversion – ASTM D2421
- Water content (Karl Fisher, coulometric) – ASTM D4928

However, a PVT model may not be capable of accurately describing *all* PVT properties with equal accuracy. Models based on equations of state often have difficulty matching retrograde phenomena (compositional variation of gas and liquid dropout), particularly when the system is near critical, or only small amounts of condensation occur just below the dew point (tail-like retrograde behavior). Oil viscosity is also difficult to predict for reservoir condensates, and measured oil viscosities are not usually available for tuning the viscosity model. Consequently, it is important to determine which PVT properties are most important to the accurate engineering of reservoir and well performance for a given field development. Different fields require different degrees of accuracy for different PVT properties, dependent on field development strategy (depletion vs. gas cycling), low or high permeability, saturated or highly undersaturated, geography (offshore or onshore), and the number of wells available for delineation and development.

The PVT properties that are important to the engineering of all gas condensate reservoirs include (i) the *Z*-factor (Hall and Yarborough, 1973), (ii) the gas viscosity (Lee et al., 1966), (iii) the variation of the condensate composition (C_{7+}) with pressure, and (iv) the viscosity of the oil as well as any liquid dropout. These properties are particularly important to reservoirs produced by pressure depletion. For gas condensate reservoirs undergoing gas cycling, it may also be important to quantify phase behavior (vaporization, condensation, and near-critical miscibility) that develops in gas cycling *below the dew point*.

Thus, obtaining preliminary values of properties such as (i) the amount of heptane and higher-boiling constituents (C_{7+}), (ii) molecular weight (MW) of the original fluid, (iii) maximum retrograde condensation (MRC), and (iv) dew-point pressure (P_d) are the required properties (Nemeth and Kennedy, 1967; Breiman and Friedman, 1985; Potsch and Bräuer, 1996; Dandekar, and Stenby, 1997; Elsharkawy, 2001; Marruffo et al., 2001). Most of these properties are extremely important to the development of a gas condensate reservoir, and the early availability of such data will allow the production engineers to carry out reservoir studies that will ensure an efficient exploitation and maximize the final recovery of the liquids present in the reservoir. The only parameter needed to use these correlations is the value of the gas–condensate ratio (GCR) of the fluid during the early stage of production (Paredes et al., 2014). These empirical equations should be valid for *any* gas condensate reservoir, although a range of usability is proposed for a better performance of the correlations.

Thus, natural-gas condensate is a low-density mixture of hydrocarbon liquids that are present as gaseous components in the raw natural gas produced from many natural gas fields. Natural-gas condensate is also called condensate, or gas condensate, or sometimes natural gasoline because it contains hydrocarbon derivatives (*n*-hydrocarbon and other hydrocarbon isomers) within the naphtha–gasoline boiling range (Tables 10.5 and 10.6). Some gas species within the raw natural gas will condense to a liquid state if the temperature is reduced to below the hydrocarbon dew-point temperature at a set pressure (Elsharkawy, 2001). Thus, condensates are a group of hydrocarbon derivatives that do not fit easily into mainstream product categories and have been defined as liquid hydrocarbon derivatives that lie between crude oil and NGLs. However, the reality is that most condensates differ significantly from crude oils and

Table 10.5 General summary of product types and distillation range.

Product	Lower carbon limit	Upper carbon limit	Lower boiling point (°C)	Upper boiling point (°C)	Lower boiling point (°F)	Upper boiling point (°F)
Refinery gas	C_1	C_4	−161	−1	−259	31
Liquefied petroleum gas	C_3	C_4	−42	−1	−44	31
Naphtha	C_5	C_{17}	36	302	97	575
Gasoline	C_4	C_{12}	−1	216	31	421
Kerosene/diesel fuel	C_8	C_{18}	126	258	302	575
Aviation turbine fuel	C_8	C_{16}	126	287	302	548
Fuel oil	C_{12}	$>C_{20}$	216	421	>343	>649
Lubricating oil	$>C_{20}$		>343		>649	
Wax	C_{17}	$>C_{20}$	302	>343	575	>649
Asphalt	$>C_{20}$		>343		>649	
Coke	$>C_{50}$[a]		>1000[a]		>1832[a]	

[a]Carbon number and boiling point difficult to assess; inserted for illustrative purposes only.

Table 10.6 Increase in the number of isomers with carbon number.

Carbon atoms	Number of isomers
1	1
2	1
3	1
4	2
5	3
6	5
7	9
8	18
9	35
10	75
15	4347
20	366 319
25	36 797 588
30	4 111 846 763
40	62 491 178 805 831

condensates are not like NGLs but are similar only to natural gasoline, the highest boiling of the NGLs. On the other hand, gas condensate and natural gasoline are often comparable to the lower-boiling fraction of naphtha.

Naphtha (*petroleum naphtha*) is a generic term applied to refined, partly refined, or unrefined petroleum products and liquid products of natural gas that distill below 240 °C (465 °F) and is the volatile fraction of the petroleum, which is used as a solvent or as a precursor to gasoline. In fact, not less than 10% v/v of material should distill below 75 °C (167 °F); not less than 95% v/v of the material should distill below 240 °C (465 °F) under standard distillation conditions, although there are different grades of naphtha within this extensive boiling range that have different boiling ranges (Hori, 2000; Parkash 2003; Pandey et al., 2004; Gary et al., 2007; Speight, 2014a, 2017; Hsu and Robinson, 2017), and the focus of this chapter is on the low-boiling liquid fraction of gas streams (gas condensate, natural gasoline), which is typically equivalent to the low-boiling fraction of naphtha (boiling point 30–200 °C, 1–392 °F).

The term *petroleum solvent* is often used synonymously with *naphtha*. However, naphtha may also be produced from tar sand bitumen, coal tar, and oil shale kerogen as well as by the destructive distillation of wood and from synthesis gas (mixtures of carbon monoxide and hydrogen produced by the gasification of coal and/or biomass or other feedstocks), which is then converted to liquid products by the Fischer–Tropsch process (Davis and Occelli, 2010; Chadeesingh, 2011; Speight, 2011, 2013, 2014a, 2014b). For that reason and to remain within the context of this book, this chapter deals only with the low-boiling naphtha fraction that is produced by the processing of crude oil in refineries (Speight, 2014a, 2017; Hsu and Robinson, 2017).

There is a third hydrocarbon product that is also somewhat interchangeable with natural gasoline and middle-of-the-road lease condensates, and that is naphtha, specifically light naphtha. Most naphtha is produced from crude oil in the first step of the refining process – distillation. The constituents of light naphtha (low-boiling naphtha similar to gas condensate) are made up predominantly of pentane and hexane hydrocarbon derivative and portions of the higher-boiling hydrocarbon derivatives. Condensate is, in fact, another source of naphtha and can be processed through a splitter (which is essentially a stand-alone, small crude oil distillation tower) that separates the naphtha range materials from lower-boiling NGLs and higher-boiling hydrocarbon derivatives to yield a clean naphtha product that can be used as a feedstock for refinery upgrading processes and for the production of petrochemicals.

All of these product categories in the condensate family – lease condensates, plant condensates (natural gasoline), and naphtha – are, in general terms, composed of the same hydrocarbon components. But only lease condensates are produced directly from the wellhead without further processing. This distinction turns out to be important when we get to the idiosyncrasies of the export rules for condensate.

In the refinery, naphtha is an unrefined or refined low-boiling distillate fraction, which usually boils below 250 °C (480 °F), but often with a fairly wide boiling range depending upon the crude oil from which the naphtha is produced as well as the process that generates the naphtha. More specifically, there is a range of special-purpose hydrocarbon fractions that can be described as *naphtha*. For example, the 0–100 °C (32–212 °F) fraction from the distillation of crude oil is referred to as *light virgin naphtha*, and the 100–200 °C (212–390 °F) fraction is referred to as *heavy virgin naphtha*. The product stream from the fluid catalytic cracker is often split into three fractions: (i) the fraction boiling <105 °C/<220 °F is *light FCC naphtha*, the fraction boiling from 105 to 160 °C (220–320 °F) is *intermediate FCC naphtha*, and the fraction boiling from 160 to 200 °C (320–390 °F) is referred to as *heavy FCC naphtha* (Occelli, 2010). These boiling ranges can vary from refinery to refinery and even within a refinery when the crude oil feedstock changes or blends of crude oil are used as refinery feedstock.

More generally, naphtha is an intermediate hydrocarbon liquid stream derived from the crude oil and is usually desulfurized and then catalytically reformed to produce high-octane naphtha before blending into the streams that make up gasoline. Because of the variations in crude oil composition and quality as well as differences in refinery operations, it is difficult (if not impossible) to provide a definitive, single definition of the word naphtha since each refinery produces a site-specific naphtha – often with a unique boiling range (unique initial point and final boiling point) as well as other physical and compositional characteristics.

On a chemical basis, (petroleum) naphtha is difficult to define precisely because it can contain varying amounts of the constituents (paraffin derivatives, naphthene derivatives, aromatic derivatives, and olefins) in different proportions, in addition to the potential isomers of the paraffin derivatives that exist in the naphtha boiling range (the C_5–C_8 or the C_5–C_{10} boiling range) (Tables 10.5 and 10.6).

The so-called *petroleum ether* solvents are specific boiling range naphtha solvents, as is *ligroin*. Thus, the term *petroleum solvent* describes a special liquid hydrocarbon fractions obtained from naphtha and used in industrial processes and formulations. These fractions are also referred to as *industrial naphtha*. Other solvents include *white spirit* that is subdivided into *industrial spirit* (distilling between 30 and 200 °C, 86 °–392 °F) and *white spirit* (light oil with a distillation range of 135–200 °C (275–392 °F)). The special value of naphtha as a solvent lies in its stability and purity.

Regulatory requirements have enhanced the need for better test methods to control manufacturing and the distribution of gasolines. The addition of alcohol and ether as important blending components to gasoline to meet air quality standards has necessitated modifying some existing test methods and the development of new procedures. The desire to reduce manufacturing costs, coupled with the regulatory requirements, has enhanced the application of more cost-effective test methods including rapid screening procedures and wider use of online analyzers. During the early 1950s, instrumental analytical techniques, such as mass spectrometry, infrared (IR) spectroscopy, and ultraviolet spectroscopy, were being explored and used for hydrocarbon composition and structural analysis. Beginning in the mid-1950s, publications on gas chromatography (GC) began to appear in the literature, and this new technique was soon used for analyzing a wide variety of hydrocarbon streams. As commercial instrumentation was developed, the application of GC grew rapidly, with volumes of information being published from its beginning up to the present time. Recently, more rapid spectrometry methods such as IR and near-infrared (NIR) and the use of hyphenated analytical techniques, for example, gas chromatography–mass spectrometry (GC-MS), have been applied with success to the characterization of low-boiling distillates.

In this chapter, reference is made to the various test methods dedicated to the determination of the amounts of carbon, hydrogen, and nitrogen (ASTM D5291, 2017) as well as the determination of oxygen, sulfur, metals, and elements such as chlorine (ASTM D808, 2017).

The amount of aromatic constituents in hydrocarbon liquids is a key characteristic that will affect a variety of properties of the oil including its boiling range, viscosity, stability, and compatibility of the oil with polymers. In the current context, this is particularly the case for gas condensate and, especially, for natural gasoline where the context of aromatic constituents may influence the compatibility of the gas condensate or the natural gasoline as a blend stock with other refinery liquids. The aromatic hydrogen and aromatic carbon contents determined by a standard test method (ASTM D5291, 2017) can be used to evaluate changes in aromatic contents of hydrocarbon oils due to changes in processing conditions and to develop processing models in which the aromatic content of the hydrocarbon oil is a key processing indicator. Existing methods for estimating the content of the aromatic constituents use physical measurements, such as refractive index, density, and number average MW (ASTM D3238, 2017) or IR absorbance, and often depend on the availability of suitable standards. These procedures do not require standards of known aromatic hydrogen or aromatic carbon contents and are applicable to a wide range of hydrocarbon liquids with the caveat that the hydrocarbon liquids must be soluble in chloroform at ambient temperature.

ASTM D5291 (2017) covers the determination of the aromatic hydrogen content and aromatic carbon content of hydrocarbon liquids using high-resolution nuclear magnetic resonance (NMR) spectrometers. For pulsed Fourier transform (FT) spectrometers, the detection limit is typically 0.1 mol% aromatic hydrogen atoms and 0.5 mol% aromatic carbon atoms. For continuous-wave (CW) spectrometers, which are suitable for measuring aromatic hydrogen contents only, the detection limit is considerably higher and typically 0.5 mol% aromatic hydrogen atoms.

On the one hand, there is the notion that the composition of complex petroleum fractions is often expressed in terms of the proportions of aromatic rings (R_A), naphthene rings (R_N), and paraffin chains (C_P) that would comprise a hypothetical mean molecule. Alternatively, the composition may be expressed in terms of a carbon distribution, that is, the percentage of the total number of carbon atoms that are present in aromatic ring structures (% C_A), naphthene ring structures (% C_N), and paraffin chains (% C_p). On the other hand, there is the notion that determination of the carbon distribution and ring content (ASTM D3238, 2017) can serve to express the gross composition of the higher-boiling fractions of crude oil that can be used as an adjunct to the bulk properties in monitoring the manufacture of lubricating oil base stocks by distillation, solvent refining, or hydrogenation, or both, and in comparing the composition of stocks from different crude sources. Furthermore, the data can often be correlated with critical product performance properties. The method covers the calculation of the carbon distribution and ring content of olefin-free crude oils from measurements of the refractive index, density, and MW (n-d-M). However, the method is limited in its application (as an example, a correction must be applied for oils containing significant quantities of sulfur) and produces an average composition of the sample under investigation, and it is up to the investigator the extent of application of average composition to the processing option for which the data will be used. Hence there is a need for the analyst to understand the methods of production and purification of the low-boiling liquid such as gas condensate and natural gasoline (Chapter 1).

10.2 Types of Condensate

The hydrocarbon products described in the following paragraphs have been used for years as feedstocks (i) for refinery upgrading processes, (ii) for gasoline blending, and (iii) for petrochemical processes, among other uses. All of these product categories in the condensate family – lease condensates, plant condensates (natural gasoline), and naphtha – are, in general terms, composed of the same hydrocarbon constituents. However, it is only the lease condensates that are produced directly from the wellhead without further processing. This distinction turns out to be important when the specifications derived by means of the application of standard analytical test methods are applied to these products.

10.2.1 Gas Condensate

Typically, condensates can be produced along with significant volumes of natural gas and are typically recovered at atmospheric temperatures and pressures from the wellhead gas production. These raw (unrefined) condensates produced from a reservoir come out of the ground as mixtures of various hydrocarbon compounds including NGLs, pentane derivatives (C_{5s}), hexane derivatives (C_{6s}) (hexanes), and, depending on the condensate, a mixture of higher MW hydrocarbon derivatives in the heptane (C_7) to decane (C_{10}) or even dodecane (C_{12}) carbon number range.

10.2.2 Lease Condensate

Lease condensate is produced at the wellhead as an unprocessed (unrefined) liquid except for stabilization at or near the wellhead. Lease condensates have a wide range of API gravity from 45 to 75°. They are produced at many sites in the United States, particularly near to the Gulf Coast of the United States from fields such as Eagle Ford and other shale basins.

A lease condensate with a high API gravity contains substantial amounts lots of NGLs (including ethane, propane, and butane) and not many of the heavier hydrocarbons. These condensates are clear or translucent in color. The condensates with a lower API gravity down at the 45° level look more like crude oil (black or near black) and have much higher concentrations of the higher MW heptane derivatives as well as octane derivatives and even higher MW hydrocarbon derivatives.

The lower MW condensates (higher API gravity) can be difficult to handle due to their high vapor pressure and are usually stabilized at the wellhead. This is achieved by passing the condensate through a pressurized vessel (sometimes referred to as a knockout pot) (Chapter 4) that allows the constituents (such as the NGLs) to be separated on the basis of relative vapor phase with some remaining in the gas phase and others condensing to the liquid phase. The result is a stabilized condensate that has a lower vapor pressure, which is easier to handle, especially when the condensate must be shipped by truck or rail.

10.2.3 Plant Condensate

Plant condensate is a product of plants that process NGLs and is almost equivalent to natural gasoline. Plant condensate is an alternate term that may be applied to natural gasoline, which is composed of hydrocarbon components that are similar to the hydrocarbon constituents in lease condensate, i.e. pentane derivatives (C_{5s}), some hexane derivatives (C_{6s}), and small quantities of higher MW hydrocarbon derivatives. Since it comes from a processing plant, plant condensate is considered to be processed product rather than a naturally occurring product.

Furthermore, since plant condensate (or natural gasoline) is the product of a processing plant, the quality of the natural gasoline (defined by the specifications) is in closer ranges than lease condensates. However, both products can be used interchangeably in some markets, such as crude blending and as a diluent for tar sand bitumen (Speight, 2014a, 2017).

10.2.4 Light Naphtha

Another hydrocarbon product that is also somewhat interchangeable with natural gasoline and lease condensates is the product designated as naphtha, specifically low-boiling naphtha. Light naphtha is similar in properties to lease condensate and natural gasoline, but it (light naphtha) is considered to be a refined product since it is produced by refinery distillation or the process of condensate splitting. In fact, most naphtha is produced from crude oil in the first step of the refining process – distillation (Speight, 2014a, 2017). Light naphtha is composed predominantly of pentane and hexane derivatives as well as smaller amounts of higher MW hydrocarbon derivatives.

Another source of naphtha is condensate, which can be processed through a stand-alone, small crude oil distillation tower (sometime referred to as a *splitter*). This allows the naphtha to be separated from the more volatile constituents of the condensate and from the higher-boiling hydrocarbon derivatives. The naphtha product can be used as a feedstock for refinery upgrading processes and for petrochemical production (Sujan et al., 2015).

10.3 Production

Briefly and by way of explanation, *gas condensate* is a mixture of light hydrocarbon liquids obtained by condensation of hydrocarbon vapors, predominately butane, propane, and pentane with some heavier hydrocarbon derivatives and relatively little methane or ethane. NGLs are the hydrocarbon liquids that condense during the processing of hydrocarbon gases that are produced from oil or gas reservoir, and *natural gasoline* is a mixture of liquid hydrocarbon derivatives extracted from natural gas that is suitable for blending with the various refinery streams that are used to blend to produce the final sales gasoline (Speight, 2014a, 2017).

To complete the definitions, condensates are a group of hydrocarbons that do not fit easily into mainstream product categories. Other definitions place condensates as liquid hydrocarbons somewhere between crude oil and NGLs. However, all condensates are similar only to natural gasoline, the highest boiling of the NGLs. However, the term *condensate* can refer to several products made up of similar hydrocarbon compounds.

On the other hand, naphtha is produced in the refinery by any one of several methods that include (i) fractionation of straight-run, cracked, and reforming distillates, or even fractionation of crude petroleum; (ii) solvent extraction; (iii) hydrogenation of cracked distillates; (iv) polymerization of unsaturated compounds (olefins); and (v) alkylation processes. In fact, naphtha may be a combination of product streams from more than one of these processes and may contain constituents (such as olefins and even diolefins) not typically found in the nonthermal gas condensate. On occasion, depending upon availability, gas condensate, NGLs, and natural gasoline are combined with the naphtha streams to complement the composition and volatility requirements of a liquid stream for gasoline manufacture.

The more common method of naphtha preparation is distillation. Depending on the design of the distillation unit, either one or two naphtha streams may be produced: (i) a single naphtha with an end point of about 205 °C (400 °F) and similar to straight-run gasoline or (ii) this same fraction divided into light naphtha and heavy naphtha. The end point of the light naphtha is varied to suit the subsequent subdivision of the naphtha into narrower boiling fractions and may be of the order of 120 °C (250 °F).

Sulfur compounds are most commonly removed or converted to a harmless form by chemical treatment with lye, doctor solution, copper chloride, or similar treating agents (Speight, 2014a, 2017). Hydrorefining processes (Speight, 2014a, 2017) are also often used in place of chemical treatment. When used as a solvent, gas condensate and natural gasoline can be blended with naphtha (subject to incompatibility restrictions) (Chapter 6) (Speight, 2014a) and are selected for low sulfur content, and the usual treatment processes remove only sulfur compounds. Such blends, with a low content of aromatic derivatives, may have a slight odor, but the aromatic derivatives increase the solvent power of the blend, and there is no need to remove the aromatic derivatives from the blend (or prior to the blending operation) unless odor-free naphtha is specified.

Generally, naphtha is composed of shorter-chain hydrocarbon derivatives (C_5–C_8), such as when gas condensate and natural gasoline are employed, can be aromatic (sweet smelling and carcinogenic), and can contain sulfuric compounds or other impurities. It has the lowest boiling point of all the petroleum-derived liquid, which gives it the highest vapor pressure. As a result, naphtha produces flammable vapors whenever it is not contained in a vapor-seal container, but those vapors will disperse quickly if well ventilated. Odor is particularly important since, unlike most other petroleum liquids, many of the manufactured products containing naphtha are used in confined spaces, in factory workshops, and in the home. As a result of the volatility, naphtha (especially naphtha containing gas condensate and natural gasoline) (i) is readily flammable, (ii) will evaporate quickly from most surfaces, and (iii) must be very carefully contained at all times. In addition, aromatic naphtha that contains a high percentage of aromatic constituents can also be smoky, toxic, and carcinogenic. Some naphtha-based fuels have a reduced aromatic content, but many are naturally high or augmented in aromatic derivatives, especially if the blend stock (gas condensate or natural gasoline) contain aromatic derivative.

In order to meet the demands of a variety of uses, certain basic naphtha grades are produced, which are identified by boiling range. The complete range of naphtha solvents may be divided, for convenience, into four general categories: (i) special boiling point spirits having overall distillation range within the limits of 30–165 °C (86–329 °F); (ii) pure aromatic compounds such as benzene, toluene, xylenes, or mixtures (BTX) thereof; (iii) white spirit, also known as mineral spirit and naphtha, usually boiling within 150–210 °C (302–410 °F); and (iv) high-boiling petroleum fractions, boiling within limits 160–325 °C (320–617 °F).

Since the end use dictates the required composition of naphtha, most grades are available in both high and low solvency categories, and the various text methods can have major significance in some applications and lesser significance in others. Hence the application and significance of tests must be considered in the light of the proposed end use. To emphasize this point, naphtha contains varying amounts of its constituents, viz. paraffin derivatives, naphthene derivatives, aromatic derivatives, and olefins, in different proportions in addition to potential isomers of paraffin that exist in naphtha boiling range. Naphtha resembles gasoline in terms of boiling range and carbon number, being a precursor to gasoline. Naphtha is used as automotive fuel, engine fuel, and jet-B (naphtha type). Broadly, naphtha is classified as *light naphtha* and *heavy naphtha*. Light naphtha is used as rubber solvent and lacquer diluent, while heavy naphtha finds its application as varnish solvent, dyer's naphtha, and cleaner's naphtha.

Volatility, solvent properties (dissolving power), purity, and odor determine the suitability of naphtha for a particular use. The use of naphtha as an incendiary device in warfare and as an illuminant dates back to 1200 AD. Naphtha is characterized as lean (high paraffin content) or rich (low paraffin content). The rich naphtha with higher proportion of naphthene content is easier to process in the Platforming unit (Parkash, 2003; Gary et al., 2007; Speight, 2014a, 2017; Hsu and Robinson, 2017).

Naphtha solvents may belong to categories such as special boiling spirits having distillation range of 30–165 °C (86–330 °F), white spirit (mineral spirit) boiling within 150–210 °C (300–410 °F), and high-boiling petroleum fractions (160–325 °C; 320–615 °F). In aromatic complexes, naphtha is converted into basic petrochemical intermediates: benzene, toluene, and xylene (BTX). Petroleum naphtha is by far the most popular feedstock for aromatic derivative production.

If spilled or discharged in the environment, naphtha represents a threat of the toxicity of the constituents to land and/or to aquatic organisms. A significant spill may cause long-term adverse effects in the aquatic environment. The constituents of naphtha predominantly fall in the C_5–C_{16} carbon range: alkanes, cycloalkanes, aromatic derivatives, and, if they are subject to a cracking process, alkenes as well. Naphtha may also contain a preponderance of aromatic constituents (up to 65%), others contain up to 40% alkenes, while all of the others are aliphatic in composition, up to 100%.

By way of definition, the term aliphatic naphtha refers to naphtha containing less than 0.1% benzene and with carbon numbers from C_5 to C_{16}. Aromatic naphtha has carbon numbers from C_6 to C_{16} and contains significant quantities of aromatic hydrocarbon derivatives such as benzene (>0.1%), toluene, and xylene. The final gasoline product as a transport fuel is a carefully blended mixture having a predetermined octane value. Thus, gasoline is a complex mixture of hydrocarbon derivatives that boils below 200 °C (390 °F). The hydrocarbon constituents in this boiling range are those that have 4–12 carbon atoms in their molecular structure, which can be unadulterated gas condensate or natural gasoline.

Water solubility ranges from very low for the longest-chain alkanes to high solubility for the simplest monoaromatic constituents. Generally, the aromatic compounds are more soluble than the same-sized alkanes, isoalkanes, and cycloalkanes. This indicates that the components likely to remain in water are the one- and two-ring aromatic derivatives (C_6–C_{12}). The C_9–C_{16} alkanes, isoalkanes, and one- and two-ring cycloalkanes are likely to be attracted to sediments based on their low water solubility and moderate to high octanol–water partition coefficient ($\log K_{ow}$) and organic carbon–water partition coefficient ($\log K_{oc}$) values.

The main uses of petroleum naphtha fall into the general areas of (i) precursor to gasoline and other liquid fuels, (ii) solvents or diluents for paints, (iii) dry-cleaning solvents, (iv) solvents for cutback asphalts, (v) solvents in rubber industry, and (vi) solvents for industrial extraction processes. Turpentine, the older and more conventional solvent for paints, has now been almost completely replaced by the cheaper and more abundant petroleum naphtha. Test methods (Section 10.3) are based on defining the suitability of naphtha for any of the aforementioned uses.

Based on the descriptions in the preceding paragraphs, it is in order to apply standard test methods to gas condensate or to natural gasoline that have been designed for application to low-boiling naphtha. This statement is based on the understanding that the difference in boiling range may require some modification of a standard test method when applied to the gas condensate or to natural gasoline.

Finally, the nature of the uses found for naphtha demands compatibility with the many other materials employed in the formulation of the naphtha, including gas condensate and natural gasoline. Thus, the solvent properties of a given fraction must be carefully measured and controlled. For most purposes volatility is important, and because of the wide use of naphtha in industrial and recovery plants, information on some other fundamental characteristics is required for plant design.

10.4 Composition and Properties

Condensate liquids are generally *water white* (colorless or near colorless) or are light in color (brown, orange, or green), with an API gravity typically on the order of 40 and 60 °API. The yields of these liquids can be as high as 300 stock-tank barrels per million cubic feet (STB/MMscf). It has been suggested (McCain, 1990) that when yields are below approximately 20 STB/MMscf, even though phase behavior considerations may show retrograde behavior, the amount of liquid dropout in the reservoir can be insignificant.

The primary difficulties in producing condensate reservoirs are as follows: (i) Liquid deposition near the wellbore causes a decrease in gas deliverability that can approach 100% in a reservoir with less than 50 md permeability, and (ii) a large amount of the most valuable hydrocarbon components is left in the reservoir rather than produced.

There are many condensate sources, and each has its own unique gas condensate composition. In general, gas condensate has a specific gravity ranging from 0.5 to 0.8 and is composed of hydrocarbons such as propane, butane, pentane, hexane, and often higher MW hydrocarbons up to decane. Natural gas compounds with more carbon atoms (e.g. pentane, or blends of butane, pentane, and other hydrocarbons with additional carbon atoms) exist as liquids at ambient temperatures. Additionally, condensate may contain other impurities such as (i) hydrogen sulfide, H_2S; (ii) thiols, traditionally called mercaptans and denoted as RSH, where R is an organic group such as methyl, ethyl, propyl, and the like; (iii) carbon dioxide, CO_2; (iv) straight-chain alkane derivatives having 2–10 carbon atoms, denoted as C_2–C_{12}; (v) cyclohexane and perhaps other naphthene derivatives; and (vi) aromatic derivatives such as benzene, toluene, xylene isomers, and ethylbenzene (Pedersen et al., 1989).

10.4.1 Chemical Properties

Ten years of operation of gas condensate reservoirs has verified the necessity for and practicability of cooperative and unitized operation. The operating method for each reservoir should be based upon the characteristics of the gas condensate at reservoir conditions. Other factors such as richness of the gas, size of the reserve, capacities of wells, nature of the reservoir, and mode of occurrence of the gas condensate must be considered. Marketing conditions and other economic factors are also are important. Due to the increasing volume of the gas market, and prospects for chemical conversion of gas to liquid fuel by the Fischer–Tropsch process, many operators will be required to make a choice or compromise between complete pressure maintenance and gas sales. This can be done intelligently with known methods of evaluation.

Gas condensate fields (also called gas distillate) are today quite commonplace on the Louisiana and Texas Gulf Coast, but are not confined to that area. The frequency with which they are being discovered and their economic importance in the petroleum industry have increased tremendously with the trend toward deeper drilling. A common understanding of best operating practices by oil operators, land or royalty owners, regulatory bodies, and legislators is necessary if all parties of interest are to cooperate in realizing the maximum potentialities from these fields. During the past decade in particular, much has been learned and written regarding each of the many phases of gas condensate operations. It is the purpose of this paper to review the accepted practices, so that the modern basic thought may be available in summary form. For the purpose of this review, all underground sources of supply from which condensate can be produced are called gas condensate reservoirs. The name is prefixed with *gas* to emphasize the fact that the condensate is associated with gas when produced. The methods used to evaluate gas condensate reservoirs, and the basic principles governing the selection of the best method of operation for a particular reservoir, are outlined in the paragraphs that follow.

The terms used in discussing gas condensates have not crystallized as yet into a standard nomenclature. For instance, both distillate and condensate are used synonymously to describe the liquid produced in the tanks. It has been necessary herein to adopt somewhat arbitrarily a set of terms and to explain the meaning of each term as introduced to avoid confusion. All are more or less familiar with the fact that some wells produce, together with large volumes of gas, a water-white or light straw-colored liquid that resembles gasoline or kerosene in appearance. This liquid has been called distillate because of the resemblance to products obtained in the refinery by distilling the volatile components from crude oil. Also, the liquid has been called condensate because it is condensed out of the gas produced by the well. The latter term shall be used throughout this paper. Condensate normally differs markedly from conventional crude oil insofar as (i) the color of crude oil typically varies from dark green to black; (ii) it contains some naphtha, often incorrectly referred to as gasoline; (iii) it usually contains dark-colored, high MW nonvolatile components; and (iv) its API gravity, which is a measure of its weight or density, is commonly less than 45° (Speight, 2014a, 2015).

10.4.2 Physical Properties

When a hydrocarbon reservoir is discovered, it is important to know the type of fluids that are present as well as their main physicochemical characteristics; this is normally obtained by performing a PVT analysis to a representative fluid sample of the reservoir. In most cases, having a PVT analysis can take several months, which limits the number and type of reservoir studies that can be carried out during this period (Paredes et al., 2014). The only parameter that can be measured almost immediately after a well is completed is conventional production measurements. In some cases, this production measurement can be obtained before completing the well by using special testing or measuring equipment such as a DST.

It is important to obtain preliminary values of properties such as molar percentage of heptane and heavier components (% mole of C_{7+}), MW of the original fluid, MRC, and dew-point pressure (P_d). Most of these properties are very important for exploitation of gas condensate reservoir, and their early availability will allow engineers to carry out reservoir studies that will ensure an efficient exploitation and maximize the final recovery of the liquids present in the reservoir.

Fluids present in an oil reservoir are the result of a series of thermodynamic changes of pressure and temperature suffered by the original mixture of hydrocarbons over time and during its migration from the container rock to the trap. Reservoir pressure and temperature increase with depth, and their relative relationship will influence the behavior of the light and heavy components that the fluid could contain. In general, the content of low-boiling constituents in a mixture of hydrocarbons increases with temperature and depth, which may result in reservoirs near the critical point; gas condensate reservoirs are included in this kind of fluids (Ovalle et al., 2007).

Certain fluid properties are required for studies related to management of gas condensate reservoirs or prediction of condensate reserves. Often these studies must begin before laboratory data become available, or possibly when laboratory data are not available. Correlations to estimate values of these properties have been developed that are based solely on commonly available field data. These properties are the dew-point pressure of the reservoir fluid, changes in the surface yield of condensate as reservoir pressure declines, and changes in the specific gravity of the reservoir gas as reservoir pressure declines (Gold et al., 1989). No correlations based solely on field data have been published for any of these properties.

Reservoir fluid properties are used to characterize the condition of a fluid at a given state. A reliable estimation and description of the properties of hydrocarbon mixtures is fundamental in petroleum and natural engineering analysis and design. Fluid properties are not independent, just as pressure, temperature, and volume are not independent of each other. Equations of state provide the means for the estimation of the PVT relationship, and from them many other thermodynamic properties can be derived. Compositions are usually required for the calculation of the properties of each phase.

The field data required are initial producing GCR from the first-stage separator, initial stock-tank liquid gravity in °API, specific gravity of the initial reservoir gas, reservoir temperature, and selected values of reservoir pressure. The dew-point pressure correlation is based on data of 615 samples of gas condensates with worldwide origins. The other two correlations are based on 851 lines of constant volume depletion (CVD) data from 190 gas condensate samples, also with worldwide origins.

Correlation equations for gas condensates based on readily available field data have been developed. The correlations can be used to predict dew-point pressures, decreases in surface condensate yields after reservoir pressure has decreased below dew-point pressure, and decreases in reservoir gas specific gravity at reservoir pressures below dew-point pressure. A value of dew-point pressure is essential data for any reservoir study. A reasonably accurate estimate of dew-point pressure for a specific reservoir fluid is necessary in situations in which laboratory data are not available or before laboratory data are obtained. Laboratory measurements of dew-point pressure and other gas properties of 615 gas condensates with worldwide origins were used to develop a dew-point pressure correlation based on initial producing GCR, initial stock-tank oil gravity, and specific gravity of the original reservoir gas. This is the first proposed dew-point pressure correlation that does not require some laboratory-measured quantity.

Estimation of decreases in producing yields after the reservoir pressure drops below the dew-point pressure is necessary for accurate prediction of condensate reserves. The reduction in surface yields can be as much as 75% during the primary production of a gas condensate. This reduction must be taken into account in the prediction of ultimate recovery of a condensate. A surface-yield correlation has been developed that is a function of a selected reservoir pressure, initial stock-tank oil gravity, specific gravity of the original reservoir gas, and reservoir temperature. The data set included laboratory studies of 190 gas condensate samples. This is the first proposal offered in petroleum literature of a correlation to estimate the decreases in surface yield.

The defining property that distinguishes crude oil and gas condensate black and volatile oils is the content of the equilibrium gases. The volatilized-oil (also called lease condensate or distillate) content of a gas represents its condensable liquid portion. Condensable refers to the portion that condenses or drops out during pressure reduction and ultimately results as stock-tank liquid. Condensation may take place within the reservoir as the gas passes through the lease separators. Physically, intermediate hydrocarbon components, typically the C_2–C_7, dominate this fraction. Gas condensates and wet gases also contain volatilized oil. Volatilized oil is reported conventionally as part of the crude oil reserves and production and should not be confused with and is distinctly different from NGLs. NGLs are derived from the gas processing plant and are, therefore, products of a gas processing plant. The volatilized-oil content of gases is quantified in terms of their volatilized-oil–gas ratio, typically expressed in units of STB/MMscf or stock-tank m^3 per standard m^3 of separator gas.

The oil formation volume factor (FVF) relates the volume of oil at stock-tank conditions to the volume of oil at elevated pressure and temperature in the reservoir. Values typically range from approximately 1.0 bbl/STB for crude oil systems containing little or no solution gas to nearly 3.0 bbl/STB for highly volatile oils.

Density is customarily defined as the amount of mass contained in a unit volume of fluid. Density is the single most important property of a fluid (Table 10.7), once it is realized that most other properties can be obtained or related to density. Both specific volume and density – which are inversely

Table 10.7 Density of various hydrocarbon liquids – possible constituents of gas condensate and natural gasoline.

Hydrocarbon (phase)	Formula	Molecular weight	Density
Benzene	C_6H_6	78.114	0.877
Decane	$C_{10}H_{18}$	142.285	0.73
Heptane	C_7H_{16}	100.204	0.684
Hexane	C_6H_{14}	86.177	0.66
Hexene	C_6H_{12}	84.161	0.673
Isopentane	C_5H_{12}	72.15	0.626
Octane	C_8H_{18}	114.231	0.703
Toluene	C_7H_8	92.141	0.867

proportionally related to each other – indicate that the molecules in a fluid are from each other. For liquids, density is high, which translates to a very high molecular concentration and short intermolecular distances. For gases, density is low, which translates to low molecular concentrations and large intermolecular distances (Rayes et al., 1992; Piper et al., 1999).

The all-important parameter to calculate density is the Z-factor, both for the liquid and vapor phases. Empirical correlations for Z-factor for natural gases were developed before the advent of digital computers. Although their use is in decline, they can still be used for fast estimates of the Z-factor. These methods are invariably based on some type of corresponding states' development. According to the theory of corresponding states, substances at corresponding states will exhibit the same behavior (and hence the same Z-factor) (Standing and Katz, 1942; Hall and Yarborough, 1973).

Specific gravity is the ratio of fluid density to the density of a reference substance, both defined at the same pressure and temperature. These densities are usually defined at standard conditions (14.7 psia and 60 °F). For a condensate, oil or liquid, the reference substance is water. By definition, the specific gravity of water is unity, and, using the API scale, water has an API gravity of 10. Light crude oils have an API greater than or equal to 45°, while the API gravities of gas condensates range between 50° and 70°API.

The FVF of a natural gas (B_g) relates the volume of 1 lb-mol of gas at reservoir conditions to the volume of the same lb-mol of gas at standard conditions:

$$B_g = (\text{volume of 1 unit of gas at reservoir conditions}) / (\text{volume of 1 unit of gas, SCF})$$

Those volumes are, evidently, the specific molar volumes of the gas at the given conditions. The FVF of a condensate (B_o) relates the volume of 1 lb-mol of liquid at reservoir conditions to the volume of that liquid once it has gone through the surface separation facility:

$$B_o = \text{(volume of 1 unit of liquid at reservoir conditions)/} \\ \text{(volume of 1 unit of liquid after separation)}$$

The oil FVF can be also seen as the volume of reservoir fluid required to produce one barrel of oil in the stock tank.

Surface tension is a measure of the surface free energy of liquids, i.e. the extent of energy stored at the surface of liquids. Although it is also known as interfacial force or interfacial tension, the name *surface tension* is usually used in systems where the liquid is in contact with gas. Qualitatively, it is described as the force acting on the surface of a liquid that tends to minimize the area of its surface, resulting in liquids forming droplets with spherical shape, for instance. Quantitatively, since its dimension is of force over length (lbfft in English units), it is expressed as the force (in lbf) required to break a film of 1 ft of length. Equivalently, it may be restated as being the amount of surface energy (in lbf-ft) per square feet.

Fluid viscosity is a measure of its internal resistance to flow, and, thus, viscosity is an indicator of flow properties of the condensate. The most commonly used unit of viscosity is the centipoise, which is related to other units as follows:

$$1c_p = 0.01\,P = 0.000672\,\text{lbm/ft–s} = 0.001\,\text{Pa–s}$$

Natural gas viscosity is typically expected to increase both with pressure and temperature (Lee et al., 1966).

Finally, a comment on the sampling and analysis of gas condensate that invokes similar principles to this used in the sampling of gas.

Before a field development starts, the primary goal of sampling is to obtain representative samples (Chapter 5) of any of the fluids (including gas and gas condensate) found in the reservoir at initial conditions. It may be difficult to obtain a representative sample because of two-phase flow effects near the wellbore. This occurs when a well is produced with the flowing bottomhole pressures below the saturation pressure of the reservoir fluids. It is also commonly thought that nonrepresentative samples of reservoir are the result if gas coning (or liquid coning) occurs during sampling.

Briefly, coning is a production problem in which gas cap gas or bottom water infiltrates the perforation zone in the near-wellbore area and reduces oil production. Gas coning is distinctly different from, and should not be confused with, free gas production caused by a naturally expanding gas cap. Likewise, water coning should not be confused with water production caused by a rising water/oil contact from water influx. Coning is a rate-sensitive phenomenon generally associated with high producing rates. It is typically a near-wellbore phenomenon and only develops once the pressure forces drawing fluids toward the wellbore overcome the natural buoyancy forces that segregate gas and water from oil.

The most representative *in situ* samples are usually obtained when the reservoir fluid is single phase at the point of sampling, be it bottomhole or at the surface. Even this condition, however, may not ensure representative sampling. And samples obtained during gas coning in an oil well can provide accurate *in situ* representative samples if a proper laboratory procedure is followed (Fevang and Whitson, 1994).

Because reservoir fluid composition can vary horizontally, between fault blocks, and as a function of depth, it is essential that a sample of reservoir fluid is obtained that is representative of the volume being drained by the well during the test. Unfortunately, the concept of a representative sample is typically *a sample that correctly reflects the composition of reservoir fluid at the depth or depths being tested.* If there is any doubt that a sample is not representative (according to the above definition), then it is often the best course of action to not use the sample. If such a sample is used for analysis, the validity of the PVT analysis done on the nonrepresentative sample and, consequently, the data should not be used when developing the PVT model.

10.5 Test Methods

The following is a review of applicable test methods that can be used to measure some of the key parameters in the range of hydrocarbon derivative that occur in gas condensate and natural gasoline. By way of clarification, in many of the test methods, the terms *gas condensate* and *natural gasoline* can be substituted in place of *naphtha.*

Because of the high standards set for naphtha and the high standards required (McCann, 1998), it is also essential to employ the same high-standard techniques to gas condensate and natural gasoline when taking samples for testing (ASTM D4057, 2017). ASTM D4057 (2017) covers procedures and equipment for manually obtaining samples of liquid petroleum, and petroleum products, crude oils, and intermediate products from the sample point into the primary container are described. The sampling of gas condensate and natural gasoline may fit into this test method, provided the potential high volatility of the gas condensate and natural gasoline is recognized by the investigator and the analyst. There are inherent limitations when performing any type of sampling, any one of which may affect the representative nature of the sample. As examples, a spot sample (Chapter 5) provides a sample from only one particular point in the tank, vessel compartment, or pipeline. In the case of running or all-level samples, the sample only represents the column of material from which it was taken.

However, based on the product and testing to be performed, ASTM D4057 (2017) provides guidance on sampling equipment, container preparation, and manual sampling procedures for petroleum and petroleum products of a liquid, semiliquid, or solid state, from the storage tanks, flow lines, pipelines, marine vessels, process vessels, drums, cans, tubes, bags, kettles, and open discharge streams into the primary sample container. However, mishandling of the sample or the slightest trace of contaminant in the sample can give rise to misleading results. Special care is necessary to ensure that containers are scrupulously clean and free from odor. Samples should be taken with the minimum of disturbance so as to avoid loss of volatile components; in the case of low-boiling naphtha, it may be necessary to chill the sample. And while awaiting examination samples should be kept in a cool dark place so as to ensure that they do not lose volatile constituents or discolor and develop odors due to oxidation.

The physical properties of gas condensate and natural gasoline depend, as should be expected, on the types of hydrocarbon derivatives present in these liquids. In general, the aromatic hydrocarbon derivatives have the highest solvent power and the straight-chain aliphatic compounds the lowest. The solvent properties can be assessed by estimating the amount of the various hydrocarbon types present. This method provides an indication of the solvent power of the naphtha on the basis that aromatic constituents and naphthenic constituents provide dissolving ability that paraffinic constituents do not. Another method for assessing the solvent properties of naphtha measures the performance of the fraction when used as a solvent under specified conditions, for example, by the kauri-butanol test method (ASTM D1133, 2017).

ASTM D1133 (2017) is used to determine the kauri-butanol value, which is used as a measure of solvent power of hydrocarbon liquids – a high kauri-butanol values indicate relatively strong solvency. However, caution is advised in the use of this methods since the method is specifically designed for application to hydrocarbon liquids that have an initial boiling point over 40 °C (104 °F) and a dry point less than 300 °C (572 °F), which may disqualify most gas condensate sample but be allowable only for higher-boiling natural gasoline samples.

Other methods that are applicable to hydrocarbon liquids typically involve determination of the surface tension from which the solubility parameter is calculated and then provides an indication of dissolving power and compatibility. A similar principle is applied to determine the amount of insoluble material in lubricating oil using *n*-pentane (ASTM D893, 2017; ASTM D4055, 2017) and can be applied to liquid fuels. The insoluble constituents measured can also assist in evaluating the performance characteristics of a liquid fuel in determining the cause of equipment failure and line blockage (chapter 13, Speight and Exall, 2014). One test method (ASTM D893, 2017) covers the determination of pentane and toluene insoluble constituents in used lubricating oils using pentane dilution and centrifugation as the method of separation. Another test method (ASTM D4055, 2017) uses pentane dilution followed by membrane filtration to remove insoluble constituents that have a size greater than 0.8 μm.

Finally, for hydrocarbon liquids, solvent tests are generally performed to ensure the quality of a given product as supplied by the producer to the consumer. In this case the purpose would be to provide data related to the properties of gas condensate and natural gasoline to the refiner and the potential benefits or adverse effects of using gas condensate or natural gasoline as blending stock in the refinery. Many solvent tests are of a somewhat empirical nature such as aniline point and mixed aniline point (ASTM D611, 2017) as well as the kauri-butanol number (ASTM D1133, 2017). The data from these test methods and the need for the test methods are frequently cited in specifications and serve a useful function as control tests. Solvent purity, however, is typically monitored mainly by GC, with individual nonstandardized tests routinely being used by the associated industry. One method that resulted from health concerns and the need to reduce the benzene content of a hydrocarbon liquid is ASTM D4367 (2017).

10.5.1 Aniline Point and Mixed Aniline Point

The aniline point for a hydrocarbon fraction is defined as the minimum temperature at which equal volumes of liquid hydrocarbon and aniline are miscible. Aniline ($C_6H_5NH_2$) is an aromatic compound with a structure of a benzene molecule where one atom of hydrogen is replaced by an amino group ($-NH_2$). The aniline point is an important parameter (ASTM D611, 2017) that can be used for the characterization of petroleum fractions and analysis of molecular type and in the analysis of mixtures of hydrocarbon derivatives. ASTM D611 (2017) includes subtest methods, each of which is suitable for the conditions under which it is to be applied, none of which may be applicable to gas condensate samples, but the potential exists for application of a subtest method (possible Method A) to natural gasoline. This test method also covers the determination of the mixed aniline point of petroleum products and hydrocarbon solvents having aniline points below the temperature at which aniline will crystallize from the aniline–sample mixture.

Test Method A (ASTM D611, 2017) is suitable for transparent samples with an initial boiling point above room temperature and where the aniline point is below the bubble point and above the solidification point of the aniline–sample mixture. Test Method B, a thin-film method, is suitable for samples too dark for testing by Test Method A. Test Methods C and D are for samples that may vaporize appreciably at the aniline point. Test Method D is particularly suitable where only small quantities of sample are available. Test Method E describes a procedure using an automatic apparatus suitable for the range covered by Test Methods A and B.

Within a hydrocarbon group, the aniline point increases with MW or carbon number, but for the same carbon number, the aniline point increases from aromatic hydrocarbon derivatives to paraffin hydrocarbon derivatives – generally, oils with higher aniline points have lower aromatic content. Aromatic derivatives have a relatively low aniline point in comparison with paraffin derivatives since aniline ($C_6H_5NH_2$) is an aromatic compound and has better miscibility with aromatic hydrocarbon derivatives. The aniline points of cycloparaffin derivatives (naphthene derivatives) and olefin derivatives tend to fall between the two extremes of the aniline points of aromatic hydrocarbon derivatives the aniline points of paraffin hydrocarbon derivatives.

10.5.2 Benzene and Aromatic Derivatives

The accurate measurements of benzene and total aromatic derivatives in gasoline are regulated test parameters in modern solvents and liquid fuels (ASTM D3606, 2017), and since gas condensate and natural gasoline are used as gasoline blending stocks, the same applies to these two products of natural gas production. ASTM D3606 (2017) covers the determination of benzene and toluene in low-boiling hydrocarbon liquids by GC. Benzene can be determined between the levels of 0.1 and 5 vol.% and toluene can be determined between the levels of 2 and 20% v/v. An Appendix, in the form of a modified test method, provides an option for modifying the test method for analyzing samples containing ethanol. The modified method uses a different internal standard and a different set of gas chromatographic columns that gives better separation of ethanol- or methanol-containing fuels.

Another test method (ASTM D5580, 2017) that is used to determine benzene, toluene, ethylbenzene, xylene isomers, and higher-boiling aromatic derivative as well as the total concentration of aromatic derivative in the sample uses GC that has been developed to include liquid fuels that contain commonly encountered alcohols and ethers. In the method, the aromatic hydrocarbon derivatives are separated without interferences from other hydrocarbons in the liquid sample. Nonaromatic hydrocarbon derivatives having a boiling point greater than *n*-dodecane (which may not be relevant when the relatively low-boiling gas condensate and natural gasoline are the samples under investigation) may cause interferences with the determination of the C_9 and higher MW aromatic derivatives. For the C_8 aromatic derivatives, *p*-xylene and *m*-xylene coelute, while ethylbenzene and *o*-xylene are separated. Any C_9 and higher-boiling aromatic derivatives are determined as a single group. This test method covers the following concentration ranges, in liquid volume %, for the aromatic derivatives: (i) benzene, 0.1–5% v/v; (ii) toluene, 1–15% v/v; (iii) individual C_8 aromatic derivatives, 0.5–10% v/v; (iv) total C_9 and higher-boiling aromatic derivatives, 5–30% v/v; and (v) total aromatic derivatives, 10–80% v/v.

In addition to GC and the instrumental methods cited above, the so-called hyphenated analytical instrumental methods (ASTM D5769, 2017) that are used to determine benzene, toluene, and total aromatic derivatives in liquid samples such as gas GC-MS and gas chromatography–Fourier transform infrared spectroscopy (GC/FTIR) are also available for accurate measurement of benzene in any liquid sample (DiSanzo and Giarrocco, 1988; De Bakker and Fredericks, 1995; ASTM D5986, 2017).

In the first test method (ASTM D5769, 2017), the determination of benzene, toluene, other specified individual aromatic compounds, and total aromatics in finished gasoline-type hydrocarbon liquids is covered. The method employs GC-MS. The method can be applied to hydrocarbon liquids that contain the analytes in the following concentrations: (i) benzene, 0.1–4% v/v; (ii) toluene, 1–13%; and (iii) total (C_6–C_{12}) aromatics, 10–42% v/v. Another test method (ASTM D5986, 2017) used to determine oxygenates, benzene, and the aromatic content of gasoline is necessary to assess product quality and to meet new fuel regulations. This test method can be used for hydrocarbon liquid mixtures that contain oxygenates (alcohols and ethers) as additives. It has been determined that the common oxygenates found in finished gasoline do not interfere with the analysis of benzene and other aromatics by this test method. The method (or an adaption of the method suitable to the properties of the sample) is not typically applicable to gas condensate or natural gasoline unless oxygenated organic derivatives have been added to the production well as an aid to recovery of the natural gas, the gas condensate, and the natural gasoline. The analyst should ascertain any such additives when receiving the sample from a member of the production team.

A test method using flow injection analysis (FIA) (an approach that is accomplished by injecting a plug of sample into a flowing carrier stream) (ASTM D1319, 2017) has traditionally been used to measure aromatic derivatives as well as olefins and saturates in low-boiling liquids. ASTM D1319 (2017) covers the determination of hydrocarbon derivatives over the concentration ranges from (i) 5 to 99% v/v aromatic derivatives, (ii) 0.3 to 55% olefins olefin derivatives, and (iii) 1 to 95% v/v saturates in hydrocarbon liquid mixtures that distill below 315 °C (599 °F). The method may apply to concentrations outside these ranges, and liquid samples (such as higher-boiling natural gasoline) that contain dark-colored components that interfere in reading the chromatographic bands cannot be analyzed.

The determination of the total volume percent of saturates, olefins, and aromatics in petroleum fractions is important in characterizing the quality of petroleum fractions as gasoline blending components and as feeds to catalytic reforming processes. This information is also important in characterizing petroleum fractions and products from catalytic reforming and from thermal and catalytic cracking as blending components for motor and aviation fuels. This information is also important as a measure of the quality of fuels, such as specified in another standard test method (ASTM D1655, 2017). However, ASTM D1655 (2017) describes the minimum property requirements for Jet A and Jet A-1 aviation turbine fuel and lists acceptable additives for use in civil operated engines and aircrafts. Thus, it may not be applicable to gas condensate or natural gasoline, and the analyst should consider all of the possible properties of gas condensate or natural gasoline before applying this test method to the sample.

10.5.3 Composition

The number of potential hydrocarbon isomers in the naphtha boiling range (Tables 10.5 and 10.6) renders complete speciation of individual hydrocarbon derivatives impossible for the naphtha distillation range, and methods are used that identify the hydrocarbon types as chemical groups rather than as individual constituents. However, it must be recognized that composition varies with depth in petroleum reservoirs (Speight, 2014a) and component "segregation" due to gravitational forces is usually given as the physical explanation for the variation in composition. The result of gravitational segregation is that a gas condensate gets richer at greater depths, with increasing C_{7+} mole fraction (and dew-point pressure) (Whitson and Belery, 1994). However, not all fields show compositional gradients with depth as predicted by the isothermal model. Some fields show practically no gradient over large depths, while other oil fields have gradients larger than predicted by the isothermal model (Høier and Whitson, 1998). Variation in C_{7+} composition with depth will obviously affect the calculation of initial surface condensate in place, compared with a calculation based on a constant composition.

The accurate determination of the density, relative density (specific gravity), or API gravity of petroleum and its products is necessary for the conversion of measured volumes to volumes or masses, or both, at the standard reference temperatures of 15 °C or 60 °F during custody transfer. Thus, the data from the density (specific gravity) test method (ASTM D1298, 2017) provides a means of identification of a grade of naphtha but is not a guarantee of composition and can only be used to indicate and evaluate product composition or quality when used in conjunction with the data from other test methods. Density data are used primarily to convert naphtha volume to a weight basis, a requirement in many of the industries concerned. For the necessary temperature corrections and also for volume corrections, the appropriate sections of the petroleum measurement tables (ASTM D1250, 2017) are used.

ASTM D1298 (2017) is most suitable for determining the density, relative density (specific gravity), or API gravity of low viscosity transparent liquids. In addition, the method can also be used for viscous liquids (not a relevant issue when the method is applied to gas condensate or natural gasoline) by allowing sufficient time for the hydrometer to reach temperature equilibrium and for opaque liquids by employing a suitable meniscus correction. Additionally, for both transparent and opaque fluids, the readings shall be corrected for the thermal glass expansion effect and alternative calibration temperature effects before correcting to the reference temperature.

The method (ASTM D1298, 2017) covers the laboratory determination using a glass hydrometer in conjunction with a series of calculations of the density, relative density, or API gravity of hydrocarbon liquids that are typically handled as liquids and having a Reid vapor pressure of 101.325 kPa (14.696 psi) or less. Values are determined at existing temperatures and corrected to 15 °C or 60 °F by means of a series of calculations and international standard tables.

The initial hydrometer readings obtained are uncorrected hydrometer readings and not density measurements. Readings are measured on a hydrometer at either the reference temperature or at another convenient temperature, and readings are corrected for the meniscus effect, the thermal glass expansion effect, and alternative calibration temperature effects and to the reference temperature by means of the petroleum measurement tables. Data determined as density, relative density, or API gravity can be converted to equivalent values in other units or alternative reference temperatures by means of interconversion procedures (ASTM D1250, 2017).

The first level of compositional information is group-type totals as deduced by adsorption chromatography (ASTM D1319, 2017) to give volume-percent saturates, olefins, and aromatic derivatives in materials that boil below 315 °C (600 °F). In ASTM D1319 (2017), a small amount of sample is introduced into a glass adsorption column packed with activated silica gel, of which a small layer contains a mixture of fluorescent dyes. When the sample has been adsorbed on the gel, alcohol is added to desorb the sample down the column and the hydrocarbon constituents are separated according to their affinities into three types (aromatic derivatives, olefins, and saturates). The fluorescent dyes also react selectively with the hydrocarbon types and make the boundary zones visible under ultraviolet light. The volume percentage of each hydrocarbon type is calculated from the length of each zone in the column.

There are other test methods available. Benzene content and other aromatic derivatives may be estimated by spectrophotometric analysis and also by gas–liquid chromatography. However, two test methods based on the adsorption concept (ASTM D2007, 2017; ASTM D2549, 2017) are used for classifying liquid samples of initial boiling point of at least 200 °C (392 °F) into the hydrocarbon types of polar compounds, aromatic derivatives, and saturates and recovery of representative fractions of these types. Such methods are unsuitable for the gas condensate sample, natural gasoline sample, and majority of naphtha samples because of volatility constraints.

A standard test method (ASTM D2427, 2017) has been designed to determine the ethane (C_2) to pentane (C_5) hydrocarbon content in gasoline and low-boiling hydrocarbon liquids. However, this test method does not cover the determination of cyclic olefins, diolefins, or acetylenes. These are usually minor components in finished gasolines. Samples to be analyzed should not contain significant amounts of material boiling lower than ethylene. On the other hand, assuming the applicability of the method to the sample in hand, the method validates samples from which the pentane derivatives have been removed (ASTM D2001, 2017). The method is also suitable to application to condensate and low-boiling naphtha. The method is a mature method that utilizes packed columns and mechanical valves for the backflush and fore-flush techniques needed to separate the volatile hydrocarbons from the rest of the gasoline sample. However, a sample containing significant concentrations of nonhydrocarbon additives (such as ethanol) may interfere with the existing method, and there may be the need to develop an alternate procedure to determine the low-boiling hydrocarbon derivatives.

The combination of modern capillary columns leads to excellent separation of the ethane to pentane aliphatic hydrocarbons, which is evident with no apparent interference from higher hydrocarbons, aromatics, or oxygenated compounds. Even though the method relies on significant carrier gas pressure and flow rate changes, the quantitative and peak retention time precisions are excellent. The method will also determine the levels of methane, if present (in other sample matrices), and the method would be suitable for the analysis of other sample types in which volatile aliphatic hydrocarbons are to be monitored in a matrix containing higher hydrocarbons, aromatics, and/or polar components. Finally, the method could be extended to allow the determination of higher MW hydrocarbon derivatives by increasing the time of the backflush point.

An indication of composition may also be obtained from the determination of aniline point (ASTM D611, 2017), freezing point (ASTM D852, 2017; ASTM D1015, 2017), cloud point (ASTM D2500, 2017), and solidification point or freezing point (ASTM D852, 2017; ASTM D1015, 2017). In the application of test method ASTM D852 (2017), the analyst should be aware that this test method is to be used for determining the purity of benzene. The closer the solidification point reaches that of pure benzene, the purer the sample. The freezing point measurement (ASTM D1015, 2017), when used in conjunction with the physical constants for the relevant hydrocarbon derivative (ASTM D1016, 2017), allows the determination of the purity of the test sample. The parameters of the test method do not allow general application of the method to bulk sample of gas condensate sample or natural gasoline sample. However, the parameters do allow purity checks (through the freezing point) for any individual sample separated from gas condensate or from natural gasoline. A knowledge of the purity of these hydrocarbons is often needed to help control their manufacture and to determine their suitability for use as reagent chemicals or for conversion to other chemical intermediates or finished products.

The alkalinity and acidity (ASTM D847, 2017; ASTM D1093, 2017; ASTM D1613, 2017; ASTM D2896, 2017) as well as olefin content should be low in all cases, but there is no guarantee that this will be the case. Gas condensate or natural gasoline separated from crude oil wells may show vestiges of acidity or alkalinity or olefin content. In addition, gases produced from other sources (refinery process gas coal gas, biogas, landfill gas, and fuel gas) may show signs of acidity or alkalinity or olefin content. Relevant tests using bromine number (ASTM D1159, 2017), bromine index (ASTM D2710, 2017), and flame ionization absorption (ASTM D1319, 2017) are necessary to ensure low levels (at the maximum) of hydrogen sulfide as well as the sulfur compounds in general (ASTM D130, 2017; ASTM D849, 2017; ASTM D1266, 2017; ASTM D3120, 2017; ASTM D4045, 2017) and especially corrosive sulfur compounds such as that are determined by the doctor test method (ASTM D4952, 2017).

Aromatic constituents can be present in gas condensate and natural gasoline, and because the aromatic constituents influence a variety of properties including boiling range (ASTM D86, 2017), viscosity (ASTM D88, 2017; ASTM D445, 2017; ASTM D2161, 2017), stability (ASTM D525, 2017), and compatibility (ASTM D1133, 2017) with a variety of solutes, the relevant test methods may need to be applied. Tests such as aniline point (ASTM D611, 2017) and kauri-butanol number (ASTM D1133, 2017) are of a somewhat empirical nature and can serve a useful function as control tests. However, gas condensate composition and natural gasoline composition are monitored mainly by GC, and although most of the methods may have been developed for gasoline or naphtha (ASTM D2427, 2017; ASTM D6296, 2017), the applicability of the test methods to gas condensate composition and natural gasoline is in order.

A multidimensional gas chromatographic method (ASTM D5443, 2017) provides for the determination of paraffin derivatives, naphthene derivatives, and aromatic derivatives by carbon number in low olefin-type hydrocarbon streams having final boiling points lower than 200 °C (392 °F). This test method (ASTM D5443, 2017) is applicable to hydrocarbon refinery streams for process control and quality assurance. The method is also applicable to a variety of hydrocarbon mixtures including virgin, catalytically converted, thermally converted, alkylated, and blended naphtha. Hydrocarbon mixtures with boiling points greater than 200 °C and less than 270 °C (392–518 °F) are reported as a single group. Olefin derivatives, if present, should be hydrogenated, and the resultant saturates are included in the paraffin and naphthene distribution. Aromatics boiling at C_9 and above are also reported as a single aromatic group. However, this test method, in contrast to other gas chromatographic-based test methods, is not intended to determine individual components except for benzene and toluene that are the only C_6 and C_7 aromatics, respectively, and cyclopentane that is the only C_5 naphthene derivative. The lower limit of detection for a single hydrocarbon component or group is 0.05% w/w. The scope of this test method (ASTM D5443, 2017) excludes hydrocarbon derivatives containing oxygenates. However, an extended version of the method can be used to measure paraffin derivatives, isoparaffin derivatives, olefin derivatives, naphthene derivatives, and aromatic derivatives (PIONA analysis) in low-boiling hydrocarbon liquids (DiSanzo and Giarrocco, 1988).

In the method ASTM D5443 (2017), the sample is injected into a gas chromatographic system that contains a series of columns and switching values. First, a polar column retains polar aromatic compounds, binaphthene derivatives, and high-boiling paraffin derivatives and naphthene derivatives. The eluate from this column goes through a platinum column that hydrogenates olefins and then to a molecular sieve column that performs a carbon number separation based on the molecular structure, that is, naphthene derivatives and paraffin derivatives. The fraction remaining on the polar column is further divided into three separate fractions that are then separated on a nonpolar column by boiling point. A flame ionization detector detects eluting compounds.

Other methods including ASTM D3257 (2017) for the determination of the amount of aromatic constituents include various types of detectors and offer alternate routes to determining aromatic derivatives in naphtha. However, the method ASTM D3257 (2017) covers the determination of ethylbenzene and total eight-carbon (C_8) and higher-boiling aromatic derivatives in the concentration range from 0.1 to 30% v/v in hydrocarbon liquids having a distillation range from 149 to 210 °C (300–410 °F) as determined by the standard test method ASTM D86 (2017). The method is only applicable to the identification of any higher-boiling constituents present in natural gasoline – gas condensate frequently does not contain such high-boiling constituents.

Hydrocarbon composition, relevant to gas condensate and natural gasoline, is determined by mass spectrometry – a technique that has been widely used for hydrocarbon-type analysis of naphtha and gasoline (ASTM D2789, 2017) as well as to the identification of hydrocarbon constituents in higher-boiling naphtha fractions (ASTM D2425, 2017).

Another method (ASTM D2789, 2017) allows condensate, naphtha, and gasoline samples to be submitted for functional group hydrocarbon analysis by mass spectrometry. This test method covers the determination by mass spectrometry of the total paraffin derivatives, monocycloparaffin derivatives, dicycloparaffin derivatives (unlikely in gas condensate and natural gasoline), alkylbenzene derivatives, and other higher MW cyclo-derivatives that are also unlikely in gas condensate and natural gasoline, provided the sample has an olefin content of less than 3% v/v and a 95% v/v distillation point of less than 210 °C (411 °F) (ASTM D86, 2017). Olefin derivatives can be determined by application of alternate test methods (ASTM D1319, 2017), which are suitable for samples of low-boiling hydrocarbon mixtures in which low-boiling hydrocarbon derivatives are to be determined in samples containing significant low volatility or polar components in the matrix. The method is also suitable to determine methane content of such mixtures and can be extended to include higher MW hydrocarbon derivatives.

On the other hand, ASTM D2425 (2017) covers an analytical scheme using the mass spectrometer to determine the hydrocarbon types present in virgin middle distillates at 204–343 °C (400–650 °F) boiling range, 5–95% v/v. Samples with average carbon number value of paraffins between C_{12} and C_{16} and containing paraffins from C_{10} to C_{18} can be analyzed. Generally, the method is not suitable for samples of low-boiling hydrocarbon mixtures in which low-boiling hydrocarbon derivatives are to be determined in samples containing significant low volatility or in the mixture. Under the parameters described, the method would require considerable modification for application to gas condensate and natural gasoline unless each sample contained a substantial amount of a higher-boiling (middle distillate-type) residue.

One method (ASTM D6379, 2017) is available that can be used to determine the monoaromatic and diaromatic hydrocarbon contents in distillates boiling in the range from 50 to 300 °C (122–572 °F). This test method covers a high-performance liquid chromatographic test method for the determination of monoaromatic and diaromatic hydrocarbon contents in hydrocarbon liquids such as distillates boiling in the range from 50 to 300 °C (122–572 °F) and may be applicable (with some modification) to gas condensate and natural gasoline. The total aromatic content is calculated from the sum of the individual aromatic hydrocarbon types. This test method is calibrated for hydrocarbon liquids containing from 10 to 25% w/w monoaromatic hydrocarbons and from 0 to 7% w/x diaromatic hydrocarbons. Constituents containing sulfur, nitrogen, and oxygen (which are not frequent constituents of gas condensate and natural gasoline) are possible interferents. Monoalkenes (as may be present in refinery gas, coal gas, biogas, landfill gas, and fuel gas) do not interfere, but conjugated di- and polyalkenes, if present, are possible interferents.

In the method ASTM D6379 (2017), the sample is diluted with an equal volume of hydrocarbon, such as heptane, and a fixed volume of this solution is injected into a high-performance liquid chromatograph fitted with a polar column where separation of the aromatic hydrocarbon derivatives from the nonaromatic hydrocarbon derivatives occurs. The separation of the aromatic constituents appears as distinct bands according to ring structure, and a refractive index detector is used to identify the components as they elute from the column. The peak areas of the aromatic constituents are compared with those obtained from previously run calibration standards to calculate the %w/w monoaromatic hydrocarbon constituents and diaromatic hydrocarbon constituents in the sample.

Another method (ASTM D2425, 2017) provides more compositional detail (in terms of molecular species) than chromatographic analysis, and the hydrocarbon types are classified in terms of a Z-series in which Z (in the empirical formula C_nH_{2n+z}) is a measure of the hydrogen deficiency of the compound. This method requires that the sample be separated into saturate and aromatic fractions before mass spectrometric analysis (ASTM D2549, 2017) and the separation is applicable to some fractions but not to others. For example, the method is applicable to high-boiling naphtha but not to the low-boiling naphtha since it is impossible to evaporate the solvent used in the separation without also losing the lower-boiling constituents of the naphtha under investigation.

The percentage of aromatic hydrogen atoms and aromatic carbon atoms can be determined by high-resolution NMR spectroscopy (ASTM D5292, 2017) that gives the mole percent of aromatic hydrogen or carbon atoms. Proton (hydrogen) magnetic resonance spectra are obtained on sample solutions in either chloroform or carbon tetrachloride using a CW or pulsed FT high-resolution magnetic resonance spectrometer. Carbon magnetic resonance spectra are obtained on the sample solution in chloroform-*d* using a pulsed FT high-resolution magnetic resonance. This test method covers the determination of the aromatic hydrogen content (procedures A and B) and aromatic carbon content (procedure C) of hydrocarbon oils using high-resolution NMR spectrometers. For pulsed FT spectrometers, the detection limit is typically 0.1 mol% aromatic hydrogen atoms and 0.5 mol% aromatic carbon atoms. For CW spectrometers, which are suitable for measuring aromatic hydrogen contents only, the detection limit is considerably higher, typically 0.5 mol% aromatic hydrogen atoms.

The determination of hydrocarbon components in petroleum naphtha is of great importance to the petrochemical industry and for process control of reforming processes, as well as for regulatory purposes. The data obtained by NMR (ASTM D5292, 2017) can be used to evaluate changes in aromatic contents in naphtha as well as kerosene, gas oil, mineral oil, and lubricating oil. However, results from this test are not equivalent to mass-percent or volume-percent aromatic derivatives determined by the chromatographic methods since the chromatographic methods determine the percent by weight or percent by volume of molecules that have one or more aromatic rings and alkyl substituents on the rings will contribute to the percentage of aromatic derivatives determined by chromatographic techniques.

Low-resolution NMR spectroscopy can also be used to determine percent by weight hydrogen in jet fuel (ASTM D3701, 2017) and in light distillate, middle distillate, and gas oil (ASTM D4808, 2017). As noted above, chromatographic methods are not applicable to naphtha where losses can occur by evaporation.

ASTM D3701 (2017) is focused on the combustion quality of aviation turbine fuel that has traditionally and may be only marginally suitable for application to gas condensate and natural gasoline. Typically, the test method has been applied to specifications along with such test methods that are relevant to the determination of the smoke point (ASTM D1322, 2017), smoke volatility index, and aromatic content of luminometer number. Other test methods (ASTM D4808, 2017 or ASTM D7171, 2017) are more suitable for the determination of hydrogen in petroleum liquids.

As an example, ASTM D4808 (2017) covers the determination of the hydrogen content of petroleum products ranging from atmospheric distillates to vacuum residua using a CW low-resolution NMR spectrometer. One of the submethods (Method A) is used for hydrocarbon liquids with the boiling range of 15–260 °C (59–500 °F) and is, therefore, suitable for gas condensate and natural gasoline (if the determination of any lower-boiling material can be accolated by another standard test method or by stabilization of the sample at 15 °C).

Although the focus of many tests is on the analysis of the hydrocarbon constituents of naphtha and other petroleum fractions, heteroatom compounds that contain sulfur and nitrogen atoms cannot be ignored, and methods for their determination are available. The combination of GC with element selective detection gives information about the distribution of the element. In addition, many individual heteroatom compounds can be determined. Thus, estimates of the purity of these products were determined in laboratories using a variety of procedures such as freezing point, flame ionization absorbance, ultraviolet absorbance, GC, and capillary GC (ASTM D850, 2017; ASTM D852, 2017; ASTM D848, 2017; ASTM D849, 2017; ASTM D1015, 2017; ASTM D1016, 2017; ASTM D1078, 2017; ASTM D1319, 2017; ASTM D2008, 2017; ASTM D2360, 2017; ASTM D5134, 2017; ASTM D5917, 2017).

GC has become a primary technique for determining hydrocarbon impurities in individual aromatic hydrocarbon derivatives and the composition of mixed aromatic hydrocarbon derivatives. Although a measure of purity by GC is often sufficient, GC is not capable of measuring absolute purity; not all possible impurities will pass through the GC column, and not all those that do will be measured by the detector. Despite some shortcomings, GC is a standard, widely used technique and is the basis of many current test methods for aromatic hydrocarbon derivatives (ASTM D2360, 2017; ASTM D3797, 2017; ASTM D4492, 2017; ASTM D4735, 2017; ASTM D5060, 2017; ASTM D5135, 2017; ASTM D5713, 2017; ASTM D5917, 2017; ASTM D6144, 2017).

Impurities other than hydrocarbon derivatives are of concern in the petroleum industry. For example, many catalytic processes are sensitive to sulfur contaminants. Consequently, there is also a series of methods to determine trace concentrations of sulfur-containing compounds (ASTM D4045, 2017; ASTM D4735, 2017). When classes of hydrocarbon derivatives, such as olefins, need to be measured, techniques such as the bromine index are used (ASTM D1492, 2017; ASTM D5776, 2017). Chloride-containing impurities are determined by various test methods (ASTM D5194, 2017; ASTM D5808, 2017; ASTM D6069, 2017) that have sensitivity to 1 $mg\,kg^{-1}$, reflecting the needs of industry to determine very low levels of these contaminants.

Water is also a potential contaminant in gas condensate and natural gasoline that should be given consideration. Water content can be measured using the Karl Fischer method (ASTM E203, 2017; ASTM D1364, 2017; ASTM D1744, 2017; ASTM D4377, 2017; ASTM D4928, 2017; ASTM D6304, 2017), by distillation (ASTM D4006, 2017), or by centrifuging and excluded by relevant drying methods.

Tests should also be carried out for sediment if the gas condensate or natural gasoline has been subjected to events (such as oxidation) that could lead to sediment formation and instability of the naphtha and the resulting products. Test methods are available for the determination of sediment by extraction (ASTM D473, 2017) or by membrane filtration (ASTM D4807, 2017) and the simultaneous determination of sediment with water by centrifugation (ASTM D1796, 2017; ASTM D2709, 2017; ASTM D4007, 2017). Whether or not these tests can be applied to gas condensate or natural gasoline depends upon the experimental protocols of each individual test method and, obviously, the need for sediment removal from the gas condensate or natural gasoline.

10.5.4 Constant Composition Expansion Test

The constant composition expansion (CCE) test, sometimes referred to as a constant mass expansion test, is used to measure dew-point pressure, single-phase gas Z-factors, and oil relative volume below the dew point ("liquid drop-out curve"). A sample of reservoir fluid is charged in a visual PVT cell and brought to reservoir temperature and a pressure sufficiently high to ensure single-phase conditions.

Pressure is lowered by increasing cell volume until a liquid phase is visually detected (through a glass window). The total cell volume and liquid volume are monitored from the initial reservoir pressure down to a low pressure (dictated by cell and sample size).

10.5.5 Constant Volume Depletion

The CVD test is an extremely important laboratory test that monitors the phase and volumetric changes of a reservoir gas sample (at reservoir temperature) as the pressure drops below the dew point and equilibrium gas phase is removed (Whitson and Torp, 1983). The test simulates closely the actual behavior of a gas condensate reservoir undergoing pressure depletion, and results from the laboratory measurements can be used directly to quantify recoveries of surface gas and condensate as a function of pressure below the dew point. Combined with single-phase Z-factors from the CCE test, a complete prediction of depletion behavior (recoveries and liquid-yield variation) can be accurately predicted from initial pressure to abandonment (Whitson et al., 1999).

The test involves stepwise lowering of the pressure below the dew point, with an associated increase in cell volume. After equilibration at each pressure, enough equilibrium gas is removed from the top of the cell to bring the cell back to the original volume occupied at the dew point. The amount of gas removed, its composition and Z-factor, and the remaining oil volume in the cell are measured and reported. A "two-phase" Z-factor is also reported for use with the "gas" material balance.

Accurate measurement of the removed gas composition is very important to the prediction of condensate recovery and liquid-yield variation – much more important than accurate measurement of retrograde oil volumes. Special laboratory procedures should be followed to ensure accurate CVD compositional measurements (e.g. appropriate heating of tubing used to remove equilibrium gas from the cell). Measurement of the final low-pressure condensate composition allows an important "backward" material balance check.

10.5.6 Correlative Methods

Correlative methods have long been used as a way of dealing with the complexity of various petroleum fractions, including naphtha. Relatively easy-to-measure physical properties such as density (or specific gravity) (ASTM D3505, 2017; ASTM D4052, 2017) are also required. Viscosity (ASTM D88, 2017; ASTM D445, 2017; ASTM D2161, 2017), density (ASTM D287, 2017; ASTM D891, 2017; ASTM D1217, 2017; ASTM D1298, 2017; ASTM D1555, 2017; ASTM D1657, 2017; ASTM D4052, 2017; ASTM D5002, 2017), and refractive index (ASTM D1218, 2017) have been correlated to hydrocarbon composition (Table 10.8).

Using density as the example, density is a fundamental physical property that can be used in conjunction with other properties to characterize the quality of crude oil and other hydrocarbon liquids. The density or relative density of crude oils is used for the conversion of measured volumes to volumes at the standard temperatures of 15 °C (or 60 °F) and for the conversion of crude mass measurements into volume units. The application of the density result obtained

Table 10.8 Refractive index of selected hydrocarbon derivatives.

Compound	Refractive index n_D^{20}
n-Pentane	1.3578
n-Hexane	1.3750
n-Hexadecane	1.4340
Cyclopentane	1.4065
Cyclopentene	1.4224
Pentene-1	1.3714
1.3-Pentadiene	1.4309
Benzene	1.5011
cis-Decahydronaphthalene	1.4814
Methylnaphthalenes	1.6150

from one of the test methods (ASTM D5002, 2017) for fiscal or custody transfer accounting calculations can require measurements of the water and sediment contents obtained on similar specimens of the crude oil parcel. The method ASTM D5002 (2017) covers the determination of the density or relative density of crude oils that can be handled in a normal fashion as liquids at test temperatures between 15 and 35 °C (59 and 95 °F) utilizing either manual or automated sample injection equipment. The method applies to crude oil (and, by inference, hydrocarbon liquids) with a high vapor pressure, provided appropriate precautions are taken to prevent vapor loss during transfer of the sample to the density analyzer.

10.5.7 Density

Density (the mass of liquid per unit volume at 15 °C) and the related terms *specific gravity* (the ratio of the mass of a given volume of liquid at 15 °C to the mass of an equal volume of pure water at the same temperature) and *relative density* (same as *specific gravity*) are important properties of petroleum products as they are a part of product sales specifications, although they only play a minor role in studies of product composition. Usually a hydrometer, pycnometer, or digital density meter is used for the determination in all these standards.

Density is an important parameter for naphtha and solvents, and the determination of density (specific gravity) (ASTM D287, 2017; ASTM D891, 2017; ASTM D1217, 2017; ASTM D1298, 2017; ASTM D1555, 2017; ASTM D1657, 2017; ASTM D4052, 2017; ASTM D5002, 2017) provides a check on the uniformity of the naphtha and permits calculation of the weight per gallon. The temperature at which the determination is carried out and for which the calculations are to be made should also be known and be in accordance with the volatility of the sample as determined by a standard test method (ASTM D86, 2017). Any such method must be subject to vapor pressure constraints and are used with appropriate precautions to prevent vapor loss during sample handling and density measurement. In addition, some test methods should not be applied if the samples are sufficiently colored and that the absence of air bubbles in the sample cell cannot be established with certainty. The presence of such bubbles can have serious consequences on the reliability of the test data.

10.5.8 Detailed Hydrocarbon Analysis

One test method (ASTM D5134, 2017), which produces a detailed analysis of petroleum naphtha through *n*-nonane by capillary GC, is applicable to olefin-free liquid hydrocarbon mixtures including raw naphthas (which include condensate and natural gasoline), reformates, and alkylates. This test method covers the determination of hydrocarbon components paraffins, naphthenes, and monoaromatics (PNA) of petroleum naphthas. Components eluting after *n*-nonane (b.p. 150.8 °C, 303.4 °F) are determined as a single group. This method is applicable to olefin-free (<2% v/v olefins) liquid hydrocarbon mixtures, but the olefin content can be determined by other test methods (ASTM D1319, 2017; ASTM D6839, 2017). To be suitable for the test, the hydrocarbon mixture must be 98% w/w distilled at 250 °C (482 °F) or less as determined by an alternate test method (ASTM D3710, 2017; ASTM D7096, 2017). A detailed distribution of the hydrocarbon components in olefin-containing samples (typically not necessary for gas condensate and natural gasoline but necessary for process gases and other gases produced by thermal processes) may be determined by other test methods (ASTM D6729, 2017; ASTM D6730, 2017; ASTM D6733, 2017).

Higher-resolution GC capillary column techniques are in routine use in petroleum laboratories today to provide a detailed analysis of most of the individual hydrocarbon derivatives in gasoline, including many of the oxygenated blending components. Software is also available that allows one to summarize the data according to hydrocarbon type and predict other parameters such as vapor pressure and distillation from the results. High-resolution GC procedures for the detailed analysis of gasoline are being considered for adoption as standard ASTM standard test methods (ASTM, 2017). Capillary GC techniques can be combined with mass spectrometry to enhance the identification of the individual components and hydrocarbon types (Teng and Williams, 1994).

10.5.9 Dew-point Pressure

The dew-point pressure is the pressure where an incipient liquid phase condenses from a gas phase. Practically, the dew point marks the pressure where (i) reservoir gas-phase composition changes and becomes leaner and (ii) condensate accumulation starts in the reservoir. These two changes can have a profound effect on reservoir and well performance – or they may have little impact.

The importance of the *actual dew-point pressure* will vary from reservoir to reservoir, but in most situations accurate dew-point determination is *not important*. First, in the context of compositional variation with pressure (and associated variation of condensate yield with pressure), accurate determination of the thermodynamic dew-point pressure is *not* of particular importance. In fact, there is no knowledge of the specific dew point at all as long as the variation of composition (C_{7+} content) with pressure is well defined "near" the thermodynamic dew point. Second, when the bottom-hole flowing pressure (BHFP) drops below the dew point and two phases start flowing near the wellbore, gas relative permeability drops and well *productivity* drops.

However, as long as BHFP is "anywhere near" the dew point, the well will have *excess deliverability* – i.e. it is a matter of reducing the BHFP to produce more gas (even though the well *productivity* is lower). Only when the BHFP reaches a minimum value (dictated by some delivery-pressure constraint) will the well no longer be able to deliver the desired rate. At this point, well productivity *is* important. However, this occurs at a BHFP much lower than the dew point.

Another (less common) need for dew-point pressure is when an underlying saturated oil zone *may exist* and a PVT model is used to predict the existence and location of the gas–oil contact (GOC). In this case, the PVT-model dew point should be tuned precisely to an accurately measured dew-point pressure. It is not uncommon that a predicted GOC may vary 10s of meters per bar of uncertainty in the (PVT-model) dew-point pressure. Thus, an accurate description of the dew-point pressure will have an impact on the prediction of initial oil and gas in place, placement of delineation wells, and potential field development strategy. In this situation, accurate dew-point measurement and equally accurate modeling of the measured dew point should be given due attention.

On the other hand, if accurate treatment of the dew-point pressure is *not* required (for estimating a GOC), consideration should be given to using little, if any, weighting of the measured dew-point pressure when tuning the PVT model. Instead, priority should be given to matching the variation of C_{7+} with pressure in the removed gas from a CVD test.

10.5.10 Distillation

The primary method specified for determining boiling range of gasoline continues to be standard test method ASTM D86 (2017). The use of automated instrumentation has been incorporated into the method ASTM D3710 (2017), which allows the method to be used for determining the boiling point properties of oxygenate-free hydrocarbon distillates. This test method (ASTM D3710, 2017) has the advantage that a smaller sample size is required, and the method can be more easily automated, but the data from the method are not directly equivalent to that obtained by distillation (ASTM D86, 2017). In addition, the data from this test method (ASTM D3710, 2017) are being used by some companies and vendors by applying correlations to predict test method (ASTM D86, 2017) distillation data for various refinery streams. Improvements in the GC simulated distillation procedures have been implemented in some laboratories, and improvements include rapid GC techniques using very narrow bore capillary GC columns that will potentially reduce analysis time to a matter of minutes.

10.5.11 Evaporation Rate

The vaporizing tendencies of petroleum products are the basis for the general characterization of liquid petroleum fuels, such as liquefied petroleum gas, natural gasoline, motor and aviation gasoline, naphtha, kerosene, gas oil, diesel fuel, and fuel oil. Standard test methods (ASTM D6, 2017; ASTM D2715, 2017) are available to determine the volatility of higher-boiling products and can, with some cautious modifications, be adapted for lower-boiling products.

ASTM D6 (2017) is useful in characterizing certain hydrocarbon liquids by the determination of the loss in weight when heated under standard conditions using the correct safety and health protocols. Another method (ASTM D2715, 2017) that is often quoted as being suitable for weight-loss-on-heating studies (ASTM D2715, 2017) covers the determination of the rates of volatilization of lubricants in a thermal-vacuum environment at pressures and temperatures necessary to obtain a measurable rate of evaporation, or evidence of decomposition. The method is unsuitable for application to gas condensate and natural gasoline.

For some purposes it is necessary to have information on the initial stage of vaporization and the potential hazards that such a property can cause. To supply this need, flash and fire, vapor pressure, and evaporation methods are available. The data from the early stages of the several distillation methods are also useful (Speight, 2015). For other uses it is important to know the tendency of a product to partially vaporize or to completely vaporize and, in some cases, to know if small quantities of high-boiling components are present. For such purposes, chief reliance is placed on the distillation methods.

Nevertheless, the evaporation rate is an important property of naphtha, and although there is a significant relation between distillation range and evaporation rate, the relationship is not straightforward. A simple procedure for determining the evaporation rate involves the use of at least a pair of weighed shallow containers, each containing a weighed amount of naphtha. The cover-free containers are placed in a temperature-controlled and humidity-controlled draft-free area. The containers are reweighed at intervals until the samples have completely evaporated or left a residue that does not evaporate further (ASTM D381, 2017; ASTM D1353, 2017). The evaporation rate can be derived either (i) by a plot of time versus weight using a solvent having a known evaporation rate for comparison or (ii) from the distillation profile (ASTM D86, 2017).

The above test method (ASTM D381, 2017) covers the determination of the existent gum content of aviation fuels and the gum content of motor gasolines or other volatile distillates in their finished form, including those containing alcohol and ether-type oxygenates and deposit control additives. Provisions are made for the determination of the heptane insoluble portion of the residue of nonaviation fuels. This test method would only be suitable for application to gas condensate and natural gasoline if suitable modifications were made to the test protocols as the necessary safety precautions were addressed. Similarly, another test method (ASTM D1353, 2017) covers the analytical measurement of residual matter in solvents that are intended to be 100% volatile at 105 ± 5 °C (221 ± 9 °F). Volatile solvents are used in the manufacture of paint, varnish, lacquer, and other related products, and the presence of any residue may affect the product quality or efficiency of the process. This test method (again with suitable modifications and safety concerns addressed) may be modifiable for application to gas condensate and natural gasoline.

10.5.12 Flash Point

The flash point is the lowest temperature at atmospheric pressure (760 mmHg, 101.3 kPa) at which application of a test flame will cause the vapor of a sample to ignite under specified test conditions. The sample is deemed to have reached the flash point when a large flame appears and instantaneously propagates itself over the surface of the sample. The flash point data is used in shipping and safety regulations to define *flammable* and *combustible* materials. This can also indicate the possible presence of highly volatile and flammable constituents in a relatively nonvolatile or nonflammable material. Since the flash point of gas condensate and the flash point of natural gasoline are low, such test method can also indicate the possible presence of even more highly volatile and flammable constituents in these two liquids.

Flash point for a hydrocarbon or a fuel is the minimum temperature at which vapor pressure of the hydrocarbon is sufficient to produce the vapor needed for spontaneous ignition of the hydrocarbon with the air with the presence of an external source, i.e. spark or flame. From this definition, it is clear that hydrocarbon derivatives with higher vapor pressures (lighter compounds) have lower flash points. Generally, flash point increases with an increase in boiling point. Flash point is an important parameter for safety considerations, especially during storage and transportation of volatile petroleum products (i.e. LPG, light naphtha, gasoline) in a high-temperature environment. The surrounding temperature around a storage tank should always be less than the flash point of the fuel to avoid possibility of ignition. Flash point is used as an indication of the fire and explosion potential of a petroleum product (ASTM D93, 2017). There is another method of measuring flash point known as *open cup* for those oils with flash point greater than 80 °C (ASTM D92, 2017). Flash point should not be mistaken with *fire point*, which is defined as the minimum temperature at which the hydrocarbon will continue to burn for at least 5 s after being ignited by a flame.

Thus, of the available test methods, the most common method of determining the flash point confines the vapor (closed cup) until the instant flame is applied (ASTM D56, 2017; ASTM D93, 2017; ASTM D3828, 2017; ASTM D6450, 2017). An alternate method that does not confine the vapor (open cup method) (ASTM D92, 2017; ASTM D1310, 2017) gives slightly higher values of the flash point. For example, ASTM D56 (2017) describes the experimental protocols for the determination of the flash point, by tag manual and automated closed testers, of hydrocarbon liquids with a viscosity below 5.5 cSt at 40 °C (104 °F), or below 9.5 cSt at 25 °C (77 °F), and a flash point below 93 °C (200 °F). Gas condensate and natural gasoline typically fit into these property specifications. However, erroneously high flash points can be obtained when precautions are not taken to avoid the loss of volatile material. Samples should not be stored in plastic bottles, since the volatile material may diffuse through the walls of the container. The containers should not be opened unnecessarily. The samples should not be transferred between containers unless the sample temperature is at least 11 °C (20 °F) below the expected flash point.

Another test (ASTM E659, 2017) is available that can be used as a complement to the flash point test and involves determination of the autoignition temperature. The autoignition temperature is the minimum temperature at which hydrocarbon vapor when mixed with air can spontaneously ignite without the presence of any external source. Values of autoignition temperature are generally higher than flash point. Values of autoignition temperature for typical hydrocarbon liquids obtained from mineral sources are in the range of 150–320 °C (300–500 °F), for gasoline it is about 350 °C (660 °F), and for alcohol it is about 500 °C (930 °F). With an increase in pressure, the autoignition temperature decreases. This is particularly important from a safety point of view when hydrocarbon derivatives are compressed. However, the flash point should not be confused with autoignition temperature that measures spontaneous combustion with no external source of ignition.

This test method (ASTM E659, 2017) is dependent on the chemical and physical properties of the liquid under examination as well as the method and apparatus employed for its determination. The autoignition temperature by a given method does not necessarily represent the minimum temperature at which a given material will self-ignite in air. The volume of the vessel used is particularly important since lower autoignition temperatures will be achieved in larger vessels and vessel material can also be an important factor. The test method is not designed for evaluating materials that are capable of exothermic decomposition. For such materials, ignition is dependent upon the thermal and kinetic properties of the decomposition, the mass of the sample, and the heat transfer characteristics of the system. Also, the method can be used, with appropriate modifications, for chemicals that are gaseous at atmospheric temperature and pressure of which gas condensate and natural gasoline are examples.

10.5.13 Kauri-Butanol Value

The kauri-butanol value (ASTM D1133, 2017) is the number of milliliters of the solvent, at 15 °C (77 °F), required to produce a defined degree of turbidity when added to 20 g of a standard solution of kauri gum resin in *n*-butyl alcohol. The kauri-butanol value of naphtha, and its constituent boiling fractions, is used to determine relative solvent power. For kauri-butanol values of 60 and higher, the standard is toluene that has an assigned value of 105, whereas for kauri-butanol values less than 60, the standard is a blend of 75% *n*-heptane and 25% toluene that has an assigned value of 40. The kauri-butanol value of products that are classified as regular mineral spirits normally varies between 34 and 44; xylene is 93, and aromatic naphtha falls in the range of 55–108. However, the results obtained by the kauri-butanol test (ASTM D1133, 2017) depend upon factors other than solvent power and are specific to the solute employed. For this reason, the aniline point is often preferred to the kauri-butanol number.

However, ASTM D1133 (2017) is suitable for use with hydrocarbon liquids having an initial boiling point in excess of 40 °C (104 °F) and a dry point under 300 °C (572 °F) when determined in accordance with the procedure. In most cases, the initial boiling point (40 °C, 104 °F) would preclude application of the method to gas condensate and natural gasoline without first stabilizing these two hydrocarbon liquids at 40 °C (104 °F) or 45 °C) and analyzing the volatile fraction by GC. Depending upon the sample, the stabilization process might leave so little of the sample (the residue) that any data obtained on the residue would have little, if any, meaning.

10.5.14 Octane Number

Octane number is a parameter defined to characterize antiknock characteristic of a fuel (gasoline) for spark-ignition engines, and a higher octane number for a liquid fuel (compared with other liquid fuels) indicates a better engine performance (ASTM D909, 2017; ASTM D2699, 2017; ASTM D2700, 2017; ASTM D2885, 2017). The octane number is a measure of the ability of a liquid fuel to resist autoignition during compression and prior to ignition. The octane number of a fuel is measured based on two reference hydrocarbon derivatives of *n*-heptane with an assigned octane number of zero and isooctane (2,2,4-trimethylpentane) with assigned octane number of 100. A mixture of 70% v/v isooctane and 30% v/v *n*-heptane has an octane number of 70. There are two methods of measuring octane number of a fuel in the laboratory. The methods are known as *motor octane number* (MON) and *research octane number* (RON).

The MON is indicative of high-speed performance (900 rpm) and is measured under heavy road conditions. The RON is indicative of normal road performance under low engine speed (600 rpm) city driving conditions. The third type of octane number is defined as *posted octane number* (PON), which is the arithmetic average of the MON and RON. Thus,

$$\left(\text{PON} = [\text{MON} + \text{RON}]/2\right)$$

Generally, isoparaffin derivatives have higher octane number than normal paraffin derivatives. Naphthene derivatives have relatively higher octane numbers than the corresponding (equal carbon number) paraffin derivatives – aromatic derivatives also have very high octane numbers than the corresponding equal carbon number paraffin derivatives.

The octane number of a fuel can be predicted by using principal component regression of chromatographic data, but the test methods based on spectroscopic techniques such as NIR, IR, and NMR are applied for the measurement and/or prediction of octane numbers, and other parameters of gasoline are preferred (Myers et al., 1975; Ichikawa et al., 1992; Welch et al., 1994; Andrade et al., 1997; Speight, 2014a, 2015).

Application of any such test methods to gas condensate or natural gasoline may have a limited value if the gas condensate or natural gasoline is a very minor constituent of the gasoline blend. In such a case, the gas condensate or natural gasoline may only be a complementary constituent to make up the final blend for sales gasoline, and the octane number would have little, if any, effect on the octane number of the final blend.

10.5.15 Odor and Color

The degree of purity of naphtha is an important aspect of naphtha properties, and strict segregation of all distribution equipment is maintained in order to ensure strict and uniform specification for the product handled. Naphtha is refined to a low level of odor to meet the specifications for use. Purified naphtha is required to have a low level of odor to meet the specifications for use.

In general, the paraffinic hydrocarbon derivatives possess the mildest odor and the aromatic derivatives the strongest, the odor level (ASTM D268, 2017; ASTM D1296, 2017) being related to chemical character and volatility of the constituents. Odors due to the presence of sulfur compounds or unsaturated constituents are excluded by specification. And apart from certain high-boiling aromatic fractions, which are usually excluded by volatility from the majority of the naphtha fractions, which may be pale yellow in color, naphtha is usually colorless (water white).

Usually, gas condensate and natural gasoline have little, if any, color (in such cases, the liquid is described as being), but gas condensate and natural gasoline containing higher amounts of aromatic constituents may be pale yellow. Measurement of color (ASTM D156, 2017; ASTM D848, 2017; ASTM D1209, 2017; ASTM D1555, 2017; ASTM D5386, 2017) provides a rapid method of checking the degree of freedom from contamination. Observation of the test for residue on evaporation (ASTM D381, 2017; ASTM D1353, 2017) provides a further guard against adventitious contamination.

Assuming that sample volatility does not have an adverse effect on the outcome of the test method, on the test method, a near-clear liquid can be tested for color by a method (ASTM D5386, 2017) that is applicable to the color measurement of clear, liquid samples, free of haze and the method is applicable to nonfluorescent liquids with light absorption characteristics similar to those of the color standard solutions The color number can be calculated from spectral measurements or can be obtained directly from a colorimeter (ASTM E1455, 2017).

10.5.16 Olefin Content

One particular method (ASTM D1319, 2017) is widely used for measuring total olefins in gasoline fractions as well as aromatic derivatives and saturates. The results of this method must be corrected for the presence of oxygenates, and the precision of the method is poor. A titration procedure (ASTM D1159, 2017) involving determination of the bromine number provides an approximation of olefin content within a sample, while determination of the bromine index (ASTM D2710, 2017) can also be valuable for determining trace olefin levels. These methods do not directly measure total olefins, and the results are affected by the type of olefinic compound present.

Other test methods include (i) the use of a gas chromatographic multidimensional procedure for oxygenates and paraffin, olefin, naphthene, and aromatic (O-PONA) hydrocarbon types in petroleum distillates and (ii) a gas chromatographic rapid analyzer for total olefin content (FTO) method. This particular method has the advantage that the analysis time is quicker. The O-PONA method is an expanded version of a standard test method (ASTM D5443, 2017) and gives a detailed breakdown of the oxygenate derivatives and hydrocarbon derivatives by carbon number.

10.5.17 Sulfur Content

Sulfur-containing components exist in some gas condensate and natural gasoline samples. Individual sulfur components can be speciated (ASTM D5623, 2017), and the method uses a gas chromatographic capillary column coupled with either a sulfur chemiluminescence detector or an atomic emission detector (AED). The total sulfur content is an important test parameter in hydrocarbon liquid that is designated as liquid fuels or is used as blending stock for sales gasoline (ASTM D2622, 2017; ASTM D4045, 2017; ASTM D5453, 2017).

10.5.18 Vapor Pressure

The vapor pressure of gasoline is a critical physical test parameter for hydrocarbon liquids (ASTM D323, 2017; ASTM D5191, 2017).

The former test method (ASTM D323, 2017) describes procedures for the determination of vapor pressure volatile hydrocarbon liquids such as gasoline, volatile crude oil, and other volatile petroleum products as well as other petroleum-related liquids that have a vapor pressure of less than 26 psi (Procedure A). However, one of the subprocedures (Procedure C) can be applied to hydrocarbon liquids with a vapor pressure of greater than 26 psi. Gas condensate and natural gasoline can fit into either of these procedures. Because the external atmospheric pressure is counteracted by the atmospheric pressure initially present in the vapor chamber, the Reid vapor pressure is an absolute pressure at 37.8 °C (100 °F) in kilopascals (pounds-force per square inch). The Reid vapor pressure differs from the true vapor pressure of the sample due to some small sample vaporization and the presence of water vapor and air in the confined space. However, this test method (ASTM D323, 2017) is not applicable to liquefied petroleum gases or fuels containing oxygenated compounds other than methyl *t*-butyl ether (MTBE), and other test methods (ASTM D1267, 2017; ASTM D6897, 2017) are recommended.

The latter test method (ASTM D5191, 2017) describes the use of automated vapor pressure instruments to determine the total vapor pressure exerted in vacuum by air-containing, volatile, liquid petroleum products, including automotive spark-ignition fuels with or without oxygenates. This test method is suitable for testing samples with boiling points above 0 °C (32 °F) that exert a vapor pressure between 7 and 130 kPa (1.0 and 18.6 psi) at 37.8 °C (100 °F) at a vapor-to-liquid ratio of 4 : 1 – these parameters are typically applicable to gas condensate and natural gasoline. Measurements are made on liquid sample sizes in the range from 1 to 10 ml. This test method is suitable for calculation of the dry vapor pressure equivalent (DVPE) of gasoline and gasoline–oxygenate blends by means of a correlation equation. The calculated DVPE very closely approximates the dry vapor pressure that would be obtained on the same material when examined by another standard test method (ASTM D4953, 2017).

10.5.19 Viscosity

Oil viscosity is important in the proper modeling of "condensate blockage" – i.e. the two-phase gas–oil flow effect on gas relative permeability in the region around the wellbore. Oil viscosity is usually low for reservoir condensates, ranging from 0.1 to 1 cp in the near-wellbore region. Measurement of condensate viscosities is not made in routine laboratory tests, and it may be difficult to obtain measurements for lean condensates (where volumes of condensate are small). Viscosity correlations are typically unreliable for predicting low oil viscosities, and some approach is needed to ensure accurate and consistent modeling of this important property.

The oil viscosity model can be tuned to measured viscosities of a separator condensate sample at reservoir temperature and pressures in the range of 1500–6000 psi. More appropriate condensate viscosity measurements can be designed (at greater expense), but having oil viscosity data from a separator oil sample to tune the viscosity correlation should ensure reasonably accurate oil viscosity predictions of the condensate actually flowing in the near-wellbore region when BHFP drop below the dew point.

The most commonly used **oil viscosity** correlations are those of Beggs–Robinson and Vasquez–Beggs. Both correlations can be used to calculate (estimate) PVT properties. Corrections must be applied for undersaturated systems and for systems where dissolved gas is present in the oil. However, in compositional simulation, where both gas and condensate compositions are known at every point of the reservoir, it is customary to calculate *condensate viscosity* using Lohrenz, Bray, and Clark correlation. In this type of simulation, it is usual to calculate *gas viscosity* based on Lohrenz, Bray, and Clark correlation as well. This serves the purpose of guaranteeing that the gas phase and condensate phase converge to the same value of viscosity as they approach near-critical conditions (Lohrenz et al., 1964; Lee et al., 1966; Pedersen and Fredenslund, 1987).

Gas viscosity for most systems will vary from 0.02 to 0.03 cp for all pressure conditions. For near-critical gas condensates and high-pressure gases, the viscosity may initially be 0.05 cp, but in most of the near-wellbore region experiencing significant pressure losses, the viscosity will be in the lower range of 0.02–0.03 cp. Consequently, the absolute value of viscosity does not vary greatly for a given gas, or from gas to gas system. Viscosity correlations are fairly reliable at predicting accurate gas viscosities, within 5–10% in most cases.

What *is* important with respect to gas viscosities is that consistent viscosities be used in all engineering applications – e.g. well test interpretation, well performance design, reservoir simulation, tubing calculations, pipeline calculations, etc. It is not uncommon that a 15–25% difference in gas viscosities may result using different correlations (by different engineering disciplines). This may result in similar inaccuracies in well performance calculations, even where all flow is single-phase gas!

10.5.20 Volatility

Distillation, as a means of determining the boiling range (hence the *volatility*) of petroleum and petroleum products, has been in use since the beginning of the petroleum industry and is an important aspect of product specifications.

Naphtha fraction (including gas condensate and natural gasoline) grades are often referred to by a boiling range, which is the defined temperature range in which the fraction distills. The ranges are determined by standard methods (ASTM D86, 2017) – it is especially necessary to use a recognized method since the initial and final boiling points that ensure conformity with volatility requirements and absence of *heavy ends* are affected by the testing procedure. A simple test for the evaporation properties of naphtha (including gas condensate and natural gasoline with any suitable *ad hoc* modifications to the test method) is available (ASTM D381, 2017). The need for vacuum distillation (ASTM D1160, 2017) as a product specification in the boiling range of naphtha is not necessary. Gas condensate and natural gasoline very rarely (if at all) have such a high boiling point.

Thus, one of the most important physical parameters of gas condensate and natural gasoline is the boiling range distribution (ASTM D86, 2017; ASTM D1078, 2017; ASTM D2268, 2017; ASTM D2887, 2017; ASTM D2892, 2017). The significance of the distillation test is the indication of volatility that dictates the evaporation rate that is an important property for gas condensate and natural gasoline used in coatings and similar application where the premise is that the gas condensate and natural gasoline evaporate over time. In the basic test method (ASTM D86, 2017), a 100 ml sample is distilled (manually or automatically) under prescribed conditions prescribed. Temperatures and volumes of the gas condensate or the natural gasoline are recorded at regular intervals from which the boiling profile is derived.

The determination of the boiling range distribution of distillates such as gas condensate and natural gasoline by GC (ASTM D3710, 2017) not only helps identify the constituents but also facilitates online controls at the refinery. ASTM D323 (2017) is designed to measure the entire boiling range of gas condensate and natural gasoline with either high or low Reid vapor pressure. In the method, the sample is injected into a gas chromatographic column that separates hydrocarbon derivatives in boiling point order. The column temperature is raised at a reproducible rate and the area under the chromatogram is recorded throughout the run. Calibration is performed using a known mixture of hydrocarbon derivatives covering the expected boiling of the sample.

Another method is described as a method for determining the carbon number distribution (ASTM D2887, 2017), and the data derived by this test method are essentially equivalent to that obtained by true boiling point (TBP) distillation (ASTM D2892, 2017). The sample is introduced into a gas chromatographic column that separates hydrocarbon derivatives in boiling point order. The column temperature is raised at a reproducible rate and the area under the chromatogram is recorded throughout the run. Boiling temperatures are assigned to the time axis from a calibration curve, obtained under the same conditions by running a known mixture of hydrocarbon derivatives covering the boiling range expected in the sample. From these data, the boiling range distribution may be obtained. However, this test method is limited to samples having a boiling range greater than 55 °C (100 °F) and having a vapor pressure (ASTM D323, 2017; ASTM D4953, 2017; ASTM D5191, 2017; ASTM D5482, 2017; ASTM D6377, 2017; ASTM D6378, 2017) sufficiently low to permit sampling at ambient temperature without contamination.

However, ASTM D2887 (2017) describes the determination of the boiling range applicable to hydrocarbon liquids (and petroleum fractions) that have a final boiling point of 538 °C (1000 °F) or lower at atmospheric pressure as measured by this test method. This test method is limited to samples having a *boiling range* greater than 55.5 °C (100 °F) and having a vapor pressure sufficiently low to permit sampling at ambient temperature. The method may not be applicable to certain sample of gas condensate and natural gasoline without some form of modification.

On the other hand, test method ASTM D2892 (2017) describes the procedure for the distillation of stabilized hydrocarbon liquids (such as crude petroleum) to a final cut temperature of 400 °C (752 °F) atmospheric equivalent temperature (AET). The method also provides details of the procedures for the production of a liquefied gas, distillate fractions, and residuum of standardized quality on which analytical data can be obtained and the determination of yields of the above fractions by both mass and volume. From the preceding information, a graph of temperature versus % w/w distilled can be produced, and the distillation curve corresponds to a laboratory technique, which is defined at 15/5 (15 theoretical plate column, 5 : 1 reflux ratio) or TBP.

Pure hydrocarbon derivatives such as pentane, hexane, heptane, benzene, toluene, xylene, and ethylbenzene may be characterized by a fixed boiling point, but the constituents of gas condensate and natural gasoline (being mixtures of many hydrocarbon derivatives) are typically less easy to identify. The distillation test does however give a useful indication of their volatility. The data obtained should include the initial and final temperatures of distillation together with sufficient temperature and volume observations to permit a characteristic distillation curve to be drawn.

This information is especially important when a formulation includes other volatile liquids since the performance of the product will be affected by the relative volatility of the constituents. An illustration of the importance of this aspect is found in the use of specifically defined boiling point naphtha in cellulose lacquers, where a mixture with ester, alcohols, and other solvents may be employed. The naphtha does not act as a solvent for the cellulose ester but is incorporated as a diluent to control the viscosity and flow properties of the mixture. If the solvent evaporates too rapidly, blistering of the surface coating may result, while if the solvent evaporates unevenly, leaving behind a higher proportion of the naphtha, precipitation of the cellulose may occur leading to a milky opaqueness known as blushing.

Although much dependence is placed on the assessment of volatility by distillation methods, some specifications include measurement of drying time by evaporation from a filter paper or dish. Laboratory measurements are expressed as *evaporation rate* either by reference to a pure compound evaporated under similar conditions as the sample under test or by constructing a time weight loss curve under standard conditions. Although the results obtained on the naphtha provide a useful guide, it is, wherever possible, better to carry out a performance test on the final product when assessing formulations.

In choosing gas condensate and/or natural gasoline for a particular purpose, it is necessary to relate volatility to the fire hazard associated with its use, storage, and transport and also with the handling of the products arising from the process. This is normally based on the characterization of the gas condensate or the natural gasoline solvent by flash point limits (ASTM D56, 2017; ASTM D93, 2017).

References

Andrade, J.M., Muniategui, S., and Prada, D. (1997). Prediction of clean octane numbers of catalytic reformed naphtha using FT-MIR and PLS. *Fuel* 76: 1035–1042.

ASTM (2017). *Annual Book of Standards*. West Conshohocken, Pennsylvania: ASTM International.

ASTM D1015 (2017). *Standard Test Method for Freezing Points of High-Purity Hydrocarbon Derivatives*, Annual Book of Standards. West Conshohocken, Pennsylvania: ASTM International.

ASTM D1016 (2017). *Standard Test Method for Purity of Hydrocarbon Derivatives from Freezing Points*, Annual Book of Standards. West Conshohocken, Pennsylvania: ASTM International.

ASTM D1078 (2017). *Standard Test Method for Hydrogen in Petroleum Fractions*, Annual Book of Standards. West Conshohocken, Pennsylvania: ASTM International.

ASTM D1093 (2017). *Standard Test Method for Acidity of Hydrocarbon Liquids and Their Distillation Residues*, Annual Book of Standards. West Conshohocken, Pennsylvania: ASTM International.

ASTM D1133 (2017). *Standard Test Method for Kauri-Butanol Value of Hydrocarbon Solvents*, Annual Book of Standards. West Conshohocken, Pennsylvania: ASTM International.

ASTM D1159 (2017). *Standard Test Method for Bromine Numbers of Petroleum Distillates and Commercial Aliphatic Olefins by Electrometric Titration*, Annual Book of Standards. West Conshohocken, Pennsylvania: ASTM International.

ASTM D1160 (2017). *Standard Test Method for Distillation of Petroleum Products at Reduced Pressure*, Annual Book of Standards. West Conshohocken, Pennsylvania: ASTM International.

ASTM D1209 (2017). *Standard Test Method for Color of Clear Liquids (Platinum-Cobalt Scale)*, Annual Book of Standards. West Conshohocken, Pennsylvania: ASTM International.

ASTM D1217 (2017). *Standard Test Method for Density and Relative Density (Specific Gravity) of Liquids by Bingham Pycnometer*, Annual Book of Standards. West Conshohocken, Pennsylvania: ASTM International.

ASTM D1218 (2017). *Standard Test Method for Refractive Index and Refractive Dispersion of Hydrocarbon Liquids*, Annual Book of Standards. West Conshohocken, Pennsylvania: ASTM International.

ASTM D1250 (2017). *Standard Guide for Use of the Petroleum Measurement Tables*, Annual Book of Standards. West Conshohocken, Pennsylvania: ASTM International.

ASTM D1266 (2017). *Standard Test Method for Sulfur in Petroleum Products (Lamp Method)*, Annual Book of Standards. West Conshohocken, Pennsylvania: ASTM International.

ASTM D1267 (2017). *Standard Test Method for Gage Vapor Pressure of Liquefied Petroleum (LP) Gases (LP-Gas Method)*, Annual Book of Standards. West Conshohocken, Pennsylvania: ASTM International.

ASTM D1296 (2017). *Standard Test Method for Odor of Volatile Solvents and Diluents*, Annual Book of Standards. West Conshohocken, Pennsylvania: ASTM International.

ASTM D1298 (2017). *Standard Test Method for Density, Relative Density, or API Gravity of Crude Petroleum and Liquid Petroleum Products by Hydrometer Method*, Annual Book of Standards. West Conshohocken, Pennsylvania: ASTM International.

ASTM D130 (2017). *Standard Test Method for Corrosiveness to Copper from Petroleum Products by Copper Strip Test*, Annual Book of Standards. West Conshohocken, Pennsylvania: ASTM International.

ASTM D1310 (2017). *Standard Test Method for Flash Point and Fire Point of Liquids by Tag Open-Cup Apparatus*, Annual Book of Standards. West Conshohocken, Pennsylvania: ASTM International.

ASTM D1319 (2017). *Standard Test Method for Hydrocarbon Types in Liquid Petroleum Products by Fluorescent Indicator Adsorption*, Annual Book of Standards. West Conshohocken, Pennsylvania: ASTM International.

ASTM D1322 (2017). *Standard Test Method for Smoke Point of Kerosine and Aviation Turbine Fuel*, Annual Book of Standards. West Conshohocken, Pennsylvania: ASTM International.

ASTM D1353 (2017). *Standard Test Method for Nonvolatile Matter in Volatile Solvents for Use in Paint, Varnish, Lacquer, and Related Products*, Annual Book of Standards. West Conshohocken, Pennsylvania: ASTM International.

ASTM D1364 (2017). *Standard Test Method for Water in Volatile Solvents (Karl Fischer Reagent Titration Method)*, Annual Book of Standards. West Conshohocken, Pennsylvania: ASTM International.

ASTM D1492 (2017). *Standard Test Method for Bromine Index of Aromatic Hydrocarbon derivatives by Coulometric Titration*, Annual Book of Standards. West Conshohocken, Pennsylvania: ASTM International.

ASTM D1555 (2017). *Standard Test Method for Calculation of Volume and Weight of Industrial Aromatic Hydrocarbon derivatives and Cyclohexane (Metric)*, Annual Book of Standards. West Conshohocken, Pennsylvania: ASTM International.

ASTM D156 (2017). *Standard Test Method for Saybolt Color of Petroleum Products (Saybolt Chromometer Method)*, Annual Book of Standards. West Conshohocken, Pennsylvania: ASTM International.
ASTM D1613 (2017). *Standard Test Method for Acidity in Volatile Solvents and Chemical Intermediates Used in Paint, Varnish, Lacquer, and Related Products*, Annual Book of Standards. West Conshohocken, Pennsylvania: ASTM International.
ASTM D1655 (2017). *Standard Specification for Aviation Turbine Fuels*, Annual Book of Standards. West Conshohocken, Pennsylvania: ASTM International.
ASTM D1657 (2017). *Standard Test Method for Density or Relative Density of Light Hydrocarbon derivatives by Pressure Hydrometer*, Annual Book of Standards. West Conshohocken, Pennsylvania: ASTM International.
ASTM D1744 (2017). *Standard Test Method for Determination of Water in Liquid Petroleum Products by Karl Fischer Reagent*, Annual Book of Standards. West Conshohocken, Pennsylvania: ASTM International.
ASTM D1796 (2017). *Standard Test Method for Water and Sediment in Fuel Oils by the Centrifuge Method (Laboratory Procedure)*, Annual Book of Standards. West Conshohocken, Pennsylvania: ASTM International.
ASTM D2001 (2017). *Standard Test Method for Depentanization of Gasoline and Naphthas*, Annual Book of Standards. West Conshohocken, Pennsylvania: ASTM International.
ASTM D2007 (2017). *Standard Test Method for Characteristic Groups in Rubber Extender and Processing Oils and Other Petroleum-Derived Oils by the Clay-Gel Absorption Chromatographic Method*, Annual Book of Standards. West Conshohocken, Pennsylvania: ASTM International.
ASTM D2008 (2017). *Standard Test Method for Ultraviolet Absorbance and Absorptivity of Petroleum Products*, Annual Book of Standards. West Conshohocken, Pennsylvania: ASTM International.
ASTM D2161 (2017). *Standard Practice for Conversion of Kinematic Viscosity to Saybolt Universal Viscosity or to Saybolt Furol Viscosity*, Annual Book of Standards. West Conshohocken, Pennsylvania: ASTM International.
ASTM D2268 (2017). *Standard Test Method for Analysis of High-Purity n-Heptane and Isooctane by Capillary Gas Chromatograph*, Annual Book of Standards. West Conshohocken, Pennsylvania: ASTM International.
ASTM D2360 (2017). *Standard Test Method for Trace Impurities in Monocyclic Aromatic Hydrocarbon Derivatives by Gas Chromatography*, Annual Book of Standards. West Conshohocken, Pennsylvania: ASTM International.
ASTM D2425 (2017). *Standard Test Method for Hydrocarbon Types in Middle Distillates by Mass Spectrometry*, Annual Book of Standards. West Conshohocken, Pennsylvania: ASTM International.
ASTM D2427 (2017). *Standard Test Method for Determination of C2 through C5 Hydrocarbon Derivatives in Gasolines by Gas Chromatography*, Annual Book of Standards. West Conshohocken, Pennsylvania: ASTM International.

ASTM D2500 (2017). *Standard Test Method for Cloud Point of Petroleum Products*, Annual Book of Standards. West Conshohocken, Pennsylvania: ASTM International.

ASTM D2549 (2017). *Standard Test Method for Separation of Representative Aromatic Derivatives and Nonaromatic Derivatives Fractions of High-Boiling Oils by Elution Chromatography*, Annual Book of Standards. West Conshohocken, Pennsylvania: ASTM International.

ASTM D2622 (2017). *Standard Test Method for Sulfur in Petroleum Products by Wavelength Dispersive X-ray Fluorescence Spectrometry*, Annual Book of Standards. West Conshohocken, Pennsylvania: ASTM International.

ASTM D268 (2017). *Standard Guide for Sampling and Testing Volatile Solvents and Chemical Intermediates for Use in Paint and Related Coatings and Materials*, Annual Book of Standards. West Conshohocken, Pennsylvania: ASTM International.

ASTM D2699 (2017). *Standard Test Method for Research Octane Number of Spark-Ignition Engine Fuel*, Annual Book of Standards. West Conshohocken, Pennsylvania: ASTM International.

ASTM D2700 (2017). *Standard Test Method for Motor Octane Number of Spark-Ignition Engine Fuel*, Annual Book of Standards. West Conshohocken, Pennsylvania: ASTM International.

ASTM D2709 (2017). *Standard Test Method for Water and Sediment in Middle Distillate Fuels by Centrifuge*, Annual Book of Standards. West Conshohocken, Pennsylvania: ASTM International.

ASTM D2710 (2017). *Standard Test Method for Bromine Index of Petroleum Hydrocarbon Derivatives by Electrometric Titration*, Annual Book of Standards. West Conshohocken, Pennsylvania: ASTM International.

ASTM D2715 (2017). *Standard Test Method for Volatilization Rates of Lubricants in Vacuum*, Annual Book of Standards. West Conshohocken, Pennsylvania: ASTM International.

ASTM D2789 (2017). *Standard Test Method for Hydrocarbon Types in Low Olefinic Gasoline by Mass Spectrometry*, Annual Book of Standards. West Conshohocken, Pennsylvania: ASTM International.

ASTM D287 (2017). *Standard Test Method for API Gravity of Crude Petroleum and Petroleum Products (Hydrometer Method)*, Annual Book of Standards. West Conshohocken, Pennsylvania: ASTM International.

ASTM D2885 (2017). *Standard Test Method for Determination of Octane Number of Spark-Ignition Engine Fuels by On-Line Direct Comparison Technique*, Annual Book of Standards. West Conshohocken, Pennsylvania: ASTM International.

ASTM D2887 (2017). *Standard Test Method for Boiling Range Distribution of Petroleum Fractions by Gas Chromatography*, Annual Book of Standards. West Conshohocken, Pennsylvania: ASTM International.

ASTM D2892 (2017). *Standard Test Method for Distillation of Crude Petroleum (15-Theoretical Plate Column)*, Annual Book of Standards. West Conshohocken, Pennsylvania: ASTM International.
ASTM D2896 (2017). *Standard Test Method for Base Number of Petroleum Products by Potentiometric Perchloric Acid Titration*, Annual Book of Standards. West Conshohocken, Pennsylvania: ASTM International.
ASTM D3120 (2017). *Standard Test Method for Trace Quantities of Sulfur in Light Liquid Petroleum Hydrocarbon Derivatives by Oxidative Microcoulometry*, Annual Book of Standards. West Conshohocken, Pennsylvania: ASTM International.
ASTM D323 (2017). *Standard Test Method for Vapor Pressure of Petroleum Products (Reid Method)*, Annual Book of Standards. West Conshohocken, Pennsylvania: ASTM International.
ASTM D3238 (2017). *Standard Test Method for Calculation of Carbon Distribution and Structural Group Analysis of Petroleum Oils by the n-d-M Method*, Annual Book of Standards. West Conshohocken, Pennsylvania: ASTM International.
ASTM D3257 (2017). *Standard Test Methods for Aromatic Derivatives in Mineral Spirits by Gas Chromatography*, Annual Book of Standards. West Conshohocken, Pennsylvania: ASTM International.
ASTM D3505 (2017). *Standard Test Method for Density or Relative Density of Pure Liquid Chemicals*, Annual Book of Standards. West Conshohocken, Pennsylvania: ASTM International.
ASTM D3606 (2017). *Standard Test Method for Determination of Benzene and Toluene in Finished Motor and Aviation Gasoline by Gas Chromatography*, Annual Book of Standards. West Conshohocken, Pennsylvania: ASTM International.
ASTM D3701 (2017). *Standard Test Method for Hydrogen Content of Aviation Turbine Fuels by Low Resolution Nuclear Magnetic Resonance Spectrometry*, Annual Book of Standards. West Conshohocken, Pennsylvania: ASTM International.
ASTM D3710 (2017). *Standard Test Method for Boiling Range Distribution of Gasoline and Gasoline Fractions by Gas Chromatography*, Annual Book of Standards. West Conshohocken, Pennsylvania: ASTM International.
ASTM D3797 (2017). *Standard Test Method for Analysis of o-Xylene by Gas Chromatography*, Annual Book of Standards. West Conshohocken, Pennsylvania: ASTM International.
ASTM D381 (2017). *Standard Test Method for Gum Content in Fuels by Jet Evaporation*, Annual Book of Standards. West Conshohocken, Pennsylvania: ASTM International.
ASTM D3828 (2017). *Standard Test Method for pH of Activated Carbon*, Annual Book of Standards. West Conshohocken, Pennsylvania: ASTM International.

ASTM D4006 (2017). *Standard Test Method for Water in Crude Oil by Distillation*, Annual Book of Standards. West Conshohocken, Pennsylvania: ASTM International.
ASTM D4007 (2017). *Standard Test Method for Water and Sediment in Crude Oil by the Centrifuge Method (Laboratory Procedure)*, Annual Book of Standards. West Conshohocken, Pennsylvania: ASTM International.
ASTM D4045 (2017). *Standard Test Method for Sulfur in Petroleum Products by Hydrogenolysis and Rateometric Colorimetry*, Annual Book of Standards. West Conshohocken, Pennsylvania: ASTM International.
ASTM D4052 (2017). *Standard Test Method for Density, Relative Density, and API Gravity of Liquids by Digital Density Meter*, Annual Book of Standards. West Conshohocken, Pennsylvania: ASTM International.
ASTM D4055 (2017). *Standard Test Method for Pentane Insolubles by Membrane Filtration*, Annual Book of Standards. West Conshohocken, Pennsylvania: ASTM International.
ASTM D4057 (2017). *Standard Practice for Manual Sampling of Petroleum and Petroleum Products*, Annual Book of Standards. West Conshohocken, Pennsylvania: ASTM International.
ASTM D4367 (2017). *Standard Test Method for Benzene in Hydrocarbon Solvents by Gas Chromatography*, Annual Book of Standards. West Conshohocken, Pennsylvania: ASTM International.
ASTM D4377 (2017). *Standard Test Method for Water in Crude Oils by Potentiometric Karl Fischer Titration*, Annual Book of Standards. West Conshohocken, Pennsylvania: ASTM International.
ASTM D445 (2017). *Standard Test Method for Kinematic Viscosity of Transparent and Opaque Liquids (and Calculation of Dynamic Viscosity)*, Annual Book of Standards. West Conshohocken, Pennsylvania: ASTM International.
ASTM D4492 (2017). *Standard Test Method for Analysis of Benzene by Gas Chromatography*, Annual Book of Standards. West Conshohocken, Pennsylvania: ASTM International.
ASTM D473 (2017). *Standard Test Method for Sediment in Crude Oils and Fuel Oils by the Extraction Method*, Annual Book of Standards. West Conshohocken, Pennsylvania: ASTM International.
ASTM D4735 (2017). *Standard Test Method for Determination of Trace Thiophene in Refined Benzene by Gas Chromatography*, Annual Book of Standards. West Conshohocken, Pennsylvania: ASTM International.
ASTM D4807 (2017). *Standard Test Method for Sediment in Crude Oil by Membrane Filtration*, Annual Book of Standards. West Conshohocken, Pennsylvania: ASTM International.
ASTM D4808 (2017). *Standard Test Methods for Hydrogen Content of Light Distillates, Middle Distillates, Gas Oils, and Residua by Low-Resolution Nuclear Magnetic Resonance Spectroscopy*, Annual Book of Standards. West Conshohocken, Pennsylvania: ASTM International.

ASTM D4928 (2017). *Standard Test Method for Water in Crude Oils by Coulometric Karl Fischer Titration*, Annual Book of Standards. West Conshohocken, Pennsylvania: ASTM International.
ASTM D4952 (2017). *Standard Test Method for Qualitative Analysis for Active Sulfur Species in Fuels and Solvents (Doctor Test)*, Annual Book of Standards. West Conshohocken, Pennsylvania: ASTM International.
ASTM D4953 (2017). *Standard Test Method for Vapor Pressure of Gasoline and Gasoline-Oxygenate Blends (Dry Method)*, Annual Book of Standards. West Conshohocken, Pennsylvania: ASTM International.
ASTM D5002 (2017). *Standard Test Method for Density and Relative Density of Crude Oils by Digital Density Analyzer*, Annual Book of Standards. West Conshohocken, Pennsylvania: ASTM International.
ASTM D5060 (2017). *Standard Test Method for Determining Impurities in High-Purity Ethylbenzene by Gas Chromatography*, Annual Book of Standards. West Conshohocken, Pennsylvania: ASTM International.
ASTM D5134 (2017). *Standard Test Method for Detailed Analysis of Petroleum Naphthas through n-Nonane by Capillary Gas Chromatography*, Annual Book of Standards. West Conshohocken, Pennsylvania: ASTM International.
ASTM D5135 (2017). *Standard Test Method for Analysis of Styrene by Capillary Gas Chromatography*, Annual Book of Standards. West Conshohocken, Pennsylvania: ASTM International.
ASTM D5191 (2017). *Standard Test Method for Vapor Pressure of Petroleum Products (Mini Method)*, Annual Book of Standards. West Conshohocken, Pennsylvania: ASTM International.
ASTM D5194 (2017). *Standard Test Method for Trace Chloride in Liquid Aromatic Hydrocarbon Derivatives*, Annual Book of Standards. West Conshohocken, Pennsylvania: ASTM International.
ASTM D525 (2017). *Standard Test Method for Oxidation Stability of Gasoline (Induction Period Method)*, Annual Book of Standards. West Conshohocken, Pennsylvania: ASTM International.
ASTM D5291 (2017). *Standard Test Methods for Instrumental Determination of Carbon, Hydrogen, and Nitrogen in Petroleum Products and Lubricants*, Annual Book of Standards. West Conshohocken, Pennsylvania: ASTM International.
ASTM D5292 (2017). *Standard Test Method for Aromatic Carbon Contents of Hydrocarbon Oils by High Resolution Nuclear Magnetic Resonance Spectroscopy*, Annual Book of Standards. West Conshohocken, Pennsylvania: ASTM International.
ASTM D5386 (2017). *Standard Test Method for Color of Liquids Using Tristimulus Colorimetry*, Annual Book of Standards. West Conshohocken, Pennsylvania: ASTM International.
ASTM D5443 (2017). *Standard Test Method for Paraffin, Naphthene, and Aromatic Hydrocarbon Type Analysis in Petroleum Distillates through 200°C by*

Multi-Dimensional Gas Chromatography, Annual Book of Standards. West Conshohocken, Pennsylvania: ASTM International.

ASTM D5453 (2017). *Standard Test Method for Determination of Total Sulfur in Light Hydrocarbons, Spark Ignition Engine Fuel, Diesel Engine Fuel, and Engine Oil by Ultraviolet Fluorescence*, Annual Book of Standards. West Conshohocken, Pennsylvania: ASTM International.

ASTM D5482 (2017). *Standard Test Method for Vapor Pressure of Petroleum Products (Mini Method – Atmospheric)*, Annual Book of Standards. West Conshohocken, Pennsylvania: ASTM International.

ASTM D5580 (2017). *Standard Test Method for Determination of Benzene, Toluene, Ethylbenzene, p/m-Xylene, o-Xylene, C9 and Heavier Aromatic Derivatives, and Total Aromatic Derivatives in Finished Gasoline by Gas Chromatography*, Annual Book of Standards. West Conshohocken, Pennsylvania: ASTM International.

ASTM D56 (2017). *Standard Test Method for Flash Point by Tag Closed Cup Tester*, Annual Book of Standards. West Conshohocken, Pennsylvania: ASTM International.

ASTM D5623 (2017). *Standard Test Method for Sulfur Compounds in Light Petroleum Liquids by Gas Chromatography and Sulfur Selective Detection*, Annual Book of Standards. West Conshohocken, Pennsylvania: ASTM International.

ASTM D5713 (2017). *Standard Test Method for Analysis of High Purity Benzene for Cyclohexane Feedstock by Capillary Gas Chromatography*, Annual Book of Standards. West Conshohocken, Pennsylvania: ASTM International.

ASTM D5769 (2017). *Standard Test Method for Determination of Benzene, Toluene, and Total Aromatic derivatives in Finished Gasolines by Gas Chromatography/Mass Spectrometry*, Annual Book of Standards. West Conshohocken, Pennsylvania: ASTM International.

ASTM D5776 (2017). *Standard Test Method for Bromine Index of Aromatic Hydrocarbon Derivatives by Electrometric Titration*, Annual Book of Standards. West Conshohocken, Pennsylvania: ASTM International.

ASTM D5808 (2017). *Standard Test Method for Determining Chloride in Aromatic Hydrocarbon Derivatives and Related Chemicals by Microcoulometry*, Annual Book of Standards. West Conshohocken, Pennsylvania: ASTM International.

ASTM D5917 (2017). *Standard Test Method for Trace Impurities in Monocyclic Aromatic Hydrocarbon Derivatives by Gas Chromatography and External Calibration*, Annual Book of Standards. West Conshohocken, Pennsylvania: ASTM International.

ASTM D5986 (2017). *Standard Test Method for Determination of Oxygenates, Benzene, Toluene, C8-C12 Aromatic Derivatives and Total Aromatic Derivatives in Finished Gasoline by Gas Chromatography/Fourier Transform Infrared Spectroscopy*, Annual Book of Standards. West Conshohocken, Pennsylvania: ASTM International.

ASTM D6 (2017). *Standard Test Method for Loss on Heating of Oil and Asphaltic Compounds*, Annual Book of Standards. West Conshohocken, Pennsylvania: ASTM International.

ASTM D6069 (2017). *Standard Test Method for Trace Nitrogen in Aromatic Hydrocarbon Derivatives by Oxidative Combustion and Reduced Pressure Chemiluminescence Detection*, Annual Book of Standards. West Conshohocken, Pennsylvania: ASTM International.

ASTM D611 (2017). *Standard Test Methods for Aniline Point and Mixed Aniline Point of Petroleum Products and Hydrocarbon Solvents*, Annual Book of Standards. West Conshohocken, Pennsylvania: ASTM International.

ASTM D6144 (2017). *Standard Test Method for Analysis of AMS (α-Methyl Styrene) by Capillary Gas Chromatography*, Annual Book of Standards. West Conshohocken, Pennsylvania: ASTM International.

ASTM D6296 (2017). *Standard Test Method for Total Olefins in Spark-Ignition Engine Fuels by Multidimensional Gas Chromatography*, Annual Book of Standards. West Conshohocken, Pennsylvania: ASTM International.

ASTM D6304 (2017). *Standard Test Method for Determination of Water in Petroleum Products, Lubricating Oils, and Additives by Coulometric Karl Fischer Titration*, Annual Book of Standards. West Conshohocken, Pennsylvania: ASTM International.

ASTM D6377 (2017). *Standard Test Method for Determination of Vapor Pressure of Crude Oil: VPCRx (Expansion Method)*, Annual Book of Standards. West Conshohocken, Pennsylvania: ASTM International.

ASTM D6378 (2017). *Standard Test Method for Determination of Vapor Pressure (VPX) of Petroleum Products, Hydrocarbon Derivatives, and Hydrocarbon-Oxygenate Mixtures (Triple Expansion Method)*, Annual Book of Standards. West Conshohocken, Pennsylvania: ASTM International.

ASTM D6379 (2017). *Standard Test Method for Determination of Aromatic Hydrocarbon Types in Aviation Fuels and Petroleum Distillates – High Performance Liquid Chromatography Method with Refractive Index Detection.*, Annual Book of Standards. West Conshohocken, Pennsylvania: ASTM International.

ASTM D6450 (2017). *Standard Test Method for Flash Point by Continuously Closed Cup (CCCFP) Tester*, Annual Book of Standards. West Conshohocken, Pennsylvania: ASTM International.

ASTM D6729 (2017). *Standard Test Method for Determination of Individual Components in Spark Ignition Engine Fuels by 100 Meter Capillary High Resolution Gas Chromatography*, Annual Book of Standards. West Conshohocken, Pennsylvania: ASTM International.

ASTM D6730 (2017). *Standard Test Method for Determination of Individual Components in Spark Ignition Engine Fuels by 100–Meter Capillary (with Precolumn) High-Resolution Gas Chromatography*, Annual Book of Standards. West Conshohocken, Pennsylvania: ASTM International.

ASTM D6733 (2017). *Standard Test Method for Determination of Individual Components in Spark Ignition Engine Fuels by 50-Metre Capillary High Resolution Gas Chromatography*, Annual Book of Standards. West Conshohocken, Pennsylvania: ASTM International.

ASTM D6839 (2017). *Standard Test Method for Hydrocarbon Types, Oxygenated Compounds, and Benzene in Spark Ignition Engine Fuels by Gas Chromatography*, Annual Book of Standards. West Conshohocken, Pennsylvania: ASTM International.

ASTM D6897 (2017). *Standard Test Method for Vapor Pressure of Liquefied Petroleum Gases (LPG) (Expansion Method)*, Annual Book of Standards. West Conshohocken, Pennsylvania: ASTM International.

ASTM D7096 (2017). *Standard Test Method for Determination of the Boiling Range Distribution of Gasoline by Wide-Bore Capillary Gas Chromatography*, Annual Book of Standards. West Conshohocken, Pennsylvania: ASTM International.

ASTM D7171 (2017). *Standard Test Method for Hydrogen Content of Middle Distillate Petroleum Products by Low-Resolution Pulsed Nuclear Magnetic Resonance Spectroscopy*, Annual Book of Standards. West Conshohocken, Pennsylvania: ASTM International.

ASTM D808 (2017). *Standard Test Method for Chlorine in New and Used Petroleum Products (High Pressure Decomposition Device Method)*, Annual Book of Standards. West Conshohocken, Pennsylvania: ASTM International.

ASTM D847 (2017). *Standard Test Method for Acidity of Benzene, Toluene, Xylenes, Solvent Naphtha, and Similar Industrial Aromatic Hydrocarbon Derivatives*, Annual Book of Standards. West Conshohocken, Pennsylvania: ASTM International.

ASTM D848 (2017). *Standard Test Method for Acid Wash Color of Industrial Aromatic Hydrocarbon Derivatives*, Annual Book of Standards. West Conshohocken, Pennsylvania: ASTM International.

ASTM D849 (2017). *Standard Test Method for Copper Strip Corrosion by Industrial Aromatic Hydrocarbon Derivatives*, Annual Book of Standards. West Conshohocken, Pennsylvania: ASTM International.

ASTM D850 (2017). *Standard Test Method for Distillation of Industrial Aromatic Hydrocarbon derivatives and Related Materials*, Annual Book of Standards. West Conshohocken, Pennsylvania: ASTM International.

ASTM D852 (2017). *Standard Test Method for Solidification Point of Benzene*, Annual Book of Standards. West Conshohocken, Pennsylvania: ASTM International.

ASTM D86 (2017). *Standard Test Method for Distillation of Petroleum Products at Atmospheric Pressure*, Annual Book of Standards. West Conshohocken, Pennsylvania: ASTM International.

ASTM D88 (2017). *Standard Test Method for Saybolt Viscosity*, Annual Book of Standards. West Conshohocken, Pennsylvania: ASTM International.

ASTM D891 (2017). *Standard Test Methods for Specific Gravity, Apparent, of Liquid Industrial Chemicals*, Annual Book of Standards. West Conshohocken, Pennsylvania: ASTM International.

ASTM D893 (2017). *Standard Test Method for Insolubles in Used Lubricating Oils*, Annual Book of Standards. West Conshohocken, Pennsylvania: ASTM International.

ASTM D909 (2017). *Standard Test Method for Supercharge Rating of Spark-Ignition Aviation Gasoline*, Annual Book of Standards. West Conshohocken, Pennsylvania: ASTM International.

ASTM D92 (2017). *Standard Test Method for Flash and Fire Points by Cleveland Open Cup Tester*, Annual Book of Standards. West Conshohocken, Pennsylvania: ASTM International.

ASTM D93 (2017). *Standard Test Methods for Flash Point by Pensky-Martens Closed Cup Tester*, Annual Book of Standards. West Conshohocken, Pennsylvania: ASTM International.

ASTM E1455 (2017). *Standard Practice for Obtaining Colorimetric Data from a Visual Display Unit Using Tristimulus Colorimeters*, Annual Book of Standards. West Conshohocken, Pennsylvania: ASTM International.

ASTM E203 (2017). *Standard Test Method for Water Using Volumetric Karl Fischer Titration*, Annual Book of Standards. West Conshohocken, Pennsylvania: ASTM International.

ASTM E659 (2017). *Standard Test Method for Autoignition Temperature of Liquid Chemicals*, Annual Book of Standards. West Conshohocken, Pennsylvania: ASTM International.

Breiman, L. and Friedman, J.H. (1985). Estimating optimal transformations for multiple regression and correlation. *Journal of the American Statistical Association* 80 (391): 580–619.

Chadeesingh, R. (2011). The Fischer-Tropsch process. In: *The Biofuels Handbook* (ed. J.G. Speight), 476–517. London, UK. Part 3, Chapter 5: The Royal Society of Chemistry.

Dandekar, A.Y. and Stenby, E.H. (1997). Measurement of phase behavior of hydrocarbon mixtures using fiber optical detection techniques. Paper No. SPE38845. *Proceedings. SPE Annual Technical Conference and Exhibition*, San Antonio, TX (5–8 October). Richardson, TX: Society of Petroleum Engineers.

Davis, B.H. and Occelli, M.L. (2010). *Advances in Fischer-Tropsch Synthesis, Catalysts, and Catalysis*. Boca Raton, Florida: CRC Press, Taylor & Francis Group.

De Bakker, C.J. and Fredericks, P.M. (1995). Determination of petroleum properties by fiber-optic Fourier transform Raman spectrometry and partial least-squares analysis. *Applied Spectroscopy* 49 (12): 1766–1771.

DiSanzo, F.P. and Giarrocco, V.J. (1988). Analysis of pressurized gasoline-range liquid hydrocarbon samples by capillary column and PIONA analyzer gas chromatography. *Journal of Chromatographic Science* 26: 258–401.

Elsharkawy, A.M. (2001). Characterization of the C7 plus fraction and prediction of the dew point pressure for gas condensate reservoirs. Paper No. SPE 68776. *Proceedings. SPE Western Regional Meeting*, Bakersfield, CA (26–29 March). Richardson, TX: Society of Petroleum Engineers.

Fevang, Ø. and Whitson, C.H. (1994). Accurate in-situ compositions in petroleum reservoirs. Paper No. SPE 28829. *Proceedings. EUROPEC Petroleum Conference*, London, UK (25–27 October). Richardson, TX: Society of Petroleum Engineers.

Gary, J.G., Handwerk, G.E., and Kaiser, M.J. (2007). *Petroleum Refining: Technology and Economics*, 5e. Boca Raton, Florida: CRC Press, Taylor & Francis Group.

Gold, D.K., McCain, W.D. Jr., and Jennings, J.W. (1989). An improved method of the determination of the reservoir gas specific gravity for retrograde gases. *Journal of Petroleum Technology* 41 (7): 747–752.

Hall, K.R. and Yarborough, L. (1973). A new equation of state for Z-factor calculations. *Oil and Gas Journal* 71 (18): 82–92.

Høier, L. and Whitson, C.H. (1998). Miscibility variation in compositional grading reservoirs. Paper No. SPE 49269. *Proceedings. SPE Annual Technical Conference and Exhibition*, New Orleans, LA (27–30 September 1998). Richardson, TX: Society of Petroleum Engineers.

Hori, Y. (2000). *In Modern Petroleum Technology. Volume 2: Downstream* (ed. A.G. Lucas). New York: Wiley Chapter 2.

Hsu, C.S. and Robinson, P.R. ed. (2017). *Handbook of Petroleum Technology*. Cham: Springer International Publishing AG.

Ichikawa, M., Nonaka, N., Amono, H. et al. (1992). Proton NMR analysis of octane number for motor gasoline: Part IV. *Applied Spectroscopy* 46 (8): 1294.

Kleyweg, D. (1989). A set of constant PVT correlations for gas condensate systems. Paper No. SPE 19509. Richardson, TX: Society of Petroleum Engineers.

Lee, A., Gonzalez, M., and Eakin, B. (1966). The viscosity of natural gases. *Journal of Petroleum Technology* 18: 997–1000.

Lohrenz, J., Bray, B.G., and Clark, C.R. (1964). Calculating viscosities of reservoir fluids from their compositions. *Journal of Petroleum Technology* 16 (10): 1171–1176.

Marruffo, I, Maita, J., Him, J., and Rojas, G. (2001). Statistical forecast models to determine retrograde dew point pressure and the C7+ percentage of gas condensate on the basis of production test data from Eastern Venezuelan Reservoirs. Paper No. SPE69393. *Proceedings. SPE Latin American and Caribbean Petroleum Engineering Conference*, Buenos Aires (25–28 March). Richardson, TX: Society of Petroleum Engineers.

McCain, W.D. Jr. (1990). *The Properties of Petroleum Fluids*, 2e. Tulsa, Oklahoma: PennWell Books.

McCann, J.M. (1998). Chapter 2. In: *Manual on Hydrocarbon Analysis*, 6e (ed. A.W. Drews). West Conshohocken, Pennsylvania: American Society for Testing and Materials.

Myers, M.E., Stollsteirner, J., and Wims, A.M. (1975). Determination of gasoline octane numbers from chemical composition. *Analytical Chemistry* 47 (13): 2301–2304.

Nemeth, L.K. and Kennedy, H.T. (1967). A correlation of dew point pressure with fluid composition and temperature. Paper No. SPE-1477-PA. Richardson, TX: Society of Petroleum Engineers.

Niemstschik, G.E., Poettmann, F.H., and Thompson, R.S. (1993). Correlation for determining gas condensate composition. Paper No. SPE 26183. *Proceedings. SPE Gas Technology Symposium*, Calgary (28–30 June). Richardson, TX: Society of Petroleum Engineers.

Occelli, M.L. (2010). *Advances in Fluid Catalytic Cracking: Testing, Characterization, and Environmental Regulations*. Boca Raton, Florida: CRC Press, Taylor & Francis Group.

Organick, E.I. and Golding, B.H. (1952). Prediction of saturation pressures for condensate-gas and volatile-oil mixtures. *Transactions AIME* 195: 135–148.

Ovalle, A.P., Lenn, C.P., and McCain, W.D. (2007) Tools to manage gas/condensate reservoirs; Novel fluid-property correlations on the basis of commonly available field data. Paper No. SPE-112977-PA. *SPE Reservoir Evaluation & Engineering Volume*. Richardson, TX: Society of Petroleum Engineers.

Pandey, S.C., Ralli, D.K., Saxena, A.K., and Alamkhan, W.K. (2004). Physicochemical characterization and application of naphtha. *Journal of Scientific & Industrial Research* 63: 276–282.

Paredes, J.E., Perez, R., Perez, L.P., and Larez, C.J. (2014). Correlations to estimate key gas condensate properties through field measurement of gas condensate ratio. Paper No. SPE-170601-MS. *Proceedings. SPE Annual Technical Conference and Exhibition*, Amsterdam, the Netherlands (27–29 October). Richardson, TX: Society of Petroleum Engineers.

Parkash, S. (2003). *Refining Processes Handbook*. Amsterdam: Gulf Professional Publishing, Elsevier.

Pedersen, K.S. and Fredenslund, A. (1987). An improved corresponding states model for the prediction of oil and gas viscosities and thermal conductivities. *Chemical Engineering Science* 42: 182–186.

Pedersen, K.S., Thomassen, P., and Fredenslund, A. (1989). Characterization of gas condensate mixtures, C_{7+} fraction characterization. In: *Advances in Thermodynamics* (ed. L.G. Chorn and G.A. Mansoori). New York: Taylor & Francis Publishers.

Piper, L.D., McCain, W.D. Jr., and Corredor, J.H. (1999). Compressibility factors for naturally occurring petroleum gases. *Gas Reservoir Engineering* 52: 23–33.

Potsch, K.T. and Bräuer, L. (1996). A novel graphical method for determining dew point pressures of gas condensates. Paper No. SPE 36919. *Proceedings. SPE European Petroleum Conference*, Milan, Italy (22–24 October). Richardson, TX: Society of Petroleum Engineers.

Rayes, D.G., Piper, L.D., McCain, W.D. Jr., and Poston, S.W. (1992). Two-phase compressibility factors for retrograde gases. Paper No. SPE-20055-PA. Richardson, TX: Society of Petroleum Engineers.

Speight, J.G. ed. (2011). *The Biofuels Handbook*. London, UK: Royal Society of Chemistry.

Speight, J.G. (2013). *The Chemistry and Technology of Coal*, 3e. Boca Raton, Florida: CRC Press, Taylor & Francis Group.

Speight, J.G. (2014a). *The Chemistry and Technology of Petroleum*, 5e. Boca Raton, Florida: CRC Press, Taylor & Francis Group.

Speight, J.G. (2014b). *Gasification of Unconventional Feedstocks*. Oxford, UK: Gulf Professional Publishing, Elsevier.

Speight, J.G. (2015). *Handbook of Petroleum Product Analysis*, 2e. Hoboken, NJ: Wiley.

Speight, J.G. (2017). *Handbook of Petroleum Refining*. Boca Raton, FL: CRC Press, Taylor & Francis Group.

Speight, J.G. and Exall, D.I. (2014). *Refining Used Lubricating Oils*. Boca Raton, FL: CRC Press, Taylor & Francis Group.

Standing, M.B. and Katz, D.L. (1942). Density of natural gases. *Transactions AIME* 146: 140–149.

Sujan, S.M.A., Jamal, M.S., Hossain, M. et al. (2015). Analysis of gas condensate and its different fractions of Bibiyana gas field to produce valuable products Bangladesh. *Journal of Scientific and Industrial Research* 50 (1): 59–64.

Teng, S.T. and Williams, A.D. (1994). Detailed hydrocarbon analysis of gasoline by GC-MS (SI-PIONA). *Journal of High Resolution Chromatography* 19: 469–475.

Thornton, O.F. (1946). Gas-condensate reservoirs-a review. Paper No API-46-150. *Proceedings. API Drilling and Production Practice*, New York (1 January). API-46-150. https://www.onepetro.org/conference-paper/API-46-150 (accessed 1 November 2017).

Welch, W.T., Bain, M.L., Russell, K., Maggard, S.M., and May, J.M. 1994. Experience leads to accurate design of a NIR gasoline analysis systems. *Oil & Gas Journal*, June 27 48–56.

Whitson, C.H. and Belery, P. (1994). Compositional gradients in petroleum reservoirs. Paper No. SPE 28000. *Proceedings. SPE Centennial Petroleum Engineering Symposium* held in Tulsa, OK (29–31August). Richardson, TX: Society of Petroleum Engineers.

Whitson, C.H. and Torp, S.B. (1983). Evaluating constant volume depletion data. *Journal of Petroleum Technology* 35: 610–620.

Whitson, C.H., Fevang, Ø., and Yang, T. (1999). Gas condensate PVT: What's really important and Why? *Proceedings. IBC Conference on the Optimization of Gas Condensate Fields*, London, UK (28–29 January). London, UK: IBC UK Conferences Ltd. http://www.ibc-uk.com (accessed 20 October 2017).

Conversion Factors

1. Concentration Conversions

1 part per million (1 ppm) = 1 microgram per liter (1 $\mu g\,l^{-1}$)

1 microgram per liter (1 $\mu g\,l^{-1}$) = 1 milligram per kilogram (1 $mg\,kg^{-1}$)

1 microgram per liter ($\mu g\,l^{-1}$) × 6.243 × 10^8 = 1 lb per cubic foot (1 lb ft^{-3})

1 microgram per liter (1 $\mu g\,l^{-1}$) × 10^{-3} = 1 milligram per liter (1 $mg\,l^{-1}$)

1 milligram per liter (1 $mg\,l^{-1}$) × 6.243 × 10^5 = 1 pound per cubic foot (1 lb ft^{-3})

1 gram mole per cubic meter (1 g mol m^{-3}) × 6.243 × 10^5 = 1 pound per cubic foot (1 lb ft^{-3})

10 000 ppm = 1% w/w

1 ppm hydrocarbon in soil × 0.002 = 1 lb of hydrocarbons per ton of contaminated soil

2. Sludge Conversions

1700 lbs wet sludge = 1 yd^3 wet sludge

1 yd^3 sludge = wet tons/0.85

Wet tons sludge × 240 = gallons sludge

1 wet ton sludge × % dry solids/100 = 1 dry ton of sludge

3. Weight Conversion

1 ounce (1 oz) = 28.3495 grams (18.2495 g)

1 pound (1 lb) = 0.454 kilogram

1 pound (1 lb) = 454 grams (454 g)

1 kilogram (1 kg) = 2.20462 pounds (2.20462 lb)

1 stone (English, 1 st) = 14 pounds (14 lb)

1 ton (US; 1 short ton) = 2000 lbs

1 ton (English; 1 long ton) = 2240 lbs

1 metric ton = 2204.62262 pounds

1 tonne = 2204.62262 pounds

Handbook of Natural Gas Analysis, First Edition. James G. Speight.

4. Temperature Conversions

°F = (°C × 1.8) + 32
°C = (°F − 32)/1.8
(°F − 32) × 0.555 = °C
Absolute zero = −273.15 °C
Absolute zero = −459.67 °F

5. Area

1 square centimeter (1 cm^2) = 0.1550 square inches
1 square meter 1 (m^2) = 1.1960 square yards
1 hectare = 2.4711 acres
1 square kilometer (1 km^2) = 0.3861 square miles
1 square inch (1 $inch^2$) = 6.4516 square centimeters
1 square foot (1 ft^2) = 0.0929 square meters
1 square yard (1 yd^2) = 0.8361 square meters
1 acre = 4046.9 square meters
1 square mile (1 mi^2) = 2.59 square kilometers

6. Nutrient Conversion Factor

1 pound phosphorus × 2.3 (1 lb P × 2.3) = 1 pound phosphorous pentoxide (1 lb P_2O_5)
1 pound potassium × 1.2 (1 lb K × 1.2) = 1 pound potassium oxide (1 lb K_2O)

7. Other Approximations

14.7 pounds per square inch (14.7 psi) = 1 atmosphere (1 atm)
1 kilopascal (kPa) × 9.8692×10^{-3} = 14.7 pounds per square inch (14.7 psi)
1 yd^3 = 27 ft^3
1 US gallon of water = 8.34 lbs
1 imperial gallon of water = 10 lbs
1 yd^3 = 0.765 m^3
1 acre-inch of liquid = 27 150 gallons = 3.630 ft^3
1 ft depth in 1 acre (in situ) = 1613 × (20–25% excavation factor) = ~2000 yd^3
1 yd^3 (clayey soils-excavated) = 1.1–1.2 tons (US)
1 yd^3 (sandy soils-excavated) = 1.2–1.3 tons (US)

Glossary

Abandoned well A well not in use because it was a dry hole originally or because it has ceased to produce. Statutes and regulations in many states require the plugging of abandoned wells to prevent the seepage of oil, gas, or water from one stratum to another.

Abiogenic gas Gas formed by inorganic means.

Absolute pressure Gauge pressure plus barometric pressure; the absolute pressure can be zero only in a perfect vacuum.

Absolute zero The zero point on the absolute temperature scale. It is equal to −273.16 °C, or 0°K (Kelvin), or −459.69 °F, or 0 °R (Rankine).

Absorbed gas Natural gas that has been dissolved into the rock and requires hydraulic fracturing to be released.

Absorbent A material that, due to an affinity for certain substances, extracts one or more such substances from a liquid or gaseous medium with which it contacts and that changes physically, or both, during the process.

Absorption The process by which the gas is distributed throughout an absorbent (liquid); depends only on physical solubility and may include chemical reactions in the liquid phase (*chemisorption*).

Absorption plant A device or unit that removes hydrocarbon compounds from natural gas, especially casinghead gas; in the plant the gas is run through absorption oil that absorbs the liquid constituents, which are then recovered by distillation.

Accumulation Pressure increase over the maximum allowable working pressure of the vessel during discharge through the pressure relief valve (expressed as a percent of that pressure) is called accumulation.

Accuracy The closeness of the data of an analytical method to the true value.

Acid deposition (acid rain) Occurs when sulfur dioxide (SO_2) and, to a lesser extent, NO_x emissions are transformed in the atmosphere and return to the earth as dry deposition or in rain, fog, or snow.

Handbook of Natural Gas Analysis, First Edition. James G. Speight.

Acid gas Impurities in a gas stream usually consisting of carbon dioxide (CO_2), hydrogen sulfide (H_2S), carbonyl sulfide (COS), thiols/mercaptans (RSH), and sulfur dioxide (SO_2) – the most common in natural gas are carbon dioxide, hydrogen sulfide, and carbonyl sulfide. See also Sour gas.

Acid gas loading The amount of acid gas, on a molar or volumetric basis, that will be picked up by a solvent.

Acid rain Abnormally acidic rainfall, most often containing dilute concentrations of sulfuric acid or nitric acid.

Adsorption The process by which the gas is concentrated on the surface of a solid or liquid to remove impurities; carbon is a common adsorbing medium that can be regenerated upon *desorption*.

Alkazid process A process for removal of hydrogen sulfide and carbon dioxide from natural gas using concentrated aqueous solutions of amino acids.

Analytical batch Consists of samples that are analyzed together with the same method sequence and the same lots of reagents and with the manipulations common to each sample within the same time period or in continuous sequential time periods.

Anticline An area of the Earth's crust where folding has made a dome-like shape in the once flat rock layers. Anticlines often provide an environment where natural gas can become trapped beneath the Earth's surface and extracted. See also Traps, Faults, Permeability, Porosity.

Aquifer An underground porous, permeable rock formation that acts as a natural water reservoir.

Aquifer storage field A subsurface facility for storing natural gas consisting of water-bearing sands topped by an impermeable cap rock.

Associated natural gas Gas that occurs as free gas in a petroleum reservoir; see Dissolved natural gas; see Nonassociated natural gas.

Atmospheric discharge The release of vapors and gases from pressure-relieving and depressurizing devices to the atmosphere.

Back pressure Pressure on the discharge side of safety relief valves is back pressure; the pressure that exists at the outlet of a pressure relief device as a result of pressure in the discharge system.

Base gas The quantity of gas needed to maintain adequate reservoir pressures and deliverability rates throughout the withdrawal season; base gas usually is not withdrawn and remains in the reservoir; all gas native to a depleted reservoir is included in the base gas volume.

Base load requirements (base load storage) Gas that is used to meet seasonal demand increases, and the facilities are capable of holding enough natural gas to satisfy long-term seasonal demand requirements.

Basement rock The impervious geological stratum that underlays the reservoir rock and retains gas or oil in a reservoir.

Bcf (billion cubic feet) Gas measurement approximately equal to one trillion (1 000 000 000 000) Btus. See also Mcf, Tcf, Quad.

Biogenic coalbed methane Methane formed in coal seams by naturally occurring bacteria that are associated with meteoric water recharge at outcrop or subcrop; see Coalbed methane.

Biomass Organic nonfossil material of biological origin constituting a renewable energy source.

Boiling point (boiling temperature) The temperature at which the vapor pressure of the substance is equal to atmospheric pressure.

Black shale A thinly bedded shale that is rich in carbon, sulfide, and organic material, formed by anaerobic (lacking oxygen) decay of organic matter; black shales occur in thin beds in many areas at various depths and are of interest both historically and economically.

Blowdown The difference between set pressure and reseating pressure of a safety valve expressed in percent of the set pressure in psi or kPa.

Bottom-simulating reflector (BSR) A seismic reflection at the sediment to clathrate stability zone interface caused by the different density between normal sediments and sediments laced with clathrates.

Brine An aqueous solution of salts that occurs with gas and crude oil; seawater and saltwater are also known as brine.

Btu (British thermal unit) A unit of measurement for energy; the amount of heat that is necessary to raise the temperature of 1 lb of water by 1 °F. See also Btu, Bcf, Tcf, Quad.

Bubblepoint pressure At a given temperature, the pressure when crude oil releases a bubble of gas from solution.

Built-up back pressure The pressure in the discharge header that develops as a result of flow after the safety relief valve opens.

Bundled service Gas sales service and transportation service packaged together in a single transaction in which the pipeline, on behalf of the utility, buys gas from producers and then delivers it to the utility.

Calibration The process of adjusting the instrument readout so that it corresponds to the actual concentration value or a reference standard; involves checking the instrument with a known concentration of a gas or vapor to see that the instrument gives the proper response; calibration results in calibration factors or functions establishing the relationship between the analyzer response and the actual gas concentration introduced to the analyzer; an important element of quality assurance in emission control.

Cap rock The impervious geological stratum that overlays the reservoir rock and retains gas or oil in a reservoir.

Carbonaceous material A material that contains carbon as well as some hydrogen and other noncarbon and nonhydrogen elements such as nitrogen, oxygen, and sulfur.

Carbonate rock A rock consisting primarily of a carbonate mineral such as calcite or dolomite, the chief minerals in limestone and dolostone, respectively.

Carbonate washing A chemical conversion process in which acid contaminants in natural gas are converted to compounds that are not objectionable or that can be removed from the stream with greater ease than the original constituents.

Carbon capture and storage A combination of a number of existing technologies, with the potential to play a major role in the management and reduction of global carbon dioxide (CO_2) levels; the process allows for carbon dioxide emissions released during energy production to be captured and stored underground.

Carbon dioxide fracturing The use of gaseous carbon dioxide to fracture a formation.

Casing Used to line the walls of a gas well to prevent collapse of the well and also to protect the surrounding earth and rock layers from being contaminated by petroleum or the drilling fluids.

Casinghead gas (casinghead gas) Natural gas produced with oil in oil wells and is usually the flash gas from the oil reservoir. Casinghead gas collects in the annular space between the well tubing and casing of an oil well. The weight of the casinghead gas contributes to reducing the bottomhole pressure and lowering well production – specifically by holding a back pressure against the formation; reducing the gas pressure on the well casing (annulus) can increase oil production. The goal is to maintain casinghead pressure as close to zero as possible. In areas or the country where it is allowed, these systems are often configured to pull a vacuum.

Catalytic oxidation A chemical conversion process that is used predominantly for destruction of volatile organic compounds and carbon monoxide.

Cathodic protection The method of preventing corrosion in metal structures that involves using electric voltage to slow or prevent corrosion; used in natural gas pipelines to resist corrosion over an extended period of time.

CFCs (chlorofluorocarbons) Gaseous compounds used for cooling; release into the atmosphere causes ozone depletion.

Chemisorption See Absorption.

Christmas tree The series of pipes and valves that sits on top of a producing gas well; used in place of a pump to extract the gas from the well.

City gate A location at which custody of gas passes from a gas pipeline company to a local distributor.

Claus process A sulfur recovery process recovering elemental sulfur from sour gas; a major producer of sulfur.

Clean Air Act Amendments of 1990 Legislation to improve the quality of the atmosphere and curb acid rain promotes the use of cleaner fuels in vehicles and stationary sources.

Coalbed methane (coal bed methane) Methane from coal seams; released or produced from the seams when the water pressure within the seam is reduced by pumping from either vertical or inclined to horizontal surface holes; see also Biogenic coalbed methane, Thermogenic coalbed methane.

Coal gas A generic term for gaseous mixture (mainly hydrogen, methane, and carbon monoxide) made from coal by the destructive distillation (i.e. heating in the absence of air) of bituminous coal; also synonymous with *blue gas, producer gas, water gas, town gas, fuel gas manufactured gas,* and *syngas* (*synthetic natural gas,* SNG).

Coke oven gas The mixture of permanent gases produced by the carbonization of coal in a coke oven at temperatures in excess of 1000 °C (1832 °F).

Composition The makeup of a gaseous stream.

Compressed natural gas (CNG) Natural gas compressed to a pressure at or above 2900–3600 psi and stored in high-pressure containers; used as a fuel for natural gas-powered vehicles.

Compressibility factor The ratio of the molar volume V_m of the gas to the molar volume V_m° of an ideal gas at the same pressure and temperature. Thus

$$Z = \frac{V_m}{V_m^\circ}.$$

The value of Z provides information on the dominant types of intermolecular forces acting in a gas. Thus, when $Z = 1$, there are no intermolecular forces, ideal gas behavior; when $Z < 1$, the attractive forces dominate, and gas occupies a smaller volume than an ideal gas; when $Z > 1$, the repulsive forces dominate, and gas occupies a larger volume than an ideal gas. All gases approach $Z = 1$ at very low pressures, when the spacing between particles is typically large.

Compression Reduction in volume of natural gas is compressed during transportation and storage.

Concentration The amount of a substance, expressed as mass, volume, or number of particles in a unit volume of a solid, liquid, or gaseous substance.

Condensate A hydrocarbon liquid stream separated from natural gas; higher molecular weight hydrocarbons that exist in the reservoir as constituents of natural gas but are recovered as liquids in separators, field facilities, or gas processing plants. See Gas condensate.

Consumption Natural gas consumed within the country, including imports but excluding amounts reinjected, flared, and lost in shrinkage.

Conventional gas Natural gas that is extracted from underground reservoirs using traditional exploration and production methods.

Critical point The pressure and temperature of a reservoir fluid where the bubblepoint pressure curve meets the retrograde dew-point pressure curve representing a unique state where all properties of the bubblepoint oil are identical to the dew-point gas.

Crude oil Petroleum.

Cryogenic process A process involving low temperatures.

CSST (corrugated stainless steel tubing) Flexible piping used to install gas service in residential and commercial areas.

Cubic foot A unit of measurement for volume; an area one foot long, by one foot wide, by one foot deep.

Cutting A piece of rock or dirt that is brought to the surface of a drilling site as debris from the bottom of well; often used to obtain data for logging.

DEA Diethanolamine.

Dead crude oil Crude oil in the reservoir with minimal or no dissolved associated gas; often difficult to produce as there is little energy to drive it.

Decline rate The rate at which the production rate of a well decreases.

DGA Diglycolamine.

Dehydration Water removal from natural gas streams.

Dekatherm (dth) A unit of energy used primarily to measure natural gas and was developed in about 1972 by the Texas Eastern Transmission Corporation – a natural gas pipeline company; the dekatherm is equal to 10 therms or 1 000 000 Btu (MMBtu) or, using the SI system, 1.055 gigajoules (GJ) and is also approximately equal to 1000 ft^3 (Mcf, Mft^3) of natural gas or exactly 1000 ft^3 of natural gas with a heating value of 1000 Btu ft^{-3}.

Delivered gas The physical transfer of natural, synthetic, and/or supplemental gas from facilities operated by the responding company to facilities operated by others or to consumers.

Delivery point (receipt point) The point where natural gas is transferred from one party to another. The city gate is the delivery point for a pipeline or transportation company because this is where the gas is transferred to the LDC.

Density The mass of a substance contained in a unit volume (mass divided by volume).

Depleted reservoirs Reservoirs that have already been tapped of all their recoverable natural gas.

Depleted storage field A subsurface natural geological reservoir, usually a depleted gas or oil field, used for storing natural gas.

Desorption See Adsorption.
Detection limit The lowest amount of analyte in a sample that can be detected but not necessarily quantitated as an exact value; often called the *limit of detection* (*LOD*), which is the lowest concentration level that can be determined statistically different from a blank at a specified level of confidence; determined from the analysis of sample blanks; see Method detection limit.
Dew-point temperature The temperature at which a vapor begins to condense and deposit as a liquid.
Dip The angle at which it lies in relation to a flat line at the surface; often helps the geologist to locate possible traps (reservoirs).
DIPA Diisopropanolamine.
Dissolved natural gas Gas that occurs in solution in the petroleum in a reservoir; see Associated natural gas.
Dry natural gas Natural gas that remains after (i) the liquefiable hydrocarbon portion has been removed from the gas stream and (ii) any volumes of nonhydrocarbon gases have been removed where they occur in sufficient quantity to render the gas unmarketable; also known as consumer-grade natural gas.
EPACT (Energy Policy Act of 1992) Comprehensive energy legislation designed to expand natural gas use by allowing wholesale electric transmission access and providing incentives to developers of clean fuel vehicles.
Estimated additional amount in place The volume additional to the proved amount in place that is of foreseeable economic interest. Speculative amounts are not included.
Estimated additional reserves recoverable The volume within the estimated additional amount in place that geological and engineering information indicates with reasonable certainty might be recovered in the future.
Exploratory well A well drilled either in search of a new and as yet undiscovered accumulation of oil or gas or in an attempt to significantly extend the limits of a known reservoir.
Extraction plant A plant in which products, such as propane, butane, oil, ethane, or natural gasoline, which are initially components of the gas stream, are extracted or removed for sale.
Fabric filter Collectors in which dust is removed from the gas stream by passing the dust-laden gas through a fabric of some type; commonly termed "bag filters" or "baghouses."
Fault When part of the Earth's crust fractures due to forces exerted on it by movement of plates on the Earth's crust; of interest because they often form traps that are natural gas reservoirs.

Fire point The temperature to which gas must be heated under prescribed conditions of the method to burn continuously when the mixture of vapor and air is ignited by a specified flame.

Fischer–Tropsch process The catalytic process by which synthesis gas (syngas; mixtures of carbon monoxide and hydrogen) is converted to hydrocarbon products.

Flare A tall stack equipped with burners used as a safety device at wellheads, refining facilities, gas processing plants, and chemical plants; used for the combustion and disposal of combustible gases.

Flash point The temperature to which gas must be heated under specified conditions to give off sufficient vapor to form a mixture with air that can be ignited momentarily by a specified flame; dependent on the composition of the gas and the presence of other hydrocarbon constituents.

Flash tank A vessel used to separate the gas evolved from liquid flashed from a higher pressure to a lower pressure.

Flexsorb process A process that uses sterically hindered amines (olamines) in aqueous solutions or other physical solvents; the molecular structure hinders the carbon dioxide approach to the amine and preferentially removes hydrogen sulfide from the gas stream.

Flow rate The rate that expresses the volume of fluid or gas passing through a given surface per unit of time (e.g. cubic feet per minute).

Formation Either a certain layer of the Earth's crust or a certain area of a layer; often refers to the area of rock where a reservoir is located.

Formation volume factor (FVF) The ratio of a phase volume (water, oil, gas, or gas plus oil) at reservoir conditions, relative to the volume of a surface phase (water, oil, or gas) at standard conditions resulting when the reservoir material is brought to the surface; denoted mathematically as B_w (bbl/STB), B_0 (bbl/STB), B_g (ft^3/SCF), and B_t (bbl/STB).

Fractionation The process of separating the various natural gas liquids present in the remaining gas stream by using the varying boiling points of the individual hydrocarbons in the gas stream.

Fracturing A method used by producers to extract more natural gas from a well by opening up rock formations using hydraulic or explosive force.

Fuel cell technology The chemical interaction of natural gas and certain other metals such as platinum, gold, and other electrolytes to produce electricity.

Gas A gas for which no liquid or solid phase can exist at the temperature of interest (e.g. gases such as nitrogen, oxygen, or helium at room temperature); see Vapor.

Gas cap The gas trapped between the liquid petroleum and the impervious cap rock of the petroleum reservoir.

Gas condensate A liquid condensed from natural gas that contains a significant amount of C_{5+} components that exhibits the phenomenon of retrograde condensation at reservoir conditions; in other words, as pressure decreases, increasing amounts of liquid condenses in the reservoir (down to about 2000 psia), which results in a significant loss of *in situ* condensate reserves that may only be partially recovered by revalorization at lower pressures. See Condensate, Gas condensate reservoir.

Gas condensate reservoir A reservoir that exhibits producing gas–oil ratios from 2500 to 50 000 SCF/STB (400 to 10 STB/MMSCF). See Condensate, Gas condensate, Gas cycling project.

Gas conditioning The removal of objectionable constituents and addition of desirable constituents.

Gas cycling project A project designed to avoid liquid loss from retrograde condensation can usually be justified for fluids with liquid content higher than about 50–100 STB/MMSCF; offshore, the minimum liquid content to justify cycling is about 100 STB/MMSCF. See Condensate, Gas condensate, Gas condensate reservoir.

Gas deliverability The deliverability of the gas from the storage facility (also referred to as the deliverability rate, withdrawal rate, or withdrawal capacity) is often expressed as a measure of the amount of gas that can be delivered (withdrawn) from a storage facility on a daily basis.

Gas holder A gastight receptacle or container in which gas is stored for future use, (i) at approximately constant pressure (low-pressure containers) in which case the volume of the container changes, and (ii) in containers of constant volume (usually high-pressure containers) in which case the quantity of gas molecules stored varies with the pressure.

Gas hydrates Solid, crystalline, waxlike substances composed of water, methane, and usually a small amount of other gases, with the gases being trapped in the interstices of a water–ice lattice; formed beneath permafrost and on the ocean floor under conditions of moderately high pressure and at temperatures near the freezing point of water.

Gas, liquefied petroleum (LPG) A gas containing certain specific hydrocarbons that are gaseous under normal atmospheric conditions but can be liquefied under moderate pressure at normal temperatures. Propane and butane are the principal examples.

Gas, manufactured A gas obtained by destructive distillation of coal, or by the thermal decomposition of oil, or by the reaction of steam passing through a bed of heated coal or coke or catalyst beds. Examples are coal gases, coke oven gases, producer gas, blast furnace gas, blue (water) gas, and carbureted water gas. Btu content varies widely.

Gas–liquid separator A vertical or horizontal separator in which gas and liquid are separated by means of gravity settling with or without a mist eliminating device.

Gasoline plant A plant in which hydrocarbon components common to the gasoline fractions are removed from "wet" natural gas, leaving a "drier" gas.

Gas processing The preparation of gas for consumer use by removal of the nonmethane constituents; synonymous with gas refining.

Gas refining See Gas processing.

Gas well A well completed for production of natural gas from one or more gas zones or reservoirs.

Geological survey The exploration for natural gas that involves a geological examination of the surface structure of the Earth to determine the areas where there is a high probability that a reservoir exists.

Geophones Equipment used to detect the reflection of seismic waves during a seismic survey (*q.v.*).

Giammarco-Vetrocoke process A process for hydrogen sulfide and/or carbon dioxide removal from natural gas.

Girdler process Amine (olamine) washing of natural gas to remove acid gases.

Global warming An environmental issue that deals with the potential for global climate change due to increased levels of atmospheric greenhouse gases.

Greenhouse effect See Global warming.

Greenhouse gases See Global warming.

Guard bed A bed (usually alumina) that serves as a protector of a more expensive bed (e.g. molecular sieve); serves by the act of attrition and may be referred to as an attrition catalyst.

Heating value The number of British thermal units per cubic foot (Btu ft^{-3}) of natural gas as determined from tests of fuel samples.

Heat of combustion (energy content) The amount of energy that is obtained from burning natural gas; measured in British thermal units (Btu).

HCFCs (hydrochlorofluorocarbons) Gaseous compounds that meet current environmental standards for minimizing stratospheric ozone depletion.

High flash stocks Liquids having a closed cup flash point of 55 °C (131 °F) or over (such as heavy fuel oil, lubricating oils, transformer oils, etc.); does not include any stock that may be stored at temperatures above or within 8 °C (14.4 °F) of the flash point.

Horizontal drilling The method that allows producers to extend horizontal shafts into areas that could not otherwise be reached; especially useful in offshore drilling; categorized as short (extending only 20–40 ft from the vertical), medium (300–700 ft from the vertical), or long (1000–4500 ft from vertical) radius.

Horsehead (balanced conventional beam, sucker rod) pump A common type of cable rod lifting equipment for recovery of oil and gas; so called because of the shape of the counterweight at the end of the beam.

Hydraulic fracturing A process through which small fractures are made in impermeable rock by a pressurized combination of water, sand, and chemical additives; the small fractures are held open by a proppant (such as grains of sand), thereby allowing the natural gas to flow out of the rock and into the wellbore.

Hydrocarbon An organic compound containing only carbon and hydrogen.

Hydrocarbon, saturated A chemical compound of carbon and hydrogen in which all the valence bonds of the carbon atoms are taken up with hydrogen atoms.

Hydrocarbon, unsaturated A chemical compound of carbon and hydrogen in which not all the valence bonds of the carbon atoms are taken up with hydrogen atoms.

Hydrogen sulfide (H_2S) A poisonous, corrosive compound consisting of two atoms of hydrogen and one of sulfur, gaseous in its natural state; found in manufactured gas or process gas made from coal or crude oil containing sulfur and must be removed; also found to some extent in some natural gas.

Ideal gas A gas in which all collisions between atoms or molecules are perfectly elastic and in which there are no intermolecular attractive forces.

IFPEXOL process A methanol-based process for water removal and hydrocarbon dew-point control; also used for acid gas removal.

Ignition, automatic A means that provides for automatic lighting of gas at the burner when the gas valve controlling flow is turned on and will affect relighting if the flame on the burner has been extinguished by means other than closing the gas burner valve.

Ignition, continuous Ignition by an energy source that is continuously maintained through the time the burner is in service, whether the main burner is firing or not.

Ignition, intermittent Ignition by an energy source that is continuously maintained through the time the burner is firing.

Ignition, interrupted Ignition by an energy source that is automatically energized each time the main burner is fired and subsequently is automatically shut off during the firing cycle.

Ignition, manual Ignition by an energy source that is manually energized and where the fuel to the pilot is lighted automatically when the ignition system is energized.

Ignition temperature The temperature at which a substance, such as gas, will ignite and continue burning with adequate air supply.

Impure natural gas Natural gas as delivered from the well and before processing (refining).

Independent producer A nonintegrated company that receives nearly all of its revenues from production at the wellhead; by the IRS definition, a firm is an independent if the refining capacity is less than 50 000 barrels per day in any given day or their retail sales are less than $5 million for the year.

Intermediate precision The within-laboratory variations, such as different days, different analysts, and different equipment; see Precision.

Interruptible service contracts Contracts that allow a distributing party to temporarily suspend delivery of gas to a buyer in order to meet the demands of customers who purchased firm service.

Iron oxide process (iron sponge process, dry box method) A process in which the gas is passed through a bed of wood chips impregnated with iron oxide to scavenge hydrogen sulfide and organic sulfur compounds (mercaptans) from natural gas streams.

Joule–Thomson effect The cooling that occurs when a compressed gas is allowed to expand in such a way that no external work is done. The effect is approximately 7 °F /100 psi for natural gas.

Joule–Thomson expansion The throttling effect produced when expanding a gas or vapor from a high pressure to a lower pressure with a corresponding drop in temperature.

Kelly The heavy square or hexagonal steel pipe that goes through the rotary table and turns the drill string (also called grief stem).

Kitchen The underground deposit of organic debris that is eventually converted to petroleum and natural gas.

Knockout pot A separator used for a bulk separation of gas and liquid, particularly when the liquid volume fraction is high.

Landfill gas Gas produced during the *in situ* maturation of landfill materials that can include municipal waste and industrial waste.

Lean gas Natural gas in which methane is the major constituent.

Lease condensate Low-boiling (low molecular weight) liquid hydrocarbons recovered from lease separators or field facilities at associated and nonassociated natural gas wells. Mostly pentanes and heavier hydrocarbons. Normally enters the crude oil stream after production.

Lease fuel Natural gas used in well, field, and lease operations, such as gas used in drilling operations, heaters, dehydrators, and field compressors.

Lease operations Any well, lease, or field operations related to the exploration for or production of natural gas prior to delivery for processing or transportation out of the field gas used in lease operations include usage such as for drilling operations, heaters, dehydrator units, and field compressors used for gas lift.

Lease separation facility A facility installed at the surface for the purpose of (i) separating gases from produced crude oil and water at the temperature and pressure conditions set by the separator and/or (ii) separating gases from that portion of the produced natural gas stream that liquefies at the temperature and pressure conditions set by the separator.

Lease separator A facility installed at the surface for the purpose of separating the full well stream volume into two or three parts at the temperature and pressure conditions set by the separator; for gas wells, these parts include produced natural gas, lease condensate, and water.

Limit of quantitation (LOQ) The level above which quantitative results may be determined with acceptable accuracy and precision.

Linearity The ability of the method to elicit results that are directly proportional to analyte concentration within a given range.

Liquefied natural gas The liquid form of natural gas.

Liquefied petroleum gas (LPG) The term applied to certain specific hydrocarbons and their mixtures, which exist in the gaseous state under atmospheric ambient conditions but can be converted to the liquid state under conditions of moderate pressure at ambient temperature.

Liquids, natural gas Those liquid hydrocarbon mixtures that are gaseous at reservoir temperatures and pressures but are recoverable by condensation or absorption; gas condensate, natural gasoline, and liquefied petroleum gases are often included in this category.

Lithology The study of rocks; important for exploration and drilling crews to have an understanding of lithology as it relates to the production of gas and oil.

Local distribution company A retail gas distribution company that delivers natural gas to end users.

LO-CAT process A wet oxidation process.

Logging Lowering of different types of measuring instruments into the wellbore and gathering and recording data on *porosity*, *permeability*, and types of fluids present near the current well after which the data are used to construct subsurface maps of a region to aid in further exploration.

Lower explosive limit The lower percent by volume of the gas vapor in air at which the gas will explode or inflame. See also Upper explosive limit.

Lower flammability limit The minimum concentration by volume of a combustible substance that is capable of propagating a flame under specified conditions.

Low flash stocks Liquids having a closed cup flash point under 55 °C (131 °F) such as gasoline, kerosene, jet fuels, some heating oils, diesel fuels, and any other stock that may be stored at temperatures above or within 8 °C (14.4 °F) of the flash point.

Magnetometer A device to measure small changes in the Earth's magnetic field at the surface, which indicates what kind of rock formations might be present underground.

Manufactured gas Gas obtained by destructive distillation of coal or by the thermal decomposition of oil or by the reaction of steam passing through a bed of heated coal or coke; examples are coal gases, coke oven gases, producer gas, blast furnace gas, blue (water) gas, and carbureted water gas; the Btu content varies widely.

Marsh gas See Natural gas.
Mass concentration The concentration expressed in terms of mass of substance per unit volume (g substance/m^3 gas volume).
Mcf (thousand cubic feet) One thousand cubic feet; a unit of measure that is more commonly used in the low volume sectors of the gas industry, such as stripper well production. See also Btu, Bcf, Tcf, Quad.
MDEA Methyldiethanolamine.
MEA Ethanolamine (monoethanolamine).
MER See Most efficient recovery rate.
Mercaptan Typically, a hydrocarbon group (usually a methane, ethane, or propane) with a hydro-sulfur group (-SH) substituted on a terminal carbon atom.
Methane (CH_4) Commonly (often incorrectly) known as natural gas; colorless and naturally odorless and burns efficiently without many by-products.
Methanogens Methane-producing microorganisms.
Method detection limit (MDL) The minimum concentration of a substance that can be measured and reported with 99% confidence that the analyte concentration is greater than zero; determined from analysis of a sample in a given matrix containing the analyte; see Detection limit.
Migration The movement of oil, gas, or water through porous and permeable rock.
Mixed gas A fuel gas in which natural or LP gas is mixed with manufactured gas.
MMBtu A thermal unit of energy equal to 1 000 000 Btu, that is, the equivalent of 1000 ft^3 of gas having a heating content of 1000 Btu per cubic foot, as provided by contract measurement terms.
MMcf A million cubic feet.
Most efficient recovery rate (MER) The rate at which the greatest amount of natural gas may be extracted without harming the formation itself.
Muds Used in drilling to lubricate the drilling bit in rotary drilling rigs.
Multiple completions The result of drilling several different depths from a single well to increase the rate of production or the amount of recoverable gas.
Naphtha An arbitrarily defined crude oil fraction containing primarily aliphatic (linear) hydrocarbons with boiling points ranging from 30 to 250 °C (86–482 °F); principal uses are for (i) gasoline blend stock, (ii) solvents, and (iii) paint thinners, (iv) as a feedstock for the production of organic chemicals, and (v) as a feedstock for the production of synthetic natural gas.
Native gas Gas in place at the time that a reservoir was converted to use as an underground storage reservoir in contrast to injected gas volumes.

Natural gas Also called *marsh gas, swamp gas,* and *landfill gas*; a gaseous fossil fuel that is found in oil fields and natural gas fields and in coalbeds; a mixture of hydrocarbon and small quantities of nonhydrocarbons that exists either in the gaseous phase or in solution in crude oil in natural underground reservoirs, which is gaseous at atmospheric conditions of pressure and temperature.

Natural Gas Act Passed in 1938 and gives the Federal Power Commission (now the Federal Energy Regulatory Commission or FERC) jurisdiction over companies engaged in interstate sale or transportation of natural gas.

Natural gas cofiring The injection of natural gas with pulverized coal or oil into the primary combustion zone of a boiler.

Natural gas hydrates Solid, crystalline, waxlike substances composed of water, methane, and usually a small amount of other gases, with the gases being trapped in the interstices of a water–ice lattice; formed beneath permafrost and on the ocean floor under conditions of moderately high pressure and at temperatures near the freezing point of water.

Natural gas liquids (NGLs) A hydrocarbon liquid stream (*gas condensate*); higher molecular weight hydrocarbons that exist in the reservoir as constituents of natural gas, which are recovered as liquids in separators, field facilities, or gas processing plants.

Natural gasoline A term used in the gas processing industry to refer to a mixture of liquid hydrocarbons (mostly pentane, including isopentane, and higher molecular weight hydrocarbons) extracted from natural gas.

Natural gas plant liquids Hydrocarbons in natural gas that are separated as liquids at natural gas processing plants, fractionating and cycling plants, and, in some instances, field facilities; the products obtained include liquefied petroleum gases (ethane, propane, and butanes), pentanes plus, and isopentane.

Natural Gas Policy Act of 1978 One of the first efforts to deregulate the gas industry and to determine the price of natural gas as dictated by market forces, rather than regulation.

Natural gas processing plant A facility designed to recover natural gas liquids from a stream of natural gas that may or may not have passed through lease separators and/or field separation facilities; the facility controls the quality of the natural gas to be marketed.

Natural gas resource base An estimate of the amount of natural gas available, based on the combination of proved reserves, and those additional volumes that have not yet been discovered, but are estimated to be "discoverable" given current technology and economics.

Natural gas shrinkage, natural gas The reduction in volume of wet natural gas due to the extraction of some of its constituents, such as hydrocarbon products, hydrogen sulfide, carbon dioxide, nitrogen, helium, and water vapor.

Natural gas vehicle (NGV) A car, bus, or truck that is powered by a natural gas, either in compressed or liquefied form, rather than the traditional gasoline or diesel fuel.

NES (National Energy Strategy) A 1991 federal proposal that focused on national security, conservation, and regulatory reform, with options that encourage natural gas use.

Nonassociated natural gas Sometimes called *gas well gas*; gas produced from geological formations that typically do not contain much, if any, crude oil or higher boiling hydrocarbons (*gas liquids*) than methane; can contain nonhydrocarbon gases such as carbon dioxide and hydrogen sulfide.

Noncombustible Material incapable of igniting or supporting combustion.

Nonhydrocarbon gases Typical nonhydrocarbon gases that may be present in reservoir natural gas, such as carbon dioxide, helium, hydrogen sulfide, and nitrogen.

Nonstandard method A method that is not taken from authoritative and validated sources; includes methods from scientific journals and unpublished laboratory-developed methods.

NO_x (nitrogen oxides) Produced during combustion; precursors to acid deposition (acid rain).

NPC (National Petroleum Council) An advisory body of appointed members whose purpose is to advise the Secretary of Energy.

Off-peak period The time during a day, week, month, or year when gas use on a particular system is not at its maximum.

Olamine process A process that used an amine derivative (an olamine) to remove acid gas form natural gas streams.

Olamines Compounds such as ethanolamine (monoethanolamine, MEA), diethanolamine (DEA), triethanolamine (TEA), methyldiethanolamine (MDEA), diisopropanolamine (DIPA), and diglycolamine (DGA) that are widely used in gas processing.

Order 636 The Federal Energy Regulatory Commission's 1992 order that required pipelines to unbundle their transportation, sales, and storage services.

Paraffin hydrocarbons Saturated hydrocarbon compounds with the general formula C_nH_{2n+2} containing only single bonds and carbon and hydrogen only; sometimes referred to as alkanes or natural gas liquids.

Part concentration The concentration expressed as number of particles of substance per a certain number of particles.

Peak load requirements (peak load storage) The design to have high deliverability for short periods of time during which the natural gas can be withdrawn from storage quickly as the need arises.

Peak shaving Use of natural gas from storage to supplement the normal amounts delivered to customers during peak-use periods.

Peak-use period The period of time when gas use on a particular system is at its maximum.

Permeability A measure of the ease that a fluid can pass through a section of rock.

Petroleum Crude oil; a flammable naturally occurring liquid that may vary from almost colorless to black; occurs in many places in the upper strata of the Earth; a complex mixture of hydrocarbons with small amounts of other substances and is prepared for use as naphtha, gasoline, and other products by various refining processes.

PIGs Robotic agents used to inspect pipeline interior walls for corrosion and defects, to measure pipeline interior diameters, to remove accumulated debris, and for other specialty tasks.

Pipeline (natural gas) A continuous pipe conduit, complete with such equipment as valves, compressor stations, communications systems, and meters for transporting natural and/or supplemental gas from one point to another, usually from a point in or beyond the producing field or processing plant to another pipeline or to points of utilization. Also refers to a company operating such facilities.

Porosity The spaces between grains of sediment in sedimentary rock.

Precision The agreement between a set of replicate measurements without assumption of knowledge of the true value; see Intermediate precision, Repeatability, Reproducibility.

Pressure, absolute (psia) Pressure in excess of a perfect vacuum; absolute pressure is obtained by algebraically adding gauge pressure to atmosphere pressure. Pressures reported in "atmospheres" are understood to be absolute. Absolute pressure must be used in equations of state and in all gas law calculations. Gauge pressures below atmospheric pressure are called "vacuum."

Producer The company generally involved in exploration, drilling, and refining of natural gas.

Producer gas A mixture of flammable gases (principally carbon monoxide and hydrogen) and nonflammable gases (mainly nitrogen and carbon dioxide) made by the partial combustion of carbonaceous substances, usually coal, in an atmosphere of air and steam; producer gas has a lower heating value than other gaseous fuels, but it can be manufactured with relatively simple equipment; it is used mainly as a fuel in large industrial furnaces.

Producing sand A rock stratum that contains recoverable oil or gas.

Producing zone The interval of rock actually producing oil or gas.

Production rate The rate of production of oil and/or gas from a well; usually given in barrels per day ($bbls\,day^{-1}$) for oil or standard cubic feet ($scft^3\,day^{-1}$) for gas.

Propping agents Sand, glass beads, epoxy, or silica sand that serve to prop open the newly widened fissures in the formation.

Proved resources Part of the resource base that includes the working inventory of natural gas; volumes that have already been discovered and are readily available for production and delivery.

Proved amount in place The volume originally occurring in known natural reservoirs that has been carefully measured and assessed as exploitable under present and expected local economic conditions with existing available technology.

Proved recoverable reserves The volume within the proved amount in place that can be recovered in the future under present and expected local economic conditions with existing available technology.

PSI (pounds per square inch) Pressure measured with respect to that of the atmosphere.

PUHCA (Public Utility Holding Company Act of 1935) Amended by EPACT to allow power generation by independent power producers (IPPs) without restrictions on corporate structure.

Quad An abbreviation for a quadrillion (1 000 000 000 000,000) Btu; roughly equivalent to one trillion (1 000 000 000 000) cubic feet or 1 Tcf. See also Bcf, Mcf, Tcf.

Quenching Cooling of a hot vapor by mixing it with another fluid or by partially vaporizing another liquid.

Range The interval between the upper and lower concentration of analyte in sample for which it has been demonstrated that the analytical procedure has an acceptable level of accuracy, precision, and linearity.

Raw natural gas Impure natural gas as delivered from the well and before processing (refining).

Recovery rate The rate at which natural gas can be removed from a reservoir.

Rectisol process A process that used a physical (nonreactive) solvent for gas cleaning.

Redox process A sulfur recovery process that involves liquid-phase oxidation; uses a dilute aqueous solution of iron or vanadium to remove hydrogen sulfide selectively by chemical absorption from sour gas streams; can be used on relatively small or dilute hydrogen sulfide stream to recover sulfur from the acid gas stream or, in some cases, they can be used in place of an acid gas removal process.

Refinery gas Noncondensable gas collected in petroleum refineries.

Reid vapor pressure (RVP) The pressure of the vapor in equilibrium with liquid at 37.8 °C (100 °F).

Repeatability The precision under the same operating conditions over a short period of time; see Precision.

Reproducibility The precision between laboratories; see Precision.
Reserves The amount of a resource available for recovery and/or production; the recoverable amount is usually tied to economic aspects of production.
Residue gas Natural gas from which the higher molecular weight hydrocarbons have been extracted; mostly methane.
Retrograde dew-point pressure At a given temperature, the pressure when a gas condenses, and the pressure drops below the dew point.
Rich gas A gaseous stream is traditionally very rich in natural gas liquids (NGLs); see Natural gas liquids.
R/P (reserves/production) ratio Calculated by dividing proved recoverable reserves by production (gross less reinjected) in a given year.
Reburning An effective and economic means of reducing NO_x emissions from all types of industrial and electric utility boilers.
Reproducibility (of an analytical method) A measure of the repeatability of the data using an analytical method.
Reserve additions Volumes of the resource base that are continuously moved from the resource category to the proved resources category.
Reservoir A geological formation that retains or *traps* the gas.
Reservoir characterization Determination of the physical properties of a reservoir (such as porosity, permeability, fluid saturation, etc.) and changes in the distribution of these properties throughout the reservoir.
Reservoir energy The underground pressure in a reservoir that will push the petroleum and natural gas up the wellbore to the surface.
Reservoir water The water found in petroleum reservoirs is usually a *brine* consisting mostly of sodium chloride (NaCl) in quantities from 10 to 350 ppt (‰); seawater has about 35 ppt. Other compounds (electrolytes) found in reservoir brines include calcium (Ca), magnesium (Mg), sulfate (SO_4), bicarbonate (HCO_3), iodide (I), and bromide (Br); sometimes referred as *brine* or *connate water*.
Rod pumping The use of a lifting (pumping) method to recover oil from a reservoir.
Rotary bit The cutting tool attached to the lower end of the drill pipe of a rotary drilling rig; the bit does the actual drilling of the hole through the formation.
Rotary drilling A method for drilling wells using a cutting bit attached to a revolving drill pipe.
Ruggedness or robustness A measure of an analytical procedure's capacity to remain unaffected by small but deliberate variations in method parameters and provides an indication of its reliability during normal usage.
Salt caverns Caverns formed out of existing salt deposits.

Saturated condition A condition where an oil and gas are in thermodynamic equilibrium, that is, the chemical force exerted by each component in the oil phase is equal to the chemical force exerted by the same component in the gas phase, thereby eliminating mass transfer of components from one phase to the other. See Undersaturated condition.

Saturation pressure An oil at its bubblepoint pressure or a gas at its dew-point pressure.

SCADA (supervisory control and data acquisition) Remote controlled equipment used by pipelines and LDCs to operate their gas systems.

SCOT (Shell Claus off-gas treating) process A tail gas-treating process.

Scrubber A unit that has been designed for the removal of contaminates from a gas stream.

Seismograph An instrument used to detect and record earthquakes and is able to pick up and record the vibrations of the earth that occur during an earthquake; when seismology is applied to the search for natural gas, seismic waves, emitted from a source, are sent into the earth, and the seismic waves interact differently with the underground formation (underground layers), each with its own properties.

Seismology The study of the movement of energy, in the form of seismic waves, through the Earth's crust.

Selexol process A process that used a physical (nonreactive) solvent for gas cleaning.

Separator tank Tanks are usually located at the well site to separate oil, gas, and water before sending each off to be processed at different locations.

Sewage gas A gas produced by the fermentation of sewage sludge low in heating value due to dilution with carbon dioxide and nitrogen.

Shale A fine-grained sedimentary rock composed of clay minerals and tiny fragments (silt-sized particles) of other materials.

Shale basins Underground rock formations that serve both as natural gas generators and reservoirs.

Shale gas Gas that occurs in low permeability shale; see Unconventional gas.

Shale oil A liquid produced by the thermal decomposition of the kerogen component of oil shale.

Shift converter A reactor in which carbon monoxide and water are catalytically converted to hydrogen and carbon dioxide.

Shrinkage, natural gas The reduction in volume of wet natural gas due to the extraction of some of its constituents, such as hydrocarbon products, hydrogen sulfide, carbon dioxide, nitrogen, helium, and water vapor.

Slurry process A process that uses a slurry of iron oxide to selectively absorb hydrogen sulfide.

SO_2 (sulfur dioxide) A precursor to acid deposition (acid rain); produced when sulfur is combusted to sulfur dioxide.

Solution gas–oil ratio (GOR) The amount of surface gas that can be dissolved in a stock tank oil when brought to a specific pressure and temperature; denoted mathematically as R_s (SCF/STB).

Solution oil–gas ratio (OGR) The amount of surface condensate that can be vaporized in a surface gas at a specific pressure and temperature; sometimes referred to as liquid content. Denoted mathematically as r_s (STB/MMSCF).

Sour gas Natural gas that contains hydrogen sulfide and/or other acid gases.

Specific gravity (API) A common measure of oil specific gravity, defined by $\gamma_{API} = (141.5/\gamma_O) - 131.5$, with units in °API (degrees API).

Specific gravity (gas) The ratio of density of any gas at standard conditions (14.7 psia and 60 °F) to the density of air at standard conditions; based on the ideal gas law ($pV = nRT$), gas gravity is also equal to the gas molecular weight divided by air molecular weight ($M_{air} = 28.97$); denoted mathematically as γ_g (where air = 1).

Specific gravity (liquid) The ratio of density of any liquid measured at standard conditions (usually 14.7 psia and 60 °F) to the density of pure water at the same standard conditions; denoted mathematically as γ_o (where water = 1).

Specificity The ability to assess unequivocally the analyte in the presence of components that may be expected to be present.

Specific volume The volume of a unit weight of a substance at specific temperature and pressure conditions.

Specific weight Weight per unit volume of a substance.

Spot market A method of contract purchasing whereby commitments by the buyer and seller are of a short duration at a single volume price.

Spot purchase Gas purchased on the spot market, which involves short-term contracts for specified amount of gas, at a one-time purchase price.

Standard condition A temperature of 15 °C (59 °F) and a pressure of 1 atm (14.7 psi, 101.325 kPa), which also is known as standard temperature and pressure (STP).

Standard method A method that is traceable to a recognized, validated method.

Stretford process A wet oxidation process.

Stripper wells Natural gas wells that produce less than 60 000 ft^3 of gas per day.

SulfaTreat process A batch-type process for the selective removal of hydrogen sulfide and mercaptans from natural gas; the process is dry, using no free liquids, and can be used for all natural gas applications where a batch process is suitable.

Sulfinol process A combination process that uses a mixture of amines and a physical solvent (an aqueous amine and sulfolane).

SulFerox process A wet oxidation process.

Supercompressibility factor A factor used to account for the following effect: Boyle's law for gases states that the specific weight of a gas is directly proportional to the absolute pressure, the temperature remaining constant. All gases deviate from this law by varying amounts, and within the range of conditions ordinarily encountered in the natural gas industry, the actual specific weight under the higher pressure is usually greater than the theoretical. The factor used to reflect this deviation from the ideal gas law in gas measurement with an orifice meter is called the *supercompressibility factor, Fpv*. The factor is used to calculate actual volumes from volumes at standard temperatures and pressures from actual volumes. The factor is of increasing importance at high pressures and low temperatures.

Supplemental gaseous fuels supplies Synthetic natural gas, propane-air, coke oven gas, refinery gas, biomass gas, air injected for Btu stabilization, and manufactured gas commingled and distributed with natural gas.

Swamp gas See Natural gas.

Sweetening process A process for the removal of hydrogen sulfide and other sulfur compounds from natural gas.

Sweet gas Natural gas that contains very little, if any, hydrogen sulfide.

Synthesis gas (syngas) Mixtures of carbon monoxide and hydrogen.

Synthetic natural gas (SNG) A manufactured product, chemically similar in most respects to natural gas, resulting from the conversion or reforming of hydrocarbons that may easily be substituted for or interchanged with pipeline quality natural gas; also referred to as substitute natural gas.

Tail gas The residue gas left after the completion of a treating process designed to remove certain liquids or liquefiable hydrocarbons.

Tail gas treating Removal of the remaining sulfur compounds from gases remaining after sulfur recovery.

Tcf (trillion cubic feet) Gas measurement approximately equal to one quadrillion (1 000 000 000 000 000) Btu's. See also Bcf, Mcf, Quad.

TEA Triethanolamine.

Temperature, dew point The temperature at which a vapor begins to condense and deposit as a liquid.

Terminal velocity (dropout velocity) The velocity at which a particle or droplet will fall under the action of gravity, when drag force just balances gravitational force and the particle (or droplet) continues to fall at constant velocity.

Therm A unit of heating value equivalent to 100 000 Btu (Btu).

Thermal maturity The amount of heat, in relative terms, to which a rock has been subjected; a thermally immature rock has not been subjected to enough heat to begin the process of converting organic material into oil and/or natural gas; a thermally overmature rock has been subjected to enough heat to convert organic material to graphite. However, these are the two extremes, and there are many intermediate stages of thermal maturity.

Thermogenic Generated or formed by heat, especially via physiological processes.

Thermogenic coalbed methane Methane formed in coal seams by the action of increasing temperature; see Coalbed methane.

Thermogenic gas Gas formed by pressure effects and temperature effects on organic debris.

Three-dimensional (3D) seismic survey Allows producers to see into the crust of the Earth to find promising formations for retrieval of gas.

Threshold limit value (TLV) The amount of a contaminant to which a person can have repeated exposure for an eight hour day without adverse effects.

Tight gas Natural gas found trapped in impermeable rock and nonporous sandstone or limestone formations, typically at depths greater than 10 000 ft below the surface.

Total organic carbon (TOC) The concentration of material derived from decaying vegetation, bacterial growth, and metabolic activities of living organisms or chemicals in the source rocks.

Town gas A generic term referring to manufactured gas produced for sale to consumers and municipalities.

Trillion cubic feet A volume measurement of natural gas. Approximately equivalent to one Quad. See also Btu's, Bcf, Mcf.

Traps A generic term for an area of the Earth's crust that has developed in such a way as to *trap* gas beneath the surface.

Ultimate analysis The determination of the elements contained in a compound, i.e. carbon, hydrogen, oxygen, nitrogen, sulfur, and other components.

Unassociated gas Natural gas unaccompanied by crude oil when produced; also called nonassociated gas or gas well gas.

Unbundled services Unbundling, or separating, pipeline transmission, sales, and storage services, along with guaranteeing space on the pipelines for all gas shippers.

Unconventional gas Gas that occurs in tight sandstones, siltstones, sandy carbonates, limestone, dolomite, and chalk; natural gas that cannot be economically produced using current technology; see also Shale gas.

Underground gas storage The use of subsurface facilities for storing gas that has been transferred from its original location for the primary purpose of load balancing; usually natural geological reservoirs, such as depleted oil or gas fields or water-bearing sands on the top by and impermeable cap rock.

Undersaturated condition A condition when an oil or a gas is in a single phase but not at its saturation point (bubblepoint or dew point), that is, the mixture is at a pressure greater than its saturation pressure. See Saturated condition.

Unsaturated compound Any compound having a double bond or triple bond between two adjacent carbon atoms.

Upper explosive limit The higher percent by volume of the gas vapor in air at which the gas will explode or inflame. See Lower explosive limit.

Upper flammability limit The maximum concentration by volume of a combustible substance that is capable of continued propagation of a flame under the specified conditions.

Vacuum A pressure less than atmospheric pressure, measured either from the base of zero pressure or from the base of atmospheric pressure.

Validation The process of establishing the performance characteristics and limitations of a method and the identification of the influences that may change these characteristics and to what extent.

Vapor When the gas phase of a substance is present under conditions when the substance would normally be a solid or liquid (e.g. below the boiling point of the substance), this is referred to as the vapor phase; see Gas.

Vapor density The density of any gas compared with the density of air with the density of air equal to unity.

Vaporizer A device other than a container that receives gas in liquid form and adds sufficient heat to convert the liquid to a gaseous state.

Verification The confirmation by examination and provision of objective evidence that specified requirements have been fulfilled.

Viscosity The measure of a fluid's thickness or how well it flows.

Volatile matter Matter that is readily vaporizable at a relatively low temperature.

Volume, specific The volume of a unit weight of a substance at specific temperature and pressure conditions.

Volume concentration The concentration expressed in terms of gaseous volume of substance per unit volume (cm^3 substance/m^3 gas volume).

Wash The removal of impurities from a gas or vapor by passing the gas through water or other liquid that retains or dissolves the impurity.

Washer A shell with internal baffler or packing, so arranged that gas to be cleaned passes up through the baffles countercurrent to the flow of scrubbing liquid down through the washer. The baffler or packing causes intimate contact and mixing of the gas with the liquid stream.

Washer-cooler A washer in the form of a tall tower in which the washing liquid is sprayed in at top is collected in the bottom of the tower and then is cooled and recycled through the tower. Serves a dual purpose of washing the gas free of impurities and also cooling the gas.

Weight, specific Weight per unit volume of a substance.

Well bore (wellbore) The channel created by the drill bit.

Well casing A series of metal tubes installed in the freshly drilled hole; serves to strengthen the sides of the well hole, to ensure that no oil or natural gas seeps out of the well hole as it is brought to the surface, and to keep other fluids or gases from seeping into the formation through the well.

Well completion The process for completion of a well to allow for the flow of petroleum or natural gas out of the formation and up to the surface; includes strengthening the well hole with casing, evaluating the pressure and temperature of the formation, and then installing the proper equipment to ensure an efficient flow of natural gas out of the well.

Well head (wellhead) The pieces of equipment mounted at the opening of the well to regulate and monitor the extraction of hydrocarbons from the underground formation; prevents leaking of oil or natural gas out of the well and prevents blowouts due to high-pressure formations.

Well logging A method used for recording rock and fluid properties to find gas- and oil-containing zones in subterranean formations.

Wet gas Natural gas that contains considerable amounts of higher molecular weight hydrocarbons other than methane.

Wet oxidation process Based on reduction–oxidation (redox) chemistry to oxidize the hydrogen sulfide to elemental sulfur in an alkaline solution containing an oxygen carrier; vanadium and iron are the two oxygen carriers that are used.

Wobbe index (Wobbe number) A number that indicates interchangeability of fuel gases and is obtained by dividing the heating value of a gas by the square root of its specific gravity.

Working gas The quantity of natural gas in the reservoir (reported in thousand cubic feet at standard temperature and pressure) that is in addition to the cushion or base gas; may or may not be completely withdrawn during any particular withdrawal season; conditions permitting, the total working capacity could be used more than once during any season; the volume of gas in the reservoir above the level of base gas and is, simply, the natural gas that is available for withdrawal and sales. I.

Working gas capacity The total gas storage capacity minus base gas.

Yield point The stress at which a material exceeds its elastic limit; below this stress, the material will recover its original size on removal of the stress;

above this stress, it will not recover its original size on removal of the stress.

Zero gas Gas at atmospheric pressure.

Zone A geographical area. A geological zone, however, means an interval of strata of the geologic column that has distinguishing characteristics from surrounding strata; also, a space or group of spaces within a building with heating and/or cooling requirements sufficiently similar so that comfort conditions can be maintained by a single controlling device.

Index

Handbook of Natural Gas Analysis, First Edition. James G. Speight.